The Subnuclear Series • Volume 50

Proceedings of the International School of Subnuclear Physics

WHAT WE WOULD LIKE LHC TO GIVE US

THE SUBNUCLEAR SERIES*

Series Editor: ANTONINO ZICHICHI, *European Physical Society, Geneva, Switzerland*

17. 1979 POINTLIKE STRUCTURES INSIDE AND OUTSIDE HADRONS
18. 1980 THE HIGH-ENERGY LIMIT
19. 1981 THE UNITY OF THE FUNDAMENTAL INTERACTIONS
20. 1982 GAUGE INTERACTIONS: Theory and Experiment
21. 1983 HOW FAR ARE WE FROM THE GAUGE FORCES?
22. 1984 QUARKS, LEPTONS, AND THEIR CONSTITUENTS
23. 1985 OLD AND NEW FORCES OF NATURE
24. 1986 THE SUPERWORLD I
25. 1987 THE SUPERWORLD II
26. 1988 THE SUPERWORLD III
27. 1989 THE CHALLENGING QUESTIONS
28. 1990 PHYSICS UP TO 200 TeV
29. 1991 PHYSICS AT THE HIGHEST ENERGY AND LUMINOSITY: To Understand the Origin of Mass
30. 1992 FROM SUPERSTRINGS TO THE REAL SUPERWORLD
31. 1993 FROM SUPERSYMMETRY TO THE ORIGIN OF SPACE–TIME
32. 1994 FROM SUPERSTRING TO PRESENT-DAY PHYSICS
33. 1995 VACUUM AND VACUA: The Physics of Nothing
34. 1996 EFFECTIVE THEORIES AND FUNDAMENTAL INTERACTIONS
35. 1997 HIGHLIGHTS OF SUBNUCLEAR PHYSICS: 50 Years Later
36. 1998 FROM THE PLANCK LENGTH TO THE HUBBLE RADIUS
37. 1999 BASICS AND HIGHLIGHTS IN FUNDAMENTAL PHYSICS
38. 2000 THEORY AND EXPERIMENT HEADING FOR NEW PHYSICS
39. 2001 NEW FIELDS AND STRINGS IN SUBNUCLEAR PHYSICS
40. 2002 FROM QUARKS AND GLUONS TO QUANTUM GRAVITY
41. 2003 FROM QUARKS TO BLACK HOLES: PROGRESS IN UNDERSTANDING THE LOGIC OF NATURE
42. 2004 HOW AND WHERE TO GO BEYOND THE STANDARD MODEL
43. 2005 TOWARDS NEW MILESTONES IN OUR QUEST TO GO BEYOND THE STANDARD MODEL
44. 2006 THE LOGIC OF NATURE, COMPLEXITY AND NEW PHYSICS: FROM QUARK-GLUON PLASMA TO SUPERSTRINGS, QUANTUM GRAVITY AND BEYOND
45. 2007 SEARCH FOR THE "TOTALLY UNEXPECTED" IN THE LHC ERA
46. 2008 PREDICTED AND TOTALLY UNEXPECTED IN THE ENERGY FRONTIER OPENED BY LHC
47. 2009 THE MOST UNEXPECTED AT LHC AND THE STATUS OF HIGH ENERGY FRONTIER
48. 2010 WHAT IS KNOWN AND UNEXPECTED AT LHC
49. 2011 SEARCHING FOR THE UNEXPECTED AT LHC AND THE STATUS OF OUR KNOWLEDGE
50. 2012 WHAT WE WOULD LIKE LHC TO GIVE US

Volume 1 was published by W. A. Benjamin, Inc., New York; 2–8 and 11–12 by Academic Press, New York and London; 9–10 by Editrice Compositori, Bologna; 13–29 by Plenum Press, New York and London; 30–50 by World Scientific, Singapore.

*For the complete list of titles in this series, please go to http://www.worldscientific.com/series/ss

The Subnuclear Series • Volume 50

Proceedings of the International School of Subnuclear Physics

WHAT WE WOULD LIKE LHC TO GIVE US

Edited by

Antonino Zichichi

European Physical Society
Geneva, Switzerland

World Scientific

NEW JERSEY • LONDON • SINGAPORE • BEIJING • SHANGHAI • HONG KONG • TAIPEI • CHENNAI

Published by

World Scientific Publishing Co. Pte. Ltd.
5 Toh Tuck Link, Singapore 596224
USA office: 27 Warren Street, Suite 401-402, Hackensack, NJ 07601
UK office: 57 Shelton Street, Covent Garden, London WC2H 9HE

The Subnuclear Series — Vol. 50
WHAT WE WOULD LIKE LHC TO GIVE US
Proceedings of the International School of Subnuclear Physics 2012

ISBN 978-981-4603-89-8

In-house Editor: Rhaimie Wahap

Printed in Singapore by B & Jo Enterprise Pte Ltd

Professor Murray Gell-Mann with the Gold–Medal awarded for his contribution

$$\bar{\theta}^0 + N \rightarrow \Lambda^0 + \pi.$$

to the establishment of the
Ettore Majorana Foundation and Centre for Scientific Culture
on the occasion of the 50th Anniversary of the Erice Subnuclear Physics School

(1963– 2012)

MURRAY GELL-MANN
82

PREFACE

In June/July 2012, a group of 150 physicists from 28 countries met in Erice to participate in the 50th Course of the International School of Subnuclear Physics. The countries represented by the participants were: Armenia, Azerbaijan, Belgium, Brasil, Canada, China, Egypt, Estonia, France, Germany, Greece, Hong Kong, Hungary, India, Iran, Italy, Latvia, Mexico, the Netherlands, Pakistan, Poland, Russia, South Africa, Switzerland, Turkey, Ukraine, United Kingdom and the United States of America.

The School was sponsored by the Academies of Sciences of Estonia, Georgia, Lithuania, Russia and Ukraine; the Chinese Academy of Sciences; the Commission of the European Communities; the European Physical Society; the Italian National Institute for Nuclear Physics; the Italian Ministry of Education, University and Scientific Research; the Sicilian Regional Government; the Weizmann Institute of Science; the World Federation of Scientists and the World Laboratory.

This year's Course was the second of the three dedicated to the celebrations of the 50th Anniversary of the School. Why three? Because Subnuclear Physics School needed three years for being established. In 1961 we started to discuss the bridge between University teaching and International Research Centres like CERN. In 1962 a formal paper for its existence was signed at CERN by Blackett, Rabi, Weisskopf and two young fellows, J.S. Bell and me. The first Course in Erice was held in 1963. The three years celebrations are coupled with big discoveries. In 2011, the negative sign of the β–function and asymptotic freedom; in 2012, QCD and in 2013, SSB and Instantons.

At this year School several lectures have been devoted to celebrate Quantum ChromoDynamics (QCD). Besides that, a strong focus was given to experimental physics with lectures on recent experiments, including the latest results from the LHC, not forgetting that, if Complexity exists at the fundamental level, a totally unexpected discovery should be given to us by LHC. What we would expect LHC to give us is the Higgs–meson and hopefully the first sign for the existence of the lightest Supersymmetric Particle. Experimental highlights from other Laboratories were also presented, while the theoretical lectures dealt with what could come out from fascinating topics such as Quantum Gravity and Black Holes Physics.

As usual, the lectures were devoted to the latest and most significant achievements in theoretical and in experimental physics. In fact, during the last half century, the School has been involved in all crucial steps of our Physics. A few examples: SU(3) flavour and SU(6) [with SU(2)-spin coupled with SU(3) flavour] dismantled by the "No-Go-Theorem", the battle between S-Matrix and Field Theory, the Universality of the weak forces [started with the ε-parameter and the non existence of the "flavour changing-neutral currents" solved by the existence of "charm"], the experimental search for the 3rd lepton in the early sixties before the discovery of CP breaking, the birth of the Electroweak Unification and the SSB (Spontaneous Symmetry Breaking) mechanism, the discovery of the negative sign of the β-function and of asymptotic freedom, the triumph of non Abelian field theories (QCD and QFD) with all consequences (including Instantons), the discovery of Supersymmetry (many years – and not few days – after the "No-Go-Theorem"). Now we are engaged with RQST (Relativistic Quantum String Theory) and LHC, without forgetting Complexity.

An original feature of the School, introduced in 1996, is a series of special sessions devoted to "New Talents", with the aim of encouraging and promoting young physicists to achieve recognition at an international level. This is a serious problem in Experimental Physics, where collaborations count several hundreds of participants and it is almost impossible for young fellows to be known. Even if with much less emphasis, the problem exists also in Theoretical Physics. So it was decided to offer 40 young fellows the possibility to give an open presentation of the results of their studies, followed by a discussion.

Among them 5 best "New Talents" have been selected for publication in this volume and were given an award, one for an original presentation in theoretical physics, one for an original presentation in experimental physics, one for an original presentation in detector physics and one for an original presentation in mathematical physics. As usual the best Student was very difficult to be chosen since the level of all the participants was very high. Diplomas were awarded to honor the memory of physicists who, during many decades, have contributed to the success of our School.

During the organization and the running of the Course, I enjoyed the collaboration of my colleague and friend, Gerardus 't Hooft, who shared with me the Directorship of the School. I would like to thank him together with the group of Lecturers and invited scientists.

I hope the reader will enjoy the book as much as the students attending the lectures and the discussion sessions, which, as every year, have been the focal point of our School activity. Thanks to the work of the Scientific Secretaries, the discussions have been reproduced as faithfully as possible. At various stages of my work I have enjoyed the collaboration of many friends whose contributions have been extremely important for the School and are highly appreciated. I thank them most warmly. A final acknowledgement to all those in Erice, Bologna and Geneva, who have helped me in so many occasions and to whom I feel very much indebted.

Antonino Zichichi

Geneva, October 2012

CONTENTS

2007 7:10 AM
2006 8:54 AM
2006 10:18 AM
2006 10:05 AM

«ETTORE MAJORANA» FOUNDATION AND CENTRE FOR SCIENTIFIC CULTURE
TO PAY A PERMANENT TRIBUTE TO GALILEO GALILEI, FOUNDER OF MODERN SCIENCE AND TO ENRICO FERMI, THE "ITALIAN NAVIGATOR", FATHER OF THE WEAK FORCES

INTERNATIONAL SCHOOL OF SUBNUCLEAR PHYSICS
50TH ANNIVERSARY CELEBRATIONS 2011-2012-2013
50th Course: ***WHAT WE WOULD LIKE LHC TO GIVE US***

ERICE-SICILY: 23 June – 2 July 2012

QCD CELEBRATIONS & GOLD–MEDAL TO MURRAY GELL-MANN

A. ZICHICHI
Opening Remarks

M. GELL-MANN
Some Reminiscences of Research Leading to QCD and Beyond

A. ZICHICHI
The Erice Centre, Gell-Mann, QCD, the Effective Energy and Complexity

H. FRITZSCH
History of QCD

H. LEUTWYLER
On the History of the Strong Interaction

D. SCHILDKNECHT
Colour Transparency and Saturation in QCD

P. MINKOWSKI
Glueballs a Key consequence of QCD

C.A. DOMINGUEZ
Quark Masses in QCD

F. CLOSE
The Quark Model and QCD

T.Y. CAO
Key Steps Towards the Creation of QCD

QCD Celebrations and Gold-Medal to Murray Gell-Mann

OPENING REMARKS

A. Zichichi
INFN and University of Bologna, Italy
CERN, Geneva, Switzerland
Enrico Fermi Centre, Rome, Italy
Ettore Majorana Foundation and Centre for Scientific Culture, Erice, Italy

The Ettore Majorana Foundation and Centre for Scientific Culture (EMFCSC) is celebrating its 50th Anniversary over the three years 2011-2013. This year will be the second of the three devoted to the celebrations of the 50th Anniversary. Why three years? The EMFCSC started in 1961 when, with my great friend John Bell, I discussed the problem of creating a bridge between university courses and activities in advanced physics laboratories such as CERN. A year later, on May 8th at CERN, Bell, Patrick Blackett, Victor Weisskopf, Isidor Rabi and Zichichi formally established the EMFCSC. The Centre's first activity was the School of Subnuclear Physics at Erice in 1963. This is why the celebrations are over three years. In 2011, the discovery of the negative sign of the β-function and of asymptotic freedom was celebrated. In 2013, the SSB plus Instantons will be celebrated. In 2012 we celebrate QCD.

The first Ceremony will be the Gold-Medal to Murray Gell-Mann for his idea that V^0–particles, discovered in 1947 by the Blackett group, had to be composed of mesons and baryons carrying the additively conserved quantum number called "strangeness. The unexpected "prediction" was that θ-meson had to be with isospin ½ and therefore the neutral piece θ^0 had to be with positive and negative "strangeness".

The Gold-Medal has engraved the symbol $\bar{\theta}$ since here is one of the vital roots of the Erice Centre, as reported at the 80th Anniversary Celebration of Gell-Mann [1]. And now let us start with Murray Gell-Mann who will recall the good old times.

Reference

[1] *Murray Gell-Mann and the Last Frontier of LHC Physics: The QGCW Project*
A. Zichichi, in Proceedings of the Conference in Honour of Murray Gell-Mann's 80th Birthday: *Quantum Mechanics, Elementary Particles, Quantum Cosmology and Complexity*, Nanyang, Institute of Advanced Studies, Nanyang Technological University, Singapore (2010), H. Fritzsch, K.K. Phua (eds), World Scientific, 2011, p. 157.

SOME REMINISCENCES OF RESEARCH LEADING TO QCD AND BEYOND

M. Gell-Mann
Santa Fe Institute, Santa Fe, New Mexico, USA

Let me thank Nino for having invited me to recall the old times when we both were many decades younger and met at the International Conference where the Blackett group announced an interesting result on the study of the production of V–particles [1]. The reaction needed was

(1) $$\theta^0 + n \rightarrow \Lambda^0 + \pi \, .$$

The chair of the Session was Butler, the member of the Blackett group who had discovered, with Rochester, the new particles, called V–particles by Blackett, since they appeared in the picture as two inverted Vs.

It turned out that the one V was a meson, θ^0, and the other V was a baryon, Λ^0. In order to explain the strong production and the weak decays of these V-particles I proposed the existence of a new quantum number, which was conserved at production and violated at decay. In order to have total zero strangeness at production I proposed to give strangeness + 1 to the θ^0 meson and strangeness – 1 to the Λ^0 baryon.

These were times when people believed that ordinary spin and isotopic spin had to be correlated. The nucleon was a fermion (with spin ½) and its isospin was ½: perfect fermion. The π meson was a boson (with spin zero) and its isospin was 1: perfect boson.

If the strange meson θ had to be with zero spin, like the π, its isospin could not be ½. But if this meson had to be with strangeness + 1 its isospin had to be ½.

At the discussion Session I pointed out that reaction (1) did not conserve "strangeness". The fellow gave no-answer. Only a young fellow made a remark pointing out that the data could be explained with the reaction

(2) $$\bar{\theta} + n \rightarrow \Lambda^0 + \pi\,.$$

The English of the young fellow was quite broken and the chairman asked him to go to the blackboard. He quoted me to explain the reasons for the existence of reaction (2), with great satisfaction of all members of the Blackett group. I was pleased to know that my idea about strangeness found experimental confirmation by the group where strange particles had been discovered. I could not imagine that $\bar{\theta}$ would have resulted in the founding of the Erice Centre. In fact, thanks to his remark on recalling the existence of reaction (2), Blackett wanted Nino to join his group. And Nino looking at old photos found two pictures which proved the existence of θ–mesons with positive and negative strangeness [2]. This was the key proof for the isospin ½ of a boson.

The second time I had good news from Nino was at CERN where I gave a Seminar about the existence of particles, as the simplest $SU(3)_{uds}$ representation, the so called quarks, with fractional charges (+ 2/3 ; – 1/3; – 1/3).

The good news from Nino was that the theoretically expected value R = 2 for the Ratio R

$$R = \frac{\sigma\,(e^+e^- \;\rightarrow hadrons)}{\sigma\,(e^+e^- \;\rightarrow \mu^+\mu^-)}$$

was coming out from the ADONE collider where Nino had a powerful experimental set-up able to identify hadronic final states in a large background of electrons and muons.

The third good news coming from Nino came when he succeeded to find that the final states produced in strong, electromagnetic and weak processes, were the same if, the nominal energy of the initial state, was replaced by the "effective energy". No matter what was the pair of interacting particles in the initial state, the multihadronic final state show "universality features", which had their source in the interaction of quarks and gluons, i.e. QCD.

The latest news from Nino is the new project [3] where he plans to study the "QCD-open-colour-states" whose number is not limited by the QCD-colourless condition which all known mesons and baryons have to obey.

Finally, his interest in "Complexity" brought him to find [4] that the experimental evidences for the existence of "Complexity" are only two: the AFB (Anderson, Feynman, Beethoven) phenomena and the Sarajevo type events. Since Complexity exists at the fundamental level, his conclusion [4] is that what LHC should discover is something impossible to be theoretically predicted.

REFERENCES

[1] *Some notes On the production of V particles*
G.D. James, in Proceedings of the 1955 International Conference on Particle Physics, Nuovo Cimento Supplemento 4, page 325 (1956).

[2] *Examples of the Production of* $(K^0, \overline{K}^0)$ *and* $(K^+, \overline{K}^0)$ *Pairs of Heavy Mesons*
W.A. Cooper, H. Filthuth, J.A. Newth, G. Petrucci, R.A. Salmeron and A. Zichichi, *Nuovo Cimento* 5, 1388-1397 (June 1957).

[3] *The QGCW Project*
A. Zichichi et al., CERN-LAA Preprint, October 2006; *see also*

Logical Reasoning in Experimental Physics: Past and Future
A. Zichichi, in *Gerardus 't Hooft Liber Amicorum to celebrate his 60th anniversary* (2006).

[4] A. Zichichi, Proceedings of the Erice Subnuclear Physics Schools, *Complexity at Elementary Level,* vol. 42 (2004); *Complexity at the Fundamental Level – Consequences for LHC*; vol. 44 (2006); vol. 45 (2007), World Scientific (2007-2008-2010); *Complexity and the QGCW Project*, Lecture in this volume.

THE ERICE CENTRE, GELL-MANN, QCD, THE EFFECTIVE ENERGY AND COMPLEXITY

A. Zichichi
INFN and University of Bologna, Italy
CERN, Geneva, Switzerland
Enrico Fermi Centre, Rome, Italy
Ettore Majorana Foundation and Centre for Scientific Culture, Erice, Italy

On the occasion of the 80th Anniversary of Murray Gell-Mann I recalled the role played by Murray in those experimental and technological activities I had been directly involved during many years, which started in 1955 up to the time of the LHC supercollider, where we propose to study the Quark–Gluon–Coloured–World (QGCW), which is totally different from our world made of QCD colourless baryons and mesons.

The Project [1a] is based on the development and the realization of a new technology able to implement the collisions between different incoming particles (p, n, π, K, μ, e, γ, ν) and the QGCW. This "new world" could not exist if Gell-Mann and his friends [2] had not discovered $SU(3)_c$ with the enormous number of QCD–open–colour–states. This number is by far higher that the number of baryons and mesons detected so far.

The present Project is the first step; the next one being linked to the ELN collider with 300 km ring and total energy in the Pev range [3]. We do not know what will be the final outcome of all theoretical developments intended to make predictions. What we know is that the world appears to be complex at every scale [1b]. Therefore we must expect a series of surprises that cannot be predicted [1b].

Now let me go back to the origin and recall how it happens that Murray played a fundamental role in the establishment of the Ettore Majorana Foundation and Centre for Scientific Culture in Erice (EMFCSC).

THE $(\theta-\bar{\theta})$ MIXING IN 1955 AND ITS CONSEQUENCES

I was the youngest member of the Italian delegation at the 1955 International Conference in Particle Physics where the most powerful group of experimental physics, led by Patrick M.S. Blackett, was expected to present its most recent cosmic rays results obtained at the Hochalpine Forschungsstation at Jungfraujoch, in Switzerland.

The report was presented by a fellow of the Blackett group and the chairman of the session was the famous Butler, co-discoverer with Rochester of the first experimental evidence for the existence of two totally unexpected particles called by Blackett V^0, since they appeared in the cloud chamber as inverted V's. These two V^0 were in fact a baryon Λ^0 and a meson θ^0. No one knew the reason for their existence. In order to describe their production and decay properties Murray proposed the existence of a new quantum number called Strangeness, conserved at production and violated at decay.

The great news presented by the Blackett group [4] was a series of observations where the following reaction was needed to explain the results:

$$\theta^0 + N \rightarrow \Lambda + \pi$$

A young fellow, at the end of the presentation, stood up and said: I am sorry but I would like to point out that this reaction violates "strangeness–conservation". Silence in the over-crowed lecture hall. The chairman saw another young fellow who wanted to say something; his English was very basic, but seemed to be defending the Blackett group results. So he was invited by the chairman to go to the blackboard and using the chalk he explained that in the "strangeness theory" proposed by Gell-Mann [5a, b, c] the θ-meson isospin was

½, therefore the θ neutral component had to be with positive and negative strangeness. Furthermore according to a very recent paper by Gell-Mann and Pais [5d] the θ^0–meson, produced with strangeness +1, becomes a mixture of strangeness +1 and -1; thus the production of Λ^0 could indeed be via the reaction

$$\overline{\theta}^0 + N \rightarrow \Lambda^0 + \pi.$$

The young fellow was me and this is how I became the pupil of the great Blackett, whose cloud chamber was a powerful instrument producing very interesting pictures. In two of them I found two examples of pair production of heavy mesons (K^0, $\overline{K}^0$) and (K^+, $\overline{K}^0$) [6], thus proving the correctness of the Gell-Mann strangeness model [5a, b, c, d], [7]. Blackett was very happy that old pictures had finally found the right interpretation and contributed to the understanding of the "strange" world started with the V^0–particles.

Let me recall two statements by Blackett to the young members of his group.

n. 1

«*We experimentalists are not like theorists: the originality of an idea is not for being printed in a paper, but for being shown in the implementation of an original experiment.*»

n. 2

«*We must know what boils in the brain of our theoretical friends. Not because they are able to predict our physics discoveries, but just for intellectual pleasure*».

These statements explain why I became one of his pupils.

Let me add a few details on these very exciting years.

The V^0–particles gave also rise to the (θ–τ) problem [8], which culminated in the discovery of the breaking of the symmetry operators C and P. The discovery of the non-invariance of these symmetry operators was suggested (1956) in a detailed analysis of all weak processes by T.D. Lee and C.N. Yang [9]; the first experimental evidence was provided by C.S. Wu and collaborators one year later [10]. This is how the θ–mesons became K–mesons.

Seven years after my arrival in the Blackett group, in 1962, the document establishing the existence of the Majorana Centre in Erice (EMFCSC) was signed by Blackett, Bell, Rabi, Weisskopf and myself at CERN in Geneva. Murray, thanks to his isospin ½ for the θ–meson and to the $(\theta\ ,\ \bar{\theta})$ mixing, contributed to the foundation of the Erice Centre, despite his signature not appearing in the 1962 document.

In Figure 1 there is a picture showing the celebration by the Minister of Home Affairs of the Italian Government of the bronze dedicated to the founding paper of the Ettore Majorana Centre (EMFCSC).

Dr. Enzo Bianco, former "best student" and Minister of Home Affairs in the Italian Government, celebrating the foundation of the Ettore Majorana Centre.

Figure 1: The Minister of Home Affairs, Dr. Enzo Bianco, unveils the bronze reproduction of the document which established the constitution of the Ettore Majorana Foundation and Centre for Scientific Culture (EMFCSC) signed by J.S. Bell, P.M.S. Blackett, I.I. Rabi, V.F. Weisskopf and A. Zichichi at CERN on May 8th of 1962.

In his inaugural address (Erice, May 8th, 2000) H.E. the Minister Enzo Bianco recalled how in 1964 he was named one of the 100 "best students" in Italy in an EMFCSC competition. At the extreme right Enzo Iarocci, President of the INFN (1998–2004).

I have many reasons to be grateful to Murray for his original ideas which gave me the correct hints to invent new technologies for my experimental work. These were times when the Bubble-chamber technology was dominating. The physics I was interested in was coming from the NBC–technology, where NBC stands for non-Bubble-chamber [11]; two examples are the invention and the implementation of the "pre-shower" technology and of the "neutron missing mass spectrometer". The result of these technological inventions is the

PAPLEP-set up (Figure 3) which is the first example of what is now standard: very large acceptance detectors. This powerful instrument was needed to detect the electromagnetic decays of mesons (pseudoscalar or vector) whose masses were measured with the missing mass spectrometer. This is how we observed for the first time the 2 γ decays of the X^0–meson, which was a candidate to have spin 2 while it was with spin zero. Those who are of my age remember that the η' is the new name of the old X^0–meson.

Figure 2: Murray Gell-Mann lecturing at Erice (1966).

Nothing of all this would exist [11] without $SU(3)_{uds}$ of Gell-Mann and collaborators. We should not forget that $SU(3)_{uds}$ started with the existence of the Isospin, when, as said already, people believed that Isofermions and Isobosons had to be also fermionic and bosonic states, as was the case for the nucleon and the π–meson. But, no-one could explain the origin of the Isotopic spin and of $SU(3)_{uds}$: why is the "strong coupling", α_3, "flavour independent"? It is the gluon-gluon interaction which guarantees the "flavour" independence of α_3. It is the non-Abelian property of SU(3)–colour of Gell-Mann and collaborators, which allowed all of us, after many decades, to finally understand the origin of the Isospin and of $SU(3)_{uds}$.

PAPLEP–set up

THIS IS THE FIRST EXAMPLE of what is now "standard" in experimental subnuclear physics: VERY LARGE ACCEPTANCE DETECTORS.

In the central part of the picture is the "pre-shower" technology invented for the study of the rare decay modes of the pseudoscalar and vector mesons in order to measure the pseudoscalar meson mixing θ_{PS} and the vector meson mixing θ_V:

$$\theta_{PS} \neq \theta_V .$$

On the rails the "neutron missing mass spectrometer" needed to measure the masses of the pseudoscalar and the vector mesons.

Figure 3

FROM THE PROLIFERATION OF MESONS AND BARYONS TO THE EFFECTIVE-ENERGY

The physics of strong interactions was characterized by two classes of phenomena, one of "static" nature, the other of "dynamic" nature. Both were affected by proliferation.

The proliferation in the "static" sector of the strong interaction was the huge number of mesons and baryons [12]. This multitude of states was reduced by an order of magnitude through the octets and decuplets of Gell-Mann and Ne'eman $SU(3)_{uds}$ [13] whose consequences for new physics needed the PAPLEP-set up, previously discussed.

The proliferation in the "dynamic" sector was the multitude of final states produced by pairs of interacting particles, in strong, electromagnetic and weak processes:

Strong	EM	Weak
$\pi\, p$	$\gamma\, p$	$\nu\, p$
$K\, p$	$e\, p$	
$p\, p$	$\mu\, p$	
$p\, n$	e^+e^-	
$\bar{p}\, p$		

The properties measured in the multihadronic final states were, for example,

i) the average charged multiplicity; $<n_{ch}>$;

ii) the fractional energy distribution; $d\sigma / dx_i$;

iii) the transverse momentum distribution $d\sigma / dp_{t_i}$; etc. ..... .

All these quantities appeared to depend on the nature of the interacting particles.

How can it be that a unique fundamental force, QCD, acting among its two very simple basic components, quarks and gluons, produces such a variety of final states as those observed when a pair of particles interact?

It is the introduction of the "Effective Energy" which allowed us to put all the different final states on the same basis.

The results are the “universality features” measured in all multihadronic final states, no matter what is the pair of interacting particles in the initial state.

It was shown (Figure 4) that the momenta of the secondary particles produced in (pp) collisions had the same distribution as those produced in (e^+e^-) annihilation, at the same “Effective Energy”, i.e. $E^{had} \equiv \left(\sqrt{s}\right)_{e^+e^-}$.

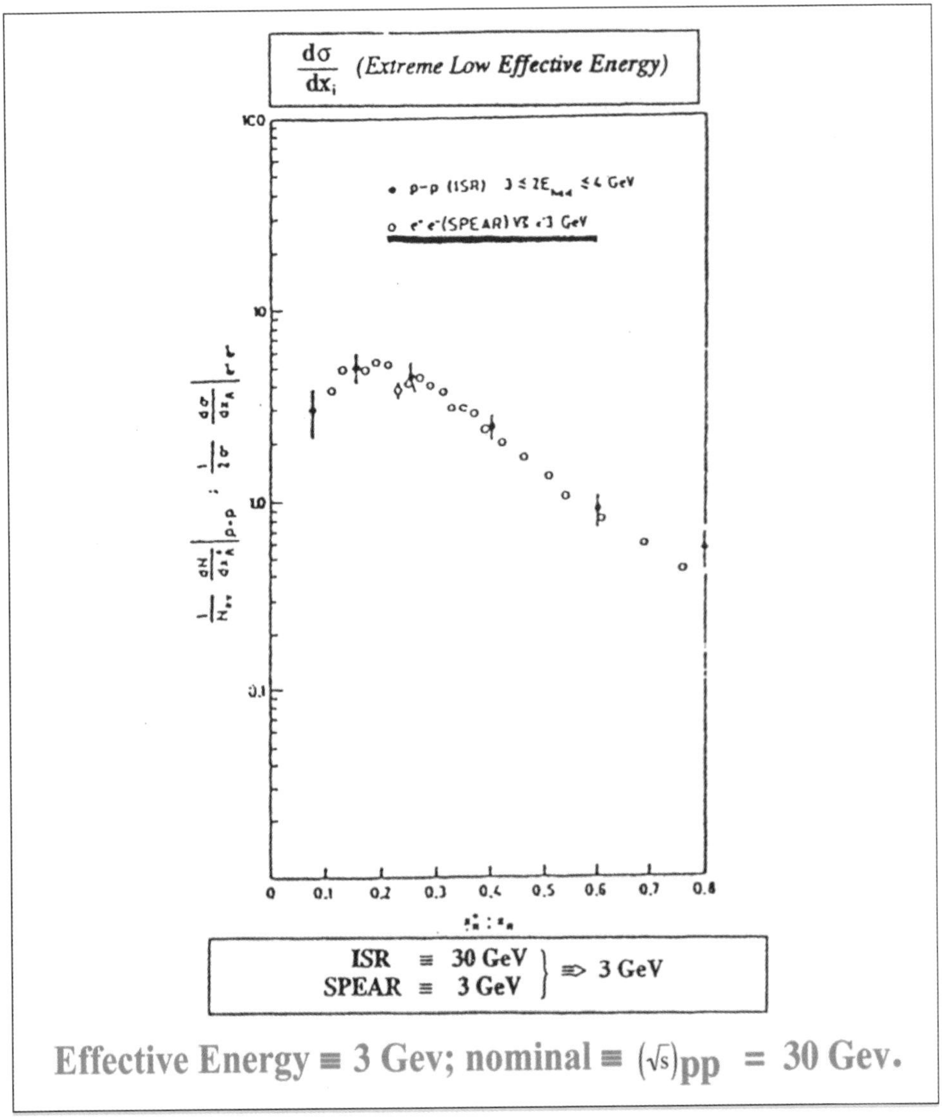

Figure 4: The inclusive single-particle fractional momentum distributions $(1/N_{ev}) \cdot (dN/dx_R)$ in the interval 3 Gev $\leq 2E^{had} \leq$ 4 Gev obtained from data at $(\sqrt{s})_{pp}$ = 30 Gev. Also shown are data from MARK I at SPEAR.

Another "universality feature", the average charged multiplicity, $< n_{ch} >$, measured in (pp) and in (e^+e^-) interactions, is reported in Figure 5.

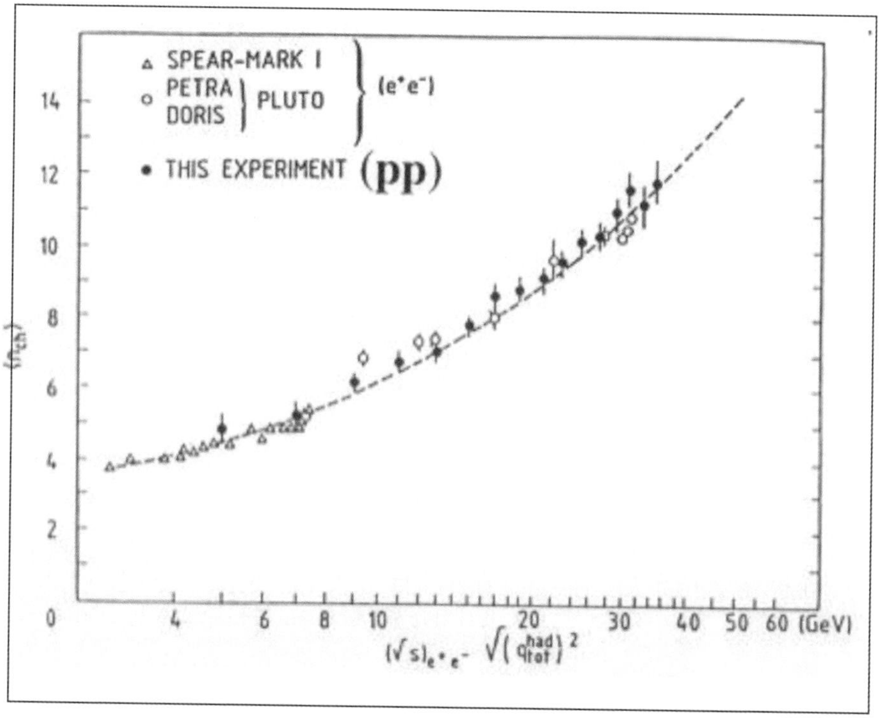

Figure 5: Direct comparison between (pp) and (e^+e^-) processes. Mean charged multiplicity as measured in e^+e^- using $(\sqrt{s})_{e^+e^-}$ and pp, using the Effective Energy E^{had}, not the nominal $(\sqrt{s})_{pp}$.

The "Effective Energy" is essential in the analysis of the multihadronic final states in order to establish the existence of a "flow" of quantum numbers from the initial to the final state.

The data for the "flow" of quantum numbers measured in strong interactions are in Figure 6, while those for the "flow" measured in electromagnetic and weak processes are in Figure 7.

None of these experiments would have ever been performed without Gell-Mann and collaborators, SU(3) – colour, QCD.

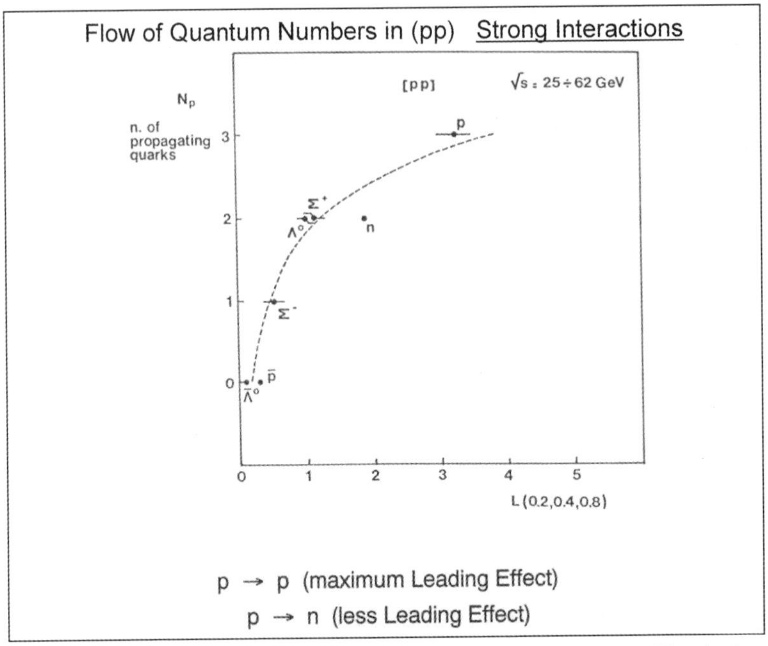

Figure 6: The centre-of-mass energy ranges from 25 to 62 Gev. The hadrons are ordered according to the number of propagating quarks. The dotted curve superimposed is obtained by using a parametrization of the single-particle inclusive cross-section, $F(x) = (1 - x)^{\alpha}$.

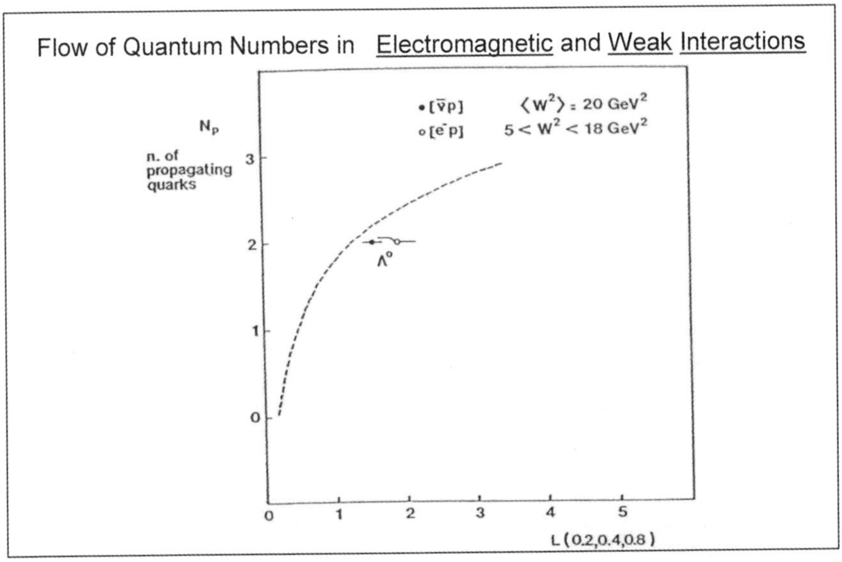

Figure 7: The "quantum number flow" from the initial to the final state is observed in an electromagnetic process ($e^- p \rightarrow \Lambda^0 x$) and in a weak process ($\bar{\nu} p \rightarrow \Lambda^0 x$).

The final states produced in (pp) interactions have properties like those produced in (e^+e^-) annihilation provided that $\left(\sqrt{s}\right)_{e^+e^-}$ is equal to E^{had}. The "nominal" ISR energy, $(\sqrt{s})_{pp}$, had to be corrected event by event in order to find the correspondent "Effective Energy". This implied the study of a very large number of events. A quotation from the first paper where these results were reported is shown in Figure 8.

Introduction of the "**Effective Energy**"

EVIDENCE OF THE SAME MULTIPARTICLE PRODUCTION MECHANISM IN p-p COLLISIONS AS IN e^+e^- ANNIHILATION

M. Basile, G. Cara Romeo, L. Cifarelli, A. Contin, G. D'Alì, P. Di Cesare, B. Esposito, P. Giusti, T. Massam, F. Palmonari, G. Sartorelli, G. Valenti and A. Zichichi.

Physics Letters 92B, 367 (1980).

"*The agreement between the momentum distributions obtained in e^+e^- annihilation and in p–p collisions suggests that the mechanism for transforming energy into particles in these two processes, so far considered very different, must be the same*".

Figure 8: The first paper where the "Effective Energy" was introduced in the study of high energy (pp) interactions at ISR.

In order to check the validity of the "Effective Energy" down to the lowest value, the ISR collider was used at its lowest nominal energy: $(\sqrt{s})_{pp} = 30$ Gev. This allowed a set of very low "Effective Energies" to be obtained using much higher energy of purely hadronic interactions.

The Future.

The next step, following the "Effective Energy" is the study [1a] of the quark-gluon-coloured-world (QGCW) whose number of QCD-open-colour-states in by far higher than the number of baryons and mesons known so far and still to discovered, but limited by the fact that they must be QCD-colourless states. This Project [1a] is our future when we will continue to pay tribute to Murray Gell-Mann, even if what will be discovered will be a series of surprises that cannot be theoretically predicted [1b].

REFERENCES

[1a] *The QGCW Project*
A. Zichichi et al., CERN-LAA Preprint, October 2006; *see also*
Logical Reasoning in Experimental Physics: Past and Future
A. Zichichi, in *Gerardus 't Hooft Liber Amicorum to celebrate his 60th anniversary* (2006).

[1b] *Complexity and the QGCW Project*
A. Zichichi, Lecture in this volume; *see also*
A. Zichichi et al., CERN-LAA Preprint, October 2006; *see also*
A. Zichichi, Proceedings of the Erice Subnuclear Physics Schools, *Complexity at Elementary Level,* vol. 42, 251 (2004); *Complexity at the Fundamental Level – Consequences for LHC*; vol. 44, 269 (2006); vol. 45, 365 (2007), World Scientific (2007-2008-2010). See also *Complexity and the QGCW Project*, Lecture in this volume.

[2] *Lectures by Fitch and Leutwyler in this volume.*

[3] *ELOISATRON (The European LOng Intersecting Storage Accelerator)*
A. Zichichi et al, INFN/AE-83/7, June 1983. INFN/AE-84/2, January 1984 (revised version 1985). Presented at the Galileo Galilei and Alfred B. Nobel Celebrations "Science for Peace", San Remo and Rome, Italy (1-11 May 1983); *see also*
A. Zichichi, *Subnuclear Physics: The first 50 Years: Highlights from Erice to ELN*, O. Barnabei, P. Pupillo, F. Roversi Monaco (eds), a joint publication by University and Academy of Sciences of Bologna, Italy (1998); 20th Century Physics Series, Vol. 24, World Scientific (2000).

[4] *Some notes On the production of V particles*
G.D. James, in Proceedings of the 1955 International Conference on Particle Physics, Nuovo Cimento Supplemento 4, page 325 (1956).

[5a] *On the Classification of Particles*
M. Gell-Mann, (circulated in preprint form, unpublished) August (1953).

[5b] M. Gell-Mann, *Phys. Rev.* 92, page 833 (1953).

[5c] M. Gell-Mann and A. Pais, *Proceedings of the Glasgow Conference on Mesons and Nuclear Physics*, page 342 (1954).

[5d] *Behavior of Neutral Particles under Charge Conjugation*
M. Gell-Mann and A. Pais, *Phys. Rev.* 97, 1387 (1955).

[6] *Examples of the Production of* (K^0, $\overline{K}^0$) *and* (K^+, $\overline{K}^0$) *Pairs of Heavy Mesons*
W.A. Cooper, H. Filthuth, J.A. Newth, G. Petrucci, R.A. Salmeron and A. Zichichi, *Nuovo Cimento* 5, 1388-1397 (June 1957).

[7] *The Interpretation of the New Particles as Displaced Charge Multiplets*
M. Gell-Mann, *Suppl. Nuovo Cimento IV, Serie X, n. 2* (1956).

[8] *On the Analysis of τ-Meson data and the Nature of the τ-Meson*
R.H. Dalitz, *Phil. Mag.* 44, 1068 (1953);

Isotopic Spin Changes in τ and θ Decay
R.H. Dalitz, *Proceedings of the Physical Society* A69, 527 (1956);

Present Status of τ Spin-Parity
R.H. Dalitz, Proceedings of the *Sixth Annual Rochester Conference on High Energy Nuclear Physics*, Interscience Publishers, Inc., New York, 19 (1956); for a detailed record of the events which led to the (θ−τ) puzzle see R.H. Dalitz "*Kaon Decays to Pions: the τ−θ Problem*" in "*History of Original Ideas and Basic Discoveries in Particle Physics*", H.B. Newman and T. Ypsilantis (eds), Plenum Press, New York and London, 163 (1994).

[9] *Question of Parity Conservation in Weak Interactions*
T.D. Lee and C.N. Yang, *Phys. Rev.* 104, 254 (1956).

[10] *Experimental Test of Parity Conservation in Beta Decay*
C.S. Wu, E. Ambler, R.W. Hayward, D.D. Hoppes, *Phys. Rev.* 105, 1413 (1957);

Observation of the Failure of Conservation of Parity and Charge Conjugation in Meson Decays: The Magnetic Moment of the Free Muon
R. Garwin, L. Lederman, and M. Weinrich, *Phys. Rev.* 105, 1415 (1957);

Nuclear Emulsion Evidence for Parity Non-Conservation in the Decay Chain $\pi^+\mu^+e^+$
J.J. Friedman and V.L. Telegdi, *Phys. Rev.* 105, 1681 (1957).

[11] *An apparatus of the NBC type and the physics results obtained*
A. Zichichi, *Annals of Physics 66,* 405 (1971).

[12] An example of the proliferation in the meson resonances is "*Meson Resonances and Related Electromagnetic Phenomena*", R.H. Dalitz and A. Zichichi (eds), Editrice Compositori, Proceedings of the EPS Conference, Bologna (1971). A record of the evolution in the multitude of baryonic and mesonic states can be found in the proceedings of the Erice Subnuclear Physics Schools.

[13] M. Gell-Mann, "*The Eightfold Way — A Theory of Strong-Interaction Symmetry*", California Institute Technology Synchrotron Lab. Report 20 (1961); "*Derivation of Strong Interactions from a Gauge Invariance*", Y. Ne'eman, *Nuclear Phys.* 26, 222 (1961). The experimental confirmation, in 1964, by Samios and collaborators of the existence of the missing member of the baryonic decuplet, Ω^-, appeared to be, at the time, a triumph for the "eightfold way". *Observation of a Hyperon with Strangeness Minus Three,* V.E. Barnes, P.L. Connolly, D.J. Crennell, B.B. Culwick, W.C. Delaney, W.B. Fowler, P.E. Hagerty, E.L. Hart, N. Horwitz, P.V.C. Hough, J.E. Jensen, J.K. Kopp, K.W. Lai, J. Leitner, J.L. Lloyd, G.W. London, T.W. Morris, Y. Oren, R.B. Palmer, A.G. Prodell, D. Radojictc, D.C. Rahm, C.R. Richardson, N.P. Samios, J.R. Sanford, R.P. Shutt, J.R. Smith, D.L. Stonehill, R.C. Strand, A.M. Thorndike, M.S. Webster, W.J. Willis and S.S. Yamamoto, *Phys. Rev. Lett.* 12, 204 (1964). The choice of the letter Ω, the last in the Greek alphabet, was due to the conviction that this particle was going to be the last ever to be discovered. See also: M. Gell-Mann and Y. Ne'eman, "*The Eightfold Way*", W.A. Benjamin inc., New York and Amsterdam (1964).

History of QCD

H. Fritzsch

Department für Physik

Ludwig-Maximilians-Universität

München, Germany

In 1961 Murray Gell-Mann and Yuval Ne'eman introduced the SU(3)-symmetry to classify the observed particles (Gell-Mann, 1961 and 1962; Ne'eman, 1961). The group SU(3) is the group of unitary 3×3 matrices. The observed hadrons are members of specific representations of SU(3). The baryons are singlets, octets and decuplets, the mesons are singlets and octets.

It remained unclear, why the members of the simplest SU(3) representation, the triplet representation, were not observed in the experiments. These particles would have non-integral electric charges: 2/3 or –1/3.

In 1964 Gell-Mann and George Zweig proposed that the baryons and mesons are bound states of those triplet particles, the quarks (Gell-Mann, 1964, Zweig, 1964). Since the quarks are an SU(3) triplet, there must be three quarks (u,d,s), with electric charges (2/3, -1/3, -1/3). The proton would be a bound state with the quark structure (uud).

Quarks with color

In the summer of 1970 Gell-Mann and I started working together. First we studied the results from SLAC on the deep-inelastic scattering of electrons and atomic nuclei. The cross-sections depend on the mass of the virtual photon and the energy transfer from the electron to the proton. However the experimental results at SLAC indicated that the cross-sections at large energies depend only on the ratio of the square of the photon mass and of the energy transfer – they showed a scaling behaviour, which had been predicted by James Bjorken.

In the SLAC experiments the nucleon matrix-element of the commutator of two electromagnetic currents is measured at nearly light-like distances. Gell-Mann and I assumed that this commutator can be abstracted from the free-quark model. We introduced the light-cone algebra of the currents (Fritzsch and Gell-

Mann, 1971). Using this algebra, we could understand the scaling behaviour. Later it turned out that the results of the light-cone current algebra are nearly correct in the theory of QCD, due to the asymptotic freedom of the theory.

The baryon Ω^- is a bound state of three strange quarks. Since it is the ground state, the space wave-function should be symmetric. The three spins of the quarks are aligned to give the spin of the omega minus. Thus the spin wave function is also symmetric, and the wave function of the Ω^- does not change, if two quarks are interchanged. However the wave function must be antisymmetric according to the Pauli principle. This was a great problem for the quark model.

In 1971 Gell-Mann and I found the solution of the statistics problem (Fritzsch and Gell-Mann, 1971). We assumed that the quarks had a new conserved quantum number, which we called "color". There would be a read quark, a green quark and a blue quark. The color symmetry SU(3) was assumed to be an exact symmetry.

The wave functions of the hadrons were assumed to be singlets of the color group. In this case the baryon wave-functions are anti-symmetric in the color indices. Thus the wave function of a baryon changes sign if two quarks are exchanged, as required by the Pauli principle.

Later we found other indications that the color quantum number must be real. The cross-section for electron–positron annihilation into hadrons at high energies depends on the squares of the electric charges of the quarks and on the number of colors. For three colors the ratio of the hadronic production cross section and of the cross section for producing a pair of muons is 2. Without color this ratio would be 2/3. The experimental data, however, were in agreement with a ratio of 2, as observed by the Zichichi group with the electron-positron collider ADONE (Zichichi, 1971).

In 1972 Gell-Mann and I worked at CERN. Together with William Bardeen we investigated the electromagnetic decay of the neutral pion into two photons. It was known that in the quark model the decay rate is about a factor nine less than the measured decay rate – another problem for the quark model.

The decay amplitude is given by a triangle diagram, in which a quark–antiquark pair is created virtually and subsequently annihilates into two photons. We found that after the introduction of color the decay amplitude increases by a factor three – each color contributes to the amplitude with the same strength. For three colors the theoretical result agrees with the experimental value.

In the spring of 1972 Gell-Mann and I started to interpret the color group as a gauge group. The resulting gauge theory is similar to quantum electrodynamics

(QED). The interaction of the quarks is generated by an octet of massless colored gauge bosons, which we called gluons (Fritzsch and Gell-Mann, 1972).

Later we introduced the name "Quantum-Chromodynamics" (QCD). We published details of this theory one year later together with Heinrich Leutwyler (Fritzsch, Gell-Mann and Leutwyler, 1972).

In QCD the gluons interact not only with the quarks, but also with themselves. This direct gluon–gluon interaction is important – it leads to the reduction of the coupling constant at increasing energy, i.e. the theory is asymptotically free, as discovered in 1972 by Gerard ’t Hooft (unpublished) and in 1973 by David Gross, David Politzer and Frank Wilczek.

Thus at high energies the quarks and gluons behave almost as free particles. This leads to the approximate "scaling behaviour" of the cross-sections in the deep-inelastic lepton–hadron scattering. The quarks behave almost as free particles at very small distances.

The logarithmic decrease of the coupling constant depends on the QCD scale parameter Λ, which is a free parameter and has to be measured in the experiments. The current experimental value of Λ is about 220 MeV.

Experiments at SLAC, at DESY, at CERN’s Large Electron-Positron-Collider and at Fermilab’s Tevatron have measured the decrease of the QCD coupling-constant. With LEP it was also possible to determine the QCD coupling-constant at the mass of the Z boson rather precisely: $\alpha_s(M_Z) = 0.1184 \pm 0.0007$.

It is useful to consider the theory of QCD with just one heavy quark Q. The ground-state meson in this hypothetical case would be a pseudoscalar meson, a quark–antiquark bound state. The effective potential between the quark and its antiquark at small distances would be a Coulomb potential proportional to 1/r, where r is the distance between the quark and the antiquark.

However at large distances the self-interaction of the gluons becomes important. The gluonic field lines at large distances do not spread out as in electrodynamics. Instead they attract each other and the quark and the antiquark are connected by a string of parallel gluonic field lines. The force between the quark and the antiquark is constant, i.e. it does not decrease as in electrodynamics. Thus the heavy quark is confined. It is still an open question whether this is true also, if light quarks are present.

In electron–positron annihilation the virtual photon creates a quark and an antiquark, which move away from each other with high speed. The quark and the antiquark "fragment" to produce two jets of particles, mostly pions. The sum of the energies and momenta of the particles in each jet should be equal to the energy of the original quark, which is equal to the energy of each colliding lepton. These quark jets were observed for the first time in 1978 at DESY. They had been predicted in 1975 by Richard Feynman.

If a quark-antiquark pair is produced in the electron–positron annihilation, then QCD predicts that sometimes a high-energy gluon should be emitted from one of the quarks. The gluon would also fragment and produce a jet of particles, thus three jets are produced. Such events were observed at DESY in 1979.

The basic quanta of QCD are the quarks and the gluons. Two color-octet gluons can form a color-singlet. Such a state would be a neutral glue-meson. In QCD with only heavy quarks this state would be stable, but in the real world it would mix with neutral quark–antiquark mesons and would decay quickly into pions. Thus far glue-mesons have not been identified clearly in the experiments.

The three quark flavours were introduced to describe the symmetry given by the flavour group SU(3). However we now know that in reality there are six quarks: the three light quarks u, d, s and the three heavy quarks c (charm), b (bottom) and t (top). These six quarks form three doublets of the electroweak symmetry group SU(2): (u,d) – (c,s) – (t,b).

The masses of the quarks are arbitrary parameters in QCD, just as the lepton masses are arbitrary parameters in QED. Since the quarks do not exist as free particles, their masses cannot be measured directly. They can, however, be estimated using the chiral symmetry and the observed hadron masses. In QCD they depend on the energy scale under consideration. Typical values of the quark masses at the energy of 2 GeV are:

u: 4 MeV	c: 1.2 GeV	t: 173 GeV
d: 6 MeV	s: 0.1 GeV	b: 4.4 GeV.

The mass of the t quark is large, similar to the mass of a gold atom. Due to this large mass the t quark decays by the weak interaction with a lifetime that is less than the time needed to form a meson or baryon. Thus there exist no hadrons containing a t quark.

The theory of QCD is the correct field theory of the strong interactions and of the nuclear forces. Both hadrons and atomic nuclei are bound states of quarks, antiquarks and gluons. It is remarkable that a simple gauge theory can describe the complicated phenomena of the strong interactions.

The physics of strong interactions is in fact characterized by two classes of phenomena. The first class discussed so far – the "static" one – was affected by the proliferation of mesonic and barionic states. The "dynamic" class was also affected by proliferation.

As stated by 't Hooft [G. 't Hooft 1998] "*The experimental results were discouraging: scattering experiments yielded different final states for each pair of interacting particles. Theoreticians were unable to prescribe what experimentalists had to look for to establish the universal nature of these final states. So it happened that these aspects of QCD had to wait until experimentalists themselves came with the right idea [A. Zichichi 1980]. The showers come with what is now called an "effective energy", and, in terms of this quantity, universality could be established*". The "effective energy" is the most spectacular consequence of QCD.

References

H. Fritzsch and M. Gell-Mann 1971 *Proceedings of the International Conference on Duality and Symmetry*, Tel Aviv

H. Fritzsch and M. Gell-Mann 1972 *Proceedings of the Int. Conference on High Energy Physics*, Chicago

H. Fritzsch, M. Gell-Mann and H. Leutwyler 1973 *Phys. Lett.* **47B,** 365

M. Gell-Mann 1964 *Phys. Lett.* **8** 214

M. Gell-Mann 1962 *Phys. Rev.* **125** 1067

M. Gell-Mann 1961 *Cal Tech. Lab. Report* **20**

Y. Ne'eman 1961 *Nucl. Phys.* **26** 222

G. 't Hooft 1998 in *The Creation of Quantum ChromoDynamics and the Effective Energy*" V.N. Gribov, G. 't Hooft, G. Veneziano and V.F. Weisskopf; N.L. Lipatov (ed), *World Scientific Series in 20th Century Physics* **25** 39

A. Zichichi (V. Alles-Borelli et al.) 1971, in: Proceedings EPS International Conference on "Meson Resonances and Related Electromagnetic Phenomena", R. H. Dalitz and A. Zichichi (eds), Editrice Compositori, Bologna, Italy

A. Zichichi et al., 1980 *Phys. Lett.* **92B** 367 – see also G. 't Hooft 1998

G. Zweig 1964 *Erice School 1964 "Symmetries in Elementary Particle Physics*", A. Zichichi (ed), *Academic Press* 192 and *CERN-TH-401* (unpublished)

On the history of the strong interaction

H. Leutwyler
Albert Einstein Center for Fundamental Physics
Institute for Theoretical Physics, University of Bern
Sidlerstr. 5, CH-3012 Bern, Switzerland

Abstract

These lecture notes recall the conceptual developments which led from the discovery of the neutron to our present understanding of strong interaction physics.

1 From nucleons to quarks

I am not a historian. The following text describes my own recollections and is mainly based on memory, which unfortunately does not appear to improve with age ... For a professional account, I refer to the book by Tian Yu Cao [1]. A few other sources are referred to below.

1.1 Beginnings

The discovery of the neutron by Chadwick in 1932 may be viewed as the birth of the strong interaction: it indicated that the nuclei consist of protons and neutrons and hence the presence of a force that holds them together, strong enough to counteract the electromagnetic repulsion. Immediately thereafter, Heisenberg introduced the notion of isospin as a symmetry of the strong interaction, in order to explain why proton and neutron nearly have the same mass [2]. In 1935, Yukawa pointed out that the nuclear force could be generated by the exchange of a hypothetical spinless particle, provided its mass is intermediate between the masses of proton and electron – a *meson* [3]. Today, we know that such a particle indeed exists: Yukawa predicted the pion. Stueckelberg pursued similar ideas, but was mainly thinking about particles of spin 1, in analogy with the particle that mediates the electromagnetic interaction [4].

In the thirties and fourties of the last century, the understanding of the force between two nucleons made considerable progress, in the framework of nonrelativistic potential models. These are much more flexible than quantum field theories. Suitable potentials that are attractive at large distances but repulsive at short distances do yield a decent understanding of nuclear structure: Paris potential, Bonn potential, shell model of the nucleus. In this framework, nuclear reactions, in particular the processes responsible for the luminosity of the sun, stellar structure, α-decay and related matters were well understood more than fifty years ago.

These phenomena concern interactions among nucleons with small relative velocities. Experimentally, it had become possible to explore relativistic collisions, but a description in terms of nonrelativistic potentials cannot cover these. In the period between 1935 and 1965, many attempts at formulating a theory of the strong interaction based on elementary fields for baryons and mesons were made. In particular, uncountable PhD theses were written, based on local interactions of the Yukawa type, using perturbation theory to analyze them, despite the fact that the coupling constants invariably turned out to be numerically large. Absolutely nothing worked even half way.

Although there was considerable progress in understanding the general principles of quantum field theory (Lorentz invariance, unitarity, crossing symmetry, causality, analyticity, dispersion relations, CPT theorem, spin and statistics) as well as in renormalization theory, faith in quantum field theory was in decline, even concerning QED (Landau-pole). To many, the renormalization procedure needed to arrive at physically meaningful results looked suspicious and it appeared doubtful that the strong interaction could at all be described by means of a local quantum field theory. Some suggested that this framework should be replaced by S-matrix theory – heated debates concerning this suggestion took place at the time [5]. Regge poles were considered as a promising alternative to the quantum fields (the Veneziano model is born in 1968 [6]). Fifty years ago, when I completed my studies, the quantum field theory of the strong interaction consisted of a collection of beliefs, prejudices and assumptions. Quite a few of these turned out to be wrong.

1.2 Flavour symmetries

Symmetries that extend isospin to a larger Lie group provided the first hints towards an understanding of the structure underneath the strong interaction phenomena. The introduction of the strangeness quantum number and the Gell-Mann-Nishijima formula [7] was a significant step in this direction. Goldberger and Treiman [8] then showed that the *axial-vector current* plays an important role, not only in the weak interaction (the pion-to-vacuum matrix element of this current – the pion decay constant F_π – determines the rate of the weak decay $\pi \longrightarrow \mu\nu$) but also in the context of the strong interaction: the nucleon matrix element of the axial-vector current, g_A, determines the strength of the interaction between pions and nucleons: $g_{\pi N} = g_A M_N / F_\pi$.

At low energies, the main characteristic of the strong interaction is that the energy gap is small: the lightest state occurring in the eigenvalue spectrum of the Hamiltonian is the pion, with $M_\pi \simeq 135$ MeV, small compared to the mass of the proton, $M_p \simeq 938$ MeV. In 1960, Nambu found out why that is so: it has to do with a hidden, approximate, continuous symmetry. Since some of its generators carry negative parity, it is referred to as a *chiral symmetry*. For this symmetry to be consistent with observation, it is essential that an analog of spontaneous magnetization occurs in particle physics: for dynamical reasons, the state of lowest energy – the vacuum – is not symmetric under chiral transformations. Consequently, the symmetry cannot be seen in the spectrum of the theory: it is *hidden* or *spontaneously broken*. Nambu realized that the spontaneous breakdown of a continuous symmetry entails massless particles analogous to the spin waves of a magnet and concluded that the pions must play this role. If the strong interaction was strictly invariant under chiral symmetry, there would be no energy gap at all – the pions would be massless.[1] Conversely, since the pions are not massless, chiral symmetry cannot be exact – unlike isospin, which at that time was taken to be an exact symmetry of the strong interaction. The spectrum does have an energy gap because chiral symmetry is not exact: the pions are not massless, only light. In fact, they represent the lightest strongly interacting particles that can be exchanged between two nucleons. This is why, at large distances, the potential between two nucleons is correctly described by the Yukawa formula.

The discovery of the *Eightfold Way* by Gell-Mann and Ne'eman paved the way to an understanding of the mass pattern of the baryons and mesons [9, 10]. Like chiral symmetry, the group SU(3) that underlies the Eightfold way represents an approximate symmetry: the spectrum of the mesons and baryons does not consist of degenerate multiplets of this group. The splitting between the energy levels, however, does exhibit a pattern that can be understood in terms of the assumption that the part of the Hamiltonian that breaks the symmetry transforms in a simple way. This led to the Gell-Mann-Okubo formula [10, 11] and to a prediction for the mass of the Ω^-, a member of the baryon decuplet which was still missing, but was soon confirmed experimentally, at the predicted place [12].

[1] A precise formulation of this statement, known as the *Goldstone theorem*, was given later [13].

1.3 Quark Model

In 1964, Gell-Mann [14] and Zweig [15] pointed out that the observed pattern of baryons can qualitatively be understood on the basis of the assumption that these particles are bound states built with three constituents, while the spectrum of the mesons indicates that they contain only two of these. Zweig called the constituents "aces". Gell-Mann coined the term "quarks", which is now commonly accepted. The Quark Model gradually evolved into a very simple and successful semi-quantitative framework, but gave rise to a fundamental puzzle: why do the constituents not show up in experiment ? For this reason, the existence of the quarks was considered doubtful: "Such particles [quarks] presumably are not real but we may use them in our field theory anyway ..." [16]. Quarks were treated like the veal used to prepare a pheasant in the royal french cuisine: the pheasant was baked between two slices of veal, which were then discarded (or left for the less royal members of the court). Conceptually, this was a shaky cuisine.

If the flavour symmetries are important, why are they not exact ? Gell-Mann found a beautiful explanation: *current algebra* [10, 16]. The charges form an exact algebra even if they do not commute with the Hamiltonian and the framework can be extended to the corresponding currents, irrespective of whether or not they are conserved. Adler and Weisberger showed that current algebra can be tested with the sum rule that follows from the nucleon matrix element of the commutator of two axial-vector charges [17]. Weinberg then demonstrated that even the strength of the interaction among the pions can be understood on the basis of current algebra: the $\pi\pi$ scattering lengths can be predicted in terms of the pion decay constant [18].

1.4 Behaviour at short distances

Bjorken had pointed out that if the nucleons contain point-like constituents, then the *ep* cross section should obey scaling laws in the deep inelastic region [19]. Indeed, the scattering experiments carried out by the MIT-SLAC collaboration in 1968/69 did show experimental evidence for such constituents [20]. Feynman called these *partons*, leaving it open whether they were the quarks or something else.

The operator product expansion turned out to be a very useful tool for the short distance analysis of the theory – the title of the paper where it was introduced [21], "Nonlagrangian models of current algebra", reflects the general skepticism towards Lagrangian quantum field theory that I mentioned at the end of section 1.1.

1.5 Colour

The Quark Model was difficult to reconcile with the spin-statistics theorem which implies that particles of spin $\frac{1}{2}$ must obey Fermi statistics. Greenberg proposed that the quarks obey neither Fermi-statistics nor Bose-statistics, but "para-statistics of order three" [22]. The proposal amounts to the introduction of a new internal quantum number. Indeed, in 1965, Bogoliubov, Struminsky and Tavkhelidze [23], Han and Nambu [24] and Miyamoto [25] independently pointed out that some of the problems encountered in the quark model disappear if the u, d and s quarks occur in 3 states. Gell-Mann coined the term "colour" for the new quantum number.

One of the possibilities considered for the interaction that binds the quarks together was an abelian gauge field analogous to the e.m. field, but this gave rise to problems, because the field would then interfere with the other degrees of freedom. Fritzsch and Gell-Mann pointed out that if the gluons carry colour, then the empirical observation that quarks appear to be confined might also apply to them: the spectrum of the theory might exclusively contain colour neutral states.

In his lectures at the Schladming Winter School in 1972 [27], Gell-Mann thoroughly discussed the role of the quarks and gluons: theorists had to navigate between Scylla and Charybdis, trying to abstract neither too much nor too little from models built with these objects. The basic tool at that time was *current algebra on the light cone.* He invited me to visit Caltech. I did that during three months in the spring break of 1973 and spent an extremely interesting period there.

1.6 QCD

Gell-Mann's talk at the High Energy Physics Conference in 1972 (Fermilab), had the title "Current algebra: Quarks and what else?" In particular, he discussed the proposal to describe the gluons in terms of a *nonabelian gauge field* coupled to colour, relying on work done with Fritzsch [26]. As it was known already that the electromagnetic and weak interactions are mediated by gauge fields, the idea that colour might be a local symmetry as well does not appear as far fetched. In the proceedings, Fritzsch and Gell-Mann mention unpublished work in this direction by Wess. For a recent outline of the development that led from the quark model to *Quantum Chromodynamics* – the quantum field theory which describes the interaction among the quarks and gluons – see [28, 29].

The main problem at the time was that for a gauge field theory to describe the hadrons and their interaction, it had to be fundamentally different from the quantum field theories encountered in nature so far: all of these, including the electroweak theory, have the spectrum indicated by the degrees of freedom occurring in the Lagrangian: photons, leptons, intermediate bosons, ... The proposal can only make sense if this need not be so, that is if the spectrum of physical states in a quantum field theory can differ from the spectrum of the fields needed to formulate it: gluons and quarks in the Lagrangian, hadrons in the spectrum. This looked like wishful thinking.

2 On the history of the gauge field concept

2.1 Electromagnetic interaction

The final form of the laws obeyed by the electromagnetic field was found by Maxwell, around 1860 – these laws survived relativity and quantum theory, unharmed.

Fock pointed out that the Schrödinger equation for electrons in an electromagnetic field,

$$\frac{1}{i}\frac{\partial\psi}{\partial t}-\frac{1}{2m_e^2}(\vec{\nabla}+i\,e\vec{A})^2\psi-e\,\varphi\,\psi=0\,,$$

is invariant under a group of local transformations:

$$\vec{A}'(x)=\vec{A}(x)+\vec{\nabla}\alpha(x),\qquad \varphi'(x)=\varphi(x)-\frac{\partial\alpha(x)}{\partial t},\qquad \psi(x)'=e^{-ie\alpha(x)}\,\psi(x)\,,$$

in the sense that the fields $\vec{A}',\varphi',\psi'$ describe the same physical situation as $\vec{A},\varphi,\psi$ [30]. Weyl termed these *gauge transformations* (with gauge group U(1) in this case). In fact, the electromagnetic interaction is fully characterized by the symmetry with respect to this group: gauge invariance is the crucial property of this interaction.

I illustrate the statement with the core of Quantum Electrodynamics: photons and electrons. Gauge invariance allows only two free parameters in the Lagrangian of this system: e, m_e. Moreover, only one of these is dimensionless: $e^2/4\pi = 1/137.035\,999\,074\,(44)$. U(1) symmetry and renormalizability fully determine the properties of the e.m. interaction, except for this number, which so far still remains unexplained.[2]

[2]As a side remark, I note that in QED, an additional term, proportional to $F_{\mu\nu}\tilde{F}^{\mu\nu}$ can be added to the Lagrangian. The term represents a total derivative, $F_{\mu\nu}\tilde{F}^{\mu\nu}=\partial_\mu f^\mu$. Since e.m. fields with a nontrivial behaviour at large distances do not appear to play a significant role (no instantons or the like), this term does not affect the physics.

2.2 Gauge fields from geometry

Kaluza [31] and Klein [32] had shown that a 5-dimensional Riemann space with a metric that is independent of the fifth coordinate is equivalent to a 4-dimensional world with *gravity*, a *gauge field* and a *scalar field*. In this framework, gauge transformations amount to a shift in the fifth direction: $x^{5'} = x^5 + \alpha(\vec{x}, t)$. In geometric terms, a metric space of this type is characterized by a group of isometries: the geometry remains the same along certain directions, indicated by Killing vectors. In the case of the 5-dimensional spaces considered by Kaluza and Klein, the isometry group is the abelian group U(1). The fifth dimension can be compactified to a circle – U(1) then generates motions on this circle. A particularly attractive feature of this theory is that it can explain the quantization of the electric charge: fields living on such a manifold necessarily carry integer multiples of a basic charge unit.

Pauli noticed that the Kaluza-Klein scenario admits a natural generalization to higher dimensions, where larger isometry groups find place. Riemann spaces of dimension > 5 admit nonabelian isometry groups that reduce the system to a 4-dimensional one with *gravity*, *nonabelian gauge fields* and several *scalar fields*. Pauli was motivated by the isospin symmetry of the meson-nucleon interaction and focused attention on a Riemann space of dimension 6, with isometry group SU(2).

Pauli did not publish the idea that the strong interaction might arise in this way, because he was convinced that the quanta of a gauge field are massless: gauge invariance does not allow one to put a mass term into the Lagrangian. He concluded that the forces mediated by gauge fields are necessarily of long range and can therefore not mediate the strong interaction, which is known to be of short range. More details concerning Pauli's thoughts can be found in [33].

2.3 Nonabelian gauge fields

The paper of Yang and Mills appeared in 1954 [34]. Ronald Shaw, a student of Salam, independently formulated nonabelian gauge field theory in his PhD thesis [35]. Ten years later, Higgs [36], Brout and Englert [37] and Guralnik, Hagen and Kibble [38] showed that Pauli's objection is not valid in general: in the presence of scalar fields, gauge fields can pick up mass, so that forces mediated by gauge fields can be of short range. The work of Glashow [39], Weinberg [40] and Salam [41] then demonstrated that nonabelian gauge fields are relevant for physics: the framework discovered by Higgs et al. does accommodate a realistic description of the e.m. and weak interactions.

2.4 Asymptotic freedom

Already in 1965, Vanyashin and Terentyev [42] found that the renormalization of the electric charge of a vector field is of opposite sign to the one of the electron. In the language of SU(2) gauge field theory, their result implies that the β-function is negative at one loop.

The first correct calculation of the β-function of a nonabelian gauge field theory was carried out by Khriplovich, for the case of SU(2), relevant for the electroweak interaction [43]. He found that β is negative and concluded that the interaction becomes weak at short distance. In his PhD thesis, 't Hooft performed the calculation of the β-function for an arbitrary gauge group, including the interaction with fermions and Higgs scalars [44]. He demonstrated that the theory is renormalizable and confirmed that, unless there are too many fermions or scalars, the β-function is negative at small coupling.

In 1973, Gross and Wilczek [45] and Politzer [46] discussed the consequences of a negative β-function and suggested that this might explain Bjorken scaling, which had been observed at SLAC in 1969 (cf. section 1.4). They pointed out that QCD predicts specific modifications of the scaling laws. In the meantime, there is strong experimental evidence for these.

A detailed account of the history of the quantum theory of gauge fields can be found in the 1998 Erice lectures of 't Hooft [47].

3 Quantum Chromodynamics

3.1 Arguments in favour of QCD

Reasons for proposing QCD as a theory of the strong interaction were given in [48]. The idea that the observed spectrum of particles can fully be understood on the basis of a theory built with quarks and gluons still looked rather questionable and was accordingly formulated in cautious terms. In the abstract, for instance, it is pointed out that "... there are several advantages in abstracting properties of hadrons and their currents from a Yang-Mills gauge model based on coloured quarks and colour octet gluons." Before the paper was completed, preprints by Gross and Wilczek [45] and Politzer [46] were circulated - they are quoted and asymptotic freedom is given as argument #4 in favour of QCD. Also, important open questions were pointed out, in particular, the U(1) problem.

Many considered QCD a wild speculation. On the other hand, several papers concerning gauge field theories that include the strong interaction appeared around the same time, for instance [49, 50].

3.2 November revolution

The discovery of the J/ψ was announced simultaneously at Brookhaven and SLAC, on November 11, 1974. Three days later, the observation was confirmed at ADONE, Frascati and ten days later, the ψ' was found at SLAC, where subsequently many further related states were discovered. We now know that these are bound states formed with the c-quark and its antiparticle and that there are two further heavy quarks: b, t.

At sufficiently high energies, quarks and gluons do manifest themselves as jets. Like the neutrini, they have left their theoretical place of birth and can now be seen flying around like ordinary, observable particles. Gradually, particle physicists abandoned their outposts in no man's and no woman's land, returned to the quantum fields and resumed discussion in the good old *Gasthaus zu Lagrange*, a term coined by Jost. The theoretical framework that describes the strong, electromagnetic and weak interactions in terms of gauge fields, leptons and quarks is now referred to as the Standard Model - this framework clarified the picture enormously.[3]

3.3 Theoretical paradise

In order to briefly discuss some of the basic properties of QCD, let me turn the electroweak interaction off, treat the three light quarks as massless and the remaining ones as infinitely heavy:

$$m_u = m_d = m_s = 0\,, \qquad m_c = m_b = m_t = \infty\,.$$

The Lagrangian then contains a single parameter: the coupling constant g, which may be viewed as the net colour of a quark. Unlike an electron, a quark cannot be isolated from the rest of the world – its colour g depends on the radius of the region considered. According to perturbation theory, the colour contained in a sphere of radius r grows logarithmically with the radius:[4]

$$\alpha_s \equiv \frac{g^2}{4\pi} = \frac{2\pi}{9\,|\ln(r\,\Lambda_{QCD})|}\,.$$

Although the classical Lagrangian of massless QCD does not contain any dimensionful parameter, the corresponding quantum field theory does: the

[3]Indeed, the success of this theory is amazing: Gauge fields are renormalizable in four dimensions, but it looks unlikely that the Standard Model is valid much beyond the explored energy range. Presumably it represents an effective theory. There is no reason, however, for an effective theory to be renormalizable. One of the most puzzling aspects of the Standard Model is that it is able to account for such a broad range of phenomena that are characterized by very different scales within one and the same renormalizable theory.

[4]The formula only holds if the radius is small, $r\,\Lambda_{QCD} \ll 1$.

strength of the interaction cannot be characterized by a number, but by a dimensionful quantity, the intrinsic scale Λ_{QCD}.

The phenomenon is referred to as *dimensional transmutation*. In perturbation theory, it manifests itself through the occurrence of divergences – contrary to what many quantum field theorists thought for many years, the divergences do not represent a disease, but are intimately connected with the structure of the theory. They are a consequence of the fact that a quantum field theory does not inherit all of the properties of the corresponding classical field theory. In the case of massless Chromodynamics, the classical version is conformally invariant, but the quantum version is not. This is crucial for QCD to be consistent with what is known about the strong interaction. In fact, massless QCD is how theories should be: it does not contain a single dimensionless parameter. In principle, the values of all quantities of physical interest are predicted without the need to tune parameters (the numerical value of the mass of the proton in kilogram units cannot be calculated, of course, because that number depends on what is meant by a kilogram, but the mass spectrum, the width of the resonances, the cross sections, the form factors, ... can be calculated in a parameter free manner from the mass of the proton, at least in principle).

3.4 Symmetries of massless QCD

The couplings of the u-, d- and s-quarks to the gauge field are identical. If the masses are set equal to zero, there is no difference at all – the Lagrangian is symmetric under SU(3) rotations in flavour space. Indeed, there is more symmetry: for massless fermions, the right- and left-handed components can be subject to independent flavour rotations. The Lagrangian of QCD with three massless flavours is invariant under $\mathrm{SU}(3)_R \times \mathrm{SU}(3)_L$. QCD thus explains the presence of the mysterious chiral symmetry discovered by Nambu: an exact symmetry of this type is present if some of the quarks are massless.

Nambu had conjectured that chiral symmetry breaks down spontaneously. Can it be demonstrated that the symmetry group $\mathrm{SU}(3)_R \times \mathrm{SU}(3)_L$ of the Lagrangian of massless QCD spontaneously break down to the subgroup $\mathrm{SU}(3)_{R+L}$? To my knowledge an analytic proof is not available, but the work done on the lattice demonstrates beyond any doubt that this does happen. In particular, for $m_u = m_d = m_s$, the states do form degenerate multiplets of $\mathrm{SU}(3)_{R+L}$ and, in the limit $m_u, m_d, m_s \to 0$, the pseudoscalar octet does become massless, as required by the Goldstone theorem.

3.5 Quark masses

The 8 lightest mesons, $\pi^+, \pi^0, \pi^-, K^+, K^0, \bar{K}^0, K^-, \eta$, do have the quantum numbers of the Nambu-Goldstone bosons, but massless they are not. The reason is that we are not living in the paradise described above: the light quark masses are different from zero. Accordingly, the Lagrangian of QCD is only approximately invariant under chiral rotations, to the extent that the symmetry breaking parameters m_u, m_d, m_s are small. Hence the multiplets are split; in particular, the Nambu-Goldstone boson multiplet is not massless.

Even before the discovery of QCD, attempts at estimating the masses of the quarks were made. In particular, nonrelativistic bound state models for mesons and baryons where constructed. In these models, the proton mass is dominated by the sum of the masses of its constituents: $m_u+m_u+m_d \simeq M_p$, $m_u \simeq m_d \simeq 300\,\mathrm{MeV}$. With the discovery of QCD, the mass of the quarks became an unambiguous concept: the quark masses occur in the Lagrangian of the theory. The first crude estimate within QCD relied on a model for the wave functions of π, K, ρ, which was based on SU(6) (spin-flavour-symmetry) and led to [51]

$$\frac{(m_u+m_d)}{2} = \frac{F_\pi M_\pi^2}{3F_\rho M_\rho} \simeq 5\,\mathrm{MeV}, \qquad m_s \simeq 135\,\mathrm{MeV}\,.$$

Similar mass patterns were found earlier, within the Nambu-Jona-Lasinio model [52] or on the basis of sum rules [53].

From the time Heisenberg had introduced isospin symmetry, it was taken for granted that the strong interaction strictly conserves isospin. QCD does have this symmetry if and only if $m_u = m_d$. If that condition were met, the mass difference between the proton and the neutron would be due exclusively to the e.m. interaction. When Jürg Gasser and I analyzed this mass difference by means of the Cottingham formula, however, we came to the conclusion that this cannot be so: evaluating the formula on the basis of Bjorken scaling, we found that the electromagnetic self energy of the proton is larger than the one of the neutron. In fact, a very strong breaking of isospin symmetry in the quark masses is needed to explain the observed value of $M_n - M_p$ within QCD [54]:

$$m_u \simeq 4\,\mathrm{MeV}, \quad m_d \simeq 7\,\mathrm{MeV}, \quad m_s \simeq 135\,\mathrm{MeV}\,.$$

In other words, QCD can be consistent with experiment only if m_u and m_d are very different. It took quite a while before this bizarre pattern was generally accepted. The Dashen theorem [55] states that, in a world where the quarks are massless, the e.m. self-energies of the kaons and pions obey the relation $M^{2\,em}_{K^+} - M^{2\,em}_{K^0} = M^{2\,em}_{\pi^+} - M^{2\,em}_{\pi^0}$. If the mass differences were dominated by the e.m. interaction, the charged kaon would be heavier than the neutral one. Hence the mass difference between the kaons cannot be due to the electromagnetic interaction, either. The estimates for the quark mass ratios obtained with the Dashen theorem confirm the above pattern [56].

3.6 Approximate symmetries are natural in QCD

At first sight, the fact that m_u strongly differs from m_d is puzzling: if this is so, why is isospin such a good quantum number ? The key observation here is the one discussed in section 3.3: QCD has an intrinsic scale, Λ_{QCD}. For isospin to represent an approximate symmetry, it is not necessary that $m_d - m_u$ is small compared to $m_u + m_d$. It suffices that the symmetry breaking parameter is small compared to the intrinsic scale, $m_d - m_u \ll \Lambda_{QCD}$.

In the case of the eightfold way, the symmetry breaking parameters are the differences between the masses of the three light quarks. If they are small compared to the intrinsic scale of QCD, then the Green functions, masses, form factors, cross sections ... are approximately invariant under the group $\mathrm{SU}(3)_{R+L}$. Isospin is an even better symmetry, because the relevant symmetry breaking parameter is smaller, $m_d - m_u \ll m_s - m_u$. The fact that $M^2_{\pi^+}$ is small compared to $M^2_{K^+}$ implies $m_u + m_d \ll m_u + m_s$. Hence all three light quark masses must be small compared to the scale of QCD.

In the framework of QCD, the presence of an approximate chiral symmetry group of the form $\mathrm{SU}(3)_R \times \mathrm{SU}(3)_L$ thus has a very simple explanation: it so happens that the masses of u, d and s are small. We do not know why, but there is not doubt that this is so. The quark masses represent a perturbation, which in first approximation can be neglected – in first approximation, the world is the paradise described above.

4 Conclusion

We now know the origin of the strong interaction. The gauge field theory that describes it, QCD, represents the first non-trivial quantum field theory that is internally consistent in four space-time dimensions. In my opinion, this constitutes the most beautiful part of the Standard Model. Apart from the scale Λ_{QCD}, which arises from dimensional transmutation, the Lagrangian of this theory contains the vacuum angle θ and the six quark mass parameters $m_u, m_d, m_s, m_c, m_b, m_t$. Their values are by now known quite accurately, but the observed pattern is bizarre and totally ungrasped. We also know that, in the basis where the quark mass matrix is diagonal, real and positive, the vacuum angle is tiny, but we do not know why that is so.

To work out the properties of QCD is a fascinating challenge. Models (AdS, CFT, superconductors, NJL ...) may help developing the intuitive understanding of QCD, but they are a meagre replacement for the real thing. Much work remains to be done to confront low energy precision experiments with our present understanding of the laws of nature: lattice simulations, effective field theory, dispersion theory, ... For an overview of ongoing work and references to the literature, I refer to [57]-[60].

Acknowledgement

I thank Tian Yu Cao, Gerhard Ecker, Lev Okun, Zurab Silagadze and Misha Vysotsky for useful comments concerning the material presented here. Also, it is a pleasure to thank Nino Zichichi for a most enjoyable stay at Erice.

References

[1] T. Y. Cao, *From Current Algebra to Quantum Chromodynamics*, Cambridge University Press, 2010.

[2] W. Heisenberg, Z. Phys. **77** (1932) 1.

[3] H. Yukawa, Proc. Phys. Math. Soc. Jap. **17** (1935) 48.

[4] E. C. G. Stueckelberg, Helv. Phys. Acta **11** (1938) 225, 299.

[5] H. Pietschmann, Eur. Phys. J. H **36** (2011) 75, arXiv:1101.2748.

[6] G. Veneziano, Nuovo Cim. A **57** (1968) 190.

[7] M. Gell-Mann, Phys. Rev. **92** (1953) 833;
T. Nakano and K. Nishijima, Prog. Theor. Phys. **10** (1953) 581.

[8] M. L. Goldberger and S. B. Treiman, Phys. Rev. **110** (1958) 1178.

[9] Y. Ne'eman, Nucl. Phys. **26** (1961) 222.

[10] M. Gell-Mann, Phys. Rev. **125** (1962) 1067.

[11] S. Okubo, Prog. Theor. Phys. **27** (1962) 949.

[12] V. E. Barnes *et al.*, Phys. Rev. Lett. **12** (1964) 204.

[13] J. Goldstone, A. Salam and S. Weinberg, Phys. Rev. **127** (1962) 965.

[14] M. Gell-Mann, Phys. Lett. **8** (1964) 214.

[15] G. Zweig, *An SU(3) model for strong interaction symmetry and its breaking*, CERN reports 8419/TH.401 and 8419/TH.412 (1964).

[16] M. Gell-Mann, Physics **1** (1964) 63.

[17] S. L. Adler, Phys. Rev. **137** (1965) B1022; **139** (1965) B1638.
W. I. Weisberger, Phys. Rev. Lett. **14** (1965) 1047.

[18] S. Weinberg, Phys. Rev. Lett. **17** (1966) 616.

[19] J. D. Bjorken, Phys. Rev. **179** (1969) 1547.

[20] For an account of the experimental developments, see the *Nobel lectures in physics 1990* by J. I. Friedman, H. W. Kendall and R. E. Taylor, SLAC-REPRINT-1991-019.

[21] K. G. Wilson, Phys. Rev. **179** (1969) 1499.

[22] O. W. Greenberg, Phys. Rev. Lett. **13** (1964) 598.

[23] N. N. Bogoliubov, B. V. Struminski and A. N. Tavkhelidze, Preprint JINR D-1968 (1964).

[24] M. Han and Y. Nambu, Phys. Rev. B139 1006 (1965).

[25] Y. Miyamoto, Prog. Theor. Phys. Suppl. Extra Number (1965) 187.

[26] H. Fritzsch and M. Gell-Mann, eConf C **720906V2** (1972) 135, hep-ph/0208010.

[27] M. Gell-Mann, Acta Phys. Austriaca Suppl. **9** (1972) 733.

[28] G. Ecker, *The colourful world of quarks and gluons I: The shaping of Quantum Chromodynamics*, Talk given at *Xth Quark Confinement and the Hadron Spectrum*, Munich, Germany (2012).

[29] H. Fritzsch, *The history of QCD*, CERN Courier, Sept. 27 (2012), http://cerncourier.com/cws/article/cern/50796

[30] V. A. Fock, Z. Phys. **39** (1926) 226 [Surveys H. E. Phys. **5** (1986) 245].
For a discussion of the significance of this paper, see
L. B. Okun, Phys. Usp. **53** (2010) 835 [Usp. Fiz. Nauk **180** (2010) 871].

[31] T. Kaluza, Sitzungsber. Preuss. Akad. Wiss. Berlin (Math. Phys.) **1921** (1921) 966.

[32] O. Klein, Z. Phys. **37** (1926) 895 [Surveys H. E. Phys. **5** (1986) 241].

[33] N. Straumann, Space Sci. Rev. **148** (2009) 25, arXiv:0810.2213.

[34] C. -N. Yang and R. L. Mills, Phys. Rev. **96** (1954) 191.

[35] R. Shaw, Cambridge Ph. E. thesis (1955) unpublished.

[36] P. W. Higgs, Phys. Lett. **12** (1964) 132.

[37] F. Englert and R. Brout, Phys. Rev. Lett. **13** (1964) 321.

[38] G. S. Guralnik, C. R. Hagen and T. W. B. Kibble, Phys. Rev. Lett. **13** (1964) 585.

[39] S. L. Glashow, Nucl. Phys. **22** (1961) 579.

[40] S. Weinberg, Phys. Rev. Lett. **19** (1967) 1264.

[41] A. Salam, Conf. Proc. C **680519** (1968) 367.

[42] V. S. Vanyashin and M. T. Terentyev, Sov. Phys. JETP **21** (1965) 375.

[43] I. B. Khriplovich, Yad. Fiz. **10** (1969) 409 [Sov. J. Nucl. Phys. 10 (1970) 235]

[44] G. 't Hooft, Nucl. Phys. B **33** (1971) 173; Nucl. Phys. B **35** (1971) 167.

[45] D. J. Gross and F. Wilczek, Phys. Rev. Lett. **30** (1973) 1343.

[46] H. D. Politzer, Phys. Rev. Lett. **30** (1973) 1346.

[47] G. 't Hooft, Proc. Int. School of Subnuclear Physics **36**, ed. A. Zichichi, World Scientific (1998), arXiv:hep-th/9812203.

[48] H. Fritzsch, M. Gell-Mann and H. Leutwyler, Phys. Lett. B **47** (1973) 365.

[49] J. C. Pati and A. Salam, Phys. Rev. D **8** (1973) 1240.

[50] S. Weinberg, Phys. Rev. Lett. **31** (1973) 494.

[51] H. Leutwyler, Phys. Lett. B **48** (1974) 431.

[52] Y. Nambu and G. Jona-Lasinio, Phys. Rev. **124** (1961) 246.

[53] S. Okubo, Phys. Rev. **188** (1969) 2300.

[54] J. Gasser and H. Leutwyler, Nucl. Phys. B **94** (1975) 269.

[55] R. F. Dashen, Phys. Rev. **183** (1969) 1245.

[56] S. Weinberg, Trans. New York Acad. Sci. **38** (1977) 185.

[57] G. Colangelo *et al.*, *Review of lattice results concerning low energy particle physics*, Eur. Phys. J. C **71** (2011) 1695, arXiv:1011.4408.

[58] U.-G. Meißner, *A walk through the world of chiral dynamics*, Talk given at Chiral Dynamics 2012, arXiv:1211.3026.

[59] H. Leutwyler, *Progress in understanding Quantum Chromodynamics*, Lectures given at the Schladming Winter School 2012.

[60] V. Cirigliano, G. Ecker, H. Neufeld, A. Pich and J. Portoles, *Kaon Decays in the Standard Model*, Rev. Mod. Phys. **84** (2012) 399, arXiv:1107.6001.

Color Transparency and Saturation in QCD

Dieter Schildknecht[1]

Fakultät für Physik, Universität Bielefeld,
Universitätsstraße 25, 33615 Bielefeld, Germany
and
Max-Planck-Institute for Physics, Föhringer Ring 6,
80805 Munich, Germany

Abstract

We review the theoretical interpretation of deep-inelastic electron-proton scattering at low values of the Bjorken variable $x \simeq Q^2/W^2 \lesssim 0.1$. The process proceeds via the interaction of quark-antiquark ($q\bar{q}$) color-dipole fluctuations of the (virtual) photon with the proton. In terms of the forward Compton scattering amplitude, two reaction channels contribute to the interaction of the $q\bar{q}$ color dipole with the gluon field in the proton. Dependent on the kinematics, there is either *color transparency*, corresponding to a cancellation of the amplitudes for the two reaction channels, or *saturation*, occuring when the process is dominated by a single interaction channel. The connection between the color-dipole picture and the pQCD improved parton model is elaborated upon.

1 Introduction

The concepts of *color transparency* and *saturation* in QCD became relevant in connection with the interpretation of the deep inelastic scattering (DIS) experiments carried out at the $e^{\pm}$-proton collider in Hamburg in the years from 1992 to 2007. The available center-of-mass energy W allowed to explore DIS in a previously non-accessible range of very low values of the Bjorken variable $x \equiv x_{bj} \simeq Q^2/W^2$,

$$5 \cdot 10^{-4} \leq x \leq 10^{-1}, \tag{1.1}$$

combined with a nevertheless appreciable range of the four-momentum squared of the virtual photon, approximately given by

$$0 \leq Q^2 \leq 100 GeV^2. \tag{1.2}$$

One of the outstanding findings from the HERA experiments is the low-x scaling behavior of the photoabsorption cross section, $\sigma_{\gamma^* p}(W^2, Q^2)$. When plotting, compare Fig. 1, the experimental data for $\sigma_{\gamma^* p}(W^2, Q^2)$ as a function of the low-x scaling variable

$$\eta(W^2, Q^2) = \frac{Q^2 + m_0^2}{\Lambda_{sat}^2(W^2)}, \tag{1.3}$$

[1]Email: schild@physik.uni-bielefeld.de

one finds a single curve [1, 2],

$$\sigma_{\gamma^* p}(W^2, Q^2) = \sigma_{\gamma^* p}(\eta(W^2, Q^2)) \sim \sigma^{(\infty)} \begin{cases} \frac{1}{\eta(W^2,Q^2)}, & \text{for } (\eta(W^2, Q^2) \gg 1), \\ \ln \frac{1}{\eta(W^2,Q^2)}, & \text{for } (\eta(W^2, Q^2) \ll 1). \end{cases} \tag{1.4}$$

In (1.3), the "saturation scale" $\Lambda^2_{sat}(W^2)$ rises with a small constant power, C_2, of the energy,

$$\Lambda^2_{sat}(W^2) \sim (W^2)^{C_2}. \tag{1.5}$$

The value of m_0 is somewhat smaller than the mass of the ρ^0 meson. The quantity $\sigma^{(\infty)}$

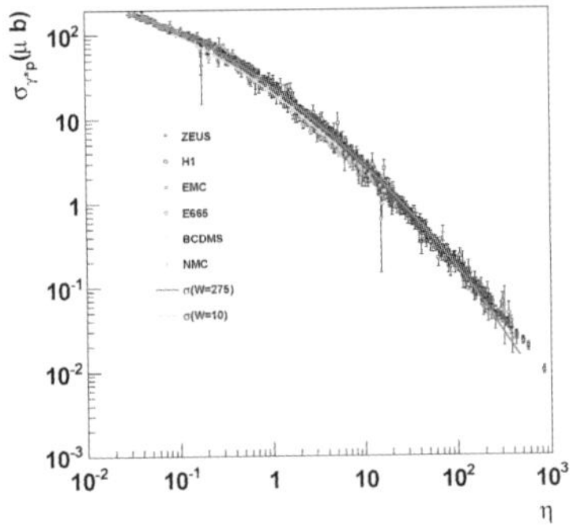

Figure 1: Scaling [1, 2] of $\sigma_{\gamma^* p}(W^2, Q^2) = \sigma_{\gamma^* p}(\eta(W^2, Q^2))$.

denotes a cross section that in good approximation is independent of the photon energy. The log-log plot of the experimental data in Fig. 1 immediately allows one to infer the $1/\eta(W^2, Q^2)$ dependence for $\eta(W^2, Q^2) \gg 1$ in (1.4), as well as the $\ln(1/\eta(W^2, Q^2)$ dependence on $\eta(W^2, Q^2)$ for $\eta(W^2, Q^2) \ll 1$.

The logarithmic behavior in the limit of $\eta(W^2, Q^2) \to 0$, or $W^2 \to \infty$ at Q^2 fixed, leads to the conclusion that the ratio of the virtual to the real photoabsorption cross section tends towards unity in this "saturation" limit [1, 2],

$$\lim_{\substack{W^2 \to \infty \\ Q^2 \text{ fixed}}} \frac{\sigma_{\gamma^* p}(\eta(W^2, Q^2))}{\sigma_{\gamma^* p}(\eta(W^2, Q^2 = 0))} = \lim_{\substack{W^2 \to \infty \\ Q^2 \text{ fixed}}} \frac{\ln\left(\frac{\Lambda^2_{sat}(W^2)}{m_0^2} \frac{m_0^2}{(Q^2+m_0^2)}\right)}{\ln \frac{\Lambda^2_{sat}(W^2)}{m_0^2}} = 1 + \lim_{\substack{W^2 \to \infty \\ Q^2 \text{ fixed}}} \frac{\ln \frac{m_0^2}{Q^2+m_0^2}}{\ln \frac{\Lambda^2_{sat}(W^2)}{m_0^2}} = 1. \tag{1.6}$$

The dependence of the photoabsorption cross section in (1.4) and (1.6) corresponds to a simple subdivision of the (Q^2, W^2) plane. It is shown in Fig. 2. The region of $\eta(W^2, Q^2) \ll 1$, corresponds to "saturation" in the sense of an approach to the $Q^2 = 0$ photoabsorption cross section, while the region of $\eta(W^2, Q^2) \gg 1$ corresponds to "color transparency". The choice of this latter terminology will be justified in Section 3 below.

It is the aim of the present lecture to show that the scaling behavior with the functional form in (1.4) arises as a consequence of the color-gauge-invariant interaction of $q\bar{q}$ color-dipole fluctuations of the photon with the proton (color-dipole picture, (CDP)). No specific parameter-dependent assumption on the dipole-proton interaction is needed to arrive at this conclusion.

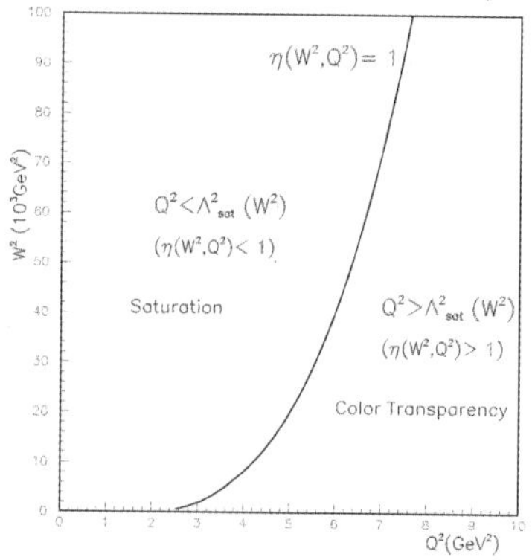

Figure 2: The (Q^2, W^2) plane corresponding to the scaling behavior of $\sigma_{\gamma^* p}(W^2, Q^2) = \sigma_{\gamma^* p}(\eta(W^2, Q^2))$ shown in Fig. 1. The line $\eta(W^2, Q^2) = 1$ subdivides the (Q^2, W^2) plane into the saturation region of $\eta(W^2, Q^2) < 1$ and the color transparency region of $\eta(W^2, Q^2) > 1$.

Again, without parameter adjustment, the CDP leads to a unique prediction [3, 4] for the ratio $R(W^2, Q^2)$ of the longitudinal to the transverse photoabsorption cross section at sufficiently large $Q^2 > \Lambda^2_{sat}(W^2)$. The definite value of $R(W^2, Q^2)$ arises as a consequence of the different transition probabilities for longitudinally polarized and transversely polarized photons into $q\bar{q}$ states, combined with the different transverse interaction size of $q\bar{q}$ states originating from longitudinally versus transversely polarized photons.

Consistency of the CDP with the pQCD-improved parton model allows one to deduce [2, 5, 6, 4] the numerical value of the exponent C_2 of the saturation scale $\Lambda^2_{sat}(W^2)$.

It is sometimes argued that the DIS data provide empirical evidence for parton recombination requiring non-linear parton evolution mechanisms (compare e.g. ref. [7] and the list of references therein). We do not see such an evidence.

2 Photon hadron interactions: Late 1960's and early 1970's

The general features of our present understanding of photon-hadron interactions at low values of $x \simeq Q^2/W^2 \ll 0.1$ date back to the pre-QCD era of the late 1960's and early 1970's. So let me briefly return to that time.

Photon-hadron interactions in the late 1960's and early 1970's were dominated by the vector-meson-dominance (VMD) picture [8]-[12]: the photon virtually dissociates, or fluctuates in modern jargon, into the low-lying vector mesons, ρ^0, ω, ϕ, that subsequently interact with e.g. the proton. Compare Fig. 3, showing the specific case of (diffractive) vector meson production in photon-proton interactions. Inserting the $\gamma(\rho^0, \omega, \phi)$ couplings that had become available from e^+e^--annihilation experiments in the late 1960's, and relating the vector-meson scattering amplitudes to the ones from pion-nucleon scattering

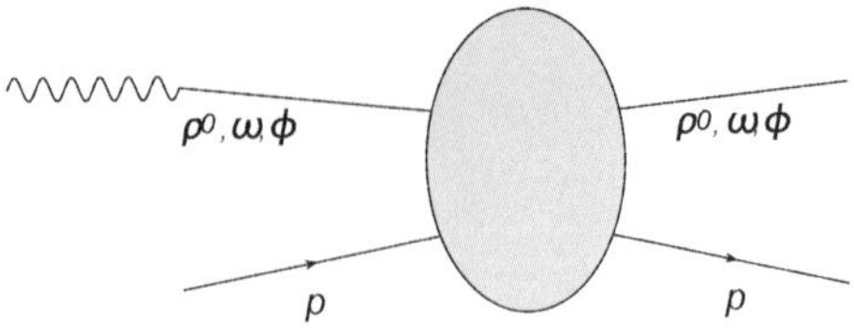

Figure 3: Vector-meson production in photon-proton interactions.

by invoking the additive quark model, vector meson production was quantitatively predicted as a function of energy and momentum transfer. The total photoproduction cross section then followed from the optical theorem by attaching a photon to the final vector mesons in Fig. 3. The experimental findings were theoretically summarized in terms of a "hadronlike behavior" [13, 14] of the photon.

Based on vector-meson dominance, hadronlike behavior of the cross section of photoproduction on heavy nuclei was predicted [15] as a consequence of the destructive interference of the one- and two-channel interactions depicted in Fig. 4. Experiments at DESY and SLAC confirmed the theoretical prediction of a hadronlike dependence on the mass number of the nucleus ("shadowing").

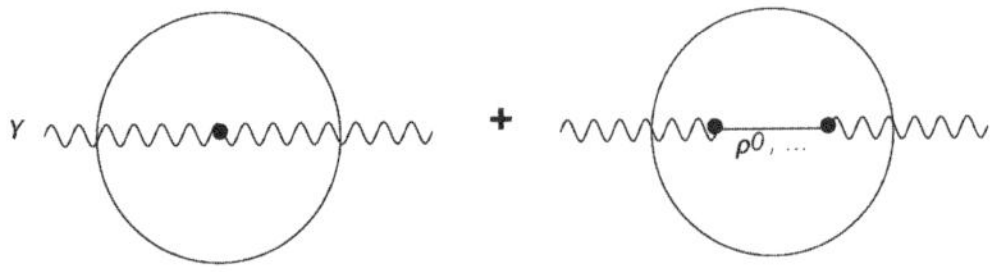

Figure 4: The forward-Compton-scattering amplitude of the interaction of a photon with a heavy nucleus.

The picture of photon-hadron interactions changed dramatically, when, starting in 1969, the first experimental results on DIS at large values of Q^2 from the SLAC-MIT collaboration became available [16]. The experimental results provided evidence for the scaling behavior of the proton structure function that had been conjectured by Bjorken [17], and they led Feynman to propose the parton-model [18].

An alternative point of view concerning the results from the SLAC-MIT experiment was advocated by Sakurai and myself in 1972 [19]. Since this point of view is of much relevance as a predecessor of our present understanding of DIS at low $x \equiv x_{bj}$ in QCD, I briefly remind you of our 1972 procedure.

We conjectured that the slow decrease of the photoabsorption cross section with increasing Q^2 observed by the SLAC-MIT collaboration at large $w' \sim 1/x$ was due to the, at that time hypothetical[2], coupling of the photon to a high-mass continuum predicted to

[2]First experimental data on e^+e^- annihilation beyond the ρ^0, ω, ϕ region had become available [20] from the Frascati e^+e^- storage ring (Adone) in 1971, providing evidence for production of multi-hadron final states at 1.3 GeV to 1.4 GeV e^+e^- center-of-mass energy.

be observed in future e^+e^- annihilation experiments beyond the ρ^0, ω, ϕ region. Compare Fig. 5 and the comparison of our generalized vector dominance (GVD) predictions [19]

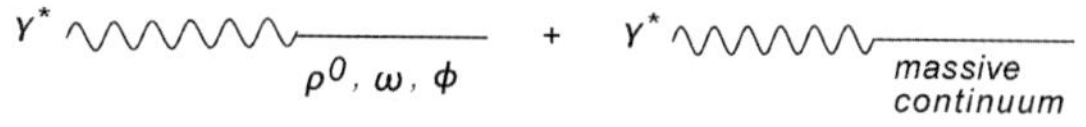

Figure 5: The coupling of the photon to a high-mass continuum.

Volume 40B, number 1 PHYSICS LETTERS 12 June 1972

GENERALIZED VECTOR DOMINANCE AND INELASTIC ELECTRON-PROTON SCATTERING *

J. J. SAKURAI and D. SCHILDKNECHT **
Department of Physics, University of California, Los Angeles, USA

Received 30 March 1972

We propose a model of inelastic electron-proton scattering which takes into account the coupling of the photon to higher-mass vector states. Both the virtual photon-proton cross section σ_T (predicted with essentially no adjustable parameters) and the q^2 dependence of R are in exceedingly good agreement with the SLAC-MIT data in the diffraction region.

with the experimental data in Fig. 6. The GVD approach was further developed in ref. [21], compare also [22] and the review in ref. [12].

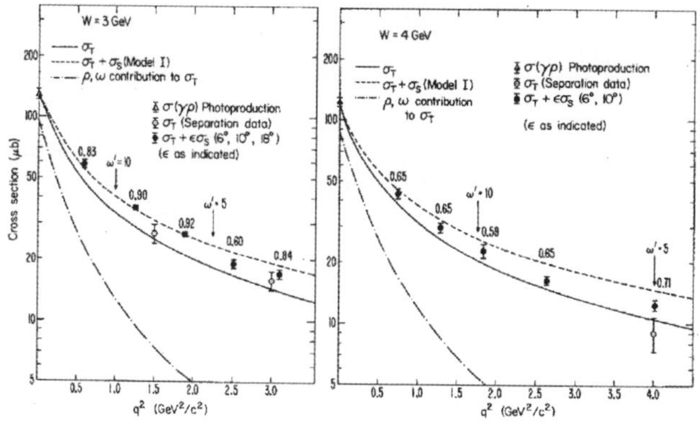

Figure 6: The SLAC-MIT experimental data compared with GVD predictions [19].

In the early 1970's, it was frequently argued that our GVD interpretation of the experimental results was incorrect, since experiments did not reveal the hadronlike shadowing in the scattering of virtual photons from complex nuclei implied by GVD. After many years of confusion, this point was clarified by the 1989 experimental results of the EMC-NMC collaboration [23], compare Fig. 7. The results indeed showed the shadowing effect expected for virtual photons due to high-mass fluctuations [24] of the virtual photon that contribute to the two-step interaction in Fig. 4.

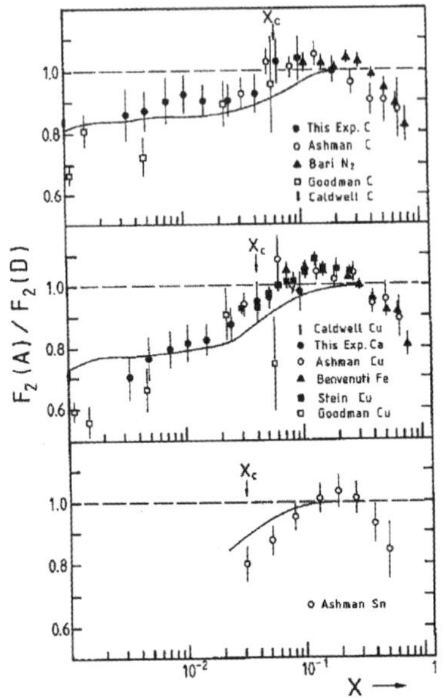

Figure 7: Shadowing in the scattering of virtual photons from complex nuclei compared with GVD predictions [24]

Since shadowing for virtual photons requires high-mass diffractive production, contrary to what is often stated, it came without surprise that one of the first results from HERA was the observation of "rap-gap" events [25] due to diffractive production of high-mass hadron states.

The 1972 approach of GVD [19] was based on postulating a mass dispersion relation as immediate generalization of (ρ^0, ω, ϕ)-dominance,

$$\sigma_{\gamma^*_T p}(W^2, Q^2) = \int dm^2 \int dm'^2 \frac{\rho(W^2, m^2, m'^2) m^2 m'^2}{(Q^2 + m^2)(Q^2 + m'^2)}. \tag{2.1}$$

The momentum space dispersion-theoretic ansatz may be equivalently interpreted in terms of a three-momentum conserving energy-conservation-violating transition to an on-shell hadronic state of finite, but sufficiently long lifetime given by [26, 24]

$$\tau = \frac{1}{\Delta E} = \frac{2M\nu}{Q^2 + M^2_{q\bar{q}}} \frac{1}{M_p} = \frac{1}{x + \frac{M^2_{q\bar{q}}}{W^2}} \frac{1}{M_p} \gg \frac{1}{M_p}, \tag{2.2}$$

followed by the interaction of the hadronic fluctuation with the nucleon. In (2.2), ν denotes the proton-rest-frame energy of the photon and $M_{q\bar{q}}$ and M_p denote the fluctuation mass and the proton mass, respectively.

3 The modern picture of DIS at low x: the color dipole picture

As in GVD, in the modern approach, the photoabsorption reaction proceeds in two steps, i) the $\gamma^* \to q\bar{q}$ transition and ii) the $(q\bar{q})$-proton interaction, whereby taking into account

i) the internal structure of the $q\bar{q}$ system via the variable $0 \leq z \leq 1$ that specifies the longitudinal momentum distribution between the quark and the antiquark the photon fluctuates into, compare Fig. 8, and

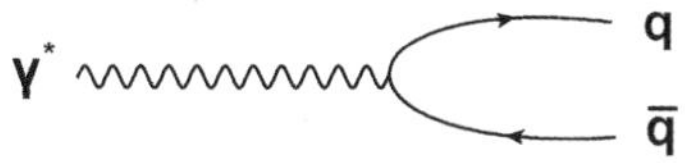

Figure 8: The $\gamma^* q\bar{q}$ transition

ii) the $q\bar{q}$ interaction with the gluon field in the nucleon [27] as a gauge-invariant color-dipole interaction, compare the (virtual) forward Compton scattering amplitude in Fig. 9.

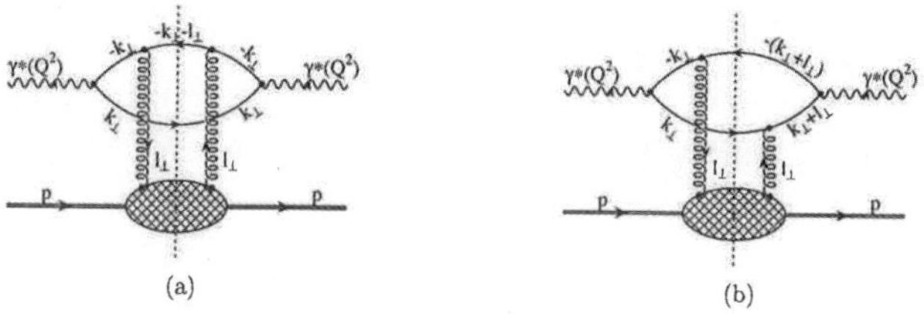

Figure 9: Two of the four diagrams for the $q\bar{q}$ dipole interaction with the gluon field in the nucleon. The diagrams (a) and (b) correspond to channel 1 and channel 2 respectively.

After Fourier transform to transverse position space, the photoabsorption cross section at low $x \simeq Q^2/W^2 < 0.1$ takes the form [28, 29]

$$\sigma_{\gamma^*_{L,T}}(W^2, Q^2) = \int dz \int d^2\vec{r}_\perp |\psi_{L,T}(\vec{r}_\perp, z(1-z), Q^2)|^2 \cdot \sigma_{(q\bar{q})p}(\vec{r}_\perp, z(1-z), W^2). \quad (3.1)$$

The quantity $|\psi_{L,T}(\vec{r}_\perp, z(1-z), Q^2)|^2$ may be interpreted as the probability for a longitudinally or a transversely polarized photon, $\gamma^*_{L,T}$, of virtuality Q^2 to undergo a transition to a $q\bar{q}$ state, $\gamma^*_{L,T} \to q\bar{q}$, being characterized by the transverse size $\vec{r}_\perp$ and by the distribution of the longitudinal momenta of quark and antiquark determined by $z(1-z)$. In the rest frame of the $q\bar{q}$ fluctuation of mass $M_{q\bar{q}}$, the quantity $z(1-z)$ determines [29] the direction of the quark (antiquark) with respect to the photon direction. The interaction cross section of the $q\bar{q}$ dipole state in (3.1) is denoted by $\sigma_{(q\bar{q})p}(\vec{r}_\perp, z(1-z), W^2)$. It depends on the energy, W, [29, 1, 30, 31] of the $(q\bar{q})p$ interaction, since the photon fluctuates into an on-shell $q\bar{q}$ state of mass $M_{q\bar{q}}$ that subsequently interacts with the nucleon. For generality, a dependence on $z(1-z)$ is allowed for in the dipole-proton cross section.

The gauge-invariant two-gluon interaction ii) of the $q\bar{q}$ dipole enters the photoabsorption cross section in (3.1) via [28, 29]

$$\sigma_{(q\bar{q})p}(\vec{r}_\perp, z(1-z), W^2) = \int d^2\vec{l}_\perp \tilde{\sigma}(\vec{l}_\perp^{\,2}, z(1-z), W^2)\left(1 - e^{-i\,\vec{l}_\perp\cdot\vec{r}_\perp}\right). \tag{3.2}$$

In (3.2), $\vec{l}_\perp$ stands for the transverse momentum of the absorbed gluon, and the first and the second term in the parenthesis on the right-hand side in (3.2), respectively, corresponds to the first and the second diagram in Fig. 9.

For the subsequent discussions, it will be useful to equivalently rewrite [32, 4] the photoabsorption cross section (3.1), in terms of dipole states, $(q\bar{q})^{J=1}_{L,T}$, that describe longitudinally and transversely polarized $q\bar{q}$ states of fixed spin $J = 1$ and polarization index L and T. In terms of the corresponding dipole cross section, $\sigma_{(q\bar{q})^{J=1}_{L,T}p}(\vec{r}_\perp^{\,\prime}, W^2)$, where $\vec{r}_\perp^{\,\prime} = \vec{r}_\perp z(1-z)$, the photoabsorption cross section (3.1) becomes [32, 4]

$$\sigma_{\gamma^*_{L,T}p}(W^2, Q^2) = \frac{\alpha}{\pi}\sum_q Q_q^2 Q^2 \int dr_\perp^{\prime 2} K_{0,1}^2(r_\perp' Q)\sigma_{(q\bar{q})^{J=1}_{L,T}p}(r_\perp^{\prime 2}, W^2). \tag{3.3}$$

In the transition from (3.1) to (3.3), assuming massless quarks, we inserted the explicit expressions for $|\psi_{L,T}(\vec{r}_\perp, z(1-z), Q^2)|^2$ in terms of the modified Bessel functions $K_0(r_\perp' Q)$ and $K_1(r_\perp' Q)$, and Q stands for $Q \equiv \sqrt{Q^2}$.The sum over the squared charges of the actively contributing quarks is given by $\sum_q Q_q^2$, and the cross section $\sigma_{(q\bar{q})^{J=1}_{L,T}}(r_\perp', W^2)$ is related to the dipole cross section in (3.1) by an appropriate projection.

In terms of the $J = 1$ projection, $\sigma_{(q\bar{q})^{J=1}_{L,T}p}(r_\perp', W^2)$, of the dipole cross section in (3.1), with $\vec{l}_\perp^{\,\prime 2} = \vec{l}_\perp^{\,2}/z(1-z)$, the two-gluon-coupling structure of the dipole cross section in (3.2) becomes [4]

$$\sigma_{(q\bar{q})^{J=1}_{L,T}p}(r_\perp', W^2) = \pi\int d\vec{l}_\perp^{\,\prime 2}\bar{\sigma}_{(q\bar{q})^{J=1}_{L,T}p}(\vec{l}_\perp^{\,\prime 2}, W^2)\cdot\left(1 - \frac{\int d\vec{l}_\perp^{\,\prime 2}\bar{\sigma}_{(q\bar{q})^{J=1}_{L,T}p}(\vec{l}_\perp^{\,\prime 2}, W^2)J_0(l_\perp' r_\perp')}{\int d\vec{l}_\perp^{\,\prime 2}\bar{\sigma}_{(q\bar{q})^{J=1}_{L,T}p}(\vec{l}_\perp^{\,\prime 2}, W^2)}\right), \tag{3.4}$$

where $J_0(l_\perp' r_\perp')$ denotes the Bessel function with index 0.

Two distinct W-dependent limits of the dipole cross section (3.4) will be relevant [4] and important for the ensuing discussions. We assume that the integrals in (3.4) exist and are determined by a W-dependent restricted range of $\vec{l}_\perp^{\,\prime 2} < \vec{l}_{Max}^{\,\prime 2}(W^2)$, in which $\bar{\sigma}_{(q\bar{q})^{J=1}_{L,T}p}(\vec{l}_\perp^{\,\prime 2}, W^2)$ is appreciably different from zero, $\vec{l}_\perp^{\,\prime 2}{}_{Max}(W^2)$ increasing with increasing W^2. The resulting cross section for a dipole of fixed transverse size, $r_\perp'$, strongly depends on the variation of the phase $l_\perp' r_\perp'$:

A) For

$$0 < l_\perp' r_\perp' < l'_{\perp\,Max}(W^2) r_\perp' \ll 1, \tag{3.5}$$

upon employing the expansion

$$J_0(l'_\perp r'_\perp) \cong 1 - \frac{1}{4}(l'_\perp r'_\perp)^2 + \frac{1}{4^3}(l'_\perp r'_\perp)^4 + \cdots, \tag{3.6}$$

we find a strong cancellation between the two additive contributions to the $J = 1$ dipole cross section (3.4) which correspond to the first and the second diagram in Fig. 9. The dipole cross section (3.4) becomes proportional to $r'^2_\perp$ ("color transparency" limit [28])

$$\sigma_{(q\bar{q})^{J=1}_{L,T}p}(r'^2_\perp, W^2) = \frac{1}{4} r'^2_\perp \sigma^{(\infty)}_L(W^2)\Lambda^2_{sat}(W^2) \begin{cases} 1, \\ \rho_W, \end{cases} \left(r'^2_\perp \ll \frac{1}{l'^2_{\perp\, Max}(W^2)} \right), \tag{3.7}$$

where by definition the W^2-dependent scale $\Lambda^2_{sat}(W^2)$ reads

$$\Lambda^2_{sat}(W^2) \equiv \frac{1}{\sigma^{(\infty)}_L(W^2)} \pi \int d\vec{l}'^2_\perp \vec{l}'^2_\perp \bar{\sigma}_{(q\bar{q})^{J=1}_L p}(\vec{l}'^2_\perp, W^2), \tag{3.8}$$

and $\sigma^{(\infty)}_L(W^2)$ is explicitly defined by (3.11) below. For vanishing size, $\vec{r}'^2_\perp$, the $q\bar{q}$ color dipole (obviously) has a vanishing cross section. The expression for the factor ρ_W in (3.7) is explicitly obtained by comparison of (3.7) with (3.4). According to (3.7), and anticipating $\rho_W > 1$, transversely polarized $(q\bar{q})^{J=1}$ states interact with enhanced transverse size [3, 4]

$$\vec{r}'^2_\perp \to \rho_W \vec{r}'^2_\perp \tag{3.9}$$

relative to longitudinal ones. The ratio ρ_W will be shown to be a W-independent constant of definite magnitude, $\rho_W = \rho = 4/3$.

B) For the case of

$$l'_{\perp Max}(W^2) r'_\perp \gg 1, \tag{3.10}$$

alternative to (3.5), for any fixed value of $r'_\perp$, rapid oscillations of the Bessel function in (3.4) lead to a vanishingly small contribution of the second term in (3.4) thus implying ("saturation" limit)

$$\lim_{\substack{r'^2 fixed \\ W^2 \to \infty}} \sigma_{(q\bar{q})^{J=1}_{L,T}p}(r'^2_\perp, W^2) = \pi \int d\vec{l}'^2_\perp \bar{\sigma}_{(q\bar{q})^{J=1}_{L,T}p}(\vec{l}'^2_\perp, W^2), \left(r'^2_\perp \gg \frac{1}{l'^2_{\perp\, Max}(W^2)} \right). \tag{3.11}$$

The high-energy limit in (3.11) of sufficiently large W at fixed dipole size $r'_\perp$, according to (3.4), coincides with the limit of sufficiently large $r'_\perp$ at fixed W i.e.

$$\lim_{\substack{r'^2 fixed \\ W^2 \to \infty}} \sigma_{(q\bar{q})^{J=1}_{L,T}p}(r'^2_\perp, W^2) = \lim_{\substack{r'^2_\perp \to \infty \\ W^2 = const}} \sigma_{(q\bar{q})^{J=1}_{L,T}p}(r'^2_\perp, W^2) = \sigma^{(\infty)}_{L,T}(W^2). \tag{3.12}$$

At fixed dipole size, at sufficiently large energy, we arrive at the large-dipole-size limit of the (at most weakly W-dependent) hadronic cross section $\sigma^{(\infty)}_{L,T}(W^2) \simeq \sigma^{(\infty)}$.

The photoabsorption cross section in (3.3), due to the strong decrease of the modified Bessel functions $K_{0,1}(r'_\perp Q)$ with increasing argument $r'_\perp Q$, is strongly dominated and actually determined at any fixed value of Q^2 by values of ${r'_\perp}^2$ such that $r'^2_\perp Q^2 < 1$. Whether color transparency of the dipole cross section according to (3.5) and (3.7) or, alternatively, saturation according to (3.10) and (3.11) is relevant for $\sigma_{\gamma^*_{L,T}p}(W^2, Q^2)$ in (3.3) depends on whether $Q^2 \gg \Lambda^2_{sat}(W^2)$ or $Q^2 \ll \Lambda^2_{sat}(W^2)$ is realized at a specific value of W. Upon substitution of (3.7) and, alternatively, of (3.11) into (3.3), for $\sigma_{\gamma^*p}(W^2, Q^2) = \sigma_{\gamma^*_Lp}(W^2, Q^2) + \sigma_{\gamma^*_Tp}(W^2, Q^2)$ one finds [1, 4]

$$\sigma_{\gamma^*p}(W^2,Q^2) = \sigma_{\gamma^*p}(\eta(W^2,Q^2)) = \frac{\alpha}{\pi}\sum_q Q_q^2 \cdot \begin{cases} \frac{1}{6}(1+2\rho)\sigma^{(\infty)}\frac{1}{\eta(W^2,Q^2)}, & (\eta(W^2,Q^2) \gg 1), \\ \sigma^{(\infty)} \ln \frac{1}{\eta(W^2,Q^2)}, & (\eta(W^2,Q^2) \ll 1), \end{cases} \tag{3.13}$$

in agreement with the experimental result (1.4) shown in Fig. 1. In (3.13), we have introduced the low-x scaling variable [1]

$$\eta(W^2, Q^2) = \frac{Q^2 + m_0^2}{\Lambda^2_{sat}(W^2)}, \tag{3.14}$$

and anticipated $\rho_W = \text{const.} = \rho$. In (3.14), via quark-hadron duality [19, 33], we introduced the lower bound $m_0^2 \lesssim m_\rho^2$ on the masses of the $q\bar{q}$ fluctuations $M^2_{q\bar{q}} \geq m_0^2$, only relevant in the limit of $Q^2 \to 0$.

From the above derivation, leading to (3.13), it has become clear that DIS at low x in terms of the Compton-forward-scattering amplitude proceeds via two different reaction channels. They correspond to the first and the second diagram in Fig. 9. For sufficiently large $Q^2 \gg \Lambda^2_{sat}(W^2)$ both channels are open, resulting in strong cancellation between them. With decreasing Q^2 at fixed W^2, or with increasing W^2 at fixed Q^2, for $Q^2 \ll \Lambda^2_{sat}(W^2)$, the second channel becomes closed, no cancellation any more. Only the first channel remains open, implying that the proportionality of the photoabsorption cross section to $\Lambda^2_{sat}(W^2)$ turns into the (soft) energy dependence proportional to $\ln \Lambda^2_{sat}(W^2)$, compare (3.13).

The longitudinal-to-transverse ratio of the photoabsorption cross sections $\sigma_{\gamma^*_Lp}(W^2, Q^2)$ and $\sigma_{\gamma^*_Tp}(W^2, Q^2)$ at large $Q^2 \gg \Lambda^2_{sat}(W^2)$ according to (3.13) is given by

$$R(W^2,Q^2)_{Q^2 \gg \Lambda^2_{sat}(W^2)} = \left.\frac{\sigma_{\gamma^*_L}(W^2,Q^2)}{\sigma_{\gamma^*_T}(W^2,Q^2)}\right|_{Q^2 \gg \Lambda^2_{sat}(W^2)} = \frac{1}{2\rho}. \tag{3.15}$$

The factor 2 in (3.15) originates from the difference in the photon wave functions in (3.3). The interaction with enhanced transverse size of $(q\bar{q})^{J=1}_T$ states relative to $(q\bar{q})^{J=1}_L$ states, ρ_W in (3.9), is a consequence of the ratio of the average transverse momenta of the quark (antiquark) in the $(q\bar{q})^{J=1}_T$ state relative to the quark (antiquark) in the $(q\bar{q})^{J=1}_L$ state. Upon applying the uncertainty principle, one obtains [3, 4]

$$\rho_W = \rho = \frac{4}{3}. \tag{3.16}$$

The longitudinal structure function, with (3.15) and (3.16), at large Q^2 is related to the transverse one via

$$F_L(x,Q^2) = \frac{1}{1+2\rho}F_2(x,Q^2) = 0.27 F_2(x,Q^2). \tag{3.17}$$

The result is consistent with the experimental data, compare Fig. 10. In Fig. 10, in addition to the experimental results on F_L from the H1 and ZEUS collaborations, we also show the result on $(F_L/F_2)/(F_L/F_2)|_{bound}$ from a very detailed analysis of the experimental data by the authors of ref. [34]. Since the value of $(F_L/F_2)|_{bound} = 0.27$ from ref. [34] (accidentally) numerically coincides with our prediction (3.17), the analysis of ref. [34] provides additional confirmation, within errors, of the prediction (3.17).

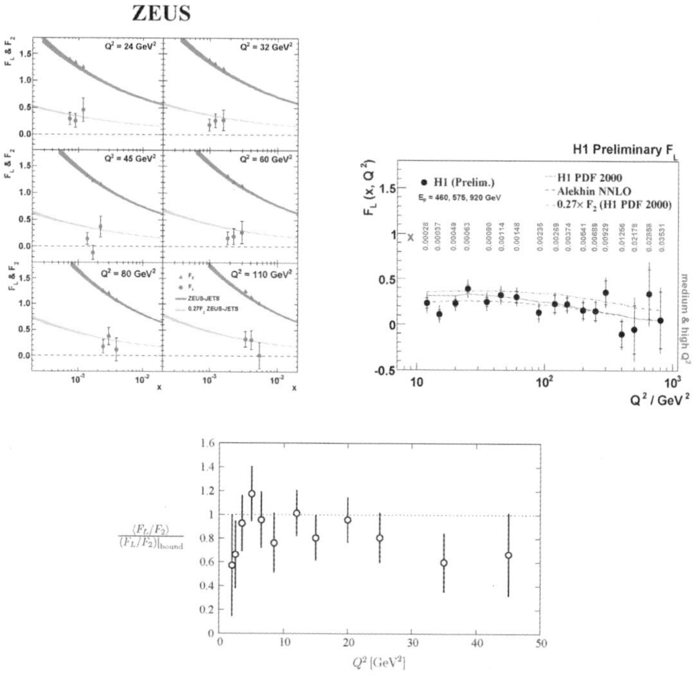

Figure 10: The prediction (3.17) compared with the experimental data for F_L and F_2.

The W-dependence in (3.1) and (3.3) of the dipole cross section, combined with the $1/Q^2$ dependence in (3.13) at large Q^2, implies that the structure function $F_2(x, Q^2) \simeq (Q^2/4\pi^2\alpha)(\sigma_{\gamma_L^* p}(W^2, Q^2) + \sigma_{\gamma_T^* p}(W^2, Q^2))$ becomes a function of the single variable W^2. The experimental data in Fig. 11, in the relevant range of $x \simeq Q^2/W^2 < 0.1$, approximately corresponding to $1/W^2 \leq 10^{-3}$, indeed show a single line [4] [3] consistent with the representation of $\sigma_{\gamma^* p}(\eta(W^2, Q^2))$ in Fig. 1. An eye-ball fit to the data in Fig. 11 yields

$$F_2(W^2) = f_2 \cdot \left(\frac{W^2}{1\mathrm{GeV}^2} \right)^{C_2 = 0.29} \tag{3.18}$$

with $f_2 = 0.063$. We shall come back to representation (3.18) of $F_2(W^2)$ below. For comparison, in Fig. 11, we also show $F_2(x, Q^2)$ as a function of the Bjorken variable x.

[3] The representation of the experimental data in Fig. 11 was kindly prepared by Prabhdeep Kaur.

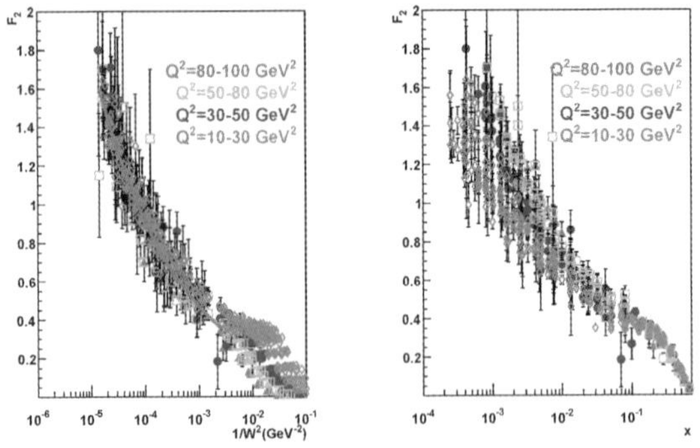

Figure 11: The structure function $F_2(x, Q^2)$ as a function of $1/W^2$ and as a function of x

We turn to the limit of $\eta(W^2, Q^2) \ll 1$, or $W^2 \to \infty$ at Q^2 fixed in (3.13), (1.4) and (1.6). The convergence to this limit is extremely slow. Compare Fig. 12, where this limit is shown in terms of the structure function $F_2(x, Q^2)$,

$$\lim_{\substack{W^2 \to \infty \\ Q^2 \text{fixed}}} \frac{F_2(x \cong Q^2/W^2, Q^2)}{\sigma_{\gamma p}(W^2)} = \frac{Q^2}{4\pi^2 \alpha}. \tag{3.19}$$

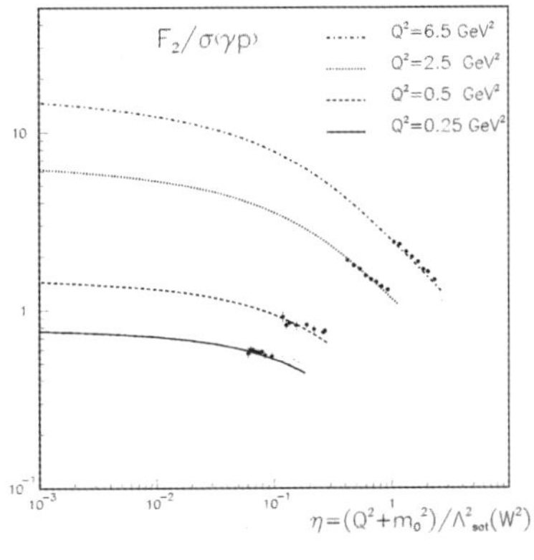

Figure 12: The approach to saturation

The theoretical curve in Fig. 12 is due to the concrete ansatz for the dipole cross section in Section 4 that interpolates between the regions of $\eta(W^2, Q^2) \gg 1$ and $\eta(W^2, Q^2) \ll 1$.

An approach to a Q^2-independent limit for $W^2 \to \infty$ at fixed Q^2 was recently independently observed by Caldwell [35] based on a purely empirical fit to the experimental data given by

$$\sigma_{\gamma^* p}(W^2, Q^2) = \sigma_0(Q^2) \left(\frac{1}{2M_p} \frac{W^2}{Q^2} \right)^{\lambda_{eff}(Q^2)}. \tag{3.20}$$

The straight lines in Fig. 13 meet at a value of W^2 approximately given by $W^2 \simeq 10^9 Q^2$, consistent with the above conclusion from the CDP.

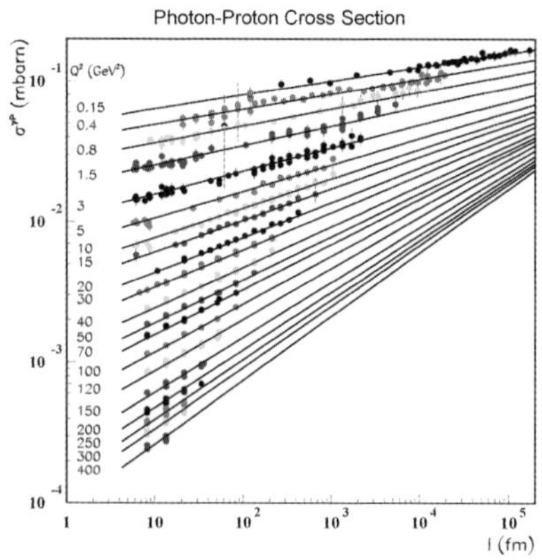

Figure 13: The Caldwell fit (3.20)[35]

The results from the above general analysis of the CDP lead to the simple structure of the (Q^2, W^2) plane shown in Fig. 2.

The (Q^2, W^2) plane is subdivided into (only) two regions as a consequence of the two interaction channels corresponding to the diagrams in Fig. 9. For $\eta(W^2, Q^2) > 1$, both interaction channels, channels 1 and 2, are open, implying color transparency of the $(q\bar{q})p$ dipole interaction and strong destructive interference in the photoabsorption cross section for $\gamma^* p \to$ hadrons. For $\eta(W^2, Q^2) \ll 1$, channel 2 becomes closed. The lack of cancellation leads to a saturation of the cross section being determined by the dipole interaction solely via channel 1. The virtual-photon-nucleon cross section approaches a Q^2-independent saturation limit that coincides with $Q^2 = 0$ photoproduction.

4 The CDP, the gluon distribution function and evolution

The CDP of DIS corresponds[4] to the low x approximation of the pQCD-improved parton model in which the interaction of the (virtual) photon occurs by interaction with the quark-antiquark sea in the proton via $\gamma^* gluon \to q\bar{q}$ fusion.

The longitudinal structure function in this approximation, for a wide range of different gluon distributions, becomes proportional to the gluon density at a rescaled value x/ξ_L [36]

$$F_L(\xi_L x, Q^2) = \frac{\alpha_s(Q^2)}{3\pi} \sum_q Q_q^2 G(x, Q^2), \tag{4.1}$$

where $G(x, Q^2) \equiv xg(x, Q^2)$ and $g(x, Q^2)$ stands for the gluon-distribution function. The rescaling factor has the preferred value of $\xi_L \simeq 0.40$.

The structure function $F_2(x, Q^2)$ at low x in this approximation is proportional to the sea-quark distribution, and again for a wide range of different gluon distributions, the

[4]With respect to this Section, compare also ref. [6]

evolution of the structure function $F_2(x, Q^2)$ with Q^2 is determined by [37, 38]

$$\frac{\partial F_2(\xi_2 x, Q^2)}{\partial \ln Q^2} = \frac{\alpha_s(Q^2)}{3\pi} \sum_q Q_q^2 G(x, Q^2), \tag{4.2}$$

where the preferred value of ξ_2 is given by $\xi_2 \simeq 0.50$.

According to the CDP, compare (3.13) and (3.18), and supported by the experimental data in Fig. 11, the structure function $F_2(x, Q^2)$ for sufficiently large Q^2 becomes a function of the single variable W^2,

$$F_2(x, Q^2) = F_2(W^2 = \frac{Q^2}{x}). \tag{4.3}$$

Employing the proportionality of $F_L(x, Q^2) = (1/(1 + 2\rho_W))F_2(x, Q^2)$ from (3.7) and (3.17), and combining (4.1) and (4.2), upon inserting (4.3) the evolution equation becomes

$$(2\rho_W + 1)\frac{\partial}{\partial \ln W^2} F_2\left(\frac{\xi_L}{\xi_2} W^2\right) = F_2(W^2). \tag{4.4}$$

A potential dependence of ρ_W on the energy W is allowed in (4.4).

We specify $F_2(W^2)$ by adopting the power law [4]

$$F_2(W^2) \sim (W^2)^{C_2} = \left(\frac{Q^2}{x}\right)^{C_2}. \tag{4.5}$$

A power law in $(1/x)^\lambda$ with $\lambda = const.$ occurs e.g. in the "hard Pomeron solution" [39] of DGLAP evolution as well as in the "hard Pomeron" part of Regge phenomenology with $(1/x)^{\epsilon_0}$ and $\epsilon_0 \simeq 0.43$ from a fit [40]. The CDP in (4.5) is more specific, however, since the W dependence of $F_2(W^2)$ implies that the x dependence and the Q^2 dependence are intimately related to each other.

Substitution of (4.5) into (4.4) implies the constraint [5, 4]

$$(2\rho_W + 1)C_2 \left(\frac{\xi_L}{\xi_2}\right)^{C_2} = 1. \tag{4.6}$$

If, and only if $\rho_W = \rho =$ const., also the exponent is constant, $C_2 =$ const. With the CDP result of $\rho = 4/3$ from (3.16), we obtain the unique value of

$$C_2 = \frac{1}{2\rho + 1}\left(\frac{\xi_2}{\xi_L}\right)^{C_2} = 0.29, \tag{4.7}$$

where $\rho = 4/3$ and $\xi_2/\xi_L = 1.25$ was inserted. The result for $C_2 = 0.29$ is fairly insensitive under variation of ξ_2/ξ_L. For $1 \leq \xi_2/\xi_L \leq 1.5$, one obtains $0.27 \leq C_2 \leq 0.31$. The value of $C_2 = 0.29$ is consistent with the experimental data, compare (3.18) and the corresponding theoretical curve in Fig. 11.

Imposing consistency between the CDP and the pQCD-improved parton model, we thus arrived at the prediction of a definite value of $C_2 = 0.29$ that coincides with the experimental findings in Fig. 11.

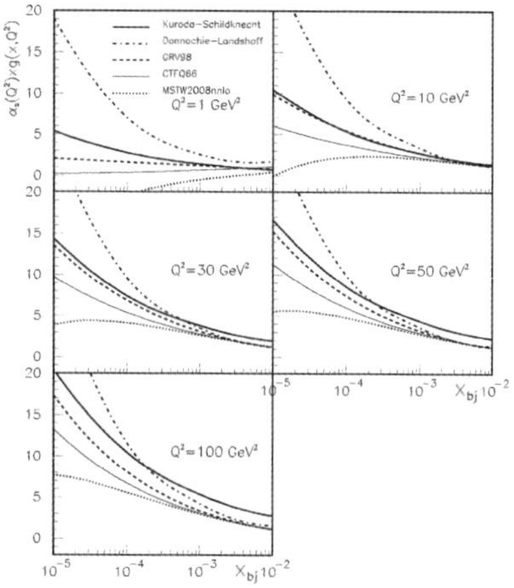

Figure 14: The gluon distribution function (4.8) compared with the results from the hard Pomeron part of a Regge fit to $F_2(x, Q^2)$ and from the fits GRV [41] CTEQ [42] and [43].

The underlying gluon distribution function can now be deduced from (4.1) by expressing the longitudinal structure function in terms of $F_2(x, Q^2)$ according to (3.17) and inserting the power law (3.18),

$$\alpha_s(Q^2)G(x,Q^2) = \frac{3\pi}{\sum_q Q_q^2(2\rho+1)} \frac{f_2}{\xi_L^{C_2=0.29}} \left(\frac{W^2}{1\text{GeV}^2}\right)^{C_2=0.29}. \tag{4.8}$$

With $\rho = 4/3$ from the CDP, compare (3.16), and $\sum_q Q_q^2 = 10/9$, the result (4.8) contains the single fitted parameter $f_2 = 0.063$ from (3.18). Inserting $W^2 = Q^2/x$ on the right-hand side in (4.8), we obtain the gluon distribution as a function of x and Q^2.

Using the next-to-leading order expression for $\alpha_s(Q^2)$, in Fig. 14 [4], we compare the gluon distribution (4.8) with various gluon distributions obtained in sophisticated global fits [41, 42, 43]to the experimental data. The consistency of our simple one-free-parameter extraction of the gluon distribution in Fig. 14 according to (4.8) may indicate that the gluon distribution is less sensitively dependent on the details of the *ggpp* structure than usually assumed, or elaborated upon and employed in the global fits to the experimental data.

5 Specific ansatz for the dipole cross section and comparison with experiment

Any specific ansatz for the dipole cross section has to interpolate between the color-transparency region of $\eta(W^2, Q^2) \gg 1$, where $\sigma_{\gamma^* p}(\eta(W^2, Q^2)) \sim 1/\eta(W^2, Q^2)$, and the saturation region of $\eta(W^2, Q^2) \ll 1$, where $\sigma_{\gamma^* p}(\eta(W^2, Q^2)) \sim \ln(1/\eta(W^2, Q^2))$, compare (3.13). For the explicit expressions for the ansatz for the dipole cross section, we refer to refs. [1] and [4] [5]. We only note that the saturation scale, $\Lambda_{sat}^2(W^2) \sim (W^2)^{C_2}$ in

[5]For deeply virtual Compton scattering and vector meson production in the CDP compare ref. [44]

the HERA energy range approximately varies between $2GeV^2 \leq \Lambda^2_{sat}(W^2) \leq 7GeV^2$, and restrict ourselves to presenting a comparison with the experimental data.

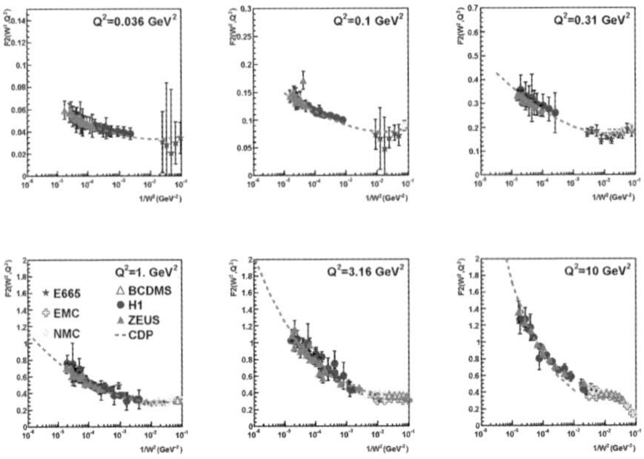

Figure 15: The predictions [4] from the CDP for the structure function $F_2(W^2, Q^2)$ compared with the experimental data for $0.036 \leq Q^2 \leq 10GeV^2$.

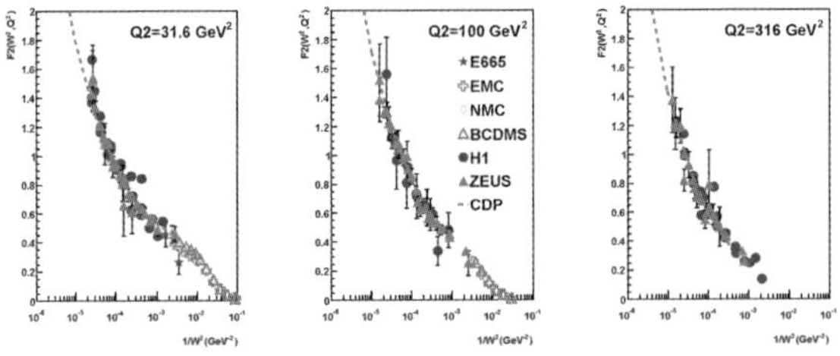

Figure 16: As in Fig. 15, but for $31.6GeV^2 \leq Q^2 \leq 316GeV^2$.

The theoretical results from the CDP in Figs. 15 and 16 show agreement with the experimental data[6] for $F_2(W^2, Q^2)$ over the full relevant region of $0.036\ GeV^2 \leq Q^2 \leq 316GeV^2$.

In Figs. 17 and 18, in addition to the theoretical results from the CDP, for comparison, we also show the results from the pQCD improved parton model based on the gluon distribution function (4.8) shown in Fig. 14.

Explicitly, by returning to (3.18) and inverting (4.8), we reinterprete (4.8) as a prediction of $F_2(W^2 = Q^2/x)$ from the (previously determined) gluon distribution according to

$$F_2(W^2 = Q^2/x) = \frac{(2\rho + 1)\sum_q Q_q^2}{3\pi} \xi_L^{C_2=0.29} \alpha_s(Q^2) G(x, Q^2), \tag{5.1}$$

where $f_2 = 0.063$ and $\xi_L = 0.40$. In Figs. 17 and 18, we see the expected consistency of the pQCD prediction (5.1) with the experimental data and the CDP in the relevant

[6]We thank Prabhdeep Kaur for providing the plots of the experimental data in Figs 15 to 18.

range of $10GeV^2 \leq Q^2 \leq 100GeV^2$. For $Q^2 < 10GeV^2$, with increasing W, gradually saturation sets in implying a breakdown of the pQCD proportionality (5.1) of the proton structure function to the gluon distribution.

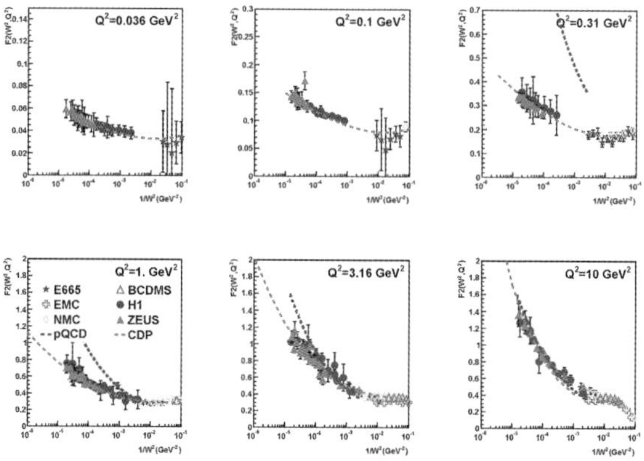

Figure 17: In addition to the prediction from the CDP, also the pQCD prediction (5.1) based on the gluon distribution (4.8) is compared with the experimental data for $F_2(W^2, Q^2)$.

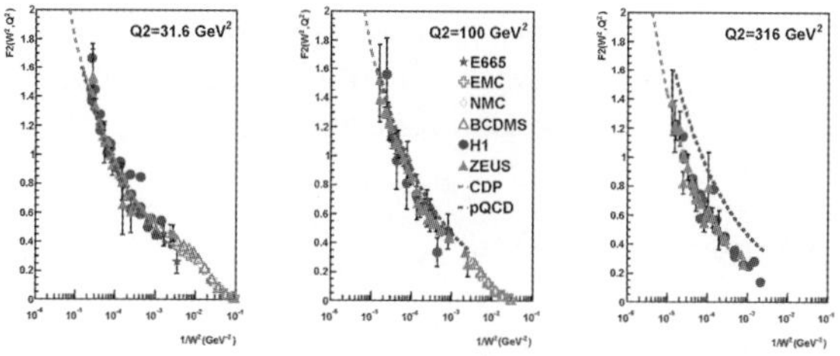

Figure 18: As in Fig. 17, but for $31.6GeV^2 \leq Q^2 \leq 316GeV^2$.

The pQCD prediction for $Q^2 = 316GeV^2$ lies above the experimental data. This is due to the breakdown of the simple form for $F_2(W^2)$ in (3.18) which is used to extract the gluon distribution. Employing the full CDP result would lead to an appropriate decrease of the gluon distribution with increasing Q^2 for $Q^2 > 100GeV^2$.

The proton structure function of the CDP, according to (3.13), in the region of $\eta(W^2, Q^2) < 1$, starts to depend logarithmically on the saturation scale, $\Lambda^2_{sat}(W^2) \sim (W^2)^{C_2}$, and with $\alpha_s(Q^2)G(x, Q^2) \sim (W^2)^{C_2}$ from (4.8) and (5.1), it depends logarithmically on the gluon distribution,

$$F_2(W^2, Q^2) \sim Q^2 \sigma_L^{(\infty)} \ln \frac{\Lambda^2_{sat}(W^2 = Q^2/x)}{Q^2 + m_0^2} \sim Q^2 \sigma_L^{(\infty)} \ln \frac{\alpha_s(Q^2)G(x, Q^2)}{\sigma_L^{(\infty)}(Q^2 + m_0^2)}, \; (Q^2 \ll \Lambda^2_{sat}(W^2)) \,. \tag{5.2}$$

The smooth transition from the color transparency region to the saturation region does not correspond to a change of the functional form of the W-dependent gluon distribution, $\alpha_s(Q^2)G(x, Q^2) \sim \Lambda^2_{sat}(W^2)$, but occurs via transition from the proportionality (5.1) to the logarithmic dependence (5.2) of $F_2(W^2, Q^2)$ on the gluon distribution function. It is the same gluon distribution that is relevant in the region of color transparency $\eta(W^2, Q^2) \gg 1$ and in the saturation region, $\eta(W^2, Q^2) \ll 1$, only the functional dependence of the proton structure function $F_2(x, Q^2)$ on the gluon distribution has changed. We disagree with the frequently expressed opinion (compare e.g. ref. [7] and the list of references given therein) that the mere existence of the saturation scale and of scaling like in Fig. 1 suggests and even requires a modification based on non-linear evolution of the gluon distribution determined in the pQCD domain of $\eta(W^2, Q^2) > 1$, when passing to the saturation region of $\eta(W^2, Q^2) < 1$.

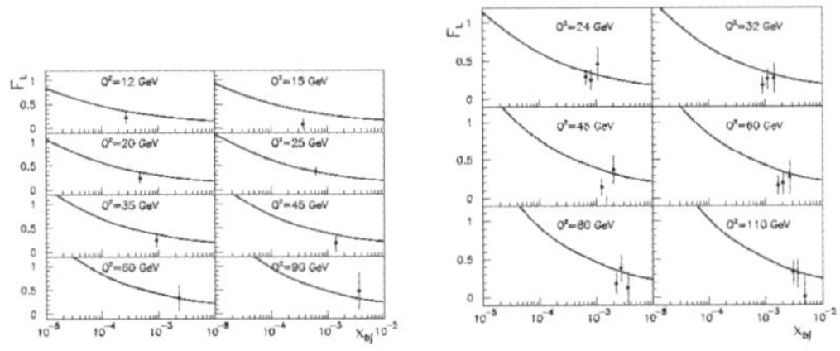

Figure 19: The longitudinal proton structure function $F_L(x, Q^2)$ from the H1 collaboration [45] and from the ZEUS collaboration [46] compared with the CDP prediction.

In Fig. 19, we show the experimental results [45, 46] for the longitudinal structure function, $F_L(x, Q^2)$, in comparison with the predictions from the CDP based on the specific ansatz for the dipole cross section that was used in Figs. 15 to 18.

6 Conclusions

The color-gauge-invariant interaction of a $q\bar{q}$ dipole state with the nucleon in terms of the forward scattering amplitude proceeds via two reaction channels. They correspond to the two diagrams shown in Fig. 9. For a given dipole size, if both reaction channels are open, the cancellation between them implies *color transparency*. In the limit in which the reaction channel corresponding to the second diagram in Fig. 9 is closed, the dipole cross section *saturates* to a cross section of standard hadronic size with at most a weak energy dependence.

In the photoabsorption cross section, the above limits of *color transparency and saturation* are realized, respectively, by the two regions in the (Q^2, W^2) plane of Fig. 2, corresponding to $\eta(W^2, Q^2) \gg 1$ and $\eta(W^2, Q^2) \ll 1$. The (Q^2, W^2) plane (under the restriction of $Q^2/W^2 \lesssim 0.1$) is accordingly simple. There are only two distinct regions in the (Q^2, W^2) plane, separated by the line $\eta(W^2, Q^2) = 1$.

The main features of the experimental data on DIS at low x have thus been recognized to follow from the color-gauge-invariant $q\bar{q}$ dipole-interaction with the gluon field of the proton without adopting a specific ansatz for the dipole cross section. Any specific ansatz has to interpolate between the model-independent restrictions on the photoabsorption cross section that hold for $\eta(W^2, Q^2) \gg 1$ and for $\eta(W^2, Q^2) \ll 1$, respectively.

Acknowledgement

The author thanks Kuroda-san for a fruitful collaboration on the color dipole picture. Thanks to Prabhdeep Kaur for providing plots of experimental data. The author thanks Allen Caldwell and Reinhart Kögerler for useful discussions.

Last not least the author thanks Professor Antonino Zichichi for the invitation to the 50th Anniversary Celebrations of his famous School of Subnuclear Physics with its lively scientific atmosphere in the magnificent environment of the "Ettore Majorana" Foundation and Centre for Scientific Culture at Erice, Sicily.

References

[1] D. Schildknecht, Contribution to Diffraction 2000, Cetraro, Italy, September 2-7, 2000, Nucl. Phys. B, Proc. Supplement **99** (2001) 121;
D. Schildknecht, B. Surrow, M. Tentyukov, Phys. Lett. **B499** (2001) 116;
G. Cvetic, D. Schildknecht, B. Surrow, M. Tentyukov, EPJC **20** (2001) 77.

[2] D. Schildknecht, Contribution to DIS 2001, The 9th International Workshop on Deep Inelastic Scattering, Bologna, Italy, 2001, G. Brassi et al. (Eds.), World Scientific, Singapore, 2002, p. 798;
D. Schildknecht, B. Surrow and M. Tentyukov, Mod. Phys. Lett. **A16** (2001) 1829.

[3] M. Kuroda and D. Schildknecht, Phys. Lett. **B670** (2008) 129.

[4] M. Kuroda and D. Schildknecht, Phys. Rev. **D85** (2012) 094001.

[5] M. Kuroda and D. Schildknecht, Phys. Lett. **B618** (2005) 84.

[6] D. Schildknecht, [arXiv:1104.0850 [hep-ph]]

[7] J. Kuokkanen, K. Rummukainen and H. Weigert, [arXiv:1108.1867 [hep-ph]].

[8] J.J. Sakurai, Ann. of Physics **11** (1960) 1; *Currents and Mesons*, The University of Chicago Press, Chicago 1969.

[9] M. Gell-Mann and F. Zachariasen, Phys. Rev. **124** (1961) 953.

[10] N.M. Kroll, T.D. Lee and B. Zumino, Phys. Rev. **157** (1967) 1376.

[11] D. Schildknecht, Springer Tracts in Modern Physics, **Vol. 63** (1972) p. 57.

[12] D. Schildknecht, in Proceedings of Photon 2005, International Conference on the Structure and Interaction of the Photon, Warsaw 2005, Acta Phys. Polon. **B37** (2006) 595 [hep-ph/0511090].

[13] J.J. Sakurai, Dynamical Models Based on Vector Meson Dominance and Universality Properties of the Fundamental Interactions, in *Properties of the Fundamental Interactions*, 1971 International School of Subnuclear Physics, Erice, Sicily (ed. A. Zichichi) Editrice Compositori, Bologna (1973), p. 243.

[14] J.J. Sakurai, Hadronlike Behaviour of the Photon, in *Laws of Hadronic Matter*, 1973 International School of Subnuclear Physics, Erice, Sicily (ed. A. Zichichi) Academic Press New York and London (1975).

[15] L. Stodolsky, Phys. Rev. Lett. **18** (1967) 135;
S.J. Brodsky and J. Pumplin, Phys. Rev. **182** (1969) 1794;
M. Nauenberg, Phys. Rev. Lett. **22** (1969) 556;
K. Gottfried and D.R. Yennie, Phys. Rev. **182** (1969) 1595.

[16] SLAC-MIT Collaboration, G. Miller et al., Phys. Rev. **D5** (1972) 528.

[17] J.D. Bjorken, Phys. Rev. **179** (1969) 1547.

[18] R.P. Feynman, *Photon Hadron Interactions*, Benjamin, New York (1972).

[19] J.J. Sakurai and D. Schildknecht, Phys. Lett. **40B** (1972) 121; Phys. Lett. **41B** (1972) 489; Phys. Lett. **42B** (1972) 216;
B. Gorczyca, D. Schildknecht, Phys. Lett. 47B, 71 (1973).

[20] C. Bernardini, in *Proc. 5th Internat. Symp. on Electron and Photon Interactions at High Energies*, Cornell University (1971)

[21] H. Fraas, B.J. Read and D. Schildknecht, Nucl. Phys. **B86** (1975) 346;
R. Devenish and D. Schildknecht, Phys. Rev. **D19** (1976) 93.

[22] D. Schildknecht, Is Vector Dominance Still Alive?, in *Lepton and Hadron Structure*, 1974 Int. School of Subnuclear Physics, Erice, Sicily, 14-31. July 1974 (ed. A. Zichichi), Academic Press New York and London, p. 448.

[23] EMC-NMC Collaboration, compare: M. Arneodo, Phys. Rep.**240** (1994) 301.

[24] D. Schildknecht, Nucl. Phys. **B66** (1973) 398;
C. Bilchak and D. Schildknecht, Phys. Lett. **B214** (1988) 441; Phys. Lett. **B233** (1989) 461.

[25] H1 Collaboration, T. Ahmed et al., Nucl. Phys. **B429** (1994) 477; ZEUS Collaboration, M. Derrick et al., Phys. Lett. **B315** (1993) 481; R. Wichmann, on behalf of the ZEUS and H1 Collaborations, Nucl. Phys. B (Proc. Suppl.) **82** (2000) 268.

[26] T. Nieh, Phys. Rev. **D1** (1970) 3161;
B.L. Joffe, Phys. Lett. **30B** (1969) 123;
V.N. Gribov, Sov. Phys.-JETP **30** (1970) 709.

[27] F.E. Low, Phys. Rev. **D12** (1075) 163;
S. Nussinov, Phys. Rev. Lett. **34** (1975) 1286; Phys. Rev. **D14** (1976) 246;
J. Gunion, D. Soper, Phys. Rev. **D15** (1977) 2617.

[28] N.N. Nikolaev, B.G. Zakharov, Z. Phys. C49, 607 (1991).

[29] G. Cvetic, D. Schildknecht, A. Shoshi, Eur. Phys. J **C13** (2000) 301.

[30] J.R. Forshaw, G. Kerley, G. Shaw, Phys. Rev. **D60** (1999) 074012.

[31] C. Ewerz and O. Nachtmann, Annals of Phys. **322** (2007) 1635; **322** (2007) 1670;
C. Ewerz, A.v. Manteuffel, O. Nachtmann, [arXiv:1101.0288 [hep-ph]].

[32] M. Kuroda and D. Schildknecht, Phys. Rev. **D66**(2002) 094005; Phys. Rev. **D67** (2003) 094008.

[33] D. Schildknecht and F. Steiner, Phys. Lett. **B56** (1975) 36;
J.J. Sakurai, Phys. Lett. **B46** (1973) 207, and *New Duality in Electromagnetic Interactions*, 1973 International School of Subnuclear Physics (ed. A. Zichichi) loc. cit. p. 291

[34] C. Ewerz, A.v. Manteuffel, O. Nachtmann, A. Schöning, arXiv: 1201.6296 [hep-ph]

[35] A. Caldwell, [arXiv:0802.0769]

[36] A.D. Martin, R.G. Roberts and W.J. Stirling, Phys. Rev. **D37** (1988) 1161;
A.M. Cooper-Sarkar et al., Z. Phys. **C39** (1988) 281.

[37] L.N. Lipatov, Sov. J. Nucl. Phys. **20** (1975) 95,
V.N. Gribov and L.N. Lipatov, Sov. J. Nucl. Phys. **15** (1972) 438,
G. Altarelli and G. Parisi, Nucl. Phys. **B126** (1977) 298;
Yu. L. Dokshitzer, Sov. Phys. JETP **46** (1977) 641.

[38] K. Prytz, Phys. Lett. **B311** (1993) 286.

[39] K. Adel, F. Barreiro, F.J. Yndurain, Nucl. Phys. **B495** (1997) 221;
F.J. Yndurain, The Theory of Quark and Gluon Interactions (Springer 1999), p.157.

[40] A. Donnachie and P.V. Landshoff, Phys. Lett. **B533** (2002) 277, [hep-ph/0111427];
Acta Physica Polonica **B34** (2003) 2989, [hep-ph/0305171].

[41] M. Glück, E. Reya, A. Vogt, Z. Phys. C67 (1994) 433; Eur. Phys. J. C5 (1998) 461

[42] CTEQ Collaboration: J. Pumplin et al., JHEP 0207 (2002) 012.

[43] A.D. Martin et al., Eur. Phys. J. C18 (2000) 117.

[44] M. Kuroda and D. Schildknecht, Eur. Phys. J. **C37** (2004) 205; Erratum, ibid. **C44** (2005) 613; Phys. Lett. **B618** (2005) 84.

[45] H1 Collaboration, F.D. Aaron et al., Phys. Lett. **B665** (2008) 139.

[46] ZEUS Collaboration, S. Chekanov et al., Phys. Lett. **B682** (2009) 8.

Glue-mesons : their conception needs all of QCD in the infrared

The trace anomaly and modification of canonical structure

Peter Minkowski
Albert Einstein Center for Fundamental Physics - ITP, University of Bern

Abstract

The regularities implied at large distances by complete gauge invariance in QCD are shown to bear nontrivial consequences for the selection among inequivalent representations of canonical (anti-) commutation rules for gauge boson - and quark fields . The trace anomaly forces a modification of the gauge boson Lagrangean and by this of the entire associated canonical stucture .

Lecture prepared for the
' INTERNATIONAL SCHOOL OF SUBNUCLEAR PHYSICS
50th Cours : WHAT WE WOULD LIKE LHC TO GIVE US '
Erice, 23. June - 2. July 2012
50th Anniversary Celebrations 2011-2012-2013
ETTORE MAJORANA FOUNDATION
AND CENTRE FOR SCIENTIFIC CULTURE

1 - Introduction

1 - 1 Assembling elements of the Lagrangean density

1-1-a – Premises

We face the theoretical abstraction of QCD with $N_{fl} = 6$, representing strong interactions – adaptable to two or three light flavors (u , d , s) of quarks and antiquarks. $\leftrightarrow$

quarks : color is counted in $\pi^{0} \rightarrow \gamma\gamma$

$\left(\begin{array}{l} \text{assuming global color- and} \\ \text{flavor-projections to commute} \end{array} \right)$ yet see ref. [1-2001]

spin and flavor are clearly seen in $q\overline{q}$ and $3q, 3\overline{q}$ spectroscopy $\left(\begin{array}{l} \text{a pre-condition} \\ \text{to count color} \end{array} \right)$.

$$\mathcal{L} = \left[\begin{array}{c} \overline{q}\,^{\dot{c}'}_{\dot{\mathcal{S}}'\,\dot{f}} \left\{ \begin{array}{c} \frac{i}{2} \overset{\leftrightarrow}{\partial}_{\mu}\, \delta_{c'\dot{c}} \\ + W^{\,r}_{\mu} \left(\frac{1}{2} \lambda_{r} \right)_{c'\dot{c}} \end{array} \right\} \gamma^{\,\mu}_{\dot{\mathcal{S}}'\mathcal{S}}\, q^{\,c}_{\mathcal{S}\,f} \\ - m_{f}\, \overline{q}\,^{\dot{c}}_{\dot{\mathcal{S}}\,\dot{f}}\, q^{\,c}_{\mathcal{S}\,f} \end{array} \right] - \frac{1}{4\,g^{2}} B^{\,\mu\nu\,r} B^{\,r}_{\mu\nu} + \Delta\,\mathcal{L} \tag{1}$$

$W^{\,r}_{\mu} \equiv -\,v^{\,r}_{\mu}$: for identification of convention for potentials

quarks : c' , $c = 1, 2, 3$ color , $f = 1, \cdots, 6$ flavor

$\mathcal{S}', \mathcal{S} = 1, \cdots, 4$ spin , m_{f} mass

In eq. 1 , the $\mathcal{D}$ associated gauge connection fields – where $\mathcal{D} = \mathcal{D}\,(\,\mathcal{G}\,)$ denotes a general , irreducible representation of the local gauge group $\mathcal{G} = SU3_{\,c}$ – appear in the form appropriate for quarks : $\mathcal{D} = \{\,3\,\}$, and antiquarks : $\mathcal{D} = \{\,\overline{3}\,\}$ respectively

$$\begin{array}{l} \left(\mathcal{W}_{\mu}\,(\,\mathcal{D}\,) \right)_{\alpha\beta}\,(\,x\,) = W^{\,r}_{\mu}\,(\,x\,)\,(\,d_{\,r}\,)_{\alpha\beta} \\ d_{\,r} = -\,d^{\,\dagger}_{\,r} = \frac{1}{i}\,J_{\,r} \in Lie\,(\,\mathcal{D}\,)\ ;\ [\,d_{\,p}\,,\,d_{\,q}\,] = f_{\,pqr}\,d_{\,r} \\ r\,,\,p\,,\,q = 1,\,\cdots,\,dim\,\mathcal{G}\ ;\ \alpha\,,\,\beta = 1,\,\cdots,\,dim\,\mathcal{D} \end{array} \tag{2}$$

For $\mathcal{D}\,(\,SU3_{\,c}\,) = \{\,3\,(\overline{3})\,\}$ the representation matrices become (the Gell-Mann matrices [2-1964])

$$\begin{array}{l} \left(d_{\,r}\,(3) = \frac{1}{i}\,\frac{1}{2}\,\lambda_{\,r} \right)_{\alpha\beta}\ ;\ r = 1,\,\cdots,\,8 \\ (\,\alpha\,,\,\beta\,) \leftrightarrow \left(c'\,,\,\dot{c} \right) = 1,\,\cdots,\,3\ ;\ d_{\,r}\,(\overline{3}) = \overline{d}_{\,r}\,(3) \end{array}$$

with the *conventional* normalization conditions : $-\ tr\ d^{\,r}\,d^{\,s} = \frac{1}{2}\,\delta^{\,rs}$

$$\tag{3}$$

The quantity proportional to the gauge potentials W_{μ}^{r} for the $\overline{q}\, q$ terms in eq. 1 is thus identified as

$$\left[W_{\mu}^{r} \left(\tfrac{1}{2} \lambda_{r} \right)_{c' \dot{c}} = i \left(\mathcal{W}_{\mu} \left(\mathcal{D} = \{ 3 \} \right) \right)_{c' \dot{c}} \right] (x) \tag{4}$$

Here we postpone the discussion of complete connections and extend the QCD Lagrangean density to include the term quadratic in the field strengths $B_{\mu\nu}^{r}$ and $\Delta \mathcal{L}$ in eq. 1, pertinent to Fermi gauges.

$$\begin{array}{c}
\text{gauge bosons :} \qquad \mathcal{L}_{B} = - \frac{1}{4 g^{2}} B^{\mu\nu\, r} B_{\mu\nu}^{r} \\
B_{\mu\nu}^{r} = \partial_{\mu} W_{\nu}^{r} - \partial_{\nu} W_{\mu}^{r} + f_{rst} W_{\mu}^{s} W_{\nu}^{t} \longleftarrow \left(W_{\mu}^{r} \equiv - v_{\mu}^{r} \right) \\
r, s, t = 1, \cdots, dim \left(G = SU3_{c} \right) = 8 \\
\text{Lie algebra labels, } \left[\tfrac{1}{2} \lambda^{r} , \tfrac{1}{2} \lambda^{s} \right] = i f_{rst} \tfrac{1}{2} \lambda^{t} \\
\text{perturbative rescaling :} \\
W_{\mu}^{r} = g W_{\mu\, pert}^{r} , \; B_{\mu\nu}^{r} = g B_{\mu\nu\, pert}^{r}
\end{array} \tag{5}$$

Degrees of freedom are seen in jets , in (e.g.) the energy momentum sum rule in deep inelastic scattering but not clearly in spectroscopy.
Completing $\Delta \mathcal{L}$ in Fermi gauges

$$\begin{array}{l}
\Delta \mathcal{L} = \left\{ \begin{array}{l} - \frac{1}{2 \eta g^{2}} \left(\partial_{\mu} W^{\mu\, r} \right)^{2} \\ + \partial^{\mu} \overline{c}^{r} \left(D_{\mu} c \right)^{r} \end{array} \right\} \; ; \; \eta \; : \; \text{gauge parameter} \\
\text{ghost fermion fields : } c \, , \, \overline{c} \; ; \; \left(D_{\mu} c \right)^{r} = \partial_{\mu} c^{r} + f_{rst} W_{\mu}^{s} c^{t} \\
\text{gauge fixing constraint : } C^{r} = \partial_{\mu} W^{\mu\, r}
\end{array} \tag{6}$$

perturbative renormalization rescaling

We begin with the renormalization of external operaters and gauge boson fields, denoting unrenormalized fields and operators by the suffix $^{(0)}$

$$\begin{array}{lll}
J_{\alpha} = \left(Z_{J} \right)^{-1} J_{\alpha}^{(0)} & , & \mathcal{O} = \left(Z_{\mathcal{O}} \right)^{-1} \mathcal{O}^{(0)} \\
g = \left(Z_{3} \right)^{3/2} \left(Z_{1} \right)^{-1} g^{(0)} & , & \eta = \left(Z_{3} \right)^{-1} \eta^{(0)} \\
W_{\mu\, pert}^{r} = \left(Z_{3} \right)^{-\frac{1}{2}} \left(W_{\mu\, pert}^{r} \right)^{(0)} & &
\end{array} \tag{7}$$

and continue with the ghost fields c^{r} , $\overline{c}^{s}$ as defined in $\Delta \mathcal{L}$ in eq. 6

$$\begin{array}{lll}
c^{r} = \left(\widetilde{Z}_{2} \right)^{-\frac{1}{2}} \left(c^{r} \right)^{(0)} & , & \overline{c}^{r} = \left(\widetilde{Z}_{2} \right)^{-\frac{1}{2}} \left(\overline{c}^{r} \right)^{(0)} \\
g = \left(Z_{3} \right)^{1/2} \widetilde{Z}_{2} \left(\widetilde{Z}_{1} \right)^{-1} g^{(0)} & &
\end{array} \tag{8}$$

The renormalization constant $Z_1 = Z_1(W-3)$ refers to the 3 gauge boson vertex of the derivative type $W\,W\,\partial\,W$ as shown in figure 1, while $\widetilde{Z}_1 = \widetilde{Z}_1(\bar{c}\,c\,W)$ refers to the ghost-W vertex, shown in figure 2, while the 4 gauge boson vertex involves the analogous renormalization constant $Z_{1(4W)}$ and the $\bar{q}\,q\,W$ vertex $Z_{1(\bar{q}qW)}$, shown in figures 3 and 4 respectively below .

4 irreducible vertex diagrams

W-3 : (i)

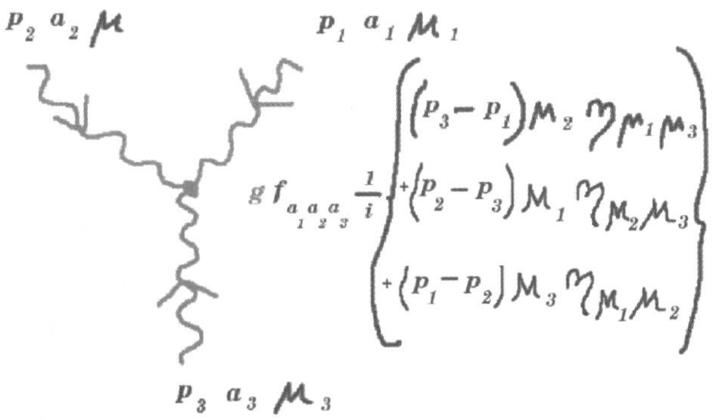

Fig. 1 : $Z_1 \leftrightarrow$ **W-3 vertex of the type** $W\,W\,\partial\,W$

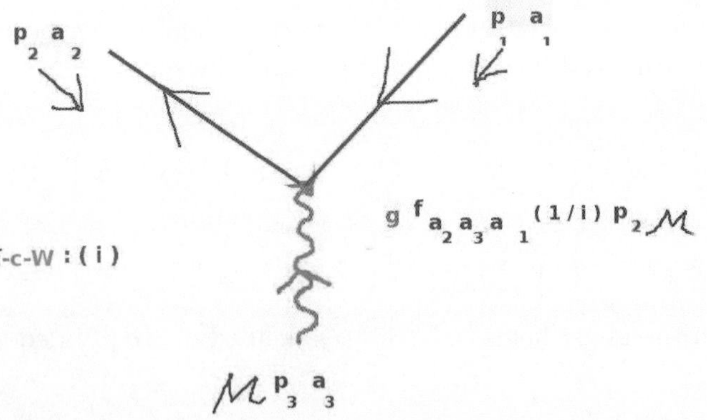

Fig. 2 : $\widetilde{Z}_1 \leftrightarrow$ **cbar-c W-3 vertex of the type** $\partial\,\bar{c}\,c\,W$

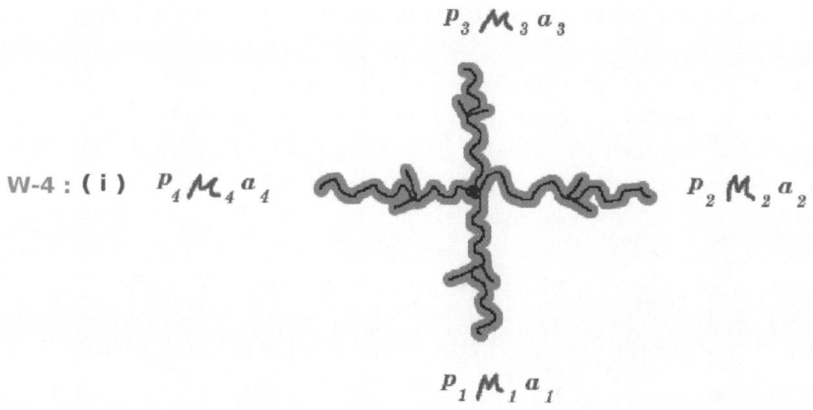

Fig. 3 : $Z_{1(4W)} \leftrightarrow$ **W-4 vertex for gauge fields**

$$g^{2}\left(\begin{array}{ll} r_{[a_1 a_2][a_3 a_4]} & \left[\begin{array}{c} \eta_{\mu_1 \mu_4}\eta_{\mu_2 \mu_3} \\ -\eta_{\mu_1 \mu_3}\eta_{\mu_2 \mu_4}\end{array}\right] \\ +r_{[a_1 a_3][a_2 a_4]} & \left[\begin{array}{c} \eta_{\mu_1 \mu_4}\eta_{\mu_2 \mu_3} \\ -\eta_{\mu_1 \mu_2}\eta_{\mu_3 \mu_4}\end{array}\right] \\ +r_{[a_1 a_4][a_2 a_3]} & \left[\begin{array}{c} \eta_{\mu_1 \mu_3}\eta_{\mu_2 \mu_4} \\ -\eta_{\mu_1 \mu_2}\eta_{\mu_3 \mu_4}\end{array}\right] \end{array}\right)^{\rightarrow r}$$

$$r_{[a_1 b_1][a_2 b_2]} = f_{d a_1 b_1} f_{d a_2 2_2}$$

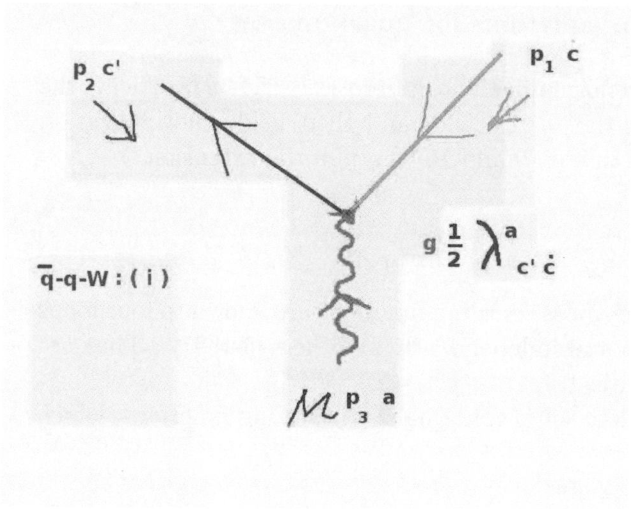

Fig. 4 : $Z_{1(\overline{q}qW)} \leftrightarrow$ **W-3 vertex for** $\overline{q}\, q$ **to gauge fields**

the complete coupling constant renormalization equations

The minimal renormalization constants defined in eqs. 7 and 8, referring beyond the vertices shown in figures 1-4 to the two point functions for gauge connections , Z_3 , and ghost fields , $\widetilde{Z}_2$, thus need extension (also) to $q, \overline{q}$ fields and their two point function, Z_2 , as shown below

$$\begin{array}{lllcl}
g = (Z_3)^{1/2} (Z_3 / Z_1) \, g^{(0)} & ; & W^{r}_{\mu \; pert} & = & (Z_3)^{-\frac{1}{2}} \left(W^{r}_{\mu \; pert} \right)^{(0)} \\
g = (Z_3)^{1/2} \left(\widetilde{Z}_2 / \widetilde{Z}_1 \right) g^{(0)} & ; & (c , \overline{c})^{r} & = & \left(\widetilde{Z}_2 \right)^{-\frac{1}{2}} \left((c , \overline{c})^{r} \right)^{(0)} \\
g^{2} = Z_3 \left(Z_3 / Z_{1(4W)} \right) \left(g^{(0)} \right)^{2} & & & & \\
\hline
g = (Z_3)^{1/2} \left(Z_2 / Z_{1(\overline{q}qW)} \right) g^{(0)} ; & & q^{\, c}_{\, \mathcal{S} \, f} & = & (Z_2)^{-\frac{1}{2}} \left(q^{\, c}_{\, \mathcal{S} \, f} \right)^{(0)} \\
 & & \overline{q}^{\, \dot{c}'}_{\, \dot{\mathcal{S}}' \, \dot{f}} & = & (Z_2)^{-\frac{1}{2}} \left(\overline{q}^{\, \dot{c}'}_{\, \dot{\mathcal{S}}' \, \dot{f}} \right)^{(0)} \\
\hline
\end{array} \tag{9}$$

Maintaining gauge invariance through renormalization thus implies the relations

$$Z_3 / Z_1 = \widetilde{Z}_2 / \widetilde{Z}_1 = Z_2 / Z_{1(\overline{q}qW)} \; ; \; Z_3 / Z_{1(4W)} = (Z_3 / Z_1)^{2} \tag{10}$$

It remains to include quark mass renormalization and associated rescaling.

renormalization equations for quark masses m_f

We perform quark mass renormalization 'at zero mass' [4-1973] , with the following relations , using the $q, \overline{q}$ fields and their field renormalization constant, Z_2 , as defined in eqs. 9 and 10, through the extension

$$(\mathcal{L}_m)^{(0)} = - \sum_f Z_2 Z_m m_f \; ; \; m_f = (Z_m)^{-1} m_f^{(0)} \tag{11}$$

The coefficients of the quark mass rescaling functions are known to four loops as calculated by Chetyrkin and independently by others [5-1997] . They are given in eq. 147 in Appendix 1.

Here we subsum the complete set of renormalized rescaling functions, which follows from the renormalization constants defined in eqs. 8 - 10

$$\left(\begin{array}{l} \mu^2 \partial_{\mu^2} + (\beta(g) / g) \, g^2 \partial_{g^2} + \\ + \gamma_m \, m_f \partial_{m_f} - 2 \gamma_3 (\eta \partial_\eta) - \gamma_{JO} \end{array} \right) C^{T\,(\Pi)}_{JO} (z ; \mu , g , m_\beta , \eta) = 0$$

$$\left\{ \begin{array}{l} \beta(g) = - g \, b(g^2) \\ \gamma_m(g^2) \equiv - \chi_m \\ \gamma_{JO}(g^2, (\eta)) \\ \gamma_3(g^2, \eta) \end{array} \right\} = \mu^2 d / d\mu^2 \left\{ \begin{array}{c} 2 \log \left((Z_3)^{3/2} (Z_1)^{-1} \right) \\ \log \left((Z_m)^{-1} \right) \\ \log \left(Z_O / Z_J^2 \right) \\ \log \left((Z_3)^{1/2} \right) \end{array} \right\} \tag{12}$$

We can always return to the standard form of the renormalization group equation

$$\left(\begin{array}{l} \mu \partial_\mu + \beta(g) \partial_g + (2 \gamma_m) \, m_f \partial_{m_f} \\ - 2 (2 \gamma_3) (\eta \partial_\eta) - (2 \gamma_{JO}) \end{array} \right) C^{T\,(\Pi)}_{JO} (z ; \mu , g , m_\beta , \eta) = 0 \tag{13}$$

without changing the definition of β, transforming the lower relation in eq. 12, which amounts to double the coefficients of the quantities γ_m, γ_3, γ_{JO}, as displayed in eq. 13

$$\left\{ \begin{array}{l} \beta(g) = - g \, b(g^2) \\ 2 \gamma_m(g^2) \equiv - 2 \chi_m \\ 2\gamma_{JO}(g^2, (\eta)) \\ 2 \gamma_3(g^2, \eta) \end{array} \right\} = \mu \, d / d\mu \left\{ \begin{array}{c} \log \left((Z_3)^{3/2} (Z_1)^{-1} \right) \\ \log \left((Z_m)^{-1} \right) \\ \log \left(Z_O / Z_J^2 \right) \\ \log \left((Z_3)^{1/2} \right) \end{array} \right\} \tag{14}$$

rescaling equations in the $\overline{\textbf{MS}}$ scheme for coupling constant g and quark masses m_f
$\beta(g)$ and $\gamma_m(\kappa)$, $\kappa = g^2 / (16 \pi^2)$

We turn to the scale (-square) evolution of the rationalized coupling constant and quark masses, introducing the variables

$$\begin{array}{l} \tau = \log \left(\mu^2 / \mu_0^2 \right) \; ; \; \kappa = g^2 / (16 \pi^2) \; \rightarrow \\ \left\{ \widehat{\beta} = (\beta / g) \, \kappa = \widehat{\beta} (\kappa) , \gamma_m (\kappa) \right\} \end{array} \tag{15}$$

The renormalization group equation in the form given in eq. 12 then becomes

$$\left(\begin{array}{l} \partial_\tau + \widehat{\beta} \, \partial_\kappa + \gamma_m \, m_f \partial_{m_f} \\ - 2 \gamma_3 \; \eta \partial_\eta \; - \gamma_{JO} \end{array} \right) C^{T\,(\Pi)}_{JO} (z ; \mu , g , m_\beta , \eta) = 0 \tag{16}$$

The partial derivative terms in brackets in eq. 16 determine the sliding scale equations with respect to the quantities $\tau, \overline{\kappa}, \overline{m}_f, \overline{\eta}$ in the sense of its initial value structure. We restrict the discussion to $\tau, \overline{\kappa}(\tau), \overline{m}_f(\tau)$ here for simplicity .

$$\begin{aligned} &\dot{\overline{\kappa}} = \widehat{\beta}(\overline{\kappa}), \dot{\overline{m}}_f = \gamma_m(\overline{\kappa})\overline{m}_f \; ; \; \bullet = d/d\tau \\ &\overline{\kappa}(\tau = 0) = \kappa, \overline{m}_f(\tau = 0) = m_f \end{aligned} \tag{17}$$

$$\text{sliding coupling constant : } \quad \dot{\overline{\kappa}} = \widehat{\beta}(\overline{\kappa})$$

We associate integration variables , initial values and endpoint variables in the following way

$$\theta \leftrightarrow [\tau, 0], \lambda \leftrightarrow [\overline{\kappa}, \kappa] \tag{18}$$

The integration of the rescaling differential equation becomes

$$\int_0^\tau d\theta = \tau = \int_\kappa^{\overline{\kappa}} d\lambda \left(\widehat{\beta}(\lambda)\right)^{-1} \longrightarrow \tau = \int_{\overline{\kappa}}^{\kappa} d\lambda \left(-\widehat{\beta}(\lambda)\right)^{-1} \tag{19}$$

The first two coefficients in the expansion of $\widehat{\beta}$ in powers of λ necessitate a twofold subtraction for $\lambda \downarrow 0$ for the integral on the right hand side of eq. 19 to converge at the lower limit of integration

$$\widehat{\beta}(\lambda) = \sum_{n=0}^{\infty} \widehat{\beta}_n \lambda^{n+2} \; ; \; \widehat{\beta}_n \equiv -b_n \tag{20}$$

The first four coefficients $b_n \; ; \; n = 0, \cdots, 3$ are known , (ref. [6-1997]) and given in eq. 146 in Appendix 1 .
We perform this subtraction splitting $\widehat{\beta}$

$$\widehat{\beta} = \widehat{\beta}_{(2)} + \Delta_{(2)}\widehat{\beta} \; ; \; \left[\begin{array}{lcl} \widehat{\beta}_{(2)} & = & -\left(b_0\lambda^2 + b_1\lambda^3\right) \\ \Delta_{(2)}\widehat{\beta} & = & -\sum_{n=2}^{\infty} b_n \lambda^{n+2} \end{array} \right. \tag{21}$$

The subtraction, displayed in eq. 21
gives rise to the substitutions

$$\begin{aligned} \zeta_{(2)} &= \frac{\Delta_{(2)}\widehat{\beta}}{\widehat{\beta}_{(2)}} = b_2 b_0^{-1} \lambda^2 (1 + O(\lambda)) \\ \left(-\widehat{\beta}\right)^{-1} &= \left(-\widehat{\beta}_{(2)}\right)^{-1} \Big/ \left(1 + \zeta_{(2)}\right) = \left(-\widehat{\beta}_{(2)}\right)^{-1} - \psi_{(2)} \\ \psi_{(2)} &= \left(-\widehat{\beta}_{(2)}\right)^{-1} \frac{\zeta_{(2)}}{1 + \zeta_{(2)}} = b_2 (b_0)^{-2} (1 + O(\lambda)) \end{aligned} \tag{22}$$

ψ_2 as defined in eq. 22 has a regular power series expansion in λ , fully determined by $\widehat{\beta}$ and hence the sought subtractions involve only $\widehat{\beta}_{(2)}$, i.e. the beta function truncated to the first two terms.
We proceed in the reduction of $\left(- \widehat{\beta} \right)^{-1}$ using the substitutions in eq. 22

$$
\begin{aligned}
\left(- \widehat{\beta} \right)^{-1} &= \left(- \widehat{\beta}_{(2)} \right)^{-1} - \psi_{(2)} \\
\left(- \widehat{\beta}_{(2)} \right)^{-1} &= b_0^{-1} \lambda^{-2} \left(1 + b_1^{(0)} \lambda \right)^{-1} ; \; b_1^{(0)} = b_1 / b_0 \\
&= b_0^{-1} \lambda^{-2} \left(1 - b_1^{(0)} \lambda \right) \left(1 - \left(b_1^{(0)} \lambda \right)^2 \right)^{-1}
\end{aligned}
\tag{23}
$$

The last expression in eq. 23 decomposes into

$$
\begin{aligned}
\left(- \widehat{\beta}_{(2)} \right)^{-1} &= b_0^{-1} \lambda^{-2} - b_0^{-2} b_1 \lambda^{-1} + \phi_{(2)} \\
\phi_{(2)} &= b_0^{-1} \left(b_1^{(0)} \right)^2 \left(1 - \left(b_1^{(0)} \lambda \right)^2 \right)^{-1} ; \; b_1^{(0)} = b_1 / b_0
\end{aligned}
\tag{24}
$$

$\phi_{(2)}$ defined in eq. 24 , as ψ_2 defined in eq. 22 , has a regular power series expansion in λ .
Thus we rearrange the expressions in eq. 23

$$
\begin{aligned}
\left(- \widehat{\beta} \right)^{-1} &= \left(- \widehat{\beta}_{sing.} \right)^{-1} + \left(- \widehat{\beta}_{reg.} \right)^{-1} ; \; \widehat{\beta} = \widehat{\beta}_{(2)} + \Delta_{(2)} \widehat{\beta} \\
\left(- \widehat{\beta}_{sing.} \right)^{-1} &= b_0^{-1} \lambda^{-2} - b_0^{-2} b_1 \lambda^{-1} , \; \left(- \widehat{\beta}_{reg.} \right)^{-1} = \phi_{(2)} - \psi_{(2)} \\
\hline
\phi_{(2)} &= b_0^{-1} \left(b_1^{(0)} \right)^2 \left(1 - \left(b_1^{(0)} \lambda \right)^2 \right)^{-1} ; \; b_1^{(0)} = b_1 / b_0 \\
\psi_{(2)} &= \left(- \widehat{\beta}_{(2)} \right)^{-1} \frac{\zeta_{(2)}}{1 + \zeta_{(2)}} = b_2 \left(b_0 \right)^{-2} \left(1 + O (\lambda) \right) \\
\zeta_{(2)} &= \frac{\Delta_{(2)} \widehat{\beta}}{\widehat{\beta}_{(2)}} = b_2 b_0^{-1} \lambda^2 \left(1 + O (\lambda) \right)
\end{aligned}
\tag{25}
$$

We recall that in the decomposition $\widehat{\beta} = \widehat{\beta}_{(2)} + \Delta_{(2)} \widehat{\beta}$, retained in eq. 25 , $\widehat{\beta}_{(2)}$ and $\Delta_{(2)} \widehat{\beta}$ denote the beta function truncated to the first wo terms and the remainder term respectively . It further follows from eq. 25

$$
\left(- \widehat{\beta}_{reg.} \right)^{-1} = b_0^{-1} \left(\left(b_1 / b_0 \right)^2 - b_2 / b_0 \right) \left(1 + O (\lambda) \right)
\tag{26}
$$

We return to eq. 19 and integrate the singular part $\left(- \widehat{\beta}_{sing.} \right)^{-1}$

$$\begin{aligned} \tau &= \int_{\overline{\kappa}}^{\kappa} d\lambda \left(- \widehat{\beta}(\lambda) \right)^{-1} \\ &= \int_{\overline{\kappa}}^{\kappa} d\lambda \left\{ \left(- \widehat{\beta}_{reg.}(\lambda) \right)^{-1} + \left(- \widehat{\beta}_{sing.}(\lambda) \right)^{-1} \right\} \end{aligned}$$

$$\int_{\overline{\kappa}}^{\kappa} d\lambda \left(- \widehat{\beta}_{sing.}(\lambda) \right)^{-1} = b_0^{-1} \left[\begin{array}{c} \overline{\kappa}^{-1} - \kappa^{-1} \\ - b_0^{-1} b_1 \left(\log\left(\overline{\kappa}^{-1} \right) - \log\left(\kappa^{-1} \right) \right) \end{array} \right] \tag{27}$$

Eq. 19 thus allows the separation of variables $\overline{\kappa}$ and κ

$$\begin{aligned} \tau &= F(\overline{\kappa}) - G_{reg.}(\overline{\kappa}) - \left(F(\kappa) - G_{reg.}(\kappa) \right) \\ F(\overline{\kappa}) &= b_0^{-1} \overline{\kappa}^{-1} - \left(b_1 / b_0^2 \right) \log\left(\overline{\kappa}^{-1} \right) \\ G_{reg}(\overline{\kappa}) &= \int_0^{\overline{\kappa}} d\lambda \left(- \widehat{\beta}_{reg.}(\lambda) \right)^{-1} \end{aligned} \tag{28}$$

the substitution $\tau \rightarrow t = \tau - \log\left(\mu_0^2 / \Lambda^2 \right)$ and inverting the functional relation $t = t(\overline{\kappa}) \longleftrightarrow \overline{\kappa} = \overline{\kappa}(t)$

We rewrite eq. 28 separating sliding scale parts and associated scale μ and initial value parts associated with scale μ_0, as indicated in eqs. 15 - 17

$$\left[\begin{array}{c} \tau + b_0^{-1} \kappa^{-1} - \left(b_1 / b_0^2 \right) \log\left(b_0^{-1} \kappa^{-1} \right) - G_{reg.}(\kappa) + \\ + \left(b_1 / b_0^2 \right) \log\left(b_0^{-1} \overline{\kappa}^{-1} \right) + G_{reg.}(\overline{\kappa}) \end{array} \right] = b_0^{-1} \overline{\kappa}^{-1} \tag{29}$$

The substitutions $\kappa \rightarrow b_0 \kappa$ and $\overline{\kappa} \rightarrow b_0 \overline{\kappa}$ in the inverse of the arguments of the logarithm terms in eq. 29 cancels out in the difference. Next we adjust the scale parameter $\mu_0 \rightarrow \Lambda$
setting

$$\begin{aligned} & - b_0^{-1} \kappa^{-1} + \left(b_1 / b_0^2 \right) \log\left(b_0^{-1} \kappa^{-1} \right) + G_{reg.}(\kappa) = \log\left(\mu_0^2 / \Lambda^2 \right) \\ & t = \tau - \log\left(\mu_0^2 / \Lambda^2 \right) = \log\left(\Lambda^2 / \mu^2 \right) \\ & x = b_0 \overline{\kappa} \; ; \; G_{reg.}(x / b_0) = G(x) \; ; \; b_1 / b_0^2 = a \end{aligned} \tag{30}$$

Further we substitute the sliding scale variable $\overline{\kappa} \longrightarrow b_0 \overline{\kappa} = x$. With these substitutions eq. 29 becomes

$$\begin{array}{c} t + a \log\left(x^{-1} \right) + G(x) = x^{-1} \\ \hline G(x) = x \sum_{n=0}^{\infty} G_n x^n \end{array} \tag{31}$$

In order to establish the asymptotic 'ultraviolet' expansion for $t \rightarrow \infty$ and $Y = x^{-1} \rightarrow \infty$ we introduce the notation , for clarity, including a change of variables $t = \exp(L)$; $L = \log(t)$

$$\begin{array}{l} \log(t) = L \; ; \; x = x(L) \longrightarrow Y = Y(L) \equiv (x(L))^{-1} \\ \log(Y) = Z(L) \equiv \log\left((x(L))^{-1} \right) \end{array} \tag{32}$$

Eq. 31 takes the form

$$\begin{array}{c} L + \log \left[1 + \dfrac{a\, Z(L) + G(x = Y^{-1})}{\exp(L)} \right] = Z(L) \\ \hline t = \exp(L) \; ; \; Y = \exp(Z) \longleftarrow Z = Z(L) \end{array} \tag{33}$$

Eq. 33 can now be solved by iteration in the asymptotic limit $L \rightarrow \infty$

$$\begin{array}{l} L = \log\left[\log\left(\Lambda^2 / \mu^2 \right) \right] \rightarrow \infty \\ Z(L) = lim_{\, n \rightarrow \infty} Z_n(L) \\ L + \log \left[1 + \dfrac{a\, Z_n(L) + G\left(x_n = Y_n^{-1} \right)}{\exp(L)} \right] = Z_{n+1}(L) \\ \hline \text{anchor} : Z_0(L) = L \end{array} \tag{34}$$

The first anchor takes care of the largest contribution to to Z , i.e. L . Thus we obtain $Z_1(L)$ and $Y_1 = \exp(Z_1)$ from rq. 34

$$\begin{array}{l} Z_1(L) = L + \log \left[1 + a\, L \exp(-L) + \dfrac{G(\exp(-L))}{\exp(L)} \right] \\ Y_1(L) = e^L + a\, L + G(\exp(-L)) \; ; \; t = e^L \end{array} \tag{35}$$

Comparing withe recursive equations (eq. 34 , repeated below

$$\begin{array}{l} L = \log\left[\log\left(\mu^2 / \Lambda^2 \right) \right] \rightarrow \infty \\ Z(L) = lim_{\, n \rightarrow \infty} Z_n(L) \\ L + \log \left[1 + \dfrac{a\, Z_n(L) + G\left(x_n = Y_n^{-1} \right)}{\exp(L)} \right] = Z_{n+1}(L) \\ \hline \text{anchor} : Z_0(L) = L \end{array} \tag{36}$$

it follows that all contributions from the function G , i.e. resulting from the subleading terms proportinal in the beta function truncated by the first to terms $b_0 \kappa^2 + b_1 \kappa^3$ give vanishing contributions to $Y(L) \equiv (b_0 \overline{\kappa})^{-1}(L)$ for $L \to \infty$. This can be verified recursively.
Thus modulo terms vanishing for $L \to \infty$ we can solve the simpler functional equation, omitting G from eqs. 31, 35 and 36

$$t + a \log \left(\widetilde{Y} \right) = \widetilde{Y} \; ; \; \widetilde{Y} = x^{-1} \; ; \; Y - \widetilde{Y} \to 0 \text{ for } t \to \infty \tag{37}$$

or equivalently

$$L + \log \left(1 + a \widetilde{Z} e^{-L} \right) = \widetilde{Z} \; ; \; \widetilde{Y} = e^{\widetilde{Z}} \; ; \; t = e^{L} \tag{38}$$

The same argument which led to the elimination of the term proportional to G in eq. 36 as far as nonvanishing contributions for Y are concerned in the limit $L \to \infty$ imply that expanding the logarithm on the left hand side of eq. 38 only the first term need be retained

$$\log \left(1 + a \widetilde{Z} e^{-L} \right) \sim a \widetilde{Z} e^{-L} \tag{39}$$

leading to the equivalent approximate functional equation simplifying eq. 38

$$\begin{aligned} Z \sim \widetilde{Z} \sim L / \left(1 - a e^{-L} \right) \sim L + a e^{-L} L \longrightarrow \\ Y \sim \widetilde{Y} \sim e^{L} \left(1 + a e^{-L} L \right) = e^{L} + a L + [0] \text{ for } L \to \infty \end{aligned} \tag{40}$$

In eq. 40 the $\sim$ symbol is meant to imply modulo additive terms to Y, vanishing for $L \to \infty$, proving the remarkable fact that the potential constant in the asymptotic expansion for Y vanishes .
We conclude this subsection comparing side by side the functional dependence of $t \equiv \log(\mu^2 / \Lambda^2) = t(\overline{\kappa}(\mu))$ and $Y \equiv (b_0 \overline{\kappa})^{-1} = Y(t)$.

The first relation is given in eq. 31

$$
\begin{array}{l}
t \ = \ x^{\ -1} \ - \ a \ \log \ \left(\ x^{\ -1} \ \right) \ - \ G \ (\ x \ ) \\
\hline
G \ (\ x \ ) \ = \ x \ \sum_{\ n=0}^{\ \infty} \ G_{\ n} \ x^{\ n} \ ; \ x \ = \ b_{\ 0} \ \overline{\kappa} \ (\ \mu^2 \ ) \\
t \ = \ \log \ (\ \mu^{\ 2} \ / \ \Lambda^{\ 2} \ ) \ \longrightarrow \\
\hline
\Lambda^{\ 2} \ = \ \mu^{\ 2} \ \left[\ \exp \ \left(- \ \dfrac{1}{b_{\ 0} \ \overline{\kappa}} \ \right) \right] \ \left[\dfrac{1}{b_{\ 0} \ \overline{\kappa}} \right]^{\ a} \ \exp \ \left[\ G \ (b_{\ 0} \ \overline{\kappa} \ ) \ \right] \\
\hline
a \ = \ b_{\ 1} \ / \ b_{\ 0}^{\ 2}
\end{array}
\tag{41}
$$

The second relation is displayed in eq. 40 , as an asymptotic expansion for $t \ = \ e^{\ L} \ \rightarrow \ \infty$

$$
\begin{array}{ll}
\hline
Y \ = & Y \ (\ L \ ) \ \equiv \ (\ b_{\ 0} \ \overline{\kappa} \ )^{\ -1} \ (\ L) \text{ for } L \ = \ \log \ \left[\ \log \ \left(\ \mu^{\ 2} \ / \Lambda^{\ 2} \ \right) \ \right] \ \rightarrow \ \infty \\
\sim & e^{\ L} \ + \ a \ L \ + \ [\ 0 \ ] \\
\hline
\end{array}
\tag{42}
$$

Eqs. 41 and 42 constitute the essence of this subsection, giving rise to some remarks :

1) The central scale Λ

with all subtle properties as defined, in the $\overline{\text{MS}}$ scheme, constitutes a renormalization group invariant finite mass scale, given a region of sliding scale coupling constant $\kappa \ = \ g^{\ 2} \ / \ \left(\ 16 \ \pi^{\ 2} \ \right)$ generically or $\overline{\kappa}$ in the perturbative region region here, nonvanishing but appropriately small.

This scale does *not* depend on quark masses , only on the number of quark flavors , while derivations from measurable quantities (in the asymptiotic region) suffer from systematic errors , which do depend on quark massses .

The combined analysis of the sliding scale coupling constant , as of 2009 by Bethke , is shown in Fig. 5 below.

I quote here a discussion of Λ for $N_{\ fl} \ = \ 3$ by Bodenstein et al., ref. [8-2011] , from data in the mass scale region of the τ lepton

$$
\begin{array}{llll}
\Lambda_{\ N_{\ fl=3}} \ = \ 382 \ \pm \ 24 \text{ MeV} & \longleftrightarrow & \alpha_{\ s} & = \ 4\pi \ \kappa \ (\ \mu \ = \ m_{\ \tau} \ ) \\
 & & & = \ 0.344 \ \pm \ 0.014
\end{array}
\tag{43}
$$

The reconstruction of Λ from the 'perturbatively accessible region' implies a hard breaking of scale invariance , and hence equivalently a nonvanishing trace of the co-renormalizable and renormalized energy momentum density operator. This shall be discussed in the next subsection.

2) sliding scale quark masses

The rescaling equations (eqs. 15 - 17) extend to the quark mass parameters , again universally and quark mass independently rescaling m_f , keeping ratios invariant

$$r_{f_1 f_2} = m_{f_1} / m_{f_2} \tag{44}$$

Here no detailed discussion of extracting sliding scale quark masses including heavy flavors c , b is given. For heavy flavor comparisons with initially the reaction $e^+ e^- \to f \overline{f}-$ flavored hadrons , the QCD sum rule approach is the main tool , as pioneered by Shifman, Vainshtein and Zakharov [7-1979] , together with lattice QCD .

The sliding quark masses as for 3 flavors are shown in figure 6 below , restricted to two loop approximation . References [8-2011] - [11-2011] are representative for present refinement to four loops in the rescaling functions and derived results , e.g.

$$\begin{aligned}
\overline{m}_b (\overline{m}_b) &= \begin{cases} 4163 \pm 16 \text{ MeV} & \text{ref. [9-2010]} \\ 4171 \pm 7 \text{ MeV} & \text{ref. [10-2011]} \\ 4177 \pm 11 \text{ MeV} & \text{ref. [11-2011]} \end{cases} \\
\overline{m}_c (3 \text{ GeV}) &= 986 \pm 13 \text{ MeV} , \text{ ref. [9-2010]} \\
\overline{m}_c (\overline{m}_c) &= 1262 \pm 17 \text{ MeV} , \text{ ref. [11-2011]}
\end{aligned} \tag{45}$$

1-1-b – Gauge boson binary bilocal and adjoint (here octet-) string operators

One goal is, to identify – not just some candidate resonance – gluonic mesons, binary and higher modes, and to relate them to the base quantities within QCD.

$$\begin{aligned}
& B_{[\mu_1 \nu_1],[\mu_2 \nu_2]} (x_1 , x_2) = \\
& = B^{\,r}_{[\mu_1 \nu_1]} (x_1) \, U (x_1 , r ; x_2 , s) \, B^{\,s}_{[\mu_2 \nu_2]} (x_2) \\
& r , s , \cdots = 1, \cdots , 8 \; ; \text{ no flavor but spin}
\end{aligned} \tag{46}$$

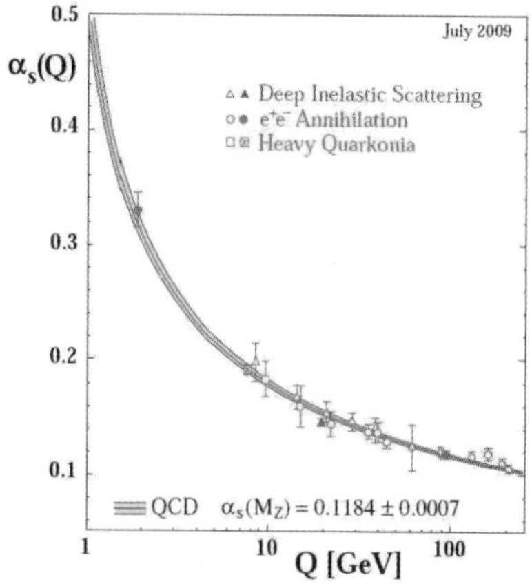

Fig. 5 : $\alpha_s (Q) = 4\pi \, \kappa (\mu = Q)$ **from ref. [A215-2009] .**

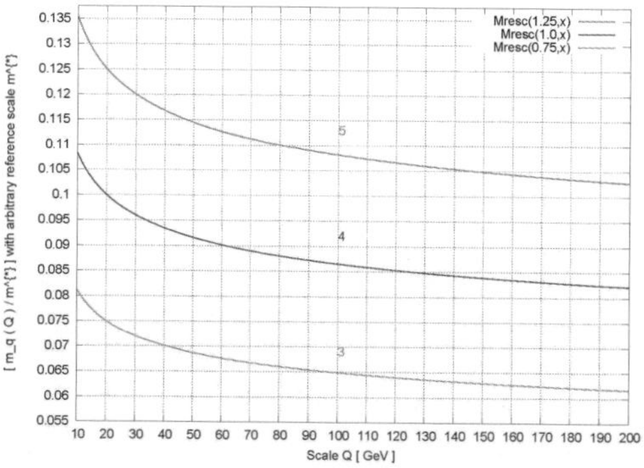

Fig. 6 : $m_q (Q) / m^*$ **with fixed ratio of rescaled quark masses**
$m_u : \frac{1}{2} (m_d + m_u) : m_d = 3 : 4 : 5$
from eqs. 15 - 17 , 146 - 147

$B^{\,r}_{[\mu \nu]} (x)$ denote the local color octet of field strengths.
The quantity $U (x , r ; y , s)$ in eq. (46) denotes the octet string operator, i. e. the path ordered exponential over a straight line path $\mathcal{C}$ from y to x

$$
\begin{array}{lll}
U (x , r ; y , s) & = & P \exp \left(\left. \int_y^x \right|_{\mathcal{C}} d z^{\mu} \, \frac{1}{i} \, v^{\,t}_{\mu} (z) \, \mathcal{F}_t \right) \Bigg|_{r s} \\
 & = & P \exp \left(- \left. \int_y^x \right|_{\mathcal{C}} d z^{\mu} \, W^{\,t}_{\mu} (z) \, (\frac{1}{i} \, \mathcal{F}_t) \right) \Bigg|_{r s} \\
(\mathcal{F}_t)_{rs} = i \, f_{r t s} \; ; \; (ad_t)_{rs} = \frac{1}{i} \, (\mathcal{F}_t)_{rs} = f_{r t s} & &
\end{array}
\tag{47}
$$

The path ordered exponential as a matrix function of the argument is to be performed before the matrix elements, denoted $|_{\,..}$ in eq. 47 , are taken. The local limit becomes

$$B_{[\,\mu_1\,\nu_1\,]\,,\,[\,\mu_2\,\nu_2\,]}\,(\,x_{\,1}\,=\,x_{\,2}\,=\,x\,)\;=\;(:)\;B^{\,r}_{\,[\,\mu_1\,\nu_1\,]}\,(\,x\,)\;B^{\,r}_{\,[\,\mu_2\,\nu_2\,]}\,(\,x\,)(:)$$

no flavor but spin

$$(48)$$

1-1-c – $\overline{q}\,q$ bilinears and triplet-string operators

$$B^{\,q}_{\,[\,\dot{\mathcal{A}}\,\dot{f}_{\,1}\,,\,\mathcal{B}\,f_{\,2}\,]}\,(\,x_{\,1}\,,\,x_{\,2}\,)\;=\;\overline{q}^{\,\dot{c}_{\,1}}_{\,\mathcal{B}\,\dot{f}_{\,2}}\,(\,x_{\,1}\,)\,U\,(\,x_{\,1}\,,\,c_{\,1}\,;\,x_{\,2}\,,\,\dot{c}_{\,2}\,)\,q^{\,c}_{\,\mathcal{A}\,f_{\,1}}\,(\,x_{\,2}\,)$$

flavor *and* spin

$$\begin{aligned} U\,(\,x\,,\,c_{\,1}\,;\,y\,,\,\dot{c}_{\,2}\,) \;&=\; P\,\exp\left(\left.\left(\int_{y}^{x}\right|_{C}\,d\,z^{\,\mu}\,\tfrac{1}{i}\,v^{\,t}_{\,\mu}\,(\,z\,)\,\tfrac{1}{2}\,\lambda_{\,t}\right)\right|_{\,c_{\,1}\,\dot{c}_{\,2}} \\ &=\; P\,\exp\left(\left.-\int_{y}^{x}\right|_{C}\,d\,z^{\,\mu}\,W^{\,t}_{\,\mu}\,(\,z\,)\,\left(\,\tfrac{1}{i}\,\tfrac{1}{2}\,\lambda_{\,t}\,\right)\right)\Bigg|_{\,c_{\,1}\,\dot{c}_{\,2}} \end{aligned} \tag{49}$$

with the local limit

$$B^{\,q}_{\,[\,\dot{\mathcal{B}}\,\dot{f}_{\,2}\,,\,\mathcal{A}\,f_{\,1}\,]}\,(\,x_{\,1}\,=\,x_{\,2}\,=\,x\,)\;=\;(:)\,\overline{q}^{\,\dot{c}}_{\,\mathcal{B}\,\dot{f}_{\,2}}\,(\,x\,)\,q^{\,c}_{\,\mathcal{A}\,f_{\,1}}\,(\,x\,)\,(:) \tag{50}$$

The symbols (:) in eqs. 48 and 50 should indicate that normal ordering of regulating the local limits is required and further that such normal ordering is *not* unique , and dependent on quark masses in the case of the $\overline{q}\,q$ bilinears .

1-1-d – Connection and curvature - forms preparing the ensueing analysis of regularity conditions

Lets begin this (sub-)section rewriting the bilocal (formally) unitary operators forming the gauge connection dependent octet- (eq. 47) and triplet (eq. 49) QCD strings , substituting an equivalent matrix oriented notation

$$\begin{array}{l} \begin{array}{c}\text{octet}\\ \text{string}\end{array} : \ U (x , r ; y , s) = P \exp \left(- \int_{y}^{x} \Big|_{C} d z^{\mu} W_{\mu}^{t} (z) (\frac{1}{i} \mathcal{F}_{t}) \right) \Bigg|_{r s} \\ \begin{array}{c}\text{triplet}\\ \text{string}\end{array} : U (x , c_{1} ; y , \dot{c}_{2}) = P \exp \left(- \int_{y}^{x} \Big|_{C} d z^{\mu} W_{\mu}^{t} (z) (\frac{1}{i} \frac{1}{2} \lambda_{t}) \right) \Bigg|_{c_{1} \dot{c}_{2}} \end{array}$$

with the substitutions $\longrightarrow$

$$\left. \begin{array}{l} \begin{array}{c}\text{octet}\\ \text{string}\end{array} : \ U (x , r ; y , s) \rightarrow \left(U \left(x \overset{C}{\leftarrow} y \right) \right)_{rs} \\ \begin{array}{c}\text{triplet}\\ \text{string}\end{array} : U (x , c_{1} ; y , \dot{c}_{2}) \rightarrow \left(U \left(x \overset{C}{\leftarrow} y \right) \right)_{c_{1} \dot{c}_{2}} \end{array} \right\} \rightarrow \begin{array}{r} (U (x , C , y ; \mathcal{D}))_{\alpha\beta} \\ \in \mathcal{D} (\mathcal{G}) \end{array}$$

with $\mathcal{G}$ = simple compact gauge group ; $\mathcal{D}$: general irreducible representation of $\mathcal{G}$

(51)

Here $\mathcal{G} = SU3_{c}$ and $\mathcal{D}$ is the adjoint- , triplet representation for the respective QCD $\mathcal{D}$ - strings.
Further let us consider matrix valued connection 1-forms , which define the bilocal matrix valued operators $(U (x , C , y ; \mathcal{D}))_{\alpha\beta} \in \mathcal{D} (\mathcal{G})$ as given in eq. 51 . To this end the form of octet and triplet strings in eq. 51 is repeated below

$$\begin{array}{l} \begin{array}{c}\text{octet}\\ \text{string}\end{array} : \ U (x , r ; y , s) = P \exp \left(- \int_{y}^{x} \Big|_{C} d z^{\mu} W_{\mu}^{t} (z) (\frac{1}{i} \mathcal{F}_{t}) \right) \Bigg|_{r s} \\ \begin{array}{c}\text{triplet}\\ \text{string}\end{array} : U (x , c_{1} ; y , \dot{c}_{2}) = P \exp \left(- \int_{y}^{x} \Big|_{C} d z^{\mu} W_{\mu}^{t} (z) (\frac{1}{i} \frac{1}{2} \lambda_{t}) \right) \Bigg|_{c_{1} \dot{c}_{2}} \end{array} \tag{52}$$

The two matrices in brackets to the right of the integrand expressions in eq. 52 form an antihermitian basis of the Lie algebra representation $Lie (\mathcal{D})$ for $\mathcal{D}$ = adjoint and $\mathcal{D}$ = triplet representations of $\mathcal{G} = SU3_{c}$ respectively

$$\begin{array}{l} d_{t} \equiv d_{t} (\mathcal{D}) \leftrightarrow (d_{t})_{\alpha\beta} = \left\{ \begin{array}{ll} (\frac{1}{i} \mathcal{F}_{t})_{rs} & \text{for } Lie (\mathcal{D}) = \text{adjoint} \\ (\frac{1}{i} \frac{1}{2} \lambda_{t})_{c_{1} \dot{c}_{2}} & \text{for } Lie (\mathcal{D}) = \text{triplet} \end{array} \right. \\ d_{t} = - d_{t}^{\dagger} \ ; \ t = 1, \cdots , dim \ \mathcal{G} \ ; \ \alpha , \beta = 1, \cdots , dim \ \mathcal{D} \text{ for general } \mathcal{D} \end{array} \tag{53}$$

From eqs. 52 and 53 we construct a – hopefully – consistent notation as appropriate for matrix valued $\mathcal{D}$ connections, 1-forms and strings, as well as derived 2- and higher forms. First eq. 53 is subject to the (matrix-)

commutation relations

$$
\begin{array}{l}
[\, d_{\,r} \, , \, d_{\,s} \,] \; = \; f_{\,rst} \, d_{\,t} \; ; \; \forall \, \mathcal{D} \, (\, \mathcal{G} \,) \; ; \; r,s,t \; = \; 1, \cdots , dim \; \mathcal{G} \quad \longrightarrow \\
(\, d_{\,t} \, (\, \mathcal{D} \; = \; \text{adjoint representation} \,) \,)_{\,sr} \; = \; (\, ad_{\,t} \,)_{\,sr} \; = \; f_{\,str} \; : \; \text{independent of } \mathcal{D} \\
f_{\,str} \; : \; \text{totally antisymmetric, real structure constants of } \; Lie \, (\, \mathcal{G} \,)
\end{array}
\tag{54}
$$

In physics the antihermitian matrix code with respect to the representations $Lie \, (\, \mathcal{D} \,)$ is (mostly) replaced by the hermitian one [1]

$$
\begin{array}{l}
(\, d_{\,t} \; \equiv \; \frac{1}{i} \, h_{\,t} \,) \; (\, Lie \, (\, \mathcal{D} \,) \,) \, \big|_{\,\alpha\beta} \; ; \; h_{\,t} \; = \; h_{\,t}^{\,\dagger} \; ; \; [\, h_{\,r} \, , \, h_{\,s} \,] \; = \; i \, f_{\,rst} \, h_{\,t} \\
\alpha , \beta \; = \; 1, \cdots , dim \; \mathcal{D}
\end{array}
\tag{55}
$$

Eq. 54 serves to define
matrix valued connections built from a basis of $Lie \, (\, \mathcal{D} \,)$ representation matrices as defined in eq. 54 for general irredcible representations $\mathcal{D} \, (\, \mathcal{G} \,)$ denoted $\mathcal{W}_{\,\mu} \, (\, z \, , \, \mathcal{D} \,)$

$$
\begin{array}{l}
\mathcal{W}_{\,\mu} \, (\, z \, , \, \mathcal{D} \,) \, \big|_{\,\alpha\beta} \; = \; W_{\,\mu}^{\,r} \, (\, z \,) \; (\, d_{\,r} \,)_{\,\alpha\beta \; Lie \; (\, \mathcal{D} \,)} \; ; \; \left[\begin{array}{l} r \; = \; 1, \cdots , dim \, (\, \mathcal{G} \,) \\ \alpha \, , \, \beta \; = \; 1, \cdots , dim \, (\, \mathcal{D} \,) \end{array} \right] \\
\mathcal{W}_{\,\mu} \, (\, z \, , \, \mathcal{D} \,) \, \big|_{\,\alpha\beta} \; \longrightarrow \; \mathcal{W}_{\,\mu} \; \text{for compact matrix notation}
\end{array}
\tag{56}
$$

In the following it is to be understood, that $\mathcal{W}_{\,\mu}$ is extended to a general collection of representations $\bigcup \mathcal{D}$ – thought to be carried by real and spurious spin $\frac{1}{2}$ fields – care beeing taken that asymptotic freedom in the ultraviolet is not upset.
From eq. 56 we define the associated matrix valued connection 1-form displayed alongside the base definition repeated from eq. 56 in eq. 57 below

$$
\begin{array}{rcll}
\mathcal{W}^{\,(1)} \, (\, z \, , \, \mathcal{D} \,) \, \big|_{\,\alpha\beta} & = & d \, z^{\,\mu} \; \mathcal{W}_{\,\mu} \, (\, z \, , \, \mathcal{D} \,) \, \big|_{\,\alpha\beta} & \longrightarrow \; \mathcal{W}^{\,(1)} \\
\mathcal{W}_{\,\mu} \, (\, z \, , \, \mathcal{D} \,) \, \big|_{\,\alpha\beta} & = & W_{\,\mu}^{\,r} \, (\, z \,) \; (\, d_{\,r} \,)_{\,\alpha\beta \; Lie \; (\, \mathcal{D} \,)} & \longrightarrow \; \mathcal{W}_{\,\mu}
\end{array}
\tag{57}
$$

and the matrix valued field-strength tensor

$$
\mathcal{W}_{\,\mu\nu} \, (\, z, \, \mathcal{D} \,) \, \big|_{\,\alpha\beta} \; = \; \left\{ \begin{array}{c} \partial_{\,\mu} \, \mathcal{W}_{\,\nu} \, (\, z, \, \mathcal{D} \,) \; - \; \partial_{\,\nu} \, \mathcal{W}_{\,\mu} \, (\, z, \, \mathcal{D} \,) \, + \\ + \; [\, \mathcal{W}_{\,\mu} \, (\, z, \, \mathcal{D} \,) \, , \, \mathcal{W}_{\,\nu} \, (\, z, \, \mathcal{D} \,) \,] \end{array} \right\}_{\alpha\beta} \; \longrightarrow \; \mathcal{W}_{\,\mu\nu}
\tag{58}
$$

[1] A (partial) collection of historical and textbook references to the topics pertaining to 'Continous transformation groups and differential geometry' is presented at the end of references and labelled by the symbols 1H , 2H $\cdots$.

together with their associated curvature 2-form

$$\begin{aligned} \mathcal{W}^{(2)}(z, \mathcal{D})\big|_{\alpha\beta} &= \tfrac{1}{2}\, d z^{\mu} \wedge d z^{\mu}\, \mathcal{W}_{\mu\nu}(z, \mathcal{D})\big|_{\alpha\beta} && \longrightarrow \mathcal{W}^{(2)} \\ \mathcal{W}_{\mu\nu}(z, \mathcal{D})\big|_{\alpha\beta} &= \left\{ \begin{array}{c} \partial_{\mu} \mathcal{W}_{\nu}(z, \mathcal{D}) - \partial_{\nu} \mathcal{W}_{\mu}(z, \mathcal{D}) + \\ + [\mathcal{W}_{\mu}(z, \mathcal{D}), \mathcal{W}_{\nu}(z, \mathcal{D})] \end{array} \right\}_{\alpha\beta} && \longrightarrow \mathcal{W}_{\mu\nu} \\ &= W^{r}_{\mu\nu}(z)\,(d_r)_{\alpha\beta\ Lie\ (\mathcal{D})} \end{aligned}$$

$$W^{r}_{\mu\nu} = \partial_{\mu} W^{r}_{\nu} - \partial_{\nu} W^{r}_{\mu} + f_{rst}\, W^{s}_{\mu} W^{t}_{\nu} \ ; \ \text{independent of } \mathcal{D} \tag{59}$$

Two remarks are in place here

1) In order to distinguish field strength from potentials (connections) the following equivalent but different notations for the field strength shall be used

$$\mathcal{W}_{\mu\nu} \equiv \mathcal{B}_{\mu\nu} \ ; \ \mathcal{W}^{(2)} \equiv \mathcal{B}^{(2)} \ ; \ W^{r}_{\mu\nu} \equiv B^{r}_{\mu\nu} \tag{60}$$

2) From the last relation in eq. 59 it may appear redundant to extend connections and curvatures to matrix valued form with respect to a wide collection of irreducible representations $\mathcal{D}(\mathcal{G})$. This however is tantamount to neglecting nontrivial global regularity conditions in the infrared .

We end this subsection (1-1-d) displaying the bilocal (parallel transport-) operators defined in eq. 51 using the shorthand notation in eq. 59

$$\begin{array}{ccc} (U(x, C, y; \mathcal{D}))_{\alpha\beta} & = & P \exp\left(- \int_{y}^{x} \Big|_{C} \mathcal{W}^{(1)}(z, \mathcal{D}) \right)\Bigg|_{\alpha\beta} \\ \downarrow & & \downarrow \\ U(x, C, y) & = & P \exp\left(- \int_{y}^{x} \Big|_{C} \mathcal{W}^{(1)} \right) \quad [\text{for } (\bigcup \mathcal{D})] \end{array} \tag{61}$$

1-1-e – The U1- or singlet axial current anomaly

The U1-axial central anomaly involves the local chiral current projections from $B^{q}_{[\dot{\mathcal{B}} \dot{f}_2 , \mathcal{A} f_1]}(x)$ in eq. 50

$$\begin{aligned} (j^{\pm}_{\mu})_{\dot{f}_2 f_1}(x) &= B^{q}_{[\dot{\mathcal{B}} \dot{f}_2 , \mathcal{A} f_1]}(x) \left(\gamma_{\mu} \tfrac{1}{2} (\P \pm \gamma_5) \right)_{\mathcal{B} \dot{\mathcal{A}}} \\ &= (:)\, \overline{q}^{\dot{c}}_{\dot{f}_2} \gamma^{\pm}_{\mu} q^{c}_{f_1}(x)\, (:) \\ \gamma_5 = \gamma_{5R} = \tfrac{1}{i} \gamma_0 \gamma_1 \gamma_2 \gamma_3 \ &; \ \gamma^{\pm}_{\mu} = \gamma_{\mu} \tfrac{1}{2} (\P \pm \gamma_5) \end{aligned} \tag{62}$$

The equations of motion for the fermion fields are *and superficially imply* (upon $f_1 \leftrightarrow f_2$)

$$\begin{aligned} \partial\!\!\!/\, q^{c}_{f_2} &= \tfrac{1}{i} \left(p\!\!\!/^{\,c \dot{c}'} + \delta^{c \dot{c}'} m_{f_2} \right) q^{c'}_{f_2} \\ \overline{q}^{\dot{c}}_{\dot{f}_1} \overleftarrow{\partial\!\!\!/} &= \overline{q}^{\dot{c}'}_{\dot{f}_1} \tfrac{1}{i} \left(- p\!\!\!/^{\,c' \dot{c}} - \delta^{c' \dot{c}} m_{f_1} \right) \end{aligned} \quad ; \text{ no sums over } \dot{f}_1 , f_2 \rightarrow$$

$$\begin{aligned} \partial^{\mu} (j^{\pm}_{\mu})_{\dot{f}_1 f_2} &= \tfrac{1}{2i} \left((m_{f_2} - m_{f_1}) S_{\dot{f}_1 f_2} \mp (m_{f_2} + m_{f_1}) P_{\dot{f}_1 f_2} \right) \\ S_{\dot{f}_1 f_2} &= (:)\, \overline{q}^{\dot{c}}_{\dot{f}_1} q^{c}_{f_2} (:) \ , \ P_{\dot{f}_1 f_2} = (:)\, \overline{q}^{\dot{c}}_{\dot{f}_1} \gamma_5 q^{c}_{f_2} (:) \end{aligned} \tag{63}$$

In eq. 63 m_f denotes the *real* , *nonnegative* quark mass for flavor f. From eq. 63 the relations for vector and axial vector currents *superficially* follow

$$\begin{aligned} (j_{\mu})_{\dot{f}_1 f_2} &= (j^{+}_{\mu})_{\dot{f}_1 f_2} + (j^{-}_{\mu})_{\dot{f}_1 f_2} \\ (j^{5}_{\mu})_{\dot{f}_1 f_2} &= (j^{+}_{\mu})_{\dot{f}_1 f_2} - (j^{-}_{\mu})_{\dot{f}_1 f_2} \\ \partial^{\mu} (j_{\mu})_{\dot{f}_1 f_2} &= \tfrac{1}{i} (m_{f_2} - m_{f_1}) S_{\dot{f}_1 f_2} \\ \partial^{\mu} (j^{5}_{\mu})_{\dot{f}_1 f_2} &= (m_{f_2} + m_{f_1})\, i P_{\dot{f}_1 f_2} \end{aligned} \tag{64}$$

As it follows from the original derivation by Adler and Bell and Jackiw [12-RR-1969] in QED, the vecor current Ward identities in eq. 64 can be implemented also in QCD , leaving the axial current ones reduced to the flavor non-singlet case , leaving the U1 axial current divergent anomalous

$$\partial^{\mu} (j_{\mu})_{\dot{f}_1 f_2} = \tfrac{1}{i} (m_{f_2} - m_{f_1}) S_{\dot{f}_1 f_2} \ \surd$$

$$\left\{ \begin{array}{c} j^{5}_{\mu} \\ P \end{array} \right\}^{NS}_{\dot{f}_1 f_2} = \left\{ \begin{array}{c} j^{5}_{\mu} \\ P \end{array} \right\}_{\dot{f}_1 f_2} - \frac{1}{N_{fl}} \delta_{\dot{f}_1 f_2} \sum_f \left\{ \begin{array}{c} j^{5}_{\mu} \\ P \end{array} \right\}_{\dot{f} f} \tag{65}$$

and similarly

$$\left\{ \begin{array}{c} j^{5}_{\mu} \\ P \end{array} \right\}^{S}_{\dot{f}_1 f_2} = \sum_f \left\{ \begin{array}{c} j^{5}_{\mu} \\ P \end{array} \right\}_{\dot{f} f} \tag{66}$$

1-1-f – Quark masses and splittings : m_f **and** $\Delta m_f = m_f - \langle m \rangle$

In the subtitle above $\langle m \rangle$ stands for the mean quark mass

$$\langle m \rangle = \frac{1}{N_{fl}} \sum_f m_f \tag{67}$$

The identities for vector currents in eqs. 64 and 65 can be extended separating the conributions proportional to Δm_f and $\langle m \rangle$

$$\begin{aligned}
\partial^{\mu} (j_{\mu})_{\dot{f}_1 f_2} &= \tfrac{1}{i} (\Delta m_{f_2} - \Delta m_{f_1}) S_{\dot{f}_1 f_2} \checkmark \\
\partial^{\mu} (j^5_{\mu})^{NS}_{\dot{f}_1 f_2} &= (\Delta m_{f_2} + \Delta m_{f_1}) i P^{NS}_{\dot{f}_1 f_2} \checkmark \\
\partial^{\mu} (j^5_{\mu})^{S}_{\dot{f}_1 f_2} &= 2 \langle m \rangle i P^S \not\checkmark \; [\longrightarrow + \delta_5] \\
\delta_5 &= (2 N_{fl}) \tfrac{1}{32\pi^2} B^r_{\mu\nu} \widetilde{B}^{\mu\nu r} \Big|_{\to ren.gr.inv} \; ; \; \widetilde{B}^r_{\mu\nu} = \tfrac{1}{2} \varepsilon_{\mu\nu\sigma\tau} B^{\sigma\tau r}
\end{aligned}$$

2

$$\tag{68}$$

We shall return to the question of how the local operator $ch_2 (B) \equiv \frac{1}{32\pi^2} (:) B^r_{\mu\nu} \widetilde{B}^{\mu\nu r} (:)$ is to be normalized and rendered renormalization group invariant [13-1991] . Here we just assume this to have been achieved and denote the U1-axial anomaly, the first of the central two, in its general form
(eq. 68)

$$\begin{aligned}
&\left\{ \partial^{\mu} (j^5_{\mu})^S = 2 \langle m \rangle i P^S + \delta_5 \right\} (x) \\
&\delta_5 = (2 N_{fl}) \tfrac{1}{32\pi^2} (:) B^r_{\mu\nu} \widetilde{B}^{\mu\nu r} (:) \Big|_{\to ren.gr.inv}
\end{aligned} \tag{69}$$

From here it is conceptually clear how the scale- (or trace-) anomaly arises but strictly within QCD . The renormalizability of a field theory in the limit of uncurved space-time gives rise to a local , symmetric and *conserved* energy momentum tensor , implying exact Poincaré invariance

$$\begin{aligned}
&\{ \vartheta_{\mu\nu} = \vartheta_{\nu\mu} \} (x) \\
&\partial^{\nu} \vartheta_{\mu\nu} = 0
\end{aligned} \tag{70}$$

In connection with the normal ordering questions it is important to admit in the precise form of the energy momentum tensor a nontrivial vacuum expected value , which

---

2 δ_5 **was – as far as I know – introduced by Murray Gell-Mann in lectures $\sim$ 1970 in Hawaii**

in view of exact Poincaré invariance must be of the form

$$\begin{array}{l} \langle \Omega \mid \vartheta_{\mu\nu}(x) \mid \Omega \rangle = \frac{1}{4} \eta_{\mu\nu} \tau \\ \left\{ \begin{array}{rcl} \eta_{\mu\nu} & = & diag(1, -1, -1, -1) \\ & & \tau \end{array} \right\} \text{ independent of x} \\ \\ \Delta \vartheta_{\mu\nu}(x) = \vartheta_{\mu\nu}(x) - \langle \Omega \mid \vartheta_{\mu\nu}(x) \mid \Omega \rangle \times \left\{ \begin{array}{l} \widehat{¶} \\ \text{or } \mid \Omega \rangle \langle \Omega \mid \end{array} \right. \\ \text{with } \partial^{\nu} \Delta \vartheta_{\mu\nu}(x) = 0 \; ; \; \langle \Omega \mid \Delta \vartheta_{\mu\nu}(x) \mid \Omega \rangle = 0 \end{array} \tag{71}$$

In eq. 71 $\widehat{¶}$ denotes the unit operator in the entire Hilbert space of states , while $P_{\Omega} = \mid \Omega \rangle \langle \Omega \mid$ stands for the projector on the ground state . Furthermore from the two local , *conserved* tensors in eq. 71 only $\Delta \vartheta_{\mu\nu}(x)$ with vanishing vacuum expected value is acceptable as representing the conserved 4 momentum *operators* in the integral form

$$\widehat{P}_{\mu} = \int_{t} d^{3} x \, \Delta \vartheta_{\mu 0}(t, \vec{x}) \tag{72}$$

All these arguments *notwithstanding* to subtract any eventual vacuum expected values of local operators , often put forward as mathematical prerequisites , it is wise *not to do so* in the presence of spontaneous parameters , the dynamical origin of spontaneous symmetry breaking, e.g. chiral symmetries in the limit or neighbourhood of some $m_f \rightarrow 0$.
Using the (classical) equations of motion pertaining to the Lagrangean in eqs. 1 - 6

$$\begin{array}{l} (D_{\nu} B^{\mu\nu})^{r} = j^{\mu \, r}(\overline{q}, q) \; ; \; B \rightarrow B_{pert} \\ (D_{\varrho} B^{\mu\nu})^{r} = \partial_{\varrho} B^{\mu\nu r} + f_{rst} W_{\varrho}^{s} B^{\mu\nu t} \\ j_{\mu}^{r}(\overline{q}, q) = g \, \overline{q}_{\dot{\mathcal{A}} f}^{\dot{c}} (\gamma_{\mu})_{\dot{\mathcal{A}} \mathcal{B}} \frac{1}{2} (\lambda^{r})_{c c'} q_{\mathcal{A} f}^{c'} \\ \\ (D_{\mu} q)_{\mathcal{A} f}^{c} = [\partial_{\mu} \delta_{c c'} + \frac{1}{i} W_{\mu}^{t} \frac{1}{2} (\lambda^{t})_{c c'}] q_{\mathcal{A} f}^{c'} \\ \hline W_{\mu}^{r} \equiv - v_{\mu}^{r} = g (W_{\mu}^{r})_{pert} \equiv - g (v_{\mu}^{r})_{pert} \end{array} \tag{73}$$

the associated form of the energy momentum becomes

$$\vartheta_{\mu\nu}^{(cl)} = \left[\begin{array}{c} \frac{1}{4} \left[B_{\mu \varrho}^{t} B_{\nu}^{\varrho t} - \frac{1}{4} \eta_{\mu\nu} B_{\sigma \varrho}^{t} B^{\varrho \sigma t} \right] + \\ + \frac{1}{2} \left[\overline{q}_{\dot{f}} \gamma_{\mu} \frac{i}{2} \overleftrightarrow{D}_{\nu} q_{f} + \mu \leftrightarrow \nu \right] \end{array} \right] \tag{74}$$

and using once more the fermion part of the equations of motion the trace of the classical energy momentum tensor becomes

$$\begin{array}{lll} \vartheta^{\mu\,(cl)}_{\ \mu} & = & \sum_f m_f S_{\dot{f} f} \\ S_{\dot{f}_1 f_2} & = & (:)\, \overline{q}^{\,\dot{c}}_{\dot{f}_1}\, q^{\,c}_{f_2}\, (:) \end{array} \tag{75}$$

1-1-g – The scale- or trace- anomaly

From the classical soft fermionic contribution to the trace of the energy momentum tensor there is a clear conjecture , also by Murray Gell-Mann, of the anomalous contribution , which subsequently became the scale- or trace- anomaly within QCD

$$\begin{array}{lll} \vartheta^{\mu}_{\ \mu} & = & \sum_f m_f S_{\dot{f} f} + \delta_0 \\ \delta_0 & = & -\left(-2\,\beta(g)\,/\,g^3 \right) \left[\frac{1}{4}\,(:)\, B^{\,t}_{\mu\nu}\, B^{\mu\nu\,t}\,(:) \right]_{\to\ ren.gr.inv} \end{array} \tag{76}$$

1-1-h – The two central anomalies alongside : scale- or trace- and U1-axial anomaly

We collect the two anomalous identities in eqs. 76 and 58

$$\begin{array}{lll} \left\{ \vartheta^{\mu}_{\ \mu} \right. & = & \left. \sum_f m_f S_{\dot{f} f} + \delta_0 \right\}(x) \\ \left\{ \partial^{\mu} \left(j^{5}_{\mu} \right)^{S} \right. & = & \left. 2\,\langle m \rangle\, i\, P^{S} + \delta_5 \right\}(x) \\ \delta_0 & = & -\left(-2\,\beta(g)\,/\,g^3 \right) \left[\frac{1}{4}\,(:)\, B^{\,t}_{\mu\nu}\, B^{\mu\nu\,t}\,(:) \right]_{\to\ ren.gr.inv} \\ \delta_5 & = & \left(2\, N_{fl} \right) \frac{1}{8\pi^2} \left[\frac{1}{4}\,(:)\, B^{\,t}_{\mu\nu}\, \widetilde{B}^{\mu\nu\,t}\,(:) \right]_{\to ren.gr.inv} \\ \hline -\beta\,/g^3 & = & \frac{1}{16\pi^2}\, b_0 + O(X)\ ;\ X = g^2\,/\,(16\,\pi^2 \\ \beta(g) & : & \text{Callan-Symanzik rescaling function in QCD} \end{array} \tag{77}$$

The qualification 'central' for the anomalies in eq. 77 stands for the property that in rendering the square coupling constant and the associated ϑ – parameter in the gauge boson *renormalized* Lagrangean density x dependent

$$\begin{array}{l} \mathcal{L}_{\,g.b.} = -\frac{1}{g^2}\,\frac{1}{4}\,(:)\, B^{\,t}_{\mu\nu}\, B^{\mu\nu\,t}\,(:) + \vartheta\,\frac{1}{8\pi^2}\,\frac{1}{4}\,(:)\, B^{\,t}_{\mu\nu}\, \widetilde{B}^{\mu\nu\,t} \\ g^2 \to g^2(x)\ ;\ \vartheta \to \vartheta(x) \end{array} \tag{78}$$

maintains perturbative renormalizability and acts together with suitable boundary- – more generally – regularity conditions as external source for the scalar and pseudoscalar local field strength bilinears

$$\frac{1}{4}\,(:)\, B^{\,t}_{\mu\nu}\, B^{\mu\nu\,t}\,(:)\ ,\ \frac{1}{4}\,(:)\, B^{\,t}_{\mu\nu}\, \widetilde{B}^{\mu\nu\,t}\,(:) \tag{79}$$

We will use the following definitions relative to the rescaling function β

$$
\begin{array}{l}
- \beta / g = X B (X) \; ; \; B (X) = b_0 A (X) \\
B (X) \sim \sum_{n=0}^{\infty} b_n X^n \; , \; A (X) \sim \sum_{n=0}^{\infty} a_n X^n \\
\kappa = g^2 / (16 \pi^2) \text{ generic } \longrightarrow X , Y \\
b_0 = \frac{1}{3} (33 - 2 N_{fl}) \; , \; a_0 = 1 \; , \; a_n = b_n / b_0 \\
b_1 = \frac{2}{3} (153 - 19 N_{fl}) \\
b_2 = \frac{1}{54} \left(77139 - 15099 N_{fl} + 325 N_{fl}^2 \right) \\
b_3 \sim 29243 - 6946.3 N_{fl} + 405.089 N_{fl}^2 + 1.49931 N_{fl}^3
\end{array}
\tag{80}
$$

References in conjunction with this section ('1-1') are presented in five (partial) collections :

1 : (R) directly related to the two central anomalies

2 : (rBsquare) establishing the one renormalization group invariant quantity of dimension $\left[M^4 \right]$

3 : (r-sp-1) a recent paper by Guido Altarelli and references cited therein

4 : (r-A2x) a selection of papers and textbooks for the entire realm of QCD

5 : (r-condx) : Condensation phenomena and field theory realizations

2 – The scale Λ , the trace of the enrgy momentum density tensor and an essential complication to establish field equations , consistent with the trace anomaly

We repeat here eq. 77 from subsection 1-1-c , subsumming the sructure of the two 'central' anomalies

$$
\begin{array}{l}
\left\{ \vartheta^{\mu}{}_{\mu} = \sum_f m_f S_{\dot{f} f} + \delta_0 \right\} (x) \\
\left\{ \partial^{\mu} (j_{\mu}^5)^S = 2 \langle m \rangle i P^S + \delta_5 \right\} (x) \\
\delta_0 = - (-2 \beta (g) / g^3) \left[\frac{1}{4} (:) B_{\mu\nu}^t B^{\mu\nu t} (:) \right]_{\rightarrow ren.gr.inv} \\
\delta_5 = (2 N_{fl}) \frac{1}{8\pi^2} \left[\frac{1}{4} (:) B_{\mu\nu}^t \widetilde{B}^{\mu\nu t} (:) \right]_{\rightarrow ren.gr.inv} \\
\hline
- \beta / g^3 = \frac{1}{16\pi^2} b_0 + O (X) \; ; \; X = g^2 / (16 \pi^2
\end{array}
\tag{81}
$$

The predicate 'central' for the anomalies in eq. 81 stands for the property that in rendering the square coupling constant and the associated ϑ –

parameter in the gauge boson *naively renormalized* Lagrangean density x dependent

$$\mathcal{L}_{g.b.} = - \frac{1}{g^2} \frac{1}{4} (:) B^t_{\mu\nu} B^{\mu\nu t} (:) + \vartheta \frac{1}{8\pi^2} \frac{1}{4} (:) B^t_{\mu\nu} \widetilde{B}^{\mu\nu t}$$
$$g^2 \rightarrow g^2 (x) \; ; \; \vartheta \rightarrow \vartheta (x) \tag{82}$$

maintains perturbative renormalizability and acts together with suitable boundary- – more generally – regularity conditions $\rightarrow$
as external source for the scalar and pseudoscalar local field strength bilinears

$$\frac{1}{4} (:) B^t_{\mu\nu} B^{\mu\nu t} (:) \; , \; \frac{1}{4} (:) B^t_{\mu\nu} \widetilde{B}^{\mu\nu t} (:) \tag{83}$$

We use the following definitions relative to the rescaling function β

$$- \beta / g = X B (X) \; ; \; B (X) = b_0 A (X)$$
$$B (X) \sim \sum_{n=0}^{\infty} b_n X^n \; , \; A (X) \sim \sum_{n=0}^{\infty} a_n X^n \tag{84}$$
$$\kappa = g^2 / (16 \pi^2) \text{ generic } \longrightarrow X \, , \, Y$$

2-2 – Consequences of the trace anomaly for the canonical structure behind compatible covariant field equations for gauge fields

The issue of consistency of equations of motion for gauge fields and the trace anomaly as shown in eq. 81 was taken up anew by the author of this outline in ref. [14-a-2012] . The associated canonical structure needs precise specifications in their own right. In addition it was endowed with *hypothetical* assigned properties, influencing the phase structure of QCD , in particular for the neighbourhood of vanishing chemical potentials and a range of (ratios of) quark masses of the light (u,d,s -) flavors including the realistic case, discussed in recent work in collaboration with Sonia Kabana , ref. [15-a-2010] . The latter work is also a reappraisal of questions left open in our earlier work, ref. [16-a-2000] .

2-2-2 – Sketch of how to construct the local, gauge invariant and conserved energy momentum density tensor, compatible with the trace anomaly

We begin studying the problem of parallel transport and the equations determining a gauge in which the time component of the connection vanishes, for complete connections with respect to a simple local compact structure group $\mathcal{G}$

$$(\mathcal{W}_\mu (\mathcal{D}))_{\alpha\beta} (x) = \mathcal{W}^r_\mu (x) (d_r)_{\alpha\beta}$$
$$d_r = - d^\dagger_r = \frac{1}{i} J_r \in Lie (\mathcal{D}) \; ; \; [d_p \, , \, d_q] = f_{pqr} \, d_r \tag{85}$$
$$r \, , \, p \, , \, q = 1, \cdots, dim \, \mathcal{G} \; ; \; \alpha \, , \, \beta = 1, \cdots, dim \, \mathcal{D}$$

The quantities defined in eq. 85 follow the notation in the notefile [17-a-2011], recapitulated below

$$\begin{array}{ll} (\mathcal{W}_{\mu} (\mathcal{D}))_{\alpha\beta} (x) & : \text{local operator } \times \ \mathcal{D} - \text{representation valued} \\ & \ \ \text{connection over flat space time } x \\ d_{r} \in Lie (\mathcal{D}) & : \text{basis of antihermitian matrices forming} \\ & \ \ \text{an irreducible representation of the} \\ & \ \ \text{Lie algebra of } \mathcal{G} \\ \mathcal{W}_{\mu}^{r} (x) & : 4 \times \ dim \ \mathcal{G} \text{ components of hermitian local} \\ & \ \ \text{connection fields} \end{array} \tag{86}$$

A connection is called complete, if all regularity- , differentiability- and integrability conditions extend to the complete ring of representations $\mathcal{R} \mid_{\mathcal{G}}$, pertaining to $\mathcal{G}$ [18-a-1968, 17-a-2011] .
The matrix-operator valued 1-form associated with the gauge connection $(\mathcal{W}_{\mu} (\mathcal{D}))_{\alpha\beta} (x)$ in eq. 86 is denoted

$$(\mathcal{W}^{(1)} (\mathcal{D})_{\alpha\beta} = (\mathcal{W}_{\mu} (\mathcal{D}))_{\alpha\beta} (x) \, d x^{\mu} \longrightarrow \mathcal{W}^{(1)} (\mathcal{D}) \tag{87}$$

parallel transport along a general and special set of curves C

We define the parallel transport operators, along an oriented curve C from base point y to target point x denoted $U \left(x \overset{C}{\leftarrow} y \right)$

$$\begin{array}{l} U \left(x \overset{C}{\leftarrow} y \right) = P \exp \left(- \int_{C} \mathcal{W}^{(1)} (\overline{x}) \right) \\ \rightarrow (U (x ; C , \{ \mathcal{W}^{(1)} \} , y))_{\alpha\beta} \in \mathcal{D} (\mathcal{G}) \\ \hline C = C \{ \overline{x} \} : \\ \overline{x} (\tau) \ ; \ 1 \geq \tau \geq 0 \ ; \ \tau : \text{ path parameter} \\ \qquad \overline{x} (\tau = 1) = x \leftarrow x_{1} \ ; \ \overline{x} (\tau = 0) = y \leftarrow x_{0} \end{array} \tag{88}$$

The substitutions $y \leftarrow x_{1} ; x \leftarrow x_{0}$ are relevant if the generic base- and target- points are specified in a particular way.
A remark concerning the operator definitions inherent in general gauges as is appropriate here.

- – the linear space on which gauge boson local operators act

  The use of Lorentz-covariant gauges and the BRST algebra of associated ghost fields , [19-a-1975, 20-a-1975] , demonstrate that the linear space on which all local fields associated with gauge fields act cannot be endowed with a positive definite norm , nor can formally associated ghost fields inherit a selfadjoint or hermitian structure. At most a Hilbert space definition necessitates a projection on gauge invariant quantities, which for local fields are composite ones. This is the price to pay for the consistent use of Poincarë covriant Feynman rules and integrals in loops, behaving consistently in the perturbatively accessible region.

$$U \left(x \overset{C}{\leftarrow} y \right) = P \exp \left(- \int_C \mathcal{W}^{(1)} (\overline{x}) \right)$$

defined in eq. 88
satisfies a first order differential equation whence the tangent vectors along the curve $\mathcal{C}$ are introduced

$$\begin{aligned} & \underline{\mathrm{v}}^{\mu} (\partial_{x\,\mu} + \mathcal{W}_{\mu} (x)) \, U (x) \, |_{\, x = \overline{x}(\tau)} = 0 \; ; \; \underline{\mathrm{v}}^{\mu} = \dot{\overline{x}}^{\mu} \\ & U (x) \longleftarrow U \left(x \, ; \, \mathcal{C} \, , \, \{ \mathcal{W}^{(1)} (\mathcal{D}) \} \, , \, y \right) \\ & U \left(y \, ; \, \mathcal{C} \, , \{ \mathcal{W}^{(1)} (\mathcal{D}) \} \, , \, y \right) = \P \end{aligned} \tag{89}$$

with the initial condition as specified in the last line of eq. 89 .
In line with the $\mathcal{D}$ projection we specify the $\mathcal{D}-$ dependent trace normalization

$$\begin{aligned} & - \, tr_{\,\mathcal{D}} \; d^{\,\mathrm{p}} \, d^{\,\mathrm{q}} = \delta^{\,pq} \, T_2 (\mathcal{D}) \\ & - \textstyle\sum_r \, (d^{\,r})^{\,2} = C_2 (\mathcal{D}) \, \P_{\, dim\, \mathcal{D} \,\times\, dim\, \mathcal{D}} \; \rightarrow \\ & T_2 (\mathcal{D}) = \frac{dim \; (\mathcal{D})}{dim \; (\mathcal{G})} \, C_2 (\mathcal{D}) \; ; \text{ for } \left. \begin{array}{rcl} \mathcal{G} & = & SU3 \\ \mathcal{D} & = & 3 \text{ or } \overline{3} \end{array} \right\} : \\ & dim \; \mathcal{D} = 3 \; , \; dim \; \mathcal{G} = 8 \; , \; T_2 (\mathcal{D}) = \tfrac{1}{2} \; C_2 (\mathcal{D}) = \tfrac{4}{3} \end{aligned} \tag{90}$$

In eq. 90 $C_2 (\mathcal{D})$ denotes the eigenvalue of the second order Casimir operator projected on the irreducible representation $\mathcal{D}$

2-2-a – Bare Lagrangean density and equations of motion in unconstrained gauges

The bare local Lagrangean density is formed by the bare field strengths for its gauge part , and the quark-antiquark matter field part, using the relations in eq. 90

$$\mathcal{L}(x) = \frac{1}{4\,T_2(\mathcal{D})\,g^2}\; tr_{\mathcal{D}}\left[\mathcal{B}_{\mu\nu}\,\mathcal{B}^{\mu\nu}\right](\mathcal{D})(x) + \mathcal{L}_{\{q\}}(x)$$
$$\mathcal{L}_{\{q\}}(x) =$$
$$\sum_{q-fl} \overline{q}^{\,\dot{c}'} \left\{ \frac{i}{2}\gamma^{\mu} \left[\begin{array}{c} \left(\overrightarrow{D}_{\mu}(3) \right)_{c'\dot{c}} - \\ - \left(\overleftarrow{D}_{\mu}(\overline{3}) \right)_{\dot{c}c'} \end{array} \right] - m_q\, \delta_{c'\dot{c}} \right\} q^{c}(x) \tag{91}$$

The Euler-Lagrange equations for gauge fields – not including the color currents of the quark-antiquark matter fields – generating an extremum of the bare action take the form

$$\begin{array}{l} \partial_{\varrho} \left\{ \dfrac{\delta\,\mathcal{L}}{\delta\,B^{\,r}_{\mu\,\nu}} \dfrac{\delta\,B^{\,r}_{\mu\nu}}{\delta\,(\partial_{\varrho} W^{\,s}_{\sigma})} \right\} - \dfrac{\delta\,\mathcal{L}}{\delta\,B^{\,r}_{\mu\,\nu}} \dfrac{\delta\,B^{\,r}_{\mu\nu}}{\delta\,W^{\,s}_{\sigma}} = 0 \\ \dfrac{\delta\,\mathcal{L}}{\delta\,B^{\,r}_{\mu\,\nu}} = \dfrac{1}{2\,g^{2}}\,B^{\,\nu\mu\,r} \;;\; \mathcal{L} = \mathcal{L}_{gauge}\ (\text{bare}) \end{array} \tag{92}$$

Including the contribution of quark-antiquark fields eq. 92 becomes

$$\begin{array}{l} \partial_{\varrho} \left\{ \dfrac{\delta\,\mathcal{L}}{\delta\,B^{\,r}_{\mu\,\nu}} \dfrac{\delta\,B^{\,r}_{\mu\nu}}{\delta\,(\partial_{\varrho} W^{\,s}_{\sigma})} \right\} - \dfrac{\delta\,\mathcal{L}}{\delta\,B^{\,r}_{\mu\,\nu}} \dfrac{\delta\,B^{\,r}_{\mu\nu}}{\delta\,W^{\,s}_{\sigma}} = \dfrac{\delta\,\mathcal{L}_{\{q\}}}{\delta\,W^{\,s}_{\sigma}} \\ \dfrac{\delta\,\mathcal{L}_{\{q\}}}{\delta\,W^{\,s}_{\sigma}} = \sum_{q-fl} \overline{q}^{\,\dot{c}'} \left\{ \gamma^{\sigma} \left(\tfrac{1}{2}\lambda^{s} \right)_{c'\dot{c}} \right\} q^{c} = \left(j^{\,\sigma\,s} \right)_{\{q\}} \end{array} \tag{93}$$

Eq. 93 becomes

$$\begin{array}{l} \left(D_{\varrho}(ad) \right)_{sr} \left\{ \dfrac{1}{g^{2}}\,B^{\,\sigma\varrho\,r} \right\} = \left(j^{\,\sigma\,s} \right)_{\{q\}} \\ \left(D_{\varrho}(ad) \right)_{sr} = \partial_{\varrho}\,\delta_{sr} + W^{\,t}_{\varrho}\,(ad_{t})_{sr} \;;\; (ad_{t})_{sr} = f_{str} \\ \left(j^{\,\sigma\,s} \right)_{\{q\}} = \sum_{q-fl} \overline{q}^{\,\dot{c}'} \left\{ \gamma^{\sigma} \left(\tfrac{1}{2}\lambda^{s} \right)_{c'\dot{c}} \right\} q^{c} \end{array} \tag{94}$$

It becomes obvious, that if we seek the energy momentum density tensor as a generalized Noether current, and compatible with the trace anomaly , the structural form of the bare Lagrangean quadratic in the gauge field strengths has to be essentially modified. This has been noted in refs. [21-a-1978, 22-a-1981] and is here taken up anew in the completion of QCD outside the perturbatively accessible region , following a first assessment in ref. [23-a-2012] .

2-2-b - Energy momentum density tensor as a conserved generalized Noether current , restricted to $\overline{\mathcal{L}}$, replacing its bare form $\mathcal{L}$, in the absence of matter fields, i.e. neglecting $\mathcal{L}_{\{q\}}$

The search for consistent equations of motion for gauge fields is initiated by comparing the bare Lagrangean density as qualified in the title of this subsection . To this end we repeat the first part of eq. 91 below

$$\mathcal{L}(x) = \mathcal{L}_{gauge} = \frac{1}{4\, T_2(\mathcal{D})\, g^2}\; tr_{\mathcal{D}} \left[\mathcal{B}_{\mu\nu}\, \mathcal{B}^{\mu\nu} \right] (\mathcal{D})(x) \tag{95}$$

The equations of motions induced by $\mathcal{L} \leftarrow \mathcal{L}_{gauge}$ are given in eqs. 91 - 94, where on the right hand side either matter quark-antiquark contributions are included or , as relevant in this subsection , replaced by 0 .
In the following the suffix label $_{gauge}$ will be suppressed , for clarity, except when it is compromised .
Replacing the gauge field Lagrangean density to achieve consistency with the trace anomaly amounts to the substitution , as derived in ref. [23-a-2012]

$$\begin{array}{l} \mathcal{L}_{gauge} = -\dfrac{1}{g^2}\,\mathcal{X} \rightarrow \overline{\mathcal{L}} = -\left[\dfrac{1}{\overline{g}^{\,2}(\mathcal{X})} - J \right] \mathcal{X} \\ \mathcal{X} = \frac{1}{4}\, B^{\;r}_{\mu\nu}\, B^{\mu\nu\; r} \end{array} \tag{96}$$

In eq. 96 J stands for a suitable constant, while $\overline{g}(\overline{l})$ denotes the scale dependent coupling constant, as discussed also in ref. [21-a-1978] , established initially in the perturbative domain of QCD

$$\begin{array}{l} \overline{l} = \log(\overline{\mu}/\Lambda) = \frac{1}{4}\log(\overline{\mu}^{\,4}/\Lambda^4) \rightarrow \frac{1}{4}\log(\mathcal{X}/\Lambda^4) \rightarrow \frac{1}{8}\log\left((\mathcal{X}/\Lambda^4)^2\right) \\ \overline{g}(\mathcal{X}) = \overline{g}\left(\overline{l} = \frac{1}{8}\log\left((\mathcal{X}/\mu^4)^2\right)\right) \\ \dfrac{d}{d\,\overline{l}}\,\overline{g} = \beta(\overline{g}) = \overline{g}\,\overline{\kappa}\left[-\left(b_0 + b_1\overline{\kappa}^{\,1} + \cdots\right)\right] ;\ \overline{\kappa} = \dfrac{\overline{g}^{\,2}}{16\,\pi^2} ;\ \overline{\alpha}_s = 4\,\pi\,\overline{\kappa} \\ \dfrac{d}{d\,\overline{l}}\,\overline{g}^{\,-2} = \dfrac{1}{8\,\pi^2}\, b(\overline{\kappa}) ;\ b(\overline{\kappa}) = -\left(\overline{g}\,\overline{\kappa}\right)^{-1}\beta(\overline{g}) = \left(b_0 + b_1\overline{\kappa}^{\,1} + \cdots\right) \end{array} \tag{97}$$

In order to substitute $\overline{\mu}^{\,4} \rightarrow \mathcal{X}$ in accordance with eq. 97 in the entire domain of the field operator $\mathcal{X}$, the functional dependance of the sliding scale

coupling constant $\overline{\kappa} \ (\mu \ / \ \Lambda)$ must be *known* beyond the perturbatively accessible region.
The variational derivatives of field strengths are

$$\frac{\delta \, B^{\,r}_{\mu\nu}}{\delta \, (\partial_{\varrho} W^{\,s}_{\sigma})} = \delta^{\,rs} \, (\delta^{\,\varrho}_{\mu} \, \delta^{\,\sigma}_{\nu} - \delta^{\,\varrho}_{\nu} \, \delta^{\,\sigma}_{\mu}) \; ; \; \frac{\delta \, B^{\,r}_{\mu\nu}}{\delta \, W^{\,s}_{\sigma}} = f_{\,srt} \, (\delta^{\,\sigma}_{\nu} \, W^{\,t}_{\mu} - \delta^{\,\sigma}_{\mu} \, W^{\,t}_{\nu}) \tag{98}$$

and obtain the variational derivatives of of the quantity $\mathcal{X}$ defined in eq. 96

$$\begin{aligned} \frac{\delta \, \mathcal{X}}{\delta \, (\partial_{\varrho} W^{\,s}_{\sigma})} &= \frac{\delta \, \mathcal{X}}{\delta \, B^{\,r}_{\mu\,\nu}} \, \frac{\delta \, B^{\,r}_{\mu\nu}}{\delta \, (\partial_{\varrho} W^{\,s}_{\sigma})} = B^{\,\varrho\,\sigma\,s} \\ \frac{\delta \, \mathcal{X}}{\delta \, B^{\,r}_{\mu\,\nu}} &= \tfrac{1}{2} \, B^{\,\mu\nu\,r} \; ; \; \frac{\delta \, \mathcal{X}}{\delta \, (W^{\,s}_{\sigma})} = \frac{\delta \, \mathcal{X}}{\delta \, B^{\,r}_{\mu\,\nu}} \, \frac{\delta \, B^{\,r}_{\mu\nu}}{\delta \, (W^{\,s}_{\sigma})} \\ \mathcal{X} &= \tfrac{1}{4} \, B^{\,r}_{\mu\nu} \, B^{\,\mu\nu\,r} = - \, g^{\,2} \, \mathcal{L}_{\,gauge} \end{aligned} \tag{99}$$

The variation of the trace anomaly induced Lagrangean density $\overline{\mathcal{L}}$, defined in eq. 96 , is obtained by repeated use of the chain rule

$$\boxed{\begin{aligned} \frac{\delta \, \overline{\mathcal{L}}}{\delta \, (\partial_{\varrho} W^{\,s}_{\sigma})} &= \left[\frac{\delta \, \overline{\mathcal{L}}}{\delta \, \mathcal{X}} \, \frac{\delta \, \mathcal{X}}{\delta \, B^{\,r}_{\mu\nu}} \right] \frac{\delta \, B^{\,r}_{\mu\nu}}{\delta \, (\partial_{\varrho} W^{\,s}_{\sigma})} \\ \frac{\delta \, \overline{\mathcal{L}}}{\delta \, (W^{\,s}_{\sigma})} &= \left[\frac{\delta \, \overline{\mathcal{L}}}{\delta \, \mathcal{X}} \, \frac{\delta \, \mathcal{X}}{\delta \, B^{\,r}_{\mu\,\nu}} \right] \frac{\delta \, B^{\,r}_{\mu\nu}}{\delta \, (W^{\,s}_{\sigma})} \\ \hline \frac{\delta \, B^{\,r}_{\mu\nu}}{\delta \, (\partial_{\varrho} W^{\,s}_{\sigma})} &= \delta^{\,rs} \, (\delta^{\,\varrho}_{\mu} \, \delta^{\,\sigma}_{\nu} - \delta^{\,\varrho}_{\nu} \, \delta^{\,\sigma}_{\mu}) \\ \frac{\delta \, B^{\,r}_{\mu\nu}}{\delta \, W^{\,s}_{\sigma}} &= f_{\,srt} \, (\delta^{\,\sigma}_{\nu} \, W^{\,t}_{\mu} - \delta^{\,\sigma}_{\mu} \, W^{\,t}_{\nu}) \end{aligned}} \tag{100}$$

We proceed to evaluate the derivatives in eq. 100 using eq, 97 .
The quantities in [.] brackets in the first two relations of eq. 100 contain expressions wherein both canonically conjugate variables appear. Hence an ordering is necessary. This is straightforward in all such situations and amounts to neglecting all local terms arising from the noncommutativity of

the latter.

$$\begin{array}{l}
\frac{\delta \overline{\mathcal{L}}}{\delta \mathcal{X}} = \overline{\mathcal{L}} / \mathcal{X} - \left(\frac{d}{d \, \overline{l}} \, \overline{g}^{\,-2} \right) \frac{\delta \, \overline{l}}{\delta \mathcal{X}} \, \mathcal{X} \\
\overline{l} = \frac{1}{8} \log \left((\mathcal{X} / \mu^{4})^{2} \right) \\
\hline
\frac{d}{d \, \overline{l}} \overline{g}^{\,-2} = \frac{1}{8 \pi^{2}} \, b (\overline{\kappa}) \\
b (\overline{\kappa}) = - (\overline{g} \, \overline{\kappa})^{-1} \beta (\overline{g}) = - \left(b_{0} + b_{1} \overline{\kappa}^{1} + \cdots \right)
\end{array} \tag{101}$$

We substitute the derivatives from eq. 99

$$\frac{\delta \mathcal{X}}{\delta B_{\mu \nu}^{\;r}} = \frac{1}{2} B^{\mu \nu \, r} \tag{102}$$

This yields the derivatives of $\mathcal{X}$, using eq. 100

$$\begin{array}{l}
\frac{\delta \mathcal{X}}{\delta (\partial_{\varrho} W_{\sigma}^{\;s})} = B^{\varrho \sigma \, s} \; ; \; \frac{\delta \mathcal{X}}{\delta W_{\sigma}^{\;s}} = \frac{1}{2} B^{\mu \nu \, r} \frac{\delta B_{\mu \nu}^{\;r}}{\delta W_{\sigma}^{\;s}} \\
\frac{\delta \mathcal{X}}{\delta W_{\sigma}^{\;s}} = - f_{str} \, W_{\varrho}^{\;t} \, B^{\varrho \sigma \, r} \\
\quad = \frac{1}{2} B^{\mu \nu \, r} \, f_{srt} \left(\delta_{\nu}^{\;\sigma} W_{\mu}^{\;t} - \delta_{\mu}^{\;\sigma} W_{\nu}^{\;t} \right) \\
\quad = \frac{1}{2} f_{srt} \left(W_{\mu}^{\;t} B^{\mu \sigma \, r} - W_{\nu}^{\;t} B^{\sigma \nu \, r} \right)
\end{array} \tag{103}$$

Next we derive the energy momentum density tensor $\vartheta_{\nu}^{\;\mu} (x)$ as a local functional of the Lagrangean density $\overline{\mathcal{L}}$ and the gauge field variables $\partial_{\varrho} W_{\sigma}^{\;s} \longleftrightarrow B_{\varrho \sigma}^{\;s}$, $W_{\sigma}^{\;s}$, as they appear e.g. in eq. 103 by variational techniques going back to Emmy Noether [24-a-1918] .
First we consider general variations of the base variables, understood to depend continuously on a family parameter, denoted f . This relates to the equations of motion

$$\begin{array}{l}
W_{\sigma}^{\;s} = W_{\sigma}^{\;s} (f \, ; x) \; ; \; \delta W_{\sigma}^{\;s} (x) = \partial_{f} \; W_{\sigma}^{\;s} (f \, ; x) \, |_{\, f=0} \; \rightarrow \\
\delta \, \partial_{\varrho} W_{\sigma}^{\;s} (x) = \partial_{\varrho} \, \delta \, W_{\sigma}^{\;s} (x)
\end{array} \tag{104}$$

and determine the conditions for the f-dependent action integral over a general 4-dimensional volume V

$$S (V , f) = \int_{V} d^{4} x \, \overline{\mathcal{L}} \left(W_{\sigma}^{\;s} (f , x) \, ; \, \partial_{\varrho} W_{\sigma}^{\;s} (f \, ; x) \right) \tag{105}$$

to acquire an extremal value for the particular member of the family of base fields corresponding to $f = 0$.
This gives rise to the condition

$$\begin{aligned} & \delta S = \partial_f \; S (V , f) \,|_{\,f=0} = 0 \;\rightarrow \\ & \delta S = \\ & = \int_V d^4 x \left(\delta W^{\;s}_{\sigma} \; \overline{\mathcal{L}}_{\,, W^{\;s}_{\sigma}} \,|_{\,f=0} + \delta \, \partial_{\varrho} W^{\;s}_{\sigma} \; \overline{\mathcal{L}}_{\,, \partial_{\varrho} W^{\;s}_{\sigma}} \,|_{\,f=0} \right) = 0 \end{aligned}$$

$$\begin{aligned} & \overline{\mathcal{L}}_{\,, W^{\;s}_{\sigma}} \,|_{\,f=0} \rightarrow \overline{\mathcal{L}}_{\,, W^{\;s}_{\sigma}} = \frac{\partial \overline{\mathcal{L}}}{\partial W^{\;s}_{\sigma} (f = 0 ; x)} \\ & \overline{\mathcal{L}}_{\,, \partial_{\varrho} W^{\;s}_{\sigma}} \,|_{\,f=0} \rightarrow \overline{\mathcal{L}}_{\,, \partial_{\varrho} W^{\;s}_{\sigma}} = \frac{\partial \overline{\mathcal{L}}}{\partial (\partial_{\varrho} W^{\;s}_{\sigma}) (f = 0 ; x)} \end{aligned} \tag{106}$$

For simplicity of notation we neglect the symbol $|_{\,f=0}$ in the quantities $\overline{\mathcal{L}}_{\,, W^{\;s}_{\sigma}}$, $\overline{\mathcal{L}}_{\,, \partial_{\varrho} W^{\;s}_{\sigma}}$ in the following as indicated in eq. 106 δS (eq. 106) becomes using eq. 104

$$\begin{aligned} & \delta S = \int_V d^4 x \left(\delta W^{\;s}_{\sigma} \overline{\mathcal{L}}_{\,, W^{\;s}_{\sigma}} + (\partial_{\varrho} \delta W^{\;s}_{\sigma}) \overline{\mathcal{L}}_{\,, \partial_{\varrho} W^{\;s}_{\sigma}} \right) \rightarrow \\ & (\partial_{\varrho} \delta W^{\;s}_{\sigma}) \overline{\mathcal{L}}_{\,, \partial_{\varrho} W^{\;s}_{\sigma}} = \partial_{\varrho} \left(\delta W^{\;s}_{\sigma} \overline{\mathcal{L}}_{\,, \partial_{\varrho} W^{\;s}_{\sigma}} \right) - \delta W^{\;s}_{\sigma} \; \partial_{\varrho} \overline{\mathcal{L}}_{\,, \partial_{\varrho} W^{\;s}_{\sigma}} \end{aligned} \tag{107}$$

Substituting the second relation in eq. 107 we obtain

$$\delta S = \int_V d^4 x \left[\begin{array}{l} \delta W^{\;s}_{\sigma} \left(\overline{\mathcal{L}}_{\,, W^{\;s}_{\sigma}} - \partial_{\varrho} \overline{\mathcal{L}}_{\,, \partial_{\varrho} W^{\;s}_{\sigma}} \right) + \\ + \partial_{\varrho} \left(\delta W^{\;s}_{\sigma} \overline{\mathcal{L}}_{\,, \partial_{\varrho} W^{\;s}_{\sigma}} \right) \end{array} \right] \tag{108}$$

The variational setting is incomplete without conditions on the boundary of the volme V – boundary conditions – which arise from the integration of the divergence in the last term in eq. 108

$$\begin{aligned} & \int_V d^4 x \, \partial_{\varrho} \left(\delta W^{\;s}_{\sigma} \overline{\mathcal{L}}_{\,, \partial_{\varrho} W^{\;s}_{\sigma}} \right) = \int_{\partial V} d^3 \sigma_{\,\varrho \,, \partial V} \left(\delta W^{\;s}_{\sigma} \overline{\mathcal{L}}_{\,, \partial_{\varrho} W^{\;s}_{\sigma}} \right) \\ & \longrightarrow \text{boundary conditions} : \delta W^{\;s}_{\sigma} (x) = 0 \text{ for } x \in \partial V \end{aligned} \tag{109}$$

With the boundary conditions as defined in eq. 109 satisfied, the extremum condition for the action integral (for all volumes V) and *otherwise* arbitrary variations $\delta W^{\;s}_{\sigma} (x)$
is equivalent to the local Euler-Lagrange equations of motion (eq. 108)

$$\left(\partial_{\varrho} \overline{\mathcal{L}}_{\,, \partial_{\varrho} W^{\;s}_{\sigma}} - \overline{\mathcal{L}}_{\,, W^{\;s}_{\sigma}} \right) (x) = 0 \tag{110}$$

Together this sets the stage for deriving associating with every additional symmetry of the Lagrangean density a conserved Nöther 'current' , but with general spin, depending on the symmetry involved.
We apply the variations as generally given in eq. 104 to the special case relating rigid space-time translations to the canonical energy momentum tensor

$$\begin{array}{lll} x^{\nu} \rightarrow f\, a^{\nu} \longleftrightarrow & \delta_{a} W^{s}_{\sigma} & = a^{\nu}\, \partial_{\nu} W^{s}_{\sigma} \\ & \delta_{a}\, \partial_{\varrho} W^{s}_{\sigma} & = a^{\nu}\, \partial_{\nu}\, \partial_{\varrho} W^{s}_{\sigma} \\ \hline \delta_{a} \overline{\mathcal{L}} = a^{\nu} \left(\begin{array}{l} (\partial_{\nu} W^{s}_{\sigma})\, \overline{\mathcal{L}}_{,\, W^{s}_{\sigma}} + \\ + (\partial_{\varrho} \partial_{\nu} W^{s}_{\sigma})\, \overline{\mathcal{L}}_{,\, \partial_{\varrho} W^{s}_{\sigma}} \end{array} \right) & & = a^{\nu}\, \partial_{\nu} \overline{\mathcal{L}} \end{array} \tag{111}$$

The underlying translation symmetry reveals itself subtracting the last term in the expression for $\delta_{a} \overline{\mathcal{L}}$ in eq. 111 from the second to yield

$$\begin{array}{l} a^{\nu} \left(\partial_{\varrho} \left[(\partial_{\nu} W^{s}_{\sigma})\, \overline{\mathcal{L}}_{,\, \partial_{\varrho} W^{s}_{\sigma}} - \delta^{\varrho}_{\nu}\, \overline{\mathcal{L}} \right] - (\partial_{\nu} W^{s}_{\sigma})\, \mathcal{E}^{\sigma s} \right) = 0 \\ \mathcal{E}^{\sigma s}(x) = \left(\partial_{\varrho} \overline{\mathcal{L}}_{,\, \partial_{\varrho} W^{s}_{\sigma}} - \overline{\mathcal{L}}_{,\, W^{s}_{\sigma}} \right)(x) \rightarrow 0 \end{array} \tag{112}$$

From eq. 112 we read off the canonical energy momentum density tensor in mixed components

$$\begin{array}{l} T_{\nu}^{\varrho} = (\partial_{\nu} W^{s}_{\sigma})\, \overline{\mathcal{L}}_{,\, \partial_{\varrho} W^{s}_{\sigma}} - \delta^{\varrho}_{\nu}\, \overline{\mathcal{L}} \; ; \; \partial_{\varrho} T_{\nu}^{\varrho} = 0 \\ T_{\nu\varrho} = \eta_{\rho\tau}\, T_{\nu}^{\tau} \; ; \; T_{\nu\varrho} - T_{\varrho\nu} \neq 0 \end{array} \tag{113}$$

$T_{\nu\varrho}$ is neither symmetric nor gauge invariant. Nevertheless a symmetric and gauge invariant energy momentum density tensor can always be achieved, by also considering variations of $\overline{\mathcal{L}}$ under Lorentz transformations.
The consistensy of the gravitational coupling of $\sqrt{g}\, \overline{\mathcal{L}}\, (g_{\mu\nu} \, ; \, B^{s}_{\sigma\tau})$, with $g = - Det\, g_{\mu\nu}$, allows to simplify the explicit symmetrization procedure , due to Belinfante [25-a-1940] , in the limit $g_{\mu\nu} \rightarrow \eta_{\mu\nu}$, i.e. in uncurved space time , after considering a variation of the metric to first order .
Here I follow my derivation in ref. [22-a-1981] , denoting the symmetric energy momentum tensor $\vartheta_{\nu\varrho}$.

$$\begin{array}{l} 2\, \delta \left(\sqrt{g}\, \overline{\mathcal{L}} \right) = \sqrt{g}\, \vartheta_{\mu\nu}\, (\delta g^{\mu\nu}) \Big|_{\, g_{\alpha\beta} \rightarrow \eta_{\alpha\beta}} \\ \hline \delta \left(\sqrt{g}\, \overline{\mathcal{L}} \right) = (\delta \sqrt{g})\, \overline{\mathcal{L}} + \sqrt{g}\, \delta \overline{\mathcal{L}} \\ \delta \, : \, g^{\mu\nu} \rightarrow g^{\mu\nu} + \delta g^{\mu\nu} \; ; \; [\, g_{\mu\nu} \, : \, \text{base metric} \,] \\ g^{\mu\nu} = (g^{-1})_{\mu\nu} \delta \sqrt{g} = - \frac{1}{2} \sqrt{g}\, g_{\mu\nu}\, (\delta g^{\mu\nu}) \end{array} \tag{114}$$

In eq. 114 $\overline{\mathcal{L}}$ is, for general metric, a scalar quantity and $\vartheta_{\mu\nu}$ a symmetric tensor .
The variation of $\overline{\mathcal{L}}$ becomes using eq. 101

$$\delta\,\overline{\mathcal{L}} = \overline{\mathcal{L}}_{,X}\,\delta\,X \; ; \; X = \tfrac{1}{4}\,B^{\;s}_{\mu\sigma}\,B_{\nu\tau}\,g^{\mu\nu}\,g^{\sigma\tau}$$
$$\overline{\mathcal{L}}_{,X} = \frac{\delta\,\overline{\mathcal{L}}}{\delta\,\mathcal{X}} = \overline{\mathcal{L}}\,/\,\mathcal{X} - \left(\frac{d}{d\,\bar{l}}\,\overline{g}^{\,-2}\right)\frac{\delta\,\bar{l}}{\delta\,\mathcal{X}}\,\mathcal{X} \tag{115}$$
$$\bar{l} = \tfrac{1}{8}\log\left(\left(\mathcal{X}\,/\,\mu^{4}\right)^{2}\right)$$

Proceeding step by step we first substitute

$$\frac{\delta\,\bar{l}}{\delta\,\mathcal{X}}\,\mathcal{X} = \tfrac{1}{4} \; ; \; \overline{\mathcal{L}}\,/\,\mathcal{X} = -\left(\overline{g}^{\,-2} - J\right) \tag{116}$$

in the second relation in eq. 115

$$\delta\,\overline{\mathcal{L}} = \left[-\left(\overline{g}^{\,-2} - J\right) - \tfrac{1}{4}\left(\frac{d}{d\,\bar{l}}\,\overline{g}^{\,-2}\right)\right]\delta\,\mathcal{X} \tag{117}$$

Next we recall eqs. 96 - 97 and 101

$$\frac{d}{d\,\bar{l}}\,\overline{g}^{\,-2} = 2\left(-\beta\left(\overline{g}\right)/\,\overline{g}^{\,3}\right) = \tfrac{1}{8\,\pi^{2}}\,b_{0}\,B\left(\overline{\kappa}\right)$$
$$\overline{\kappa} = \frac{\overline{g}^{\,2}}{16\,\pi^{2}} \; ; \; \overline{\alpha}_{s} = 4\pi\,\overline{\kappa} \tag{118}$$

$$B\left(\overline{\kappa}\right) = b\left(\overline{\kappa}\right)/\,b_{0} = \left[-\beta\left(\overline{g}\right)/\,\overline{g}\right]\left(b_{0}\,\overline{\kappa}\right)^{-1}$$
$$b_{0} = 11 - \tfrac{2}{3}\,N_{fl}$$
$$B\left(\overline{\kappa}\right) = B_{0} + B_{1}\,\overline{\kappa} + \cdots)$$
$$B_{0} = 1\;,\;B_{n} = b_{n}\,/\,b_{0}\;;\;n = 0.1.2\,\cdots$$

The differential equation in eq. 118 is to be solved for suitable initial conditions as if l , $\overline{\kappa}$ were c-numbers and the functional dependence as indicated in eq. 119 below , evaluated with the operator valued (local field valued) substitution, defined in eq. 101 .

$$\overline{\kappa} = \overline{\kappa}\left(l\right) = F\left(l\right) \; ; \; l \longrightarrow \tfrac{1}{8}\log\left(\left(\mathcal{X}\,/\,\mu^{4}\right)^{2}\right) \tag{119}$$

In the substition, defined in eqs. 101 and 119 , local products of local operators are involved, which again requires a regularization in the sense of normal ordering of canonical variables . We insert the relation in eq. 118 in eq. 117 and obtain

$$\delta \overline{\mathcal{L}} = \left[- \left(\overline{g}^{\,-2} - J \right) - \frac{1}{32\,\pi^{\,2}}\, b_{\,0}\, B\left(\overline{\kappa} \right) \right] \delta\, \mathcal{X} \tag{120}$$

Next we evaluate $\delta\, \mathcal{X}$ using the first relation in eq. 101

$$\begin{array}{rcl} \delta\, \overline{\mathcal{L}} & = & \overline{\mathcal{L}}_{\,,\,X}\, \delta\, X \ ; \ X \ = \ \frac{1}{4}\, B_{\,\mu\sigma}^{\,s}\, B_{\,\nu\tau}\, g^{\,\mu\,\nu}\, g^{\,\sigma\,\tau} \ \rightarrow \\ \delta\, X & = & \frac{1}{2}\, B_{\,\mu\sigma}^{\,s}\, g^{\,\sigma\,\tau}\, B_{\,\nu\tau}^{\,s} \Big|_{\,g^{\,\alpha\,\beta} \rightarrow \eta^{\,\alpha\,\beta}}\, \delta\, g^{\,\mu\,\nu} \\ & = & -\,\frac{1}{2} \left(B_{\,\mu\sigma}^{\,s}\, B_{\,\nu}^{\,\sigma\,s} \right) \delta\, g^{\,\mu\,\nu} \end{array} \tag{121}$$

The - sign in the last expression in eq. 121 together with a transposition of the tensor indices in $B_{\,\nu}^{\,\sigma\,s}$ is chosen to offset the two - signs in the expression inside [.] brackets in eq. 120 as well as to comply with the analogous expressions in QED [15-B2-1976] .
Inserting the expressions in eq. 121 in eq. 120 the final form of $\delta\, \overline{\mathcal{L}}$ becomes

$$\delta\, \overline{\mathcal{L}} = \frac{1}{2} \left[\left(\overline{g}^{\,-2} - J \right) + \frac{1}{32\,\pi^{\,2}}\, b_{\,0}\, B\left(\overline{\kappa} \right) \right] \left(B_{\,\mu\sigma}^{\,s}\, B_{\,\nu}^{\,\sigma\,s} \right) \delta\, g^{\,\mu\,\nu} \tag{122}$$

Finally we complete the expression for the symmetric energy momentum tensor $\vartheta_{\,\mu\,\nu}$ in eq. 114 , which is conserved in the limit of uncurved space time

$$\begin{array}{l} 2\,\delta \left(\sqrt{g}\, \overline{\mathcal{L}} \right) \ = \\ = \ \sqrt{g}\ \vartheta_{\,\mu\,\nu} \left(\delta\, g^{\,\mu\,\nu} \right) \Big|_{\,g_{\,\alpha\,\beta} \rightarrow \eta_{\,\alpha\,\beta}} \qquad \rightarrow \\ \hline 2\,\delta \left(\sqrt{g}\, \overline{\mathcal{L}} \right) \ = \\ = \ \sqrt{g} \left[\begin{array}{l} \left[\left(\overline{g}^{\,-2} - J \right) + \frac{1}{32\,\pi^{\,2}}\, b_{\,0}\, B\left(\overline{\kappa} \right) \right] B_{\,\mu\sigma}^{\,s}\, B_{\,\nu}^{\,\sigma\,s} \\ -\,\frac{1}{4}\, g_{\,\mu\nu} \left(\overline{g}^{\,-2} - J \right) B_{\,\varrho\sigma}^{\,s}\, B^{\,\sigma\,\varrho\,s} \end{array} \right] \delta\, g^{\,\mu\,\nu} \end{array} \tag{123}$$

Combining the two terms proportional to $\overline{g}^{\,-2} - J$ in eq. 123 we obtain two characteristic contributions to $\sqrt{g}\ \vartheta_{\,\mu\,\nu} \left(\delta\, g^{\,\mu\,\nu} \right)$, even before the limit $g_{\,\mu\,\nu} \rightarrow \eta_{\,\mu\,\nu}$ is taken .
It follows

$$\vartheta_{\,\mu\,\nu} = \left[\begin{array}{c} \left(\overline{g}^{\,-2} - J \right) \left[B_{\,\mu\sigma}^{\,s}\, B_{\,\nu}^{\,\sigma\,s} - \frac{1}{4}\, g_{\,\mu\nu}\, B_{\,\varrho\sigma}^{\,s}\, B^{\,\sigma\,\varrho\,s} \right] \\ +\,\frac{1}{8\,\pi^{\,2}}\, b_{\,0}\, B\left(\overline{\kappa} \right) \left[\frac{1}{4}\, B_{\,\mu\sigma}^{\,s}\, B_{\,\nu}^{\,\sigma\,s} \right] \end{array} \right] \tag{124}$$

The upper term in the outer [.] brackets in eq. 124 , proportional to $\overline{g}^{\,-2} - J$, is traceless , whereas the anomalous trace is contained in the lower term , proportional to $\frac{1}{8\,\pi^{\,2}}\, b_{\,0}\, B\,(\,\overline{\kappa}\,)$.
In the limit of vanishing gravitational interactions , i.e. $g_{\,\alpha\,\beta} \rightarrow \eta_{\,\alpha\,\beta}$, the energy momentum tensor $\vartheta_{\,\mu\,\nu}$, defined in eq. 124 , is not only symmetric but conserved , provided the *modified* equations of motion can be enforced, as well as gauge invariant with respect to the local gauge group of QCD $SU3_{\,c}$

$$\partial^{\,\mu}\,\vartheta_{\,\mu\,\nu}\,(\,x\,) = \,0\;;\;\partial^{\,\mu} \,=\, \eta^{\,\mu\,\beta}\,\partial_{\,\beta} \,=\, \eta^{\,\mu\,\beta}\,\partial\,/\,\partial_{\,x\,\beta} \tag{125}$$

The trace anomaly allows precisely the general form of the modified Lagrangean $\overline{\mathcal{L}}$, as introduced in eq. 96 including the a priori free constant J. It takes the form using eq. 124 , independent of J

$$\begin{array}{l} \overline{\mathcal{L}} \,=\, -\,\left[\,\dfrac{1}{\overline{g}^{\,2}\,(\,\mathcal{X}\,)}\,-\,J\,\right]\,\mathcal{X}\;\longrightarrow \\ \vartheta^{\,\mu}_{\;\;\mu} \,=\, \frac{1}{8\,\pi^{\,2}}\,b_{\,0}\,B\,(\,\overline{\kappa}\,)\,\left[\,\frac{1}{4}\,B^{\;s}_{\,\mu\sigma}\,B^{\,\sigma\mu\;s}\,\right] \,=\, -\,\frac{1}{8\,\pi^{\,2}}\,b_{\,0}\;:\;B\,(\,\overline{\kappa}\,(\mathcal{X}\,)\,)\,\mathcal{X}\;: \end{array} \tag{126}$$

In the last expression in eq. 126 the need of (normal) ordering of local products of quantized fields is indicated by the : symbols , omitted before and also hereafter whenever not explicitely necessary .
The negative sign in the same expression is significant: the factor $B\,(\,\overline{\kappa}\,)\,(\,\mathcal{X}\,)$ is a positive operator at least for $\overline{\kappa}$ in the perturbative regime, wile $\mathcal{X}$ becomes a positive operator in the Euclidean region

$$\begin{array}{l} \mathcal{X} \,=\, \frac{1}{2}\,\left[\,\vec{B}^{\,s}\,\vec{B}^{\,s} \,-\, \vec{E}^{\,s}\,\vec{E}^{\,s}\,\right]\,\rightarrow \\ \rightarrow\,\mathcal{X}_{\,Euc} \,=\, \frac{1}{2}\,\left[\,\vec{B}^{\,s}\,\vec{B}^{\,s} \,+\, \vec{E}^{\,s}\,\vec{E}^{\,s}\,\right]_{\,Euc} \end{array} \tag{127}$$

This completes the construction of the trace anomaly induced canonical variables and modified Lagrangean $\overline{\mathcal{L}}$. Consequences are discussed in the next section .

3 - Consequences arising from derivations in the previous section

We collect the reults contained in eqs. 98 - 103 , 110 - 113 , 115 - 127 in eq.

128 below

$$\vartheta_{\mu\nu} = \left[\begin{array}{l} (\overline{g}^{-2} - J) \left[B^{\ s}_{\mu\sigma} B^{\ \sigma\ s}_{\nu} - \frac{1}{4} g_{\mu\nu} B^{\ s}_{\varrho\sigma} B^{\sigma\varrho s} \right] \\ + \frac{1}{8\pi^2} b_0 B(\overline{\kappa}) \left[\frac{1}{4} B^{\ s}_{\mu\sigma} B^{\ \sigma\ s}_{\nu} \right] \end{array} \right]$$

$$B(\overline{\kappa}) = \leftarrow : B(\overline{\kappa}(\mathcal{X}/\Lambda^4)) : \rightarrow ; \mathcal{X} =: \tfrac{1}{4} B^{\ r}_{\mu\nu} B^{\mu\nu\ r} := -g^2 \mathcal{L}_{gauge}$$

$$\begin{array}{l} B(\overline{\kappa}) = b(\overline{\kappa}) / b_0 = [-\beta(\overline{g}) / \overline{g}] (b_0 \overline{\kappa})^{-1} ; b_0 = 11 - \tfrac{2}{3} N_{fl} \\ B(\overline{\kappa}) = B_0 + B_1 \overline{\kappa} + \cdots) ; B_0 = 1 , B_n = b_n / b_0 ; n = 0.1.2 \cdots \end{array} \tag{128}$$

The symbol $\leftarrow : B(\overline{\kappa}(\mathcal{X}/\Lambda^4)) : \rightarrow$ specifically the external arrows , shall indicate that all normal orderings which have to be applied to products of gauge field operators and functions thereof , as they appear in the various expressions in eq. 128 have to extend from the exterior leftmost to the exterior rightmost operators .
We distinguish both the canonical energy momentum (density) tensors (eq. 113) and their symmetric equivalents according to the two choices for the gauge field Lagrangean (density) (eq. 123) , abbreviated to $\mathcal{L}_{gauge} = \mathcal{L}$ and $\overline{\mathcal{L}}$ respectively

$$\begin{array}{l} T_{\nu}^{\ \varrho} = (\partial_{\nu} W^{\ s}_{\sigma}) \mathcal{L}^{\ genaral}_{\ ,\partial_{\varrho} W^{\ s}_{\sigma}} - \delta^{\ \varrho}_{\nu} \mathcal{L}^{\ general} \\ 2\delta \left(\sqrt{g}\, \mathcal{L}^{\ general} \right) = \sqrt{g}\, \vartheta_{\mu\nu} (\delta g^{\mu\nu}) \big|_{g_{\alpha\beta} \rightarrow \eta_{\alpha\beta}} \rightarrow \\ \hline T_{\nu}^{\ \varrho} = T_{\nu}^{\ \varrho}(\,.\,) ; \vartheta_{\mu\nu} = \vartheta_{\mu\nu}(\,.\,) ; \text{ with } . = \{ \overline{\mathcal{L}} , \mathcal{L} \} \end{array} \tag{129}$$

The Belinfante construction [25-a-1940] links uniquely $T_{\nu}^{\ \mu}(\,.\,) \leftrightarrow \vartheta_{\mu\nu}(\,.\,)$, separately for both values of $(\,.\,)$. The respective canonical energy momentum tensors are neither symmetric nor gauge invariant also for both choices of $(\,.\,)$, whereas both $\vartheta_{\mu\nu}(\,.\,)$ are gauge invariant.
Furthermore all energy momentum tensors are conserved , provided the equations of motion are enforced

$$\begin{array}{l} \partial^{\mu} T_{\nu\mu}(\,.\,) = 0 ; T_{\nu\mu}(\,.\,) = \eta_{\mu\varrho} T_{\nu}^{\ \varrho}(\,.\,) \\ \partial^{\mu} \vartheta_{\nu\mu}(\,.\,) = 0 ; \vartheta_{\nu\mu}(\,.\,) = \vartheta_{\mu\nu}(\,.\,) \end{array} \tag{130}$$

Notwithstanding the two cases

it is $\overline{\mathcal{L}}$ which needs to be chosen.

With respect to the trace anomaly (eq. 126) and thereby with respect to dilatation transformations, for which the symmetric tensor(s) $\vartheta_{\nu\mu}(\,.\,)$ are the relevant ones, it follows

$$\begin{aligned} \vartheta^{\mu}_{\ \mu} (\overline{\mathcal{L}}) &= \frac{1}{8 \pi^2} b_0 : B \left(\overline{\kappa} \left(\mathcal{X} / \Lambda^4 \right) \right) \left[\tfrac{1}{4} B^{s}_{\mu\sigma} B^{\sigma\mu\, s} \right] : \\ &= - \frac{1}{8 \pi^2} b_0 : B \left(\overline{\kappa} \left(\mathcal{X} / \Lambda^4 \right) \right) \mathcal{X} : \neq 0 \\ \vartheta^{\mu}_{\ \mu} (\mathcal{L}) &= 0 \end{aligned} \tag{131}$$

Concerning redundancy of normal order proedures it shall be noted , that here the normal ordering is restricted to obey chain rules of (covariant) differentiation . This is exemplified by the relations in eq. 100 and as another example the relation

$$\begin{aligned} \partial_{\varrho} : B^{s}_{\mu\nu} (x) B^{s}_{\sigma\tau} (x) :=: & \left[\begin{array}{l} \left(\left(D_{\varrho} (ad) \right)_{sr} B^{r}_{\mu\nu} \right) B^{s}_{\sigma\tau} \\ + B^{s}_{\mu\nu} \left(\left(D_{\varrho} (ad) \right)_{sr} B^{r}_{\mu\nu} \right) \end{array} \right] : \\ \left(D_{\varrho} (ad) \right)_{sr} = \partial_{\varrho} \delta_{sr} + W^{t}_{\varrho} (ad_t)_{sr} \ ; & \ (ad_t)_{sr} = f_{str} \end{aligned} \tag{132}$$

In the following normal ordering symbols : ( ) : shall be omitted whenever possible , for shortness of notation , except if crucial to avoid misinterpretation .

3-1 - Equations of motion and canonically conjugate variables pertaining to $\overline{\mathcal{L}} + \mathcal{L}_{\{q\}}$

The conflict between the canonical forms of obviously inequivalent Lagrangean densities in the gauge field sector

$$\overline{\mathcal{L}} \longleftrightarrow \mathcal{L} \,\Big|_{gaugefields} \tag{133}$$

can be traced – in the perturbative sector – to maintaining Poincaré invariance and full gauge invariance , which necessitates the use of Fermi like gauges and thus a nontrivial coupling of ghost fields, which do contribute to the energy momentum tensor , as shown e.g. in ref. [20-1975] . At this stage we turn to the equations of motion for gauge fields as induced by the Lagrangean $\overline{\mathcal{L}} + \mathcal{L}_{\{q\}}$ defined in eqs. 1 - 6 and 96 - 97 , merging eqs. 103 and 120

$$\begin{aligned} &\delta \overline{\mathcal{L}} = - \left[\left(\overline{g}^{\,-2} - J \right) + \frac{1}{32 \pi^2} b_0 B (\overline{\kappa}) \right] \delta \mathcal{X} \\ &\frac{\delta \mathcal{X}}{\delta \left(\partial_{\varrho} W^{s}_{\sigma} \right)} = B^{\varrho \sigma s} \ ; \ \frac{\delta \mathcal{X}}{\delta W^{s}_{\sigma}} = - f_{str} W^{t}_{\varrho} B^{\varrho \sigma r} \end{aligned} \tag{134}$$

→

It follows for the partial derivatives of $\overline{\mathcal{L}}$

$$\begin{array}{lll} \overline{\mathcal{L}}_{\,,\,\partial_{\varrho} W_{\sigma}^{s}} & = + & \left[\left(\overline{g}^{\,-2} - J \right) + \frac{1}{32\,\pi^{\,2}}\, b_{\,0}\, B\left(\overline{\kappa} \right) \right] B^{\,\sigma\,\varrho\,s} \\ \overline{\mathcal{L}}_{\,,\,W_{\sigma}^{s}} & = - & \left[\left(\overline{g}^{\,-2} - J \right) + \frac{1}{32\,\pi^{\,2}}\, b_{\,0}\, B\left(\overline{\kappa} \right) \right] f_{\,str} \times W_{\varrho}^{\,t}\, B^{\,\sigma\varrho\,r} \end{array} \tag{135}$$

Using eq. 132

$$\left(D_{\varrho} \left(ad \right) \right)_{sr} \left\{ B_{\sigma\tau}^{\,r} \right\} = \partial_{\varrho} \left\{ B_{\sigma\tau}^{\,s} \right\} + f_{\,str}\; W_{\varrho}^{\,t} \left\{ B_{\sigma\tau}^{\,r} \right\} \tag{136}$$

the Euler-Lagrange derivative of $\overline{\mathcal{L}}$ on the left hand side of eq. 110 take the form

$$\begin{array}{l} \left(\partial_{\varrho} \overline{\mathcal{L}}_{\,,\,\partial_{\varrho} W_{\sigma}^{s}} - \overline{\mathcal{L}}_{\,,\,W_{\sigma}^{s}} \right) \left(x \right) = \\ \quad = D_{\varrho} \left(ad \right) \left\{ \left[\left(\overline{g}^{\,-2} - J \right) + \frac{1}{32\,\pi^{\,2}}\, b_{\,0}\, B\left(\overline{\kappa} \right) \right] B^{\,\sigma\,\varrho} \right\}^{s} \left(x \right) \end{array} \tag{137}$$

The Euler-Lagrange equations including quark flavors according to $\mathcal{L}_{\{q\}}$ (eqs. 1 , 94 , 110) become

$$\begin{array}{l} D_{\varrho} \left(ad \right) \left\{ \left[\left(\overline{g}^{\,-2} - J \right) + \frac{1}{32\,\pi^{\,2}}\, b_{\,0}\, B\left(\overline{\kappa} \right) \right] B^{\,\sigma\,\varrho} \right\}^{s} \left(x \right) = \\ \quad = \left(j^{\,\sigma\,s} \right)_{\{q\}} \left(x \right) \end{array} \tag{138}$$

$$\left(j^{\,\sigma\,s} \right)_{\{q\}} = \sum_{q-fl} \overline{q}^{\,\dot{c}'} \left\{ \gamma^{\,\sigma} \left(\tfrac{1}{2} \lambda^{\,s} \right)_{c'\dot{c}} \right\} q^{\,c} = \left(\mathcal{L}_{\{q\}} \right)_{,\,W_{\sigma}^{s}}$$

The quantity on the gauge field side of the 'divergence' on the right hand side of eq. 137 and on the left hand side of the first relation in eq. 138 differs in an essential way from the bare Lagrangean expressions corresponding to $\mathcal{L}_{\,gauge}$ (eqs. 94 , 139 below)

$$\begin{array}{l} D_{\varrho} = \partial_{\varrho} + \mathcal{W}_{\varrho} \; ; \; \mathcal{W}_{\varrho} = W_{\varrho}^{\,r}\, ad_{\,r} \\ D^{\,\varrho} \left\{ \dfrac{1}{g^{\,2}}\, B_{\,\sigma\varrho} \right\}^{s} = j_{\,\sigma}^{\,s} \end{array} \tag{139}$$

To make this more transparent lets introduce the notation

$$\begin{array}{lcl} \overline{\mathcal{L}} \leftrightarrow \overline{G} & \leftrightarrow & \mathcal{L} \leftrightarrow G \\ \overline{G} = \text{local, scalar, color neutral field} & \leftrightarrow\; G = & g^{\,-2} \\ & & \text{const. c-number} \\ \overline{G} = : \left[\left(\overline{g}^{\,-2} - J \right) + \frac{1}{32\,\pi^{\,2}}\, b_{\,0}\, B\left(\overline{\kappa} \right) \right] : \left(x \right) & & \end{array} \tag{140}$$

The precise way , $\overline{G}$ and associated quantities, in particular the energy momentum (density) tensor pertaining to gauge fields, defined in eq. 124, are determined as local field operators , shall be (re-) specified below

$$\begin{array}{l} (a) \; : \; \overline{G} \; = \; \overline{G} \; (\; l \;) \; ; \; l \; = \; \log \; (\; \overline{\mu} \; / \; \mu \;) \; \rightarrow \\ (b) \; : \; l \; \rightarrow \; \frac{1}{8} \; \log \; \left((\; \mathcal{X} \; (\; x \;) \; / \; \Lambda^{\;4} \;)^{\;2} \right) \; \leftrightarrow \; e^{\;8 l} \; = \; (\; \mathcal{X} \; (\; x \;) \; / \; \Lambda^{\;4} \;)^{\;2} \\ \hline \mathcal{X} \; (\; x \;) \; = : \; \frac{1}{4} \; B^{\;r}_{\;\mu\nu} \; B^{\;\mu\nu\;r} \; : \; (\; x \;) \end{array} \tag{141}$$

The functional dependence $\overline{G} = \overline{G} \; (\; l \;)$ in step (a) in eq. 141 can be determined in the perturbative regime, through the renormalization group equation(s) , e.g. in the $\overline{\text{MS}}$ renormalization scheme .
Through 4 loop order $\overline{\text{MS}}$ is renormalization group invariant, through the substitutions pertaining to a general sliding scale μ , $\mu = \Lambda_{QCD}$, and e.g. through the moments of deep inelastic scattering amplitudes , $\overline{\mu}^{\;2} = Q^{\;2}$, where Q^2 is the (positive) momentum transfer square in the deep inelastic reaction studied [29-1988, 30-1997] . In any renormalization scheme, where at least in the perturbative regime full renormalization group invariance is verified , it follows that the substitution in the second step (b) in eq. 141 can equally be performed in a renormalization group invariant manner . This is tantamount to resolve all ambiguities in the definition of the composite local field $\mathcal{X} \; (\; x \;)$ by a normalization in terms of renormalization group invariant , i.e. measurable quantities .
Within QCD sum rules introduced by Shifman, Vainshtain and Zakharov [40-1979] , the vacuum expected value of the multiplicatively related field , denoted $\alpha_{\;s} \; G^{\;2}$ has been intensively studied .
I cite here a recent paper and result(s) by Stephan Narison [11-2011, 45-2011] in partcular with respect to the renormalization group invariant setting – in principle – of composite local field normalization

$$\alpha_{\;s} \; G^{\;2} \; = \; \pi^{\;-1} \; \mathcal{X} \; ; \quad \begin{array}{rcl} \langle \; \Omega \; | \alpha_{\;s} \; G^{\;2} \; | \; \Omega \; \rangle & = & (\; 7.0 \; \pm \; 1.3 \;) \; 10^{\;-2} \; \text{GeV}^{\;4} \\ & = & \pi^{\;-1} \; (\; 0.22 \; \pm \; 0.04 \;) \; \text{GeV}^{\;4} \end{array} \tag{142}$$

However balancing of short distance ordered contributions from local operators of increasing mass dimension in a finite region of reference energies corresponding to characteristic, i.e. nonasymptotic hadronic ones, with contributions in the same energy interval from hadrons , proves to be less straightforward, than initially thought. The invoked compensations , traditionally called duality relations, may well not hold upon truncation of the ultraviolet and normal scale contributions to a few leading terms, translating into increased errors relative to initial estimates, exemplified by the errors in eq. 142 . I quote here a derivation of the ratio of up , down and strange quark masses by Cesareo Dominguez and Eduardo de Rafael [27-a-1991] in which subtle analytic methods are presented.
Having defined the local field structure of two inequivalent such fields $\mathcal{X} \; (\; x \;)$

and *separately* $\overline{G}\,(\,x\,)$ in eqs. 140 and 141, the symmetric, gauge invariant energy momentum tensor incompletely defined in eqs. 123 and 124 is represented as follows

$$\begin{array}{lll} \vartheta_{\,\mu\,\nu}\,(\,x) & = & \left[\begin{array}{c} :\,\overline{G}\,\left[\,B^{\;s}_{\;\mu\sigma}\;B^{\;\sigma\;s}_{\;\nu}\;+\;g_{\;\mu\nu}\;\mathcal{X}\,\right]\;: \\ -\;g_{\;\mu\nu}\;:\;\frac{1}{32\;\pi^{\;2}}\;b_{\;0}\;B\;(\,\overline{\kappa}\,(\,\mathcal{X}\,)\;\mathcal{X}\;: \end{array}\right](\,x\,) \\ \vartheta^{\;\mu}_{\;\mu}\,(\,x\,) & = & -\;\frac{1}{8\;\pi^{\;2}}\;b_{\;0}\;:\;B\;(\,\overline{\kappa}\,(\mathcal{X}\,)\,)\;\mathcal{X}\;:\;(\,x\,) \\ \hline \overline{G}\,(\,x\,) & = & :\;\left[\,(\,\overline{g}^{\;-2}\;-\,J\,)\;+\;\frac{1}{32\;\pi^{\;2}}\;b_{\;0}\;B\;(\,\overline{\kappa}\,(\,\mathcal{X}\,)\,)\,\right]:\;(\,x\,) \\ B\;(\,\overline{\kappa}\,(\,\mathcal{X}\,)\,)\;(\,x\,) & \rightarrow\equiv & \overline{B}\;(\,\mathcal{X}\,)\;(\,x\,) \end{array} \tag{143}$$

We complete the Euler-Lagrange equations of motion defined in eqs. 137 and 138 , which introduce a local, color neutral, hermitian scalar field , denoted $\varphi\,(\,x\,)$ below , depending implicitely on the gauge field strengths bilinear $\mathcal{X}\,(\,x\,)$ as specified in eqs. 140 - 143

$$\begin{array}{l} (\,\partial_{\,\varrho}\,\overline{\mathcal{L}}\,,\,\partial_{\,\varrho}\,W^{\;s}_{\;\sigma}\;-\;\overline{\mathcal{L}}\,,\,W^{\;s}_{\;\sigma}\,)\;(\,x\,)\;= \\ =\;D_{\;\varrho}\,(\,ad\,)\;:\;\left\{\left[\left(\begin{array}{c}\overline{g}^{\;-2}\;-\,J\,+ \\ +\;\frac{1}{32\;\pi^{\;2}}\;b_{\;0}\;\overline{B}\end{array}\right)(\,\mathcal{X}\,)\right]\;B^{\;\sigma\;\varrho}\right\}^{s}\;:\;(\,x\,)\;\rightarrow \\ \varphi\,(\,x\,)\;=\;:\;(\,\overline{g}^{\;-2}\;-\,J\;+\;\frac{1}{32\;\pi^{\;2}}\;b_{\;0}\;\overline{B}\,)\;(\,\mathcal{X}\,)\;:\;(\,x\,)\;\equiv\;\overline{G}\,(\,x\,) \end{array} \tag{144}$$

The field φ introduced in eq. 144 is *dimensionless* , which in itself is a consequence of the violation of dilatation invariance, or the trace anomaly . The equations of motion for the gauge fields (eqs. 138 , 143) become , suppressing the ordering signs : ( ) :

$$\begin{array}{l} D_{\;\varrho}\,(\,ad\,)\;\{\quad B^{\;\sigma\;\varrho}\,\}^{\;s}\;(\,x\,)\;=\;(\,j^{\;\sigma\;s}\,)_{\;\{q\}}\;(\,x\,) \\ \hline (\,j^{\;\sigma\;s}\,)_{\;\{q\}}\;=\;\sum_{\;q-fl}\;\overline{q}^{\;\dot{c}'}\;\{\,\gamma^{\;\sigma}\;(\,\frac{1}{2}\,\lambda^{\;s}\,)_{\;c'\dot{c}}\,\}\;q^{\;c}\;=\;(\,\mathcal{L}_{\;\{q\}}\,)\,,\,W^{\;s}_{\;\sigma} \\ \left[\begin{array}{c} D_{\;\varrho}\,(\,ad\,) \\ \{\quad B^{\;\sigma\;\varrho}\,\}^{\;s} \\ (\,x\,) \end{array}\right]\;=\;\left[\begin{array}{l} \partial_{\;x\;\varrho}\;\{\qquad B^{\;\sigma\;\varrho\;s}\,(\,x\,)\,\}\;+ \\ +\;f_{\;str}\;\left\{\,W^{\;t}_{\;\varrho}\,(\,x\,)\qquad B^{\;\sigma\varrho\;\;r}\,(\,x\,)\,\right\} \end{array}\right] \\ \qquad\qquad\qquad =\;\left[\begin{array}{l} \{\,\partial_{\;x\;\varrho}\qquad\}\;\{B^{\;\sigma\;\varrho\;s}\,(\,x\,)\,\}\;+ \\ +\qquad\{\,D_{\;\varrho}\,(\,ad\,)\;B^{\;\sigma\;\varrho}\,\}^{\;s}\;(\,x\,) \end{array}\right] \\ \hline \qquad =\;:\;(\,\overline{g}^{\;-2}\;-\,J\;+\;\frac{1}{32\;\pi^{\;2}}\;b_{\;0}\;\overline{B}\,)\;(\,\mathcal{X}\,)\;:\;(\,x\,) \\ \hline \end{array} \tag{145}$$

In deriving eq. 145 we assumed that chain rules for normal partial derivatives and covariant derivatives are maintained through the ordering processes , suppressed for simplicity of notation .

3-2 - Concluding remarks and outlook

To set the present results in perspective I quote from ref. [28-a-2010] (PM , November 2010, Singapore)

"... The conventional form for the bosonic Bogoliubov transformation is the one corresponding to $| \vartheta | < 1$.

Let me remark , that neither the real form nor the conventinal normalization ($= 1$) are representing the full structure associated with *one* bosonic pair of oscillators .

Epilogue

> The embedding of chiral symmetry depends in a nontrivial way on the strength of the *gauge field strength pair*- Bose condensate as does the excitation of binary and higher gauge boson compounds ('glueballs') *and* the phase structure of QCD .
>
> There is some way to go. I hope to come back to this theme soon. ... "

and further

"...

1) Perturbative accessibility of renormalization in asymptotically free theories

   While the entire renormalization procedure thus (eqs. 9 - 16) comes within *perturbative accessibility* – as explained in textbooks [21], [22-1982] – the associated renormalization group equation serves to restore renormalization group invariant properties, in particular such definitions of operators .

2) Infrared instability

   is associated with all physical scales *not* accessible to perturbative approximations .

3) Quark mass dependence

   We neglect in the considerations followed here the quark mass dependence of all Green functions in the deep Euclidean region , the quark masses also *to be renormalized* and thus *not* renormalization group independent [23-1975] . This is in line with the main short distance contributions , which are sorting out by the twist characteristic leading contributions *modulo less dominant ones modulo powers of inverse Euclidean distance* . These dimensional hierarchies *also* break down whence the region of perturbative accessibility is transgressed . For small quark masses at a generic scale of $\sim\ 1\ GeV$ the quark mass associated mixing of operators with different dimensions sets in in subtle ways governed by approximate chiral symmetry also outside the deep Euclidean region . $\cdots$ "

4) Canonical structure

The step performed here brings in line the structure of the trace anomaly with the derivation of consistent equations of motion for the gauge fields , displayed in eq. 145 . It shows that to transcend the perturbatively accessible region the canonical structure of gauge boson associated 'p'-s and 'q'-s has to be modified accordingly .

5) Outlook

The completion of QCD as a gauge field theory in uncurved space time remains a far goal .

Along the way let us keep in mind that (for all we know) the physical reality transcends much further, in particular to curved space time of unknown dimensionality .

— **Thank you** —

Appendix 1 - expansion coefficients of the rescaling functions $\widehat{\beta}$, γ to four loops

$$
\begin{aligned}
& -\beta / g = X B(X) \; ; \; B(X) = b_0 A(X) \\
& B(X) \sim \sum_{n=0}^{\infty} b_n X^n \; , \; A(X) \sim \sum_{n=0}^{\infty} a_n X^n \\
& \kappa = g^2 / (16 \pi^2) \text{ generic } X \\
& b_0 = \tfrac{1}{3} (33 - 2 N_{fl}) \; , \; a_0 = 1 \; , \; a_n = b_n / b_0 \\
& b_1 = \tfrac{2}{3} (153 - 19 N_{fl}) \\
& b_2 = \tfrac{1}{54} \left(77139 - 15099 N_{fl} + 325 N_{fl}^2 \right) \\
& b_3 \sim 29243 - 6946.3 N_{fl} + 405.089 N_{fl}^2 + 1.49931 N_{fl}^3
\end{aligned}
\tag{146}
$$

$$
\begin{aligned}
& -\gamma_m^0 = 4 \, , \, -\gamma_m^1 = \frac{202}{3} - n_{fl} \frac{20}{9} \qquad | - \gamma_m^l \equiv \chi_m^l \\
& -\gamma_m^2 = 1249 - \left[\frac{2216}{27} + \frac{160}{3} \zeta(3) \right] N_{fl} - \frac{140}{81} N_{fl}^2 \\
& -\gamma_m^3 = \left\{ \begin{array}{l}
\dfrac{4603055}{162} + \dfrac{135680}{27} \zeta(3) - 8800 \zeta(5) + \\
+ \left[\begin{array}{l} - \dfrac{91723}{27} - \dfrac{34192}{9} \zeta(3) + \\ + 880 \zeta(4) + \dfrac{18400}{9} \zeta(5) \end{array} \right] N_{fl} + \\
+ \left[\dfrac{5242}{243} + \dfrac{800}{9} \zeta(3) - \dfrac{160}{3} \zeta(4) \right] N_{fl}^2 + \\
+ \left[- \dfrac{332}{243} + \dfrac{64}{27} \zeta(3) \right] N_{fl}^3
\end{array} \right\}
\end{aligned}
\tag{147}
$$

References

[1-2001] O. Bär and U.J. Wiese, 'Can one see the number of colors?', MIT-CTP-3144, May 2001, 24 pp., Nucl.Phys. B609 (2001) 225-246, hep-ph/0105258 .

[2-1964] M. Gell-Mann and Y. Neeman, 'The eightfold way', W. A. Benjamin, Inc., New York 1964 .

[3-1975] H. Fritzsch and P. Minkowski, 'Psi Resonances, Gluons and the Zweig Rule',
March 1975, 61 pp., CALT-68-492, Nuovo Cim. A30 (1975) 393 .

[4-1973] S. Weinberg, 'New approach to the renormalization group', Phys. Rev. D8 (1973) 3497 .

[A215-2009] S. Bethke, 'The 2009 Wolrd Average of $\alpha_s(M_Z)$ ', Eur.Phys.J.C64 (2009) 689-703, arXiv:0908.113 [hep-ph] .

[5-1997] K.G. Chetyrkin, 'Quark mass anomalous dimension to $O\ (\alpha\ _s^4\ )$ ', March 1997, 7 pp., MPI-PHT-97-019, Phys.Lett. B404 (1997) 161-165, hep-ph/9703278 , and references cited therein .

[6-1997] T. van Ritbergen, J.A.M. Vermaseren and S.A. Larin, 'The Four loop beta function in quantum chromodynamics', UM-TH-97-01, NIKHEF-97-001, Jan 1997. 9pp.,
Phys.Lett.B400 (1997) 379 , hep-ph/9701390 , and references cited therein .

[7-1979] M.A. Shifman, A.I. Vainshtein and V.I. Zakharov, 'QCD and resonance physics. Theoretical Foundations.' , Nucl. Phys. B147 (1979) 385, 448 .

[8-2011] S. Bodenstein, C.A. Dominguez, S.I. Eidelman, H. Spiesberger and K. Schilcher, 'Confronting electron-positron annihilation into hadrons with QCD: An Operator product expansion analysis', Oct 2011, 17 pp., UCT-TP-282-10,MZ-TH-10-40, JHEP 1201 (2012) 039, arXiv:1110.2026 [hep-ph] , and references cited therein .

[9-2010] K. Chetyrkin, J.H. Kühn, A. Maier, P. Maierhöfer, P. Marquard, M. Steinhauser and C. Sturm, 'Precise Charm- and Bottom-Quark Masses: Theoretical and Experimental Uncertainties', Theor.Math.Phys. 170 (2012) 217-228, arXiv:1010.6157v2 [hep-ph] , and references cited therein .

[10-2011] S. Bodenstein, J. Bordes, C.A. Dominguez, J. Penarrocha and K. Schilcher, 'Bottom-quark mass from finite energy QCD sum rules', Nov 2011, UCT-TP-288-11, MZ-TH-11-37, Phys.Rev. D85 (2012) 034003, arXiv:1111.5742 [hep-ph] .

[11-2011] S. Narison, 'Gluon Condensates and $m_{c,b}$ from QCD-Moments and their ratios to Order α_s^3 and $< G^4 >$ ', Phys.Lett. B706 (2012) 412-422, arXiv:1105.2922 [hep-ph] .

[12-RR-1969] S. L. Adler, 'Axial vector vertex in spinor electrodynamics', Phys.Rev.177 (1969) 2426-2438.

J. Bell and R. Jackiw, 'A PCAC puzzle: $\pi^0 \rightarrow \gamma\gamma$ in the sigma model', Nuovo Cim.A60 (1969) 47-61 .

[13-1991] P. Minkowski, 'Proton spin and quark spin', in Proceedings of the "Workshop on Effective Field Theories of the Standard Model", Dobogókõ, Hungary August 22 - 26, 1991 ; U. Meissner edt., World Scientific Singapore 1992 , BUTP-91-2, Aug 1991. 31pp. .

[14-a-2012] P. Minkowski, 'Canonical (anti-) commutation rules in QCD abd unbroken gauge invariance, QCD - the two central anomalies and canonical structure', Feb 2012, arXiv:1202.2488 [hep-th] .

[15-a-2010] S. Kabana and P. Minkowski, 'On the thermal phase structure of QCD at vanishing chemical potentials', Jan 2010, 7 pp., CERN-PH-TH-2010-002, Int.J.Mod.Phys. A26 (2011) 3035-3050, arXiv:1001.0707 [hep-ph] .

[16-a-2000] S. Kabana and P. Minkowski, 'Mapping out the QCD phase transition in multiparticle production', Oct 2000, 53 pp., New J.Phys. 3 (2001) 4, hep-ph/0010247 .

[17-a-2011] P. Minkowski, 'QCD and the incomplete links between mass and gauge', massgauge-kyoto2011.pdf,
(URL) http://www.mink.itp.unibe.ch/lectures.html .

[18-a-1968] G. Segal, 'The representation ring of a compact Lie group', Publications Mathématiques de L'IHS, Volume 34, Number 1 (1968) 113-128, DOI: 10.1007/BF02684592 , and references cited therein .

[19-a-1975] C. Becchi, A. Rouet and R. Stora, Comm. Math. Phys. 42 (1975) 127 ,
Ann. Phys. 98 (1976) 287 , and in 'Renormalization Theory', G. Velo and A. S. Wightman eds.,
Reidel , Dordrecht 1976 .

[20-a-1975] I. V. Tyutin, Lebedev Institute preprint N39 , 1975 .

[21-a-1978] H. Pagels and E. Tomboulis, 'Vacuum of the quantum Yang-Mills theory and magnetostatics', Nuclear Physics B143 (1978) 485-502.

[22-a-1981] P. Minkowski, 'On The Ground State Expectation Value Of The Field Strength Bilinear In Gauge Theories And Constant Classical Fields', Nuclear Physics B 177 (1981) 203-217 , and references cited therein .

[23-a-2012] P. Minkowski, 'Canonical (anti-)commutation rules in QCD abd unbroken gauge invariance, QCD - the two central anomalies and canonical structure', Feb 2012, arXiv:1202.2488 [hep-th] .

[24-a-1918] E. Noether, "Invariante Variationsprobleme",
Nachrichten der Königlichen Gesellschaft der Wissenschaften zu Göttingen, mathematisch-physikalische Klasse 1918 , 235-257 ; english translation :
Emmy Noether, M. A. Tavel, 'Invariant Variation Problems', (Submitted on 8 Mar 2005), arXiv:physics/0503066v1 [physics.hist-ph] .

[25-a-1940] F. J. Belinfante, 'On the current and the density of the electric charge, the energy, the linear momentum and the angular momentum of arbitrary fields', Physica 7 (1940) 449-474 , a full derivation can be found in the textbook by Res Jost [26-a-1965] .

[26-a-1965] R. Jost, 'The general theory of quantzed fields', American Mathematical Society, Providence, Rhode Island , 1965 .

[27-a-1991] C. Dominguez and E. de Rafael, 'Light Quark Masses in QCD from Local Duality', Annals of Physics 174 (1987) 272 , and references cited therein .

[28-a-2010] P. Minkowski, 'Oscillations of three Majorana neutrinos and a possible connection to QCD', URL : http://www.mink.itp.unibe.ch/publications.html
(file : singafl2010.pdf) .

[14-B2-1975] H. Kluberg-Stern and J. B. Zuber, 'Ward Identities and Some Clues to the Renormalization of Gauge Invariant Operators', Phys.Rev. D12 (1975) 467 ,
'Renormalization of Nonabelian Gauge Theories in a Background Field Gauge. 1. Green Functions', Phys.Rev. D12 (1975) 482 and
'Renormalization of Nonabelian Gauge Theories in a Background Field Gauge. 2. Gauge Invariant Operators', Phys.Rev.D12 (1975) 3159-3180 .

[15-B2-1976] P. Minkowski, 'On the anomalous divergence of the dilatation current in gauge theories', Bern preprint 1976 , unpublished , URL : http://www.mink.itp.unibe.ch/publications.html [file : all.pdf] .

For fairness of citation let me quote the papers published in 1977 on the trace anomaly

S. L. Adler, J. C. Collins and A. Duncan, Phys. Rev. D15 (1977) 1712 ,

N. K. Nielsen. Nucl. Phys. B 120 (1977) 212 ,

S. L. Adler, A. Duncan and S. D. Joglekar, Phys. Rev. D 16 (1977) 438 .

[16-1972] H. Fritzsch and M. Gell-Mann, 'Current algebra: Quarks and what else?', Proceedings of 16th International Conference on High-Energy Physics, Batavia, Illinois, 6-13 Sep 1972, published in eConf C720906V2 (1972) 135-165, also in Physics, Proceedings of the XVI International Conference on High Chicago 1972 p.135 (J. D. Jackson, A. Roberts, eds.), hep-ph/0208010 .

[17-1973] H. Fritzsch, Murray Gell-Mann and H. Leutwyler , 'Advantages of the Color Octet Gluon Picture', CALT-68-409, 1973, Phys.Lett.B47 (1973) 365-368 .

[18-1973(4)] D.J. Gross and F. Wilczek,
'Asymptotically free gauge theories. 1', Phys.Rev.D8 (1973) 3633-3652 ,
'Asymptotically free gauge theories. 2', Phys.Rev.D9 (1974) 980-993 .

[19-1973(4)] D. Politzer, ' Reliable Perturbative Results for Strong Interactions?', Phys. Rev. Lett. 30 (1973) 1346 ,
H. Georgi and D. Politzer, 'Electroproduction scaling in an asymptotically free theory of strong interactions', Phys. Rev. D9 (1974) 416 .

[20-1975] H. Kluberg-Stern and J. B. Zuber, 'Ward Identities and Some Clues to the Renormalization of Gauge Invariant Operators', Phys.Rev. D12 (1975) 467 ,
'Renormalization of Nonabelian Gauge Theories in a Background Field Gauge. 1. Green Functions', Phys.Rev. D12 (1975) 482 and
'Renormalization of Nonabelian Gauge Theories in a Background Field Gauge. 2. Gauge Invariant Operators', Phys.Rev.D12 (1975) 3159-3180 .

[21] C. Itzykson and J. B. Zuber, 'Quantum field theory', McGraw-Hill, New York 1980.

M. E. Peskin, D. V. Schroeder, 'An Introduction to quantum field theory', Addison-Wesley, Reading, MA 1995 .

J. Zinn-Justin, 'Quantum field theory and critical phenomena', The International Series of Monographs on Physics, Vol. 113 , Oxford University Press, Oxford 2002 .

[22-1982] G. Altarelli, 'Partons in quantum chromodynamics', NFN-ROME-783, Jul 1981. 211pp., Phys.Rept.81 (1982) .

[23-1975] H. Fritzsch and P. Minkowski, 'Psi Resonances, Gluons and the Zweig Rule', Caltech preprint CALT-68-492, Mar 1975. 61pp., Nuovo Cim.A30 (1975) 393 .

[24-1970] C. G. Callan, 'Broken scale invariance in scalar field theory', Phys.Rev.D2 (1970) 1541-1547 .

[25-1970] K. Symanzik, 'Small distance behavior in field theory and power counting', Commun.Math.Phys.18 (1970) 227-246 .

[26-1971(2)] G. 't Hooft, 'Renormalizable Lagrangians for Massive Yang-Mills Fields', Nucl.Phys.B35 (1971) 167-188 ,
G. 't Hooft and M.J.G. Veltman, 'Regularization and Renormalization of Gauge Fields', Nucl.Phys.B44 (1972) 189-213 .

[27-1969] I. B. Khriplovich, 'Greens functions in theories with a non-abelian gauge group' , Yad. Fiz. 10 (1969) 409 , engl. transl. Sov. J. Nucl. Phys. 10 (1970) 235 .

[28-1972] G. 't Hooft, Remarks on the ultraviolet stable sign of the beta function at the Colloquium on renormalization of Yang-Mills fields and applications to particle physics, Marseille 1972 .

[29-1988] S. G. Gorishnii , A. L. Kataev and S. A. Larin, 'Next-To-Leading $O(\alpha_s^3)$ QCD Correction to Σ_t ($e^+ e^- \rightarrow$ Hadrons) : Analytical Calculation and Estimation of the Parameter Lambda (MS)', JINR-E2-88-254, Apr 1988. 18pp., Phys.Lett.B212 (1988) 238-244 .

[30-1997] T. van Ritbergen, J.A.M. Vermaseren and S.A. Larin, 'The Four loop beta function in quantum chromodynamics', UM-TH-97-01, NIKHEF-97-001, Jan 1997. 9pp.,
Phys.Lett.B400 (1997) 379 , hep-ph/9701390 .

[31-2009] S. Bethke, 'The 2009 Wolrd Average of $\alpha_s(M_Z)$ ', arXiv:0908.1135 [hep-ph] .

[32-1969] K. Wilson, ' Nonlagrangian models of current algebra', Phys. Rev. 179 (1969) 1499 .

[33-1954] M. Gell-Mann and F. Low, 'Quantum electrodynamics at small distances', Phys. Rev. 95 (1954) 1300 .

[34-1973] S. Weinberg, 'New approach to the renormalization group', Phys. Rev. D8 (1973) 3497 .

[35-1982] O. V. Tarasov, Dubna report JINR P2-82-900 (1982) , unpublished .

[36-1994] K. G. Chetyrkin, C. A. Dominguez, D. Pirjol and K. Schilcher, 'Mass singularities in light quark correlators : the strange quark case', Phys.Rev.D51 (1995) 5090-5100 , hep-ph/9409371 .

[37-2006] P. A. Baikov, K. G. Chetyrkin and J. H. Kühn, Phys. Rev. Lett. 96 (2006) 012003 .
P. A. Baikov, K. G. Chetyrkin and J. H. Kühn, 'Massless propagators: Applications in QCD and QED', SFB-CPP-08-06, TTP08-03, Oct 2008. 7pp., Contributed to 8th International Symposium on Radiative Corrections (RADCOR 2007): Application of Quantum Field Theory to Phenomenology, Florence, Italy, 1-6 Oct 2007, PoS RADCOR2007 (2007) 023 , arXiv:0810.4048 [hep-ph] .

[38-2008] P. A. Baikov, K. G. Chetyrkin and J. H. Kühn, 'Hadronic Z- and tau-decays in order α_s^4 ', Phys.Rev.Lett.101 (2008) 012002 , arXiv:0801.1821v2 [hep-ph] .

[39-2008] C.A. Dominguez, N.F. Nasrallah, R.H. Röntsch and K. Schilcher, 'Up and down quark masses from Finite Energy QCD sum rules to five loops', UCT-TP-271-08, Jun 2008. 4pp. , Phys.Rev.D79 (2009) 014009 , arXiv:0806.0467 [hep-ph] .

[40-1979] M. A. Shifman , A. I. Vainshtein and V. I. Zakharov , 'QCD and resonance physics. Theoretical Foundations.' , Nucl. Phys. B147 (1979) 385 , 448.

[41-2001] J. Gasser and H. Leutwyler, Annals Phys. 158 (1984) 142,
H. Leutwyler, 'Nonlattice determinations of the light quark masses', Nucl.Phys.Proc.Suppl.94 (2001) 108, hep-ph/0011049.

[42-2008] B. Blossier et al., European Twisted Mass Collaboration, 'Light quark masses and pseudoscalar decay constants from N(f)=2 Lattice QCD with twisted mass fermions', DESY-07-148, FTUV-07-2709, IFIC-07-57, MS-TP-07-23, RM3-TH-07-11, ROM2F-2007-16, SFB-CPP-07-58, TUM-HEP-676-07, Sep 2007. 20pp., JHEP 0804 (2008) 020 , arXiv:0709.4574 [hep-lat].

[43-2008] C. Amsler et al. (Particle Data Group), Physics Letters B667 (2008) 1.

[44-2008] T. Becher and M. D. Schwartz, 'A Precise determination of alpha(s) from LEP thrust data using effective field theory', FERMILAB-PUB-08-048-T, Mar 2008, 34pp., JHEP 0807 (2008) 034 , arXiv:0803.0342 [hep-ph] .

[45-2011] S. Narison, 'Gluon Condensates and $m_b(m_b)$ from QCD-Exponential Moments at Higher Orders', May 2011. 4pp., arXiv:1105.5070 [hep-ph] .

[46-2011] S. Borsanyi, G. Endrodi, Z. Fodor, A. Jakovac, S. D. Katz, S. Krieg, C. Ratti, K. K. Szabo, 'The QCD equation of state with dynamical quarks', WUB-10-12, Jul 2010. 31pp., JHEP 1011 (2010) 077, arXiv:1007.2580 [hep-lat] .

Historical and textbook references to 'Continous transformation groups and differential geometry'

[1H-1893] S. Lie and G. Scheffers, 'Vorlesungen über continuierliche Gruppen, mit geometrischen und anderen Anwendungen', Leipzig 1893 .

[2H-1890] W. Killing, 'Die Zusammensetzung der stetigen endlichen Transformationsgruppen', Mathematische Annalen 31 (1888) 252-290; 33 (1888) 1-48; 34 (1889) 57-122; 36 (1890) 161-189 .

[3H-1894] É. Cartan, 'Sur la structure des groupes de transformations finis et continues', Thèse, Paris 1884 , IIème édition 1933 .

[4H-1897] F. Klein and A. Sommerfeld, 'Über die Theorie des Kreisels', 4 volumes, Teubner Verlag 1897 .

[5H-1903] L. Bianchi, 'Lezioni sulla teoria dei gruppi continui finiti di trasformazioni', Pisa 1903 , IIa edizione 1918 .

[6H-1927] F. Peter und H. Weyl, 'Die Vollständigkeit der primitiven Darstellungen einer geschlossenen kontinuierlichen Gruppe' , Ann. Math. 97 (1927) 737-755 .

[7H-1933] A. Haar, 'Der Massbegriff in der Theorie der kontinuierlichen Gruppen', Ann. Math. (2) 34 (1933) 147-169 .

[8H-1933] H. Weyl, 'The structure and representation of continous groups', I and II, Princeton 1933 and 1934-35.

[9H-1933] L.P. Eisenhart, 'Continuous groups of tranformations', Princeton 1933.

[10H-1941-1986] L. S. Pontryagin, 'Topological Groups', Gordon and Breach Sci. Publ., New York/London/Paris 1986 , first edition in russian : 1941 .

[11H-1951-1961] G. Racah, 'Group theory and spectroscopy', CERN Yellow report 61-8, 1961, URL : http://doc.cern.ch/cernrep/1961/1961-008/1961-008.html , reprinted from lectures dlivered (by the author) at the Institute for Advanced Study in Princeton, 1951 ; and references therein quoted below.

[12H-1953] G. W. Mackey, 'Induced Representations of Locally Compact Groups II. The Frobenius Reciprocity Theorem', The Annals of Mathematics, Second Series, Vol. 58, No. 2 (Sep., 1953) 193-221 .

[13H-1962] S. Helgason, 'Differential geometry and Symmetric Spaces', Academic Press New York and London 1962 .

[14H-1963] S. Kobayashi and K. Nomizu, 'Foundations of Differential Geometry', 2 volumes, John Wiley and Sons, Inc., Hoboken, NJ 1963, Wiley Classics Library 2009 .

End of historical and textbook references to
'Continous transformation groups and differential geometry'

Quark Masses in QCD

C. A. Dominguez

Centre for Theoretical and Mathematical Physics, University of Cape Town, Rondebosch 7700, South Africa

Abstract

Recent progress on QCD sum rule determinations of the light and heavy quark masses is reported. In the light quark sector a major breakthrough has been made recently in connection with the historical systematic uncertainties due to a lack of experimental information on the pseudoscalar resonance spectral functions. It is now possible to suppress this contribution to the 1% level by using suitable integration kernels in Finite Energy QCD sum rules. This allows to determine the up-, down-, and strange-quark masses with an unprecedented precision of some $8 - 10\,\%$. In the heavy quark sector, the availability of experimental data in the vector channel, and the use of suitable multipurpose integration kernels allows to increase the accuracy of the charm- and bottom-quarks masses to the 1% level.

1 Introduction

Quark and gluon confinement in Quantum Chromodynamics (QCD) precludes direct experimental measurements of the fundamental QCD parameters, i.e. the strong interaction coupling and the quark masses. Hence, in order to determine these parameters analytically one needs to relate them to experimentally measurable quantities. Alternatively, simulations of QCD on a lattice provide increasingly accurate numerical values for these parameters, but little if any insight into their origin. The first approach relies on the intimate relation between QCD Green functions, in particular their Operator Product Expansion (OPE) beyond perturbation theory, and their hadronic counterparts. This relation follows from Cauchy's theorem in the complex energy plane, and is known as the QCD sum rule technique [1]. In addition to producing numerical values for the QCD parameters, this method provides a detailed breakdown of the relative impact of the various dynamical contributions. For instance, the strong coupling at the scale of the τ-lepton mass essentially follows from the relation between the experimentally measured τ ratio, R_τ, and a contour integral involving the perturbative QCD (PQCD) expression of the $V + A$ correlator. This is the cleanest, most transparent, and model independent determination of the strong coupling [2]-[3]. It also allows to gauge the impact of each individual term in PQCD, up to the currently known five-loop order. Similarly, in the case of the quark masses one considers a

QCD correlation function which on the one hand involves the quark masses and other QCD parameters, and on the other hand it involves a measurable (hadronic) spectral function. Using Cauchy's theorem to relate both representations, the quark masses become a function of QCD parameters, e.g. the strong coupling, some vacuum condensates reflecting confinement, etc., and measurable hadronic parameters. The virtue of this approach is that it provides a breakdown of each contribution to the final value of the quark masses. More importantly, it allows to tune the relative weight of each of these contributions by introducing suitable integration kernels. This last feature has been used recently to solve the historical problem of the systematic uncertainties affecting light quark mass determinations, to be discussed in this report.

In the case of the light quark masses the ideal Green function is the pseudoscalar current correlator. This contains the square of the quark masses as an overall factor multiplying the PQCD expansion, and the leading power corrections in the OPE. Unfortunately, this correlator is not realistically accessible experimentally beyond the pseudoscalar meson pole. While the existence of at least two radial excitations of the pion and the kaon are known from hadronic interaction data, this information is hardly enough to reconstruct the full spectral functions. In spite of many attempts over the years to model them, there remains an unknown systematic uncertainty that has plagued light quark mass determinations from QCD sum rules. The use of the vector current correlator, for which there is plenty of experimental data from τ decays and e^+e^- annihilation, is not a realistic option for the light quarks as their masses enter as sub-leading terms in the OPE. The scalar correlator, involving the square of quark mass differences, at some stage offered some promise for determining the strange quark mass with reduced systematic uncertainties. This was due to the availability of data on $K - \pi$ phase shifts. Unfortunately, these data do not fully determine the hadronic spectral function. The latter can be reconstructed from phase shift data only after substantial theoretical manipulations, implying a large unknown systematic uncertainty. A breakthrough has been made recently by introducing an integration kernel in the contour integral in the complex energy plane. This allows to suppress substantially the unknown hadronic resonance contribution to the pseudoscalar current correlator. As it follows from Cauchy's theorem, this suppression implies that the quarks masses are determined essentially from the well known pseudoscalar meson pole and PQCD (well known up to five-loop level). In this way it has been possible to reduce the hadronic resonance contribution to the 1% level, allowing for an unprecedented accuracy of some $8 - 10\,\%$ in the values of the up-, down-, and strange-quark masses. Further improvement on this accuracy will be possible with further reduction of the uncertainty in the strong coupling, now the main source of error.

The determination of the charm- and bottom-quark masses has been free of systematic uncertainties due to the hadronic resonance sector, as there is plenty of experimental information in the vector channel from e^+e^- annihilation into hadrons. One problem, though, is that the massive vector current correlator is not known in PQCD to the same level as the light pseudoscalar correlation function. Nevertheless, substantial theoretical progress has been made over the years leading to extremely accurate charm- and bottom-quark masses. The novel idea of introducing suitable integration kernels in Cauchy's contour integrals, as described above, has also been used recently as a way of improving accuracy in the heavy-quark sector. For instance, kernels can be used to suppress regions where the data is either not as accurate, or simply unavailable. This will also be reported here.

The paper is organized as follows. First, determinations of quark-mass ratios from various hadronic data, as well as from chiral perturbation theory, will be reviewed in Section 2. These ratios are quite useful as consistency checks for results from QCD sum rules. Section 3 describes the OPE beyond perturbation theory, one of the two pillars of QCD sum rules. Section 4 discusses quark-hadron duality and finite energy sum rules. These sum rules weighted by suitable integration kernels will be analyzed in the light quark sector in Section 5. In particular it will be shown how this technique unveils the subjacent hadronic systematic uncertainty plaguing light quark mass determinations for the past thirty years. In Section 6 recent progress on charm- and bottom-quark mass determinations will be reported. Comparison with lattice QCD results for all quark masses will also be made. Finally, Section 7 provides a very short summary of this report.
As an important disclaimer, this paper is not a comprehensive review of past quark mass determinations from QCD sum rules. It is, rather, a report on recent progress on the subject. Given that past determinations of light quark masses were affected by unknown systematic uncertainties, essentially from the hadronic resonance sector, it makes little or no sense to review them once the main uncertainty has been exposed. Any agreement between values affected by this uncertainty and current results, free of it, would only be fortuitous. In fact, once this hadronic resonance uncertainty is removed, the values of all three light quark masses get reduced by some 15 - 20 %, a clear sign of a systematic uncertainty acting in only one direction. Last but not least, light quark masses from QCD sum rules before 2006 employed correlators up to at most four-loop level in PQCD, together with superseded values of the strong coupling.

2 Quark mass ratios

Quark masses actually precede QCD by a number of years, albeit under the guise of *current algebra quark masses*, which clearly lacked today's detailed understanding of quark-mass renormalization. In fact, the study of global SU(3) × SU(3) chiral symmetry realized á la Nambu-Goldstone, and its breaking down to SU(2) × SU(2), followed by a breaking down to SU(2), and finally to U(1) was first done using the strong interaction Hamiltonian [4]-[7]

$$H(x) = H_0(x) + \epsilon_0\, u_0(x) + \epsilon_3\, u_3(x) + \epsilon_8\, u_8(x)\,. \tag{1}$$

The term $H_0(x)$ above is SU(3) × SU(3) invariant, the $\epsilon_{0,3,8}$ are symmetry breaking parameters, and the scalar densities $u_{0,3,8}(x)$ transform according to the $3\bar{3} + \bar{3}3$ representation of SU(3) × SU(3). In modern language, ϵ_8 is related to the strange quark mass m_s, and ϵ_3 to the difference between the down- and the up-quark masses $m_d - m_u$, while the scalar densities are related to products of quark-anti-quark field operators. For instance, the ratio of SU(3) breaking to SU(2) breaking is given by

$$R \equiv \frac{m_s - m_{ud}}{m_d - m_u} = \frac{\sqrt{3}}{2}\,\frac{\epsilon_8}{\epsilon_3}\,, \tag{2}$$

where $m_{ud} \equiv (m_u + m_d)/2$. In the pre-QCD era many relations for quark-mass ratios were obtained from hadron mass ratios, as well as from other hadronic information, e.g.

$\eta \to 3\pi$, K_{l3} decay, etc. [7]. To mention a pioneering determination of the ratio R above, from a solution to the $\eta \to 3\pi$ puzzle proposed in [8] it followed [9] $R^{-1} = 0.020 \pm 0.002$, in remarkable agreement with a later determination based on baryon mass splittings [10] $R^{-1} = 0.021 \pm 0.003$. With the advent of chiral perturbation theory (CHPT) [6]-[7], [11]-[12], certain quark mass ratios turned out to be renormalization scale independent to leading order, and could be expressed in terms of pseudoscalar meson mass ratios [7],[11]-[13], e.g.

$$\frac{m_u}{m_d} = \frac{M_{K^+}^2 - M_{K^0}^2 + 2M_{\pi^0}^2 - M_{\pi^+}^2}{M_{K^0}^2 - M_{K^+}^2 + M_{\pi^+}^2} = 0.56 \tag{3}$$

$$\frac{m_s}{m_d} = \frac{M_{K^+}^2 + M_{K^0}^2 - M_{\pi^+}^2}{M_{K^0}^2 - M_{K^+}^2 + M_{\pi^+}^2} = 20.2\ , \tag{4}$$

where the numerical results follow after some subtle corrections due to electromagnetic self energies [12]. Beyond leading order in CHPT things become complicated. At next to leading order (NLO) the only parameter free relation is

$$Q^2 \equiv \frac{m_s^2 - m_{ud}^2}{m_d^2 - m_u^2} = \frac{M_K^2 - M_\pi^2}{M_{K^0}^2 - M_{K^+}^2}\,\frac{M_K^2}{M_\pi^2}\ . \tag{5}$$

Other quark mass ratios at NLO and beyond depend on the renormalization scale, as well as on some CHPT low energy constants which need to be determined independently [11]-[12]. After taking into account electromagnetic self energies, Eq.(5) gives [11] $Q = 24.3$, while a recent analysis of $\eta \to 3\pi$ [11], [14] gives

$$Q = 22.3 \pm 0.8\ . \tag{6}$$

The ratios R, Eq.(2), and Q, Eq.(5), together with the leading order ratios Eqs.(3)-(4), will prove useful for comparisons with QCD sum rule results. An additional useful quark mass ratio involving the ratios Eqs.(3)-(4) is

$$r_s \equiv \frac{m_s}{m_{ud}} = \frac{2\ m_s/m_d}{1 + m_u/m_d} = 28.1 \pm 1.3\ , \tag{7}$$

where the numerical value follows from the NLO CHPT relation [11], to be compared with the LO result from Eqs.(3)-(4), $r_s = 25.9$, and a large N_c estimate [15] $r_s = 26.6 \pm 1.6$.

3 Operator product expansion beyond perturbation theory

The OPE beyond perturbation theory in QCD, one of the two pillars of the sum rule technique, is an effective tool to introduce quark-gluon confinement dynamics. It is not a model, but rather a parametrization of quark and gluon propagator corrections due to confinement, done in a rigorous renormalizable quantum field theory framework. Let us consider a typical object in QCD in the form of the two-point function, or current correlator

$$\Pi(q^2) = i \int d^4x\, e^{iqx}\ < 0|\, T(J(x)\, J(0))\, |0 >, \tag{8}$$

where the local current $J(x)$ is built from the quark and gluon fields entering the QCD Lagrangian. Equivalently, this current can also be written in terms of hadronic fields with the same quantum numbers. A relation between the two representations follows from Cauchy's theorem in the complex energy (squared) plane. This is often referred to as quark-hadron duality, the second pillar of the QCD sum rules method to be discussed in the next section. The QCD correlator, Eq.(8), contains a perturbative piece (PQCD), and a non perturbative one mostly reflecting quark-gluon confinement. Since confinement has not been proven analytically in QCD, its effects can only be introduced effectively, e.g. by parameterizing quark and gluon propagator corrections in terms of vacuum condensates. This is done as follows. In the case of the quark propagator

$$S_F(p) = \frac{i}{\not p - m} \Longrightarrow \frac{i}{\not p - m + \Sigma(p^2)} \,, \tag{9}$$

the propagator correction $\Sigma(p^2)$ contains the information on confinement, a purely non-perturbative effect. One expects this correction to peak at and near the quark mass-shell, e.g. for $p \simeq 0$ in the case of light quarks. This effect is then parameterized in terms of the quark condensate $\langle 0|\bar{q}(0)q(0)|0\rangle$. Similarly, in the case of the gluon propagator

$$D_F(k) = \frac{i}{k^2} \Longrightarrow \frac{i}{k^2 + \Lambda(k^2)} \,, \tag{10}$$

the propagator correction $\Lambda(k^2)$ will peak at $k \simeq 0$, and the effect of confinement in this case can be parameterized by the gluon condensate $\langle 0|\alpha_s\, \vec{G}^{\mu\nu} \cdot \vec{G}_{\mu\nu}|0\rangle$. In addition to the quark and the gluon condensate there is a plethora of higher order condensates entering the OPE of the current correlator at short distances, i.e.

$$\Pi(q^2)|_{QCD} = C_0\, \hat{I} + \sum_{N=0} C_{2N+2}(q^2, \mu^2)\, \langle 0|\hat{O}_{2N+2}(\mu^2)|0\rangle \,, \tag{11}$$

where μ^2 is the renormalization scale, and where the Wilson coefficients in this expansion, $C_{2N+2}(q^2, \mu^2)$, depend on the Lorentz indixes and quantum numbers of $J(x)$ and of the local gauge invariant operators $\hat{O}_N$ built from the quark and gluon fields. These operators are ordered by increasing dimensionality and the Wilson coefficients, calculable in PQCD, fall off by corresponding powers of $-q^2$. In other words, this OPE achieves a factorization of short distance effects encapsulated in the Wilson coefficients, and long distance dynamics present in the vacuum condensates. Since there are no gauge invariant operators of dimension $d = 2$ involving the quark and gluon fields in QCD, it is normally assumed that the OPE starts at dimension $d = 4$. This is supported by results from QCD sum rule analyses of τ-lepton decay data, which show no evidence of $d = 2$ operators [16]-[17]. The unit operator $\hat{I}$ in Eq.(11) has dimension $d = 0$ and $C_0\hat{I}$ stands for the purely perturbative contribution. The Wilson coefficients as well as the vacuum condensates depend on the renormalization scale. For light quarks, and for the leading $d = 4$ terms in Eq.(11), the μ^2 dependence of the quark mass cancels the corresponding dependence of the quark condensate, so that this contribution is a renormalization group (RG) invariant. Similarly, the gluon condensate is also a RG invariant, hence once determined in some channel these condensates can be used throughout. The numerical values of the vacuum condensates cannot be calculated analytically from first principles as this would be tantamount to solving QCD exactly. One exception is that of the quark condensate which enters in the Gell-Mann-Oakes-Renner relation, a QCD low energy theorem following from the global chiral symmetry of the QCD Lagrangian [18]. Otherwise, it is possible to extract values for the leading vacuum condensates using QCD sum rules together with experimental data, e.g. e^+e^- annihilation into hadrons, and hadronic decays of the τ-lepton. Alternatively, as lattice QCD improves in accuracy it should become a valuable source of information on these condensates.

4 Quark-hadron duality and finite energy QCD sum rules

Turning to the hadronic sector, bound states and resonances appear in the complex energy (squared) plane (s-plane) as poles on the real axis, and singularities in the second Riemann sheet, respectively. All these singularities lead to a discontinuity across the positive real axis. Choosing a circular integration contour, and given that there are no other singularities in the complex s-plane, Cauchy's theorem leads to the finite energy sum rule (FESR)

$$\int_{\mathrm{sth}}^{s_0} ds\, \frac{1}{\pi}\, f(s)\, Im\,\Pi(s)|_{HAD} = -\frac{1}{2\pi i} \oint_{C(|s_0|)} ds\, f(s)\, \Pi(s)|_{QCD}\,, \tag{12}$$

where $f(s)$ is an arbitrary (analytic) function, s_{th} is the hadronic threshold, and the finite radius of the circle, s_0, is large enough for QCD and the OPE to be used on the circle. Physical observables determined from FESR should be independent of s_0. In practice, though, this is not exact, and there is usually a region of stability where observables are fairly independent of s_0, typically somewhere inside the range $s_0 \simeq 1 - 4\ \mathrm{GeV}^2$. Equation (12) is the mathematical statement of what is usually referred to as quark-hadron duality. Since QCD is not valid in the time-like region ($s \geq 0$), in principle there is a possibility of problems on the circle near the real axis (duality violations), an issue identified very early in [19] long before the present formulation of QCD sum rules. At present this is a contentious issue, so I shall not comment any further. The right hand side of this FESR involves the QCD correlator which is expressed in terms of the OPE as in Eq.(11). The left hand side involves the hadronic spectral function which is written as

$$Im\,\Pi(s)|_{HAD} = Im\,\Pi(s)|_{POLE} + Im\,\Pi(s)|_{RES}\,\theta(s_0 - s) + Im\,\Pi(s)|_{PQCD}\,\theta(s - s_0)\,, \tag{13}$$

where the ground state pole is followed by the resonances which merge smoothly into the hadronic continuum above some threshold s_0. This continuum is expected to be well represented by PQCD if s_0 is large enough. Hence, if one were to consider an integration contour in Eq.(12) extending to infinity, the cancellation between the hadronic continuum on the left hand side and the PQCD contribution on the right hand side, would render the sum rule a FESR. The integration in the complex s-plane of the QCD correlator is usually carried out in two different ways, Fixed Order Perturbation Theory (FOPT) and Contour Improved Perturbation Theory (CIPT). The first method treats running quark masses and the strong coupling as fixed at a given value of s_0. After integrating all logarithmic terms ($\ln(-s/\mu^2)$) the RG improvement is achieved by setting the renormalization scale to $\mu^2 = -s_0$. In CIPT the RG improvement is performed before integration, thus eliminating logarithmic terms, and the running quark masses and strong coupling are integrated around the circle. This requires solving numerically the RGE for the quark masses and the coupling at each point on the circle.

5 Light quark masses

Traditionally, the light quark masses have been determined using the correlator, Eq.(8), involving the pseudoscalar currents $J(x) \equiv \partial_\mu A^\mu(x)|^i_j = [\overline{m}_i(\mu) + \overline{m}_j(\mu)] : \overline{q}_j(x) i\gamma_5 q_i(x) :$, where $A_\mu(x)$ is the axial vector current of flavours i and j, $\overline{m}_i(\mu)$ the quark mass in the $\overline{MS}$ scheme, μ the renormalization scale and $q_i(x)$ are the quark fields. An issue of major concern in the past was the presence of logarithmic quark-mass singularities in these correlators. This problem has been satisfactorily resolved some time ago in [20]-[21]. These correlators are now known to five-loop order in PQCD [22], and free of logarithmic quark mass singularities. The Wilson coefficients of the leading power corrections, i.e. the gluon and the quark condensates, are also known up to two-loop level [20]-[21]. Higher dimensional condensates, as well as quark mass corrections of order $\mathcal{O}(m_i^4)$ (with respect to the one-loop term) and higher turn out to be negligible. From Cauchy's theorem, Eq.(12), the FESR to determine the quark masses can be written as

$$
\begin{aligned}
-\frac{1}{2\pi i}\oint_{C(|s_0|)} ds\, \psi_5^{QCD}(s)\, \Delta_5(s) &= 2\, f_P^2\, M_P^4\, \Delta_5(M_P^2) \\
&+ \int_{s_{th}}^{s_0} ds\, \frac{1}{\pi}\, Im\, \psi_5(s)|_{RES}\, \Delta_5(s)\,, \qquad (14)
\end{aligned}
$$

where $\Delta_5(s)$ is an (analytic) integration kernel to be introduced shortly, the first term on the right hand side is the pseudoscalar meson pole contribution ($P = \pi, K$), s_{th} is the hadronic threshold, and $Im\,\psi_5(s)|_{RES}$ is the hadronic resonance spectral function. The radius of integration s_0 is assumed to be large enough for QCD to be valid on the circle. For later convenience this FESR can be rewritten as

$$
\delta_5(s_0)|_{QCD} = \delta_5|_{POLE} + \delta_5(s_0)|_{RES}\,, \qquad (15)
$$

where the meaning of each term is self evident. Historically, the problem with the pseudoscalar correlator has been the lack of direct experimental information on the hadronic resonance spectral functions. Two radial excitations of the pion and of the kaon, with known masses and widths, have been observed in hadronic interactions [23]. However, this information is hardly enough to reconstruct the full spectral function. In fact, inelasticity, non-resonant background and resonance interference are impossible to guess, leaving no choice but to model these functions. This introduces an unknown systematic uncertainty which has been present in all past QCD sum rule determinations of the light quark masses. Since the FESR Eq.(14) is valid for any analytic $\Delta_5(s)$ one can choose this kernel in such a way as to suppress $\delta_5(s_0)|_{RES}$ as much as possible. An example of such a function is the second degree polynomial [24]-[27]

$$\Delta_5(s)|_{RES} = 1 - a_0\, s - a_1\, s^2 \,, \tag{16}$$

where a_0 and a_1 are constants fixed by the requirement $\Delta_5(M_1^2) = \Delta_5(M_2^2) = 0$, where $M_{1,2}$ are the masses of the first two radial excitations of the pion or kaon. This simple kernel suppresses enormously the resonance contribution, which becomes only a couple of a percent of the pole contribution, and well below the current uncertainty due to the strong coupling. This welcome feature is essentially independent of the model chosen to parametrize the resonances. A practical parametrization consists of two Breit-Wigner forms normalized at threshold according to chiral perturbation theory, as first proposed in [28] for the pionic channel, and in [29] for the kaonic channel. Detailed results for $\delta_5(s_0)|_{QCD}$, to five-loop order in PQCD and up to dimension $d = 4$ in the OPE, after integrating in FOPT may be found in [25]. In the case of CIPT the FESR must be written in terms of the second derivative of the current correlator. This is in order to eliminate the unphysical first degree polynomial present in $\psi_5(s)$, which unlike the case of FOPT would otherwise contribute to the FESR which then becomes

$$-\frac{1}{2\pi i}\oint_{C(|s_0|)} ds\;\; \psi_5^{''QCD}(s)\,[F(s) - F(s_0)] = 2\, f_P^2\, M_P^4\, \Delta_5(M_P^2)$$

$$+\;\frac{1}{\pi}\int_{s_{th}}^{s_0} ds\, Im\, \psi_5(s)|_{RES}\, \Delta_5(s)\,, \tag{17}$$

where

$$F(s) = -s\left(s_0 - a_0\,\frac{s_0^2}{2} - a_1\,\frac{s_0^3}{3}\right) + \frac{s^2}{2} - a_0\,\frac{s^3}{6} - a_1\,\frac{s^4}{12}\,, \tag{18}$$

and

$$F(s_0) = -\frac{s_0^2}{2} + a_0\,\frac{s_0^3}{3} + a_1\,\frac{s_0^4}{4}\,. \tag{19}$$

The RG improvement is used before integration, so that all logarithmic terms vanish. The running coupling as well as the running quark masses are no longer frozen as in FOPT, but must be integrated. This can be done by solving numerically the respective RG equations at each point on the integration circle in the complex s-plane. Detailed expressions are given in [25]-[26].

The parameters of the integration kernel, Eq.(16), are $a_0 = 0.897\ \text{GeV}^{-2}$, and $a_1 = -\,0.1806\ \text{GeV}^{-4}$ for the pionic channel, and $a_0 = 0.768\ \text{GeV}^{-2}$, and $a_1 = -\,0.140\ \text{GeV}^{-4}$ for the kaonic channel. These values correspond to the radial excitations $\pi(1300)$, $\pi(1800)$, $K(1460)$ and $K(1830)$. The pion and kaon decay constants are [23] $f_\pi = 92.21 \pm 0.14$ MeV, and $f_K = (1.22 \pm 0.01) f_\pi$. In the QCD sector it is best to use the value of the strong coupling determined at the scale of the τ-mass, as this is close to the scale in current use for the light quark masses, i.e. $\mu = 2$ GeV. The extraction of $\alpha_s(M_\tau)$ from the R_τ ratio involves an integral with a natural kinematical integration kernel that eliminates the contribution of the $d = 4$ term in the OPE. This welcome feature improves the accuracy of the determination, and it makes little sense to introduce additional spurious integration kernels which would artificially recover this $d = 4$ contribution. The different values obtained from τ

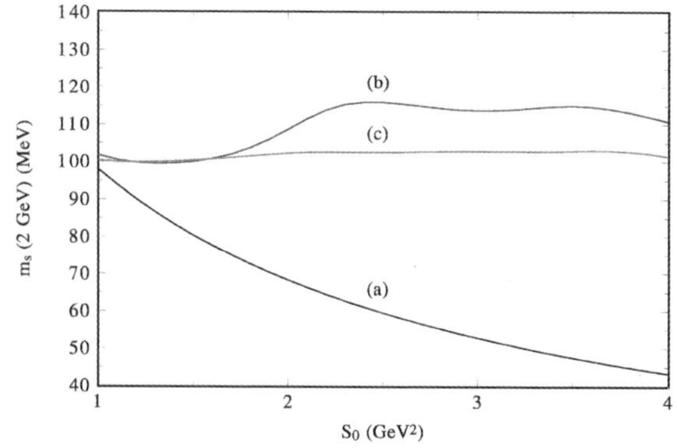

Figure 1: The strange quark mass $\overline{m}_s(2 \text{ GeV})$ in the $\overline{MS}$ scheme taking into account only the kaon pole with $\Delta_5(s) = 1$ (curve(a)), and the two Breit-Wigner resonance spectral function with a threshold constraint from CHPT [29], with $\Delta_5(s) = 1$ (curve(b)), and $\Delta_5(s)$ as in Eq.(16) (curve(c)). A systematic uncertainty of some 20 % due to the resonance sector is dramatically unveiled.

decay using CIPT are all in agreement with each other, i.e. $\alpha_s(M_\tau) = 0.338 \pm 0.012$ [3], $\alpha_s(M_\tau) = 0.341 \pm 0.008$ [30], $\alpha_s(M_\tau) = 0.344 \pm 0.009$ [31], and $\alpha_s(M_\tau) = 0.332 \pm 0.016$ [32]. These determinations are model independent and extremely transparent, with α_s obtained essentially by confronting PQCD with the single experimental number R_τ. The method of FOPT is known to give rise to a pathological non-convergent perturbative series in this application [2], so it will not be considered here. The $d = 4$ gluon condensate has been extracted from τ decays [17], but one can conservatively consider the wide range$< \alpha_s G^2 >= 0.01 - 0.12$ GeV^4. The impact of the light quark condensate is at the level of 1% in the quark masses. A $\pm$ 30% uncertainty in the resonance contribution $\delta_5(s_0)|_{RES}$ in Eq.(15) translates into a safe 1% change in the quark masses. Finally, it has been assumed that the unknown six-loop PQCD contribution is equal to the five-loop result, an extreme but very conservative estimate of higher orders in PQCD.

Beginning with the strange quark mass, Fig. 1 shows the results for $\overline{m}_s(2 \text{ GeV})|_{\overline{MS}}$ with no integration kernel, $\Delta_5(s) = 1$, and taking into account only the kaon pole, curve (a), and the kaon pole plus a two Breit-Wigner resonance model with a threshold constraint from CHPT [29], curve (b) (a misprint in the formula for the spectral function in [29] has been corrected in [35]). These curves are for the central value of $\alpha_s(M_\tau)$ whose uncertainties will be considered afterwards. The latter result is reasonably stable in the wide region $s_0 = 2 - 4$ GeV^2, so that it could lead us to conclude that $\overline{m}_s(2 \text{ GeV})|_{\overline{MS}} \simeq 100 - 120$ MeV, albeit with a yet unknown systematic uncertainty arising from the resonance sector. Introducing the kernel, Eq.(16), leads to curve (c) and to a dramatic unveiling of this systematic uncertainty. In fact, the *real* value of the quark mass is $\overline{m}_s(2 \text{ GeV})|_{\overline{MS}} = 102 \pm 8$ MeV, or some 20% below the former result (this error now includes the uncertainty in α_s). In

Source	$\overline{m}_u$	$\overline{m}_d$	$\overline{m}_s$	$\overline{m}_{ud}$	$\overline{m}_u/\overline{m}_d$	$\overline{m}_s/\overline{m}_{ud}$	R	Q
QCDSR [26]	2.6 ± 0.3	5.6 ± 0.4	-	4.1 ± 0.3	0.46 ± 0.06	-	-	-
QCDSR [25]	-	-	102 ± 8	-	-	24.9 ± 2.7	33 ± 6	21 ± 3
FLAG [12]	2.2 ± 0.3	4.6 ± 0.6	95 ± 10	3.4 ± 0.4	0.47 ± 0.04	27.8 ± 1.0	37.2 ± 4.1	23.1 ± 1.5

Table 1: The running quark masses in the $\overline{MS}$ scheme at a scale $\mu = 2$ GeV in units of MeV from QCD sum rules (QCDSR) (first two rows), and from the FLAG lattice QCD analysis [12]. The ratios $\overline{m}_u/\overline{m}_d$ and $\overline{m}_s/\overline{m}_{ud}$ are an input in the QCDSR (see text). The ratios R and Q are defined on the l.h.s. of Eqs.(2) and (3). The QCDSR results for the up- and down- quark masses and ratios differ slightly from [26]-[27] due to the input value of $\overline{m}_u/\overline{m}_d$ used here.

addition, and as a bonus the systematic uncertainty-free result is remarkably stable in the unusually wide region $s_0 \simeq 1 - 4$ GeV^2 (typical stability regions are only half as wide). It must be recalled that the pseudoscalar correlator involves the overal factor $(m_s + m_{ud})^2$. Hence, in order to determine m_s an input value for the ratio m_s/m_{ud} is needed in the result from the sum rule, which is

$$\overline{m}_s(2 \text{ GeV})|_{\overline{MS}} = \frac{105.5 \pm 8.2 \text{ MeV}}{1 + m_{ud}/m_s} \,. \tag{20}$$

Using the wide range $m_s/m_{ud} = 24 - 29$ leads to $\overline{m}_s(2 \text{ GeV})|_{\overline{MS}} = 102 \pm 8$ MeV. In this case the impact of the uncertainty in the quark mass ratio is small. However, in the case of the up- and down-quark masses the corresponding ratio m_u/m_d plays a more important role in the result from the sum rule, which is

$$\overline{m}_d(2 \text{ GeV})|_{\overline{MS}} = \frac{8.2 \pm 0.6 \text{ MeV}}{1 + m_u/m_d} \,. \tag{21}$$

The input used in Eq.(21) for the ratio $m_u/m_d = 0.47 \pm 0.04$ is from the overall lattice QCD analysis of the FLAG group [12]. Once m_d is determined from Eq.(21), m_u follows. Using these results for the individual masses one obtains the ratios m_u/m_d and m_s/m_{ud} shown in Table 1. The quark masses $\overline{m}_u(2 \text{ GeV})|_{\overline{MS}}$ and $\overline{m}_d(2 \text{ GeV})|_{\overline{MS}}$ also exhibit a remarkably wide stability region $s_0 \simeq 1-4$ GeV^2 [26]-[27]. In Table 1 one finds a summary of the results for the light quark masses, and the ratios R and Q defined on the left hand side of Eqs. (2) and (5), together with the results of the Flag group [12]. The values of the up- and down-quark masses and their ratios in Table 1 are slightly different from those in [26]-[27] due to the input value for the ratio m_u/m_d (in [26]-[27] the value $m_u/m_d = 0.55$ was used). The various sources of errors in the quark masses discussed earlier combine into the final values given in Table 1. Having all but eliminated the systematic uncertainty from the hadronic resonance sector, the main source of error is now due to the strong coupling. Improved accuracy in the determination of α_s would then allow for a reduction of the uncertainties in the light quark masses.

6 Heavy quark masses

Determinations of the charm- and bottom-quark masses are not affected by a lack of data, as there is plenty of experimental information from e^+e^- annihilation into hadrons [23], except for a gap in the region 25 GeV$^2 \lesssim s \lesssim 50$ GeV^2 . On the theoretical side there has been very good progress on PQCD up to four-loop level [33]-[51]. The leading power correction in the OPE is due to the gluon condensate with its Wilson coefficient known at the two-loop level [52]. The correlator, Eq.(8), involves the vector current $J(x) \equiv V_\mu(x) = \bar{Q}(x)\gamma_\mu Q(x)$, where $Q(x)$ is the charm- or bottom-quark field. The experimental data is in the form of the R_Q-ratio for charm (bottom) production, which determines the hadronic spectral function. Modern determinations of the heavy-quark masses have been based on inverse moment (Hilbert-type) QCD sum rules, e.g. Eq.(12) with $f(s) = 1/s^n$. These sum rules require QCD knowledge of the vector correlator in the low energy region, around the open charm (bottom) threshold, as well as in the high energy region. A recent update [53] of earlier determinations [33]-[36], [38]-[40] reports a charm-quark mass in the $\overline{MS}$ scheme accurate to 1%, and half this uncertainty for the bottom-quark mass. However, the analysis of [51] claims an error a factor two larger for the charm-quark mass. It appears that the discrepancy arises from the treatment of PQCD. In fact, in [51] two different renormalization scales were used, one for the strong coupling and another one for the quark mass. This unconventional choice results in a much larger error in the charm-quark mass obtained from inverse (Hilbert) moment QCD sum rules. It does not affect, though, sum rules involving positive powers of s. In any case, the philosophy in current use is to choose the result from the method leading to the smallest uncertainty.
Beginning with the charm-quark mass, an alternative procedure was proposed some years ago based only on the high energy expansion of the heavy-quark vector correlator [54]-[55]. This method was followed recently [56], but with updated PQCD information and the inclusion of integration kernels in the FESR, Eq.(12), tuned to enhance/suppress contributions from data in certain regions. The first such kernel is the so-called *pinched* kernel [57]-[59]

$$f(s) = 1 - \frac{s}{s_0}\,, \tag{22}$$

which is supposed to suppress potential duality violations close to the real s-axis in the complex s-plane. In connection with the charm-quark mass application, this kernel enhances the contribution from the first two narrow resonances, J/ψ and $\psi(2S)$, and reduces the weight of the broad resonance region, particularly near the onset of the continuum. The latter feature is better achieved with the alternative kernel [60]

$$f(s) = 1 - \left(\frac{s_0}{s}\right)^2\,, \tag{23}$$

which produces an obvious larger enhancement of the narrow resonances, and a larger quenching of the broad resonance region. This kernel has been used together with both the high and the low energy expansion of the vector correlator in [60]. Since there are no data in the region 25 GeV$^2 \lesssim s \lesssim 50$ GeV^2, while the data for $s \gtrsim 50$ GeV^2 agrees with PQCD, it is useful to introduce a kernel that will allow a suppression in the former region, e.g.

$$f(s) = \mathcal{P}_n[x(s)]\,, \tag{24}$$

where

$$x(s) = \frac{2s - (s_0 + s_1)}{s_0 - s_1}\ , \tag{25}$$

with $s_0 > s_1$, and $\mathcal{P}_n(x)$ are the standard Legendre polynomials, i.e. $\mathcal{P}_1(x) = x$, $\mathcal{P}_2(x) = (5x^3 - 3x)/2$, etc., which satisfy the constraint

$$\int_{s_1}^{s_0} s^k\ \mathcal{P}_n[x(s)]\ ds = 0\ , \tag{26}$$

where $s_1 \simeq 24$ GeV^2, and s_0 varies in the region where there is no data. These Legendre-type kernels provide extra weight to the well known resonance region on account of their rapid growth for $s < s_1$. The resulting charm-quark masses are essentially insensitive to the order of these polynomials, with $n = 3 - 6$ giving answers differing at the 0.1% level. The merits of each kernel may be judged by its ability to minimize the uncertainties, in particular those that might be most affected by systematic errors, such as e.g. the experimental data. The kernel $f(s) = 1 - (s_0/s)^2$ appears to be optimal as it produces the smallest uncertainty due to the data, and is very stable against changes in s_0. Some recent determinations of the charm-quark mass are based on Hilbert moments with no s_0-dependent kernel. While there is no explicit s_0-dependence in Hilbert moments (the integrals extend to infinity), there is definitely a residual dependence when choosing the threshold for the onset of PQCD. From a FESR perspective, the major drawback of the kernel $1/s^2$ is clearly the poor stability against changes in s_0. In this regard, the Legendre polynomial kernels would be optimal, except that they have the largest uncertainty due to the data. An important remark is in order concerning the uncertainty due to changes in s_0. From current data it is not totally clear where does PQCD start. This problem not only affects FESR, with their explicitly obvious s_0-dependence, but also Hilbert moments with an implicit s_0-dependence, as there is no data all the way up to infinity. Two of the most recent results for $\overline{m}_c(3\,\mathrm{GeV})$ [51], [53], together with the weighted FESR value [60] are

$$\overline{m}_c(3\,\mathrm{GeV}) \;=\; \begin{cases} 986\ \pm 13\ \mathrm{MeV}\ [53] \\ 998\ \pm 29\ \mathrm{MeV}\ [51] \\ 987\ \pm\ \ 9\ \mathrm{MeV}\ [60]\ , \end{cases} \tag{27}$$

in very good agreement with each other, except for the errors. The small uncertainty from [60] is due in part to improved quenching of the data in the broad resonance region, but mostly due to a strong reduction in the sensitivity to s_0, i.e. the onset of PQCD. For comparison, a recent lattice QCD determination gives [61]

$$\overline{m}_c(3\,\mathrm{GeV}) \;=\; 986\ \pm\ 6\ \mathrm{MeV}\ . \tag{28}$$

Turning to the bottom-quark mass, one of the most recent determination from Hilbert moment QCD sum rules [53] is $\overline{m}_b(10\,\mathrm{GeV}) = 3610 \pm 16$ MeV. The error is due to uncertainties in the data, the strong coupling and the renormalization scale. It does not include, though, an uncertainty due to the onset of PQCD. There is an appreciable difference if PQCD were to start at the end point of the (BABAR) data ($\sqrt{s_0} \simeq 11.2$ GeV), or rather at higher energies ($\sqrt{s_0} \simeq 13.0$ GeV). If one were to adopt the procedure of [56], i.e. add in quadrature the first three errors, but only linearly the one due to s_0, then using the information in [53] the result would change to $\overline{m}_b(10 \text{ GeV}) = 3600 \pm 47$ MeV. Presumably, the error would be larger if a broader region in s_0 were to be considered. One of the virtues of weighted sum rules is to reduce this uncertainty, as may appreciated from the results for the charm-quark mass. Following a similar strategy as for the charm-quark mass, and considering inverse moments, as well as Legendre polynomial kernels, the following result was obtained [62]

$$\overline{m}_b(10\,\text{GeV}) = 3623 \pm 9\,\text{MeV}\,, \tag{29}$$

which corresponds to $\overline{m}_b(\overline{m}_b) = 4171 \pm 9$ MeV. These values are fully consistent with the lattice QCD result [61] $\overline{m}_b(10\,\text{GeV}) = 3617 \pm 25$ MeV.

7 Acknowledgements

The author wishes to thank Professor Antonino Zichichi for the invitation to the School, and his collaborators in the various projects on quark masses reported here: S. Bodenstein, J. Bordes, N. Nasrallah, J. Peñarrocha, R. Röntsch and K. Schilcher. Enlightening correspondence with H. Leutwyler and P. Minkowski is greatly appreciated. This work was supported in part by NRF (South Africa), and the Alexander von Humboldt Foundation (Germany).

References

[1] P. Colangelo and A. Khodjamirian, in: "At the Frontier of Particle Physics/ Handbook of QCD"', M. A. Shifman, ed. (World Scientific, Singapore 2001), Vol. 3, 1495.

[2] A. Pich, Acta Phys. Polon. Supp. **3**, 165 (2010).

[3] A. Pich, Nucl. Phys. B (Proc. Suppl.) **218**, 89 (2011).

[4] S. Glashow and S. Weinberg, Phys. Rev. Lett. **20**, 224 (1968).

[5] M. Gell-Mann, R. Oakes and B. Renner, Phys. Rev. **175**, 2195 (1968).

[6] H. Pagels, Phys. Rep. C **16**, 219 (1975).

[7] J. Gasser and H. Leutwyler, Phys. Rep. C **87**, 77 (1982).

[8] C. A. Dominguez and A. Zepeda, Phys. Rev. D **18**, 884 (1978).

[9] C. A. Dominguez, Phys. Lett. B **86**, 171 (1979).

[10] P. Minkowski and A. Zepeda, Nucl. Phys. B **164**, 25 (1980).

[11] H. Leutwyler, PoS CD **09**, 005 (2009).

[12] G. Colangelo *et al.*, arXiv:1011.4408.

[13] S. Weinberg, Trans. New York Acad. Sci. **38**, 185 (1977).

[14] G. Colangelo, S. Lanz, and E. Passemar, PoS CD **09**, 047 (2009).

[15] H. Leutwyler, Nucl. Phys. B Proc. Suppl. **64**, 223 (1998).

[16] C. A. Dominguez and K. Schilcher, Phys. Rev. D **61**, 114020 (2000).

[17] C. A. Dominguez and K. Schilcher, J. High Energy Phys. **0701**, 093 (2007).

[18] J. Bordes, C. A. Dominguez, P. Moodley, J. Peñarrocha, and K. Schilcher, J. High Energy Phys. **1005**, 064 (2010).

[19] R. Shankar, Phys. Rev. D **15**, 755 (1977).

[20] K. G. Chetyrkin, C. A. Dominguez, D. Pirjol and K. Schilcher, Phys. Rev. D **51**, 5090 (1995).

[21] M. Jamin and M. Münz, Z. Physik, C **66**, 633 (1995).

[22] P. A. Baikov, K. G. Chetyrkin and J. H. Kühn, Phys. Rev. Lett. **96**, 012003 (2006).

[23] K. Nakamura *et al.*, Particle Data Group, J. Phys. G **37**, 075021 (2010).

[24] C. A. Dominguez, N. F. Nasrallah, and K. Schilcher, J. High Energy Phys. **0802**, 072 (2008).

[25] C. A. Dominguez, N. F. Nasrallah, R. Röntsch and K. Schilcher, J. High Energy Phys. **0805**, 020 (2008).

[26] C. A. Dominguez, N. F. Nasrallah, R. Röntsch and and K. Schilcher, Phys. Rev. D **79**, 014009 (2009).

[27] C. A. Dominguez, Int. J. Mod. Phys. A **29**, 5223 (2010); Mod. Phys. Lett. A **26**, 691 (2011).

[28] C. A. Dominguez, Z. Phys. C **26**, 269 (1984).

[29] C. A. Dominguez, L. Pirovano and K. Schilcher, Phys. Lett. B **425**, 193 (1998).

[30] G. Cvetic *et al.*, Phys. Rev. D **82**, 093007 (2010).

[31] M. Davier *et al.*, Eur. Phys. J. C **56**, 305 (2006).

[32] P. A. Baikov, K. G. Chetyrkin and J. H. Kühn, Phys. Rev. Lett. **101**, 012002 (2008).

[33] K. G. Chetyrkin, R. Harlander, J. H. Kühn, and M. Steinhauser, Nucl. Phys. B **503**, 339 (1997).

[34] P. A. Baikov, K. G. Chetyrkin, and J. H. Kühn, Nucl. Phys. B (Proc. Suppl.) **189**, 49 (2009).

[35] C. A. Dominguez, A. Ramlakan and K. Schilcher, Phys. Lett. B **511**, 59 (1998).

[36] K. G. Chetyrkin, R. Harlander, J. H. Kühn, Nucl. Phys. B **586**, 56 (2000).

[37] Y. Kiyo, A. Maier, P. Maierhöfer, and P. Marquard, Nucl. Phys. B **823**, 269 (2009).

[38] P. A. Baikov, K. G. Chetyrkin, and J. H. Kühn, Phys. Rev. Lett. **101**, 012002 (2008).

[39] P. A. Baikov, K. G. Chetyrkin, and J. H. Kühn, Nucl. Phys. B (Proc. Suppl.) **135**, 243 (2004).

[40] K. G. Chetyrkin, J. H. Kühn, and M. Steinhauser, Phys. Lett. B **371**, 93 (1996).

[41] K. G. Chetyrkin, J. H. Kühn, and M. Steinhauser, Nucl. Phys. B **482**, 213 (1996)

[42] K. G. Chetyrkin, J. H. Kühn, and M. Steinhauser, Nucl. Phys. B **505**, 40 (1997).

[43] R. Boughezal, M. Czakon, and T. Schutzmeier, Phys. Rev. D **74**, 074006 (2006)

[44] R. Boughezal, M. Czakon, and T. Schutzmeier, Nucl. Phys. B (Proc. Suppl.) **160**, 164 (2006).

[45] A. Maier, P. Maieröfer, and P. Marquard, Nucl. Phys. B **797**, 218 (2008)

[46] A. Maier, P. Maieröfer, and P. Marquard, Phys. Lett. B **669**, 88 (2008).

[47] K. G. Chetyrkin, J. H. Kühn, and C. Sturm, Eur. Phys. J. C **48**, 107 (2006).

[48] A. Maier, P. Maierhöfer, P. Marquard, and A. V. Smirnov, Nucl. Phys. B **824**, 1 (2010).

[49] A. H. Hoang, V. Mateu, and S. Mohammad Zebarjad, Nucl. Phys. B **813**, 349 (2009).

[50] J. Hoff and M. Steinhauser, Nucl. Phys. B **849**, 610 (2011).

[51] B. Dehnadi, A. H. Hoang, V. Mateu and S. Mohammad Zebarjad, arXiv: 1102.2264.

[52] D. J. Broadhurst *et al.* Phys. Lett. B **329**, 103 (1994).

[53] K. G. Chetyrkin, *et al.*, Phys. Rev. D **80**, 074010 (2009); Theor. Math. Phys. **170**, 217 (2012).

[54] J. Peñarrocha, and K. Schilcher, Phys. Lett. B **515**, 291 (2001).

[55] J. Bordes, J. Peñarrocha, and K. Schilcher, ibid. B **562**, 81 (2003).

[56] S. Bodenstein, J. Bordes, C. A. Dominguez, J. Peñarrocha, and K. Schilcher, Phys. Rev. D **82**, 114013 (2010).

[57] K. Maltman, Phys. Lett. B **440**, 367 (1998).

[58] C. A. Dominguez and K. Schilcher, Physics Letters B **448**, 93 (1999).

[59] C. A. Dominguez and K. Schilcher, Physics Letters B **581**, 193 (2004).

[60] S. Bodenstein, J. Bordes, C. A. Dominguez, J. Peñarrocha, and K. Schilcher, Phys. Rev. D **83**, 074014 (2011).

[61] C. McNeile, C. T. H. Davies, E. Follana,. K. Hornbostel and G. P. Lepage, Phys. Rev. D **82**, 034512 (2010).

[62] S. Bodenstein, J. Bordes, C. A. Dominguez, J. Peñarrocha, and K. Schilcher, Phys. Rev. D **85**, 034003 (2012).

The Quark Model and QCD

Frank Close*
Rudolf Peierls Centre for Theoretical Physics, University of Oxford,
1 Keble Road, Oxford, OX1 3NP, UK

(Dated: Version 1.0 - 1 October 2012)

The early history of the quark model, and the nature of constituent quarks are discussed from a personal perspective.

I. INTRODUCTION

Forty three years ago I was a student at Erice. I was working at my doctorate on the constituent quark model, under the guidance of Dick Dalitz. At that school, along with Morpurgo and Thirring who had discussed constituent quarks, we heard Sid Drell talk about the new phenomenon of Bjorken scaling in deep inelastic scattering, which was the first hint of the "parton" substructure being revealed in the proton. Today we recognise that as the beginning of the acceptance of quarks, leading to the emergence of QCD and the standard model.

What is also remarkable in my opinion is that four decades on, the constituent quark model remains a successful phenomenological description of hadrons, in the relatively low energy regime, and yet has no firm foundation in the fundamental Lagrangian of QCD. So my talk today is as more a statement of my personal ignorance than an answer to the puzzle of why the constituent model works so well, let alone what is the precise relation between constituent quarks and those revealed in inelastic scattering.

The emergence of quarks has some similarities with the discovery of substructure at other periods in science. However, in those previous cases, the new layer was uniquely identified. Only at the quark layer have two superficially identical, but fundamentally different, constituents emerged. It is the background to trying to understand this duality that is the motivation for this talk.

In the 19th century, the first clues that there is a deeper layer of structure within atoms came from the discovery of patterns among the atomic elements: Mendeleev's periodic table. This was accompanied by spectra and followed by Rutherford's direct observation of the atomic nucleus and Thomson's discovery of the electron. In the middle of the 20th century an analogous set of circumstances led to the revelation of the quark layer of reality. Strongly interacting particles hadrons were found to be built from quarks, the inter-quark forces being transmitted by gluons.

The analogue of Mendeleev's table was the recognition of patterns known as the Eightfold Way - among families of hadrons[1]. These arise from the various ways that the underlying quarks, which form those hadrons, combine [2, 3]. In the late 1960s the presence of quarks within hadrons was directly confirmed by experiments scattering high energy beams of electrons and neutrinos from targets of protons and neutrons [4]. Discovery of a spectroscopy of hadrons, analogous to the spectra of atomic elements, completed the parallel.

Today the quark layer of reality is established. However, the quarks revealed by spectroscopy are not easily identified with those that are manifested by the high energy deep inelastic scattering experiments. The latter have almost no mass, and as such have the attributes of the basic entities in the fundamental Lagrangian of QCD, the relativistic quantum field theory of the strong force. Their light masses, which are presumed to arise courtesy of spontaneous symmetry breaking via the Higgs mechanism, break chiral symmetry, which gives the pion its mass and the strong nuclear force a finite range. The constituent quarks were described non relativistically, implicitly assumed by Dalitz, Morpurgo and others initially to have large masses and deep binding [5–7]. What are these constituent quarks?

II. NON-RELATIVISTIC CONSTITUENT QUARKS

The non-relativistic constituent quark model works empirically even though there is no well-defined fundamental theoretical rationale for the approach in general. For massive flavors, such as bottom and charm, the model has

*E-mail: f.close1@physics.ox.ac.uk

some legitimacy. Today, the established spectroscopies of charmonium and bottomonium are well described by non-relativistic models, based on analogy with well established ideas rooted in atomic and nuclear physics. Sets of orbitally and radially excited states are known, and radiative transitions among them fit well with the classic electric and magnetic multipole transitions calculated using techniques that have been long established in those other disciplines. Applying these ideas to lighter flavours has no obvious justification, and yet empirically succeeds. The most dramatic example, known since about 1970, is the photoproduction of baryon resonances, where over 100 amplitudes were well described, both in relative magnitudes and phases, a feat which no other approach then or since has matched [8, 9].

Trying to understand precisely what features of this model were being tested, was what interested me as a student and young postdoc. The success of this unfounded non relativistic model was, in many cases at least, revealed to be due to its underlying structure rather than detailed dynamics[10]. My belief, today, is that it is these rather general features that are the cause of its success, and although they are realised in explicit non-relativistic models, it is their generality rather than specific, and unjustified, non-relativistic dynamics that is being revealed.

A hadron has total spin **J**, made up of quarks with total spin **S** and the left over defined by **L**. In a specific non relativistic model, **L** was identified with the orbital angular momentum. However, as far as the mathematical algebra was concerned it is not necessary to make this further assumption. This reduces the amount of predictions, but makes the survivors stronger. In particular relations among amplitudes survive, and have proved empirically successful. Therefore it was clear very early in the history of the quark model that its success was due to more general features, driven by the couplings of spins, rather than detailed and ill-founded dynamics.

To illustrate this, I shall look at a specific example: the photoproduction amplitudes of the prominent resonance $D_{13}(1520)$. From proton targets a specific amplitude - when the combined helicity of photon and proton are anti-aligned to total $\frac{1}{2}$ rather than aligned to $\frac{3}{2}$ - empirically vanishes. This was described in the model as a cancellation between electric and magnetic multipoles for this transition [8, 9]. Initially the parameters of the model, such as quark effective mass and the strength of the excitation in the (assumed) harmonic oscillator potential, were fixed to enforce this vanishing[8]. The resulting values then turned out to give an excellent description of a host of other amplitudes, hence the interest in the model for Feynman[9] who showed that these parameters arise rather naturally.

I noticed that although the models of refs [8] and [9] appeared rather different in detail, their successes shared a common algebraic pattern. The details of this are explained over several pages in my textbook [10]. They are due to a common algebraic structure in the models, ultimately arising from the general assumptions outlined above - namely, that a hadron has total spin **J**, made up of quarks with total spin **S** and the left over defined by **L**, and that a single quark interacts with the photon. The transversely polarised photon can flip the projection of the quark's spin or "orbital" momentum while the remaining spectators play no direct role.

If this was the answer, there was no evidence for non-relativistic constituent quarks in these data. Instead the correlations revealed some algebraic pattern whose dynamical origin remained (and remains!) to be identified. However, Fred Gilman and I then realized that, while the data could be accounted for by this general mathematical algebra, without specific recourse to non-relativistic models, a change from photo to electroproduction would provide a sensitive dynamic test [11]. Real quarks would respond more strongly to the magnetic rather than to the electric multipoles as the photon became more virtual; in an algebraic approach there was no reason for such an effect.

Empirically this was verified, and remarkably so [10]. The change to magnetic dominance took place almost immediately as the photon became virtual. The speed of this change was in line with the non-relativistic model. Why the model worked so well was a mystery, and remains so. However, this result proved that constituent quarks are real dynamical degrees of freedom within hadrons and are more than simply book-keeping devices for the mathematical algebra. The challenge was to make a dynamical model of light quarks, which move relativistically, and yet have interactions with photons (at least) that mimic algebraically those of non-relativistic constituents.

III. BAG MODELS: RELATIVISTIC QUARKS; GLUEBALLS AND HYBRIDS.

A resolution to the dichotomy came with the Bag models, notably that of MIT [12]. Quarks with the characteristics of QCD, possibly even massless, were confined in a simple model where, in effect, an infinitely high potential imprisoned them within a femto-universe whose extent is of the order of 1 fm. Applying boundary conditions on the fermion wavefunction to nullify current at the boundary gives the fermion (quark) an energy of around 350MeV. This kinetic energy of the lightweight quark could, by the mass-energy equivalence, be identified as an effective mass, namely the mass of the constituent quark. The phenomenology of this is discussed in detail in [10]

The model had considerable successes in justifying the dynamics of quark models, at least for the ground state hadrons, but was never developed enough to deal with the excited states that Gilman and I had investigated. However, it had another implication: massless gluons, similarly confined, would themselves gain an energy. This enabled a tentative spectroscopy of glueballs, containing two or more gluons, and also of hybrids, where a gluon accompanies quarks and/or antiquarks inside the confinement bag. In the 1980s this stimulated considerable interest in the spectroscopy and dynamics of these novel hadrons, which are consequences of QCD, and in finding experimental evidence of them.

This simple picture of confined gluons seems remarkably robust in comparison with modern understanding of the spectrum. The equations of QCD, when studied in computer simulations, which assume space-time to consist of a discrete lattice rather than a continuum (so called Lattice QCD), reveal a spectrum of states with remarkable parallels to those of the simple bag model of massless quarks and gluons. Naive though the models are, they clearly carry some general features of nature.

In 1983 Barnes and I computed the spectrum of the lightest hybrid mesons and baryons, including the perturbations of energy levels (hadron masses) arising from single gluon exchanges among the constituents [13, 14] For the lightest hybrid mesons this implied that the lightest such state would be $J^{PC} = 0^{-+}$, the next being the exotic 1^{-+}, with 1^{--} and 2^{-+} heavier and completing the group. This is what modern lattice QCD finds.

We also studied the spectrum of hybrid baryons. There is a rich set of N* and Delta excitations with positive parity and spins ranging from $\frac{1}{2}$ to $\frac{5}{2}$. The pattern here too seems to be confirmed in recent studies of lattice QCD by a group at Jefferson Lab[15]. The pattern in the bag model is driven by the three-gluon coupling, where a gluon is exchanged between a gluon and a quark. This gives a hyperfine spin-dependent mass shift. The message therefore may be that the effects of this three-gluon perturbation remain robust during the detailed confinement, whose dynamics are "known" to the lattice computer.

The insights from lattice QCD stimulated the creation of more sophisticated models where quarks are held to one another by a linear tube of coloured flux. These so called flux-tube models [16] have been vigorously pursued, for both spectorscopy and decays[17], and their implications for the photoproduction of hybrid mesons investigated[18]. While they contain features of hadrons, at least if lattice QCD is a guide, it still surprises me that the naive bag models remain so robust empirically. The details of the recent lattice calculations seem to agree with the spectrum of the bag models remarkably well.

The message seems to be that the simple bag picture agrees with the pattern of the real world (lattice QCD being the nearest to this that we can simulate at present.) The fact that quarks and gluons act as quasi-free particles until they hit the wall of confinement is a remarkably simple, and empirically successful, description of hadrons. Nature has turned out to be simple enough for us to decode its patterns. Obtaining a robust theoretical description though remains a challenge.

IV. GLUONIC HADRONS: A NEW PICTURE.

One thing is missing from all of the above: the structure of a hadron depends on the resolving power of the microscope with which it is examined. In a nutshell: the larger the resolution, or in effect, the shorter the wavelength of the probe, so the more structure is revealed. What appears as a quark at low resolution is revealed as a seed surrounded by virtual quarks and antiquarks, and gluons, at high resolution. This poses a challenge: what would a glueball look like as you power up the microscope?

Scattering electrons from a state of pure glue at zero momentum transfer would not occur, as the electron needs a source of electric charge to couple to. However, as the momentum transferred grows, the resolution improves and the target is increasingly perceived to contain quarks and antiquarks, from which the electron can scatter. The probability of scattering thereby grows.

The quantitative measure is the amount of momentum carried by electrically charged constituents technically this is given by the integral of the inelastic structure function, $F_2(x, q^2)$, where x is the Bjorken variable and q^2 is the square of the four-momentum transfer [10]. The integral will grow with q^2. As q^2 tends to ∞ , the integral tends to an asymptotic value, whose value depends on the number of flavours. Physically, this extreme is where the hadron is perceived as a plasma of quarks, antiquarks and gluons in equilibrium.

This behaviour contrasts what occurs for a hadron such as a proton, where there is electric charge available even as q^2 is small. In this case the integral of $F_2(x, q^2)$ falls to its asymptotic value.

Thus we have a well-defined way of classifying hadrons as gluonic or flavoured: the two classes correspond to the said integral rising or falling to the asymptotic limit. In other words, at small q^2 the hadrons momentum is dominantly carried by gluons or flavoured fermions.

This applies to hadrons made of the constituents as defined by the Lagrangian of QCD and revealed in deep inelastic experiments: lightweight fermions and gluons. There is an interesting question of counting of states relative to classification in the old constituent picture. In the latter, mesons can be glueball, hybrid or conventional. This involves three classes, whereas the novel approach only admits two: gluonic or fermionic. This is a question that I am currently investigating, together with that of decays, and how the concept of multiquark hadrons fits in with this scheme.

The fundamental issue is whether there are "unbalanced" flavors within the hadron, which can never be removed by annihilation into gluons. For example, a $u\bar{d}$ will carry electric charge at any q^2; $ud\bar{s}\bar{c}$ is also indestructible or "exotic", unless it decays into a pair of color singlet hadrons. By contrast, $u\bar{u} + d\bar{d} + s\bar{s}$ etc. in flavor singlet can appears as gluons at some different value of q^2. Thus a "glueball" is in effect a hybrid with flavor singlet quantum numbers. They are the same; it is the value of q^2 that resolves the state. The naming is thus terminology and not physics. Even if we ignored mixing of quark-gluon states at the level of color singlet hadrons, to which they couple virtually even below decay threshold,there remains the fundamental issue of the scale of resolution at which the hadron is observed.

Thus arguments about whether glueballs exist or not and whether this or that state is an example have something of the angels on heads of pins about them. At the very least, one must specify the q^2 scale. Looking for scalar glueballs is subject to a further problem, namely the coupling to color singlet hadrons. This can occur in S-wave and as such provides a long-range "cloak" that obscures the underlying "parton" structure. All is not lost however. The number of scalar states, with I=0, empirically exceeds what one expect from the simple constituent quark model. Below 1.7GeV one finds scalars that fit exactly with a nonet of $qq\bar{q}\bar{q}$ in an S-wave; a nonet of $q\bar{q}$ in a P-wave and an extra isoscalar. This is consistent with a lattice glueball mixing with nearby states. Quantifying the amount of the mixing of the glueball with these other states remains empirically unsettled, and is in any event a q^2 dependent quantity. That there is an excess of isoscalars, in line with the presence of the predicted extra state arising from the presence of glue is, I believe, established, though whether you call it a glueball or isoscalar hybrid is a matter of words more than physics. Devising more experimental tests to shed further light on this question seems to me to be a worthwhile strategy for hadron spectroscopy and dynamics. However, I doubt that it will answer the question that started all this: what are constituent quarks?

Acknowledgements

I am indebted to my collaborators in quark models over several decades for their inspiration and tolerance. I am grateful to Nino Zichichi and the Ettore Majorana Centre for enabling me to return and lecture some four decades after I was here as a student.

---

[1] Gell-Mann M., Symmetries of Baryons and Mesons, Physical Review 125, 1067-1084
[2] Gell-Mann M., A Schematic Model of Baryons and Mesons, Physics Letters 8,
[3] Zweig G., CERN report TH 402 (1964) unpublished
[4] Panofsky W., Proceedings of 14th International Conference on High Energy Physics, Vienna, 1968 (ed J Prentki and J Steinberger), (CERN)
[5] Dalitz R.H., in High Energy Physics (eds. C. de Witt and M.Jacob) (Gordon and Breach, New York)
[6] Morpurgo G., Physics 2, 95
[7] Morpurgo G., Proceedings of Erice school on Subnuclear Physics (ed A.Zichichi) (Academic Press, New York and London)
[8] Copley L.A., G. Karl and E.Obryk, Single Pion Photoproduction in the Quark Model, Nuclear Physics B13, 303-312
[9] Feynman R., M Kislinger and F Ravndal, Current Matrix Elements from a Relativistic Quark Model, Physical Review D3, 2706-2732
[10] Close F.E. Introduction to Quarks and Partons (Academic Press; New York and London)
[11] Close F.E. and F.J.Gilman, Helicity Structure of Nucleon Resonance Electroproduction and the Symmetric Quark Model, Physics Letters 38B, 541-543
[12] Jaffe R.L. Proceedings of Erice School on Subnuclear Physics (ed A.Zichichi) (Academic Press, New York and London)
[13] Barnes T., F.E.Close and F.de Viron, Hybrid mesons in the MIT Bag Model, Nuclear Physics B224, 241-264
[14] Barnes T. and F.E.Close, Where are Hermaphrodite Baryons? Physics Letters 123B, 89-92
[15] Dudek J. and R. Edwards, Hybrid Baryons in QCD http://arxiv.org/pdf/1201.2349
[16] Isgur N. and Paton J., Phys. Rev. D31, 2910 (1985)
[17] F. E. Close and P. R. Page, Phys. Lett. B**578**, 199 (2004). [arXiv:hep-ph/0309253].
[18] F. E. Close and J. J. Dudek, Phys. Rev. Lett. **91**, 142001 (2003). [arXiv:hep-ph/0304243]; Phys.Rev **D 69** 034010 (2004).

KEY STEPS TOWARD THE CREATION OF QCD —
NOTES ON THE LOGIC AND HISTORY OF THE GENESIS OF QCD

Tian Yu Cao
Boston University

The creation of QCD is one of greatest achievements in the history of science. It has radically changed our conception of the fundamental ontology of the physical world and its underlying dynamics. What it has discovered are more than new particles and a new force, but rather a deeper level of physical reality, a new kind of entities. Dynamically, the strong nuclear forces are no longer taken to be fundamental, but are relegated to the status of the un-cancelled residue of far stronger long range force mediated by gluon. From a long term perspective, perhaps more important than these discoveries is that it breaks a new path to explore many novel features of an unknown layer of the physical world, such as instanton , the theta vacuum and effective energy. The notion of effective energy allows us to conceptualize the mysterious quantum number flow and hadronization. Although the metaphysical status of effective energy is not quite clear at this moment, but a window is open for conceptualizing the physical world that is in complex processes of transmutations in a unified way.

A naive logic of the genesis of QCD

In retrospect, we know clearly that QCD is a non-abelian gauge theory of quarks and gluons (the gauge bosons). Naively we would conceive that the genesis of QCD had two components: the introduction or construction or discovery of quarks, and that of SU(3) colour gauge symmetry. Quarks were introduced and explored by George Zweig (1964), Yoichiro Nambu (1965), Richard Dalitz (1966) and many others since 1964; the SU(3) colour gauge symmetry was introduced explicitly by Nambu in 1965. Thus Harry J. Lipkin, a prominent theoretical physicist, declared at the 1992 SLAC conference, *The Third International Symposium on the History of Particle Physics*, that

> ***The Nobel prize for QCD as a description of strong interactions might have been awarded to Sakharov, Zeldovich, and Nambu. They had it all figured out in 1966: the Balmer formula, the Bohr atom, and the Schrodinger equation of strong interactions. All subsequent developments leading to QCD were just mathematics and public relations with no new physics.*** (Lipkin, 1997)

What about Murray Gell-Mann who was the first physicist to suggest the idea of quarks in (1964a) and, together with Harald Fritzsch, first formulated a viable colour octet gluon picture in 1972? In terms of SU(3) colour gauge symmetry, Nambu could easily claim his priority over Gell-Mann and Fritzsch. In terms of quarks, however, the situation is much more complicated. Gell-Mann had been notorious in his frequent and consistent claim that fractionally charged quarks are mathematical and fictitious device; they could not be real in the sense that they could never be detected in isolation in laboratory. This was in sharp contrast with Nambu's conception of quarks as integrally charged and thus observable and real. Gell-Mann's rejection of the reality of quarks was based on two reasons. First, nobody had any idea of forces that would hold quarks together to be hadrons, and without gluing forces, the very notion of quarks as the constituents of hadrons would make no sense at all.

Second, constrained by the idea of bootstrap, the fractionally charged quarks would have to be hidden forever, never be observable in isolation in laboratory. As late as 1971, Gell-Mann and Fritzsch still claimed that "if quarks are real, then we cannot assign them para-Fermi statistics of rank 3, since that is said to violate the factoring of the S-matrix for distant subsystems"; without assigning such a para-statistics to quarks, quark theorists would be in deep trouble, facing a direct conflict with numerous empirical facts, from pion → two gamma decay to the ratio of total cross sections for muon and hadron production. The later success of the idea of quark and the universal acceptance of quarks' reality have given ground to many physicists, historians and philosophers of physics to severely criticise Gell-Mann for his being a mathematical instrumentalist, and to give credit to quark realists for their discovery of quarks.

Scientific discovery, however, is a very complicated and difficult-to-settle issue, the discovery of quarks included. In terms of physical entities, Priestley is not credited as the discoverer of oxygen although he observed many of the crucial effects of oxygen which he labelled as "deflogisticated air"; but J. J. Thomson is still regarded as the discoverer of the electron even though his conception of electron has been radically revised by Rutherford and quantum physicists. What is the criterion for justifying such a different recognition? The criterion, in my view, is that the core of the conception of the hypothetical entity must contain the **identifying features** of the entity that are going to be discovered later by experiments. If the core remains the same through radical changes in identifying the entity, then the initial discovery would be respected, but not otherwise.

Now how about quarks? Experimental discoveries of quarks in fact were just the experimental confirmation of the previous theoretical discoveries. Then who discovered quarks theoretically? Quarks are now conceived to be the ingredients of hadrons with a holistic set of structural properties, such as fractionally charged, coloured and coupled with coloured vector gluons, which always cohere together serving to identify something to be a quark. Judging from such a criterion, the conceptions of quarks by Gell-Mann and Zweig 1964, and by Nambu and many others afterwards, may be credited more or less as a step toward the discovery of quarks as real physical entities, but not the discovery itself. We will come back to this interesting and important issue at the end of this talk.

Context for the creation of QCD

Apparently, the Nobel committee has not listened to Lipkin. But it is also clear that it still has difficulties in appreciating the crucial steps leading to the creation of QCD. However, it is understandable if we take into account the rich structure of scientific developments leading to the birth of QCD, which happened in the context in which conflicting ideas, confusions and false ideas were abundant.

Theoretically, the matrix from which QCD was born had three components: S-matrix theory, current algebra, and constituent quark model.

The irrevocable failure of the meson theory and Lev Laudau's severe criticism against QFT that it was standing on the false conviction of micro-spacetime and micro-causality which were in fact inaccessible to human cognition, led to the rise of S-matrix theory, in the form of dispersion relations, Regge trajectories, bootstrap scheme or nuclear democracy, initiated by Gell-Mann and Goldberger and then led by Geoffrey Chew (cf. sections 8.2 and 8.5 of Cao, 1997, in which relevant references to the works mentioned in this paragraph can be found). Motivated by avoiding infinity that tarnished QFT, bootstrappers stick to empirically accessible asymptotic states, which are outside the region of interaction. This move from fundamental inquiry about what is going on in hadron interactions to phenomenological one was guided by general principles (such as causality and unitarity) in understanding the scattering processes, and the interactions are conceptualized in terms of the poles in scattering amplitudes, using dispersion relations as Cauchy-Riemann formulae to express an S-matrix element in terms of its singularities. They claimed that all hadrons as poles of the S-matrix lie on Regge trajectories, which are characteristic for compositeness. A corollary was that all hadrons enjoy the equal status as composites, there is no end to division and no elementary particles in the atomistic sense at the hadron level, let alone at a deeper, un-accessible level. Thus SMT can be characterized, most importantly, by its rejection of elementarity.

The move from fundamental to phenomenological had another direction, guided by the idea of symmetry. The colliders with higher and higher energies supported lavishly by the US government after the World War II led to the proliferation of new particles and resonances. The recognition of their properties gave clues to their underlying structures, which were soon to be encoded in the idea of a hierarchy of symmetries: each kind of interaction is characterized by one or more symmetries from the hierarchy and is mediated by its own vector bosons. (Gell-Mann and Pais, 1954)

The symmetry approach had two versions: the composite models and current algebra. The most prominent among composite models were the Sakata model (1956) and the SU(6) quark model (Gursey and Radicati, 1964; Pais, 1964; Sakita, 1964). It took a global symmetry to characterize the structure of the hadron system, explored it with respect to hadron's static properties. In speculating, nonrelativistically, the material basis and underlying mechanism for the observed behaviour pattern (symmetry) of hadrons, composite models enjoyed some successes in spectrum and selection rules. The successful SU(6) prediction of -3/2 for the ratio of the proton and neutron magnetic moments was taken to be striking evidence for the compositeness, since only a composite model gave a simple ratio for total moments. Yet lack of knowledge of how the composites are dynamically formed betrayed its speculative nature.

Current algebra (Gell-Mann, 1964b), in contrast, was the result of abstracting from underlying relativistic field theory models. Hadronic currents derived from underlying models for conceptualizing interactions were taken to be the basic entities in hadronic processes, with their unknown constitution bracketed. This signalled a subtle shift in the ontological commitment from hadrons to currents, and thus a door was open for the exploration of experimentally accessible external and internal constraints on the currents, which would be followed by another ontological shift back to hadrons from currents, as we will see shortly. The basic assumption was that hadronic currents are the representations of a global symmetry of strong interactions, suggested by the successes of CVC and PCAC, and thus obey the algebraic relations derived from the symmetry group, which are closed under the equal time commutation relations. This conjectured symmetry constraint allowed exact algebraic relations among measurable quantities to be derived from the global symmetry of the underlying field theory. But the door was not closed to the exploration of dynamic symmetry of the underlying field theoretical model, although some stimulus from data and motivation from theoretical developments would be needed, as we will see shortly.

In order to give a fundamental explanation of the supposed symmetry, Gell-Mann in 1964 proposed the idea of quarks. The proposal opened a door for investigating current quarks that are constrained by current algebra, and thus broke a new path for further fruitful explorations. Note that Gell-Mann's current quarks were radically different from the constituent quarks in that they were the bare excitations of the quark field, they were fictitious because there was no force to hold them together to be a hadron. Most importantly, they were not hadron's ingredients independently existing, but were assumed to be enmeshed in a network of relations, merely as placeholders in currents (as structures) and satisfy all the structural constraints posed by current algebra.

The conflict between the two symmetry approaches in their underlying assumptions concerning the nature of quarks was clear, but not as radical as that between them both with a built-in notion of elementarity on the one hand and bootstrap which rejects any version of elementarity on the other, as we will see shortly. But more disturbing in any attempt to understand and appreciate what were crucial in the conceptual developments leading to the birth of QCD were some confusions and some false beliefs. Downright false beliefs, seen with the benefit of hindsight, about the charges and asymptotic states of quarks and gluons, about massive gluons mediating colour interactions that cause large mass splitting between different triplets of quarks (Han and Nambu, 1965), or about the flavour gauge invariance at the hadron level (Pauli, 1953; Yang and Mills, 1954; Sakurai, 1960; Veltman, 1966; and Bell, 1967), have been dismissed and forgotten now. But some fundamental confusions concerning the ontological basis of gauge principle, or the validity and reality of the quark models with or without QCD as their underpinning, remain widely held even now. Thus the Noble committee is not to be blamed in its hesitation to award its prestigious prize to one of the greatest achievements in the 20th century physics, the creation of QCD.

Key steps toward the creation of QCD

Physicists had no epistemic access to hypothetical entities and their dynamics for explaining phenomena, except for some hints from the constraints posed by existing theories and data. In the case of QCD, theories as outlined above had conflicting assumptions; while data, such as hadron spectrum, Regge trajectories, pion-two gamma decay rate or the ratio R of total cross sections for hadron and muon production in the electron positron annihilation, make sense only when they are interpreted by theory. Thus how to proceed from these hints, the meaning of which were up to interpretation, through conceptual syntheses, to a formulation of the conceptual foundation for a new fundamental theory, was a formidable challenge to theoreticians; and the conceptual developments, happened in the battle field of conflicting ideas, leading to the genesis of QCD and its early justifications were extremely complicated and difficult to grasp.

To make the task of judging which steps were crucial easier, I suggest that the criterion should not be that a step should have original ideas without predecessors. Originality might be the case for current algebra or renormalization group analysis based on scale anomaly, but certainly is not the case for sum rules or gauge invariance. Rather, the criterion should be the historical effectiveness of a step. That is, a step is crucial only when it has provided a new vision, broken a new direction for explorations, started a new research program, and thus effectively moved the scientific explorations ahead.

With such a criterion in mind, at the risk of oversimplification, I would suggest that the journey toward the creation of QCD started from current algebra proposed in 1962. After three breakthroughs: local current algebra sum rules, scaling and partons, and light cone current algebra, it ended with a great synthesis which, announced in 1972, was in fact the first articulation of QCD.[1]

Current algebra

As we outlined above, it was a general framework for exploring the hadron system. Initially, it focused on exploring the structure of the hadron system by posing algebraic constraints on the external relations among hadrons in hadronic processes. But it had the potential for exploring the inside of hadrons, which potential was realized in the works of Adler and Bjorken. Without the structuralist approach moving from external to internal constraints initiated by current algebra, developments along the paths of bootstrapping among hadrons without going deeper, or along the paths of directly speculating on the inside ingredients without any means of dynamically exploring the inside of hadrons, would never lead to the genesis of QCD.

Local current algebra sum rules

Applying the Fubini-Furlan method (1965; using infinite momentum frame and a complete set of intermediate states) to the axial-vector currents, together with the assumption of PCAC, Stephen Adler (1965a) and Willain Weisberger (1965) derived their sum rule for expressing the axial form factor in terms of the pion-proton scattering total cross sections, and thus delivered the first big success to current algebra. By dropping PCAC, Adler (1965b) was able to derive a new sum rule in which the pion-proton scattering was replaced by the high energy neutrino-nucleon inelastic reactions.

Mobilized by Gell-Mann, Stephen Adler (1965c, 1966) took a further step to generalize his result, in which the charge algebra sum rules tested in forward inelastic neutrino reactions, to non-forward neutrino reactions and to derive from local current algebra a sum rule involving the elastic and the inelastic form factors measurable in high energy neutrino reactions.

In such semi-leptonic processes, form factors, as scalar function of momentum transfer which is a measure of the ability to probe the structure of the nucleon, indicate the modification of weak interaction by all unknown effects caused by virtual strong interactions. Traditionally, form factors were investigated in terms of the pole structure of the S-matrix, which indicates the possible contributions of virtual particles to them. In fact, Nambu (1957) in his mass-spectral representation of nucleon's form factor took form factors as a sum in terms of the masses of the intermediate states. But in Adler's sum rules, all effects from unknown intermediate states were bracketed. This move helped him to circumvent all the difficulties caused by the implicit introduction of the built-in unobservable structures that was inherited from its underlying local field theoretical framework.

Adler's local current algebra sum rules created a new situation, opened a new direction for investigating the new behaviour pattern of form factors, which soon led to a new picture of nucleons. It was more restrictive than his charge algebra rules. It brought the localized behaviour of hadron scattering neutrinos to the forefront, and thus opened a window for peeking into the inside of hadron. One may conceive the process as lepton locally probing

---

[1] A more detailed account of these key steps than what is given below, including all relevant references, can be found in Cao (2010).

hadrons with its localized momentum transfer. But more properly it should be conceived as the hadronic current density locally probing hadron with its own localized momentum transferred to it from the leptonic current coupled to it, either directly as conceived then or indirectly through intermediate bosons. Once the focus was shifted to hadronic current and its constitution which was unknown then, a place in the hadronic current was created for current quarks to take. Such a move, manifested in Bjorken's scaling, solved the saturation problem for the sum rules. With the benefit of using leptons (and electrons and photons later), with which physicists had more knowledge about them and their interactions than with hadrons and strong interactions, Adler, and later Bjorken, can probe form factors more successfully than Nambu, who relied solely on the pole structure of the S-matrix. The forthcoming SLAC experiments showed that it is the contributions from the ingredients of hadrons (partons or current quarks), rather than resonances, that make overwhelming contribution to form factors.

Adler's work on local current algebra sum rules was resolutely rejected by Geoffrey Chew, the leader of the nuclear democracy movement. At the 14th Solvay conference held in Brussels in October 1967, Chew rejected whole project of current algebra and its sum rules because "it follows from a completeness relation, but without an elementary substructure, particle or field, no meaning is possible for completeness." Yet according to the strict bootstrap point of view, there is no end to dividing, and thus there can be no fundamental substructure. Chew acknowledged that we can "have on mass shell completeness, but there is no complete set of states which could give meaning to the sum rules." For example, one can define what is fundamental substructure in terms of pion's Compton wavelength "for the nonrelativistic physics of hadrons." But once quarks and gluons reveal their own internal structure, which is conceivable in the future, it would be unclear how to define the sum rule in that novel situation. (Chew, 1967)

The path-breaking role of Adler's work, however, was duly recognized by other active researchers. For example, at the 1992 SLAC conference on the history of particle physics, James Bjorken commented on Adler's work: "His work provided a most important basis for what was to follow when the ideas were applied to electroproduction." (Bjorken, 1997)

Scaling and partons

Bjorken's inequality (1966) for electro-production, obtained by adapting Adler's sum rule and performing an isotopic rotation, changed the landscape of high energy physics because it brought Gell-Mann—Adler line of work to close contact with feasible experiments at SLAC. More importantly, because it gave him enough background knowledge, accumulated at SLAC over decades, to fruitfully explore saturation mechanism for sum rules. That is, he could appeal to the historical analogy with nonrelativistic quantum mechanics, in which the elastic scattering from atoms or nuclei measures the time averaged charge distribution, while inelastic scattering with coarse energy resolution measures the instantaneous distribution of charge, and use the insight thus gained to construct models for saturation of Adler's sum rule and his own inequalities, in terms quasi elastic scattering between photon and point-like constituents of hadron. The result was the famous scaling (Bjorken, 1969).

Scaling was a turning point. As a bridge connecting the esoteric current algebra–sum rule project with experimental investigations of the pattern of deep inelastic scattering, scaling reoriented the phenomenological exploration of external relations between observable quantities toward the clarifications of hadron's constituents and their dynamics that underlie the observed hadronic phenomena. The expectations of experimenters were no longer restricted to the improvement of the scope and precision of their measurements of the phenomena that was shaped by the structureless picture of nucleon, and the concerns of theoreticians were no longer restricted to the verification of their predictions, derived from their model-independent esoteric schemes of current algebra or analytic s-matrix theory.

Bjorken's point-like constituent was soon to be named by Feynman and Bjorken himself as parton. As fundamental bare particles of an unspecified underlying field theory of the strong interactions, partons had no mass in the infinite momentum frame. Constrained by scaling, partons were also assumed to have no interaction to each other. Thus partons' behaviour was supposed to be described by a scale invariant free field theory, which, as we will see shortly, posed a severe constraint on theory construction for years to come. But for a while, it was quite puzzling for physicists to calculate strong interactions with a free field theory. In addition, although each parton was supposed to carry a fraction of the total momentum of the proton, the momentum distribution function of partons, which was supposed to contain all the effects of strong interactions, was not calculable from any dynamic scheme, but had to be empirically adjusted. Lack of a proper dynamical theory also forced parton theorists to keep silent on parton's hadronization, thus could not claim to have offered a complete picture for hadron physics in terms of parton's behaviour.

All these shortcomings notwithstanding, the achievement of the scaling-parton project was great. It irreversibly consolidated the constituent picture of the hadrons. More concretely, massless partons were soon to be identified with current quarks. Detailed parton model calculations also indicated that there were half non-quark partons in proton, and thus had brought the notion of gluon back to the picture of hadrons, which played no role in the constituent quark model after Gell-Mann first introduced it in 1962, and would take the central place in the subsequent conceptual development.

From a historical perspective, one may say that this line of work, by moving from exploring external structural constraints to internal ones, had revealed the potential of current algebra in exploring the structure of hadrons in terms of their constituents. With the structural knowledge about the ingredients accumulated, such as point-likeness, quasi-free and half-half proportion for quarks and gluons, the way was paved for identifying what they are and what a dynamic theory should be for describing their behaviour. Without what was achieved by this line of work, no further development in this direction would be conceivable.

Light cone current algebra

Both the free field theory assumed by the parton model and the scale invariance entailed by the model were seriously questioned and challenged within the framework of renormalized perturbation theory, once the gluons were back and their interactions with quarks had to be taken into account. The challenge took the form of anomaly and scaling violations, which were caused by renormalization effects when the interaction was not invariant to the symmetry of the theory. Scale anomaly changed the scale of current, and thus destroyed scale invariance as well as the canonical equal time commutators. The foundation of current algebra was collapsing. Then the whole situation was reconceptualised in terms of operator product expansions and the scaling law version of renormalization equations, on the basis of a new idea of broken scale invariance. One of the most significant outcomes of these developments was the rise of a new scheme of light cone current algebra.

As a response to the failure of current algebra, formulated in terms of equal time commutators, in dealing with the short distance behaviour of currents in strong interactions, such as scaling violation, K. G. Wilson (1969) suggested a new framework, in which operator product expansions (OPE) for product of two or more local fields or currents near the same point, as an extension of current algebra, was supposed to exist and contain singular functions when the operators were defined at the same point or one defined on the light cone through the other. The importance of light cone field theories was in fact recognized earlier than the appearance of OPE scheme, by Heinrich Leutwyler (1968) and Leonard Susskind (1968). They noticed that the intuitive meaning of the infinite momentum limit adopted by Adler and others in their current algebra sum rules is just to replace the equal time charges by the corresponding charges contained in a light like surface. Since the equal time commutation rules for the charges in the infinite momentum limit are equivalent to the same form for the light like generators, all the results obtained in the infinite momentum frame can be expressed in terms of the behaviour of the commutators of light like charges in coordinate space, which are integrated over a light-like plane instead of an equal time plane.

The light cone current algebra intensively explored by Fritzsch and Murray Gell-Mann at the beginning of 1970s was a algebraic system obtained by abstracting the leading singularity in the commutator of two currents on the light cone from a field theoretical quark-gluon model, with the resulting properties similar to current algebra defined on equal time commutators abstracted from the free quark model. It synthesized all the good results from scaling and partons with those from broken scale invariance and OPE, and extended to all the local operators occurring in the light cone expansions of commutators of all local operators. Such a generality enabled it to deal with processes far beyond what could be dealt with by canonical current algebra, such as pion-two gamma decay and the ratio R of total cross section for hadron and muon production in the electron positron annihilation. (Fritzsch and Gell-Mann, 1971a)

But there was a serious problem or threat to the scheme. Interactions would change the operator commutators inside the light cone, but the scheme would be unaffected if operators were formally manipulated rather than treated by renormalized perturbation theory order by order. Since the underlying model was based on vector gluon rather than scalar or pseudoscalar gluon, for the purpose of using "the scalar and pseudoscaalar densities describe the divergences of the vector and axial vector currents respectively", Fritzsch and Gell-Mann realized that the Adler anomaly would cause various pathologies in commutators of currents, and thus ruin the basic algebra as a closed system. Although the Adler anomaly was well established only for low energy regime, but there was a mathematical relation between the low energy anomalous divergence and the high energy anomalous singularity which would spoil the generalized algebra. In order to preserve their scheme, Fritzsch and Gell-Mann argued that order by order evaluations of scaling based on a renormalized quark-gluon theory had to be rejected. The renormalized perturbation theory revealed no scale invariance or free field theory behaviour at high energy, while, they claimed, "Nature reads books on free field theory, as far as the Bjorken limit is concerned." (Fritzsch and Gell-Mann, 1971b)

But they were cautious enough to realize that such an attitude "risk making some unwise generalizations of the basic algebra." Methodologically puzzled, they said: "it would be desirable to have some definite point of view about the relation of the abstracted results to the renormalized perturbation theory." Together with this methodological puzzlement they were also puzzled over the nature of their underlying model, in particular over the nature of interactions and the nature of gluons mediating the interactions. Such a desire drove them to search intensively for a clarification which, however, required other inputs and a great synthesis. (Fritzsch and Gell-Mann, 1971b)

A Great synthesis

So far, in my survey of conceptual developments, no contribution from the constituent quark model was mentioned. It was not completely fair. The model did contribute to the development, although not as a fundamental framework, but mainly in ways of giving stimulus. One of its stimuli was related with the idea of parastatistics (Greenbeerg, 1964), which was later translated into the notion of colour (Han and Nambu, 1965), although it was shown that the two notions are equivalent only when colour is neutral to electromagnetism. (Greenberg and Zwanziger, 1966) For this reason, the notion of colour adopted by Han and Nambu in their scheme with integral charged quarks could not be justified by the then accepted idea of parastatistics, although conceptually the notion of parastatistics can only be fully justified by the new quantum number of color as physicists later realized. This feature of Nambu's thinking was consequential as we will see shortly.

Colour made its first serious appearance in Adler's treatment of pion → two gamma decay within a relativistic renormalized quark gluon field theory, using the current quark picture rather than the constituent quark picture. But it was not acceptable to Fritzsch and Gell-Mann (1971b) since the way Adler's result was derived was inconsistent with the light cone current algebra in general and with scaling in particular. But the situation was dramatically changed in the late February of 1972, when one of Gell-Mann's students, Rodney Crewther, completed a work in which Adler's result was rederived from short distance behaviour of operator product expansion without assuming perturbation theory (Crewther, 1972). Once this was clear to Gell-Mann, he decided that "we need color!" the reason was that he would like to carry over the parastatistics for the current quarks from constituent quarks, and eventually to suggest a "transformation which took one into the other, conserving the statistics but changing a lot of other things." (Gell-Mann, 1972a) The transformation was done by two of Gell-Mann's students, Jay Melosh and Ken Young connecting two different algebras, underlain by two physical pictures. Although the two pictures were physically totally different, one was nonrelativistic with hadrons being conceived as composed of just a few constituent quarks, while the other was conceived in terms of relativistic quark fields, mathematically they were similar. One interesting result noticed by Young was that "the constituent quark looks like a current quark dressed up with current quark pairs." (The results of Melosh and Young were reported in Gell-Mann, 1972a)

The robustness of the notion of colour was further confirmed by the calculation of the ratio R of total cross section for hadron and muon production in the electron positron annihilation. At this juncture, with the notion of gluon firmly integrated into his underlying field theoretical framework and the notion of colour appeared so robust waiting to be integrated into his framework, Gell-Mann clearly felt the time was ripe to set a new agenda for himself and for the whole high energy physics community to construct a consistent conceptual framework, or a unified theory, to deal with hadron physics as a whole by "incorporating all the respectable ideas now being studied: constituent quarks, current quarks, the bootstrap,... scaling and current algebra. They are all or nearly all compatible, if we get our language straight." (Gell-Mann 1972b).

Crucial to the agenda of great synthesis was to clarify, and justify rather than declare, the relationship between the notion of gluon (the supposed carrier of the strong force interacting with quarks and holding them together) , which was strongly supported by evidence from parton model calculations, and the notion of colour, which as a new quantum number indicating a global symmetry among quarks was robust enough to be incorporated into the underlying quark-vector-gluon field theoretical framework for the light cone current algebra. All the evidence for quarks to have colour came from weak or electromagnetic processes. But gluons could never get involved in these processes. Then how can we know if gluons have colour or not?

There could be no simple ways to answer this central question, except for indirectly reasoning from constraints posed by the requirement of consistency and from aesthetic consideration. If the vector gluon carries no colour, Fritzsch and Gell-Mann had been worried for a long time, then it and some vector baryon currents might exist in the same channel and coupled. In addition, this would indicate a fundamental asymmetry between quarks, which were supposed to be colour triplets and thus not real in the sense that they cannot be detected in laboratory in isolation, and gluons, which as colour singlets presumable would be real in the sense they could be detected in isolation in laboratory. Thus their answer was simple and clear once they accepted colour as an indispensable component of the whole scheme in the Spring of 1972, namely, gluons should be colour non-singlets. Once the gluons as the strong force mediators were taken to be the colour non-singlets, the seductive power for taking gluons as the gauge potentials of the SU(3) colour symmetry, which should be correspondingly transmuted from a global symmetry to a local gauge symmetry, was difficult to resist. Mathematically this implied that gluons should be colour octet.

Some from the audience might say, which I have heard, ever since or even before I have published a book on this subject (Cao, 2010), from some prominent physicists and historians of physics, such as David Olive and Laurie Brown (private communications). Their point is simple: It is not fair to give all the credit to Fritzsch and Gell-Mann in this regard while all what they did was already clearly formulated and published in the mid1960s by Yoichiro Nambu, although he did not use the name of colour. My response is as follows.

Gauge principle has no universal free-floating magic power. It is useless unless it is anchored at, or applied to, certain proper system of entities at certain proper level. Otherwise one simply could not understand why so many physicists, including Pauli, Yang and Mills, Schwinger, Sakurai, and also Veltman and Bell, who tried hard to use gauge principle to analyse strong interactions at the hadron level, were all failed. The reason for their failure was that they did not get the level right.

Nambu was different. He got the level right. But he got the entities wrong. His assumption of integrally charged quarks caused serious problems. Most serious was that it led to wrong conceptions about the nature of gluons and the nature of the colour symmetry. The ultimate cause of his mistake was that his reasoning was speculative and *ad hoc* in nature. He applied the gauge principle *a priori* to some subsystem of hadron, solely based on the statistics problem in nonrelativistic constituent quark model, without any inputs from other places. In contrast, the gluons treated by Fritzsch and Gell-Mann in the early 1970s came from parton model calculations, and thus carried with them all the constraints from experiments, most important among them was that gluons could not be involved in any electromagnetic or weak interactions.

In addition, the context matters. Nambu's gauge theory for quarks and gluons was proposed in the mid-1960s, well before the quantization and renormalization problems of non-abelian gauge theory were properly dealt with. It would not be an exaggeration to say that only when 't Hooft and Veltman published their world-changing works in the early 1970s, physicists began to obtain the conviction for seriously developing non-abelian gauge theoretical models to treat nuclear forces. For these reasons, Nambu's two sets of triplets left no trace in the thinking of gauging the inter-quark forces, although it had been well known as an alternative building scheme,

Come back to Fritzsch and Gell-Mann again. After having set the agenda in his plenary talk at the 16th International Conference on High Energy Physics held at Batavia in September 1972, Gell-Mann presented a joint paper with Fritzsch (1972) at a session chaired by David Gross. The paper made a fundamental assumption about the color singlet restriction within their general framework of the quark vector-gluon field theory model: "real particles are required to be singlets with respect to the SU(3) of colour." The implication of this restriction for quarks was clear: "we assume here that quarks do not have real counterparts that are detectable in isolation in the laboratory – they are supposed to be permanently bound inside the mesons and baryons." Their conviction about the power of the singlet restriction was clearly displayed:

> *We might eventually abstract from the quark-vector-gluon field theory model enough algebraic information about the colour singlet operators in the model to describe all the degrees of freedom that are present...We should have a complete theory of the hadrons and their currents, and we need never mention any operators other than colour singlets.*

Gell-Mann acknowledged that the singlet restriction for observable states, which had its origin in the bootstrap philosophy that he respected greatly as far as hadron physics was concerned, amounted to nothing but the requirement of gauge invariance. But this in fact begged the question. At a deeper level, Schwinger once warned that the general requirement of gauge invariance could not be used to dispose of essentially dynamical questions. The dynamical question in this context was the miracle of confinement, which should be and was attacked by Wilson and others in dynamical ways.

As to gluons, the paper declared that "they could form a colour octet of neutral vector fields obeying the Yang-Mills equations." The move was justified by informing the audience of a significant advantage in terms of conceptual consistency:

> *If the gluons of the model are to be turned into colour octets, then an annoying asymmetry between quarks and gluons is removed, namely that there is no physical channel with quark quantum numbers, while gluons communicate freely with the channel containing the ω and φ mesons. (In fact, this communication of an elementary gluon potential with the real current of baryon number makes it very difficult to believe that all the formal relations of light cone current algebra could be true even in "finite" version of singlet neutral vector/gluon field theory.)*

This was enough to specify the QCD Lagrangian. Although this was only a verbal description of the QCD Lagrangian without using the word Lagrangian, and Gell-Mann's nickname QCD was not coined till 1973, the implication of this description to knowledgeable physicists was perfectly clear: a new non-abelian gauge theory of quark-gluon interactions was born.

But a widely circulated opinion holds that QCD started with the discovery of asymptotic freedom. (Pais, 1986; Pickering, 1984) At a 1996 round table discussion, participated by Sidney Coleman, Sheldon Glashow, David Gross, Steven Weinberg and Arthur Wightman, and chaired by Stanley Deser, Gross was introduced as the Mr. QCD. (Deser, 1999) Surely the discovery of asymptotic freedom (Gross and Wilczek, 1973a, b, c)clearly showed that QCD satisfied the constraint posed by scaling, and thus contributed to its justification. But the discovery itself was not decisive and enough for the emergence of QCD, which in fact existed already. More serious was that the discovery had no implication to the nature of the colour symmetry: exact symmetry or broken, which, as we know now, is crucial for identifying QCD. In fact, Gross and his pupil wanted broken symmetry to avoid the infrared problems (massless gluons, long-range force, which would cause problems with the spectrum of states and spoil S-matrix in perturbation theory). Thus they declared that in the construction of realistic physical models of the strong interactions, "*the major problem remaining in these gauge theories is how to break the gauge symmetry and provide masses for the vector mesons*." (Gross and Wilczek, 1973b) This is clearly not QCD. In addition, when they discussed the work of Fritzsch and Gell-Mann (1972), they commented,

> *the proponents of red, white and blue quarks as a mathematical abstraction argue that the colour SU(3) group should be exact, and that all non-colour-singlets should be suppressed completely. One clearly requires a dynamical explanation of such a miracle*. (Gross and Wilczek, 1973b)

The requirement for giving confinement a dynamical explanation is respectable. But the statement also clearly betrayed a lack of understanding of what was the identifying feature of QCD, as late as 1973. As Fritzsch, Gell-Mann and Leutwyler (1973) indicated immediately afterwards, asymptotic freedom was only a consequence, or an advantage of QCD, which they at that time called the colour octet gluon picture.

Gell-Mann and the reality of quarks

No serious attempt to properly understand and appreciate the logic of, and contributions to, the genesis of QCD would be possible without a proper understanding of Gell-Mann's changing conception of the reality of quarks. But the latter would be impossible without a proper philosophical understanding of the complicated relationship between physics and metaphysics.

Since the meaning of scientists' fundamental concepts is dictated by an established meaning structure provided by the prevailing metaphysical framework, it would be very difficult for physicists to digest the radical changes of the meaning of such terms as "constituents" and "reality" that accompanied the profound development in physics, namely the colour confinement in QCD, without at the same time having realized the necessity of revising the underlying metaphysical scheme, and thus the meaning structure they are using. But the latter realization is quite difficult task even for philosophers, let alone physicists. For example, even if the notion of a running coupling and the consequent confinement were accepted in physical reasoning, it still took times for physicists to realize that the idea that constituents had to be capable of existing in isolation (a reductionist sense) was valid only for weakly coupled filed theories, but had to be revised when the coupling was strong. That is, they may exist permanently in the confined rather than asymptotic states (a reconstructive sense).

Thus Gell-Mann's painful struggle with the question of whether permanently confined quarks are real cannot be frivolously dismissed as mathematical instrumentalism, but has to be put in the context that when the idea of quark was proposed, the very idea of reality, according to the prevailing metaphysics of the time, was inseparably tied with observability, which was defined in terms of asymptotic states. For this reason, Gell-Mann understandably did not know now to handle the complicated situation when, on the one hand, the progress in hadron physics gave him good reasons to believe that hadrons consisted of quark constituents, and on the other hand, he had deep reasons to believe that quarks could not be observed asymptotically. Were quarks real? The question had frequently tortured Gell-Mann. But later, he realized that he could not take quarks realistically because the use of quarks in his original theory with scalar gluon was wrong; if he had a theory with vector gluons he would take then realistically. (Gell-Mann, 1997) This implies that he took the reality of quarks as depending on the validity of a quark-gluon theory. However, quarks and gluons in a correct theory remained unobservable, where they real?

Constrained by various logical inferences, Gell-Mann and the whole community soon realized that when the notion *real* applies to any physically hidden entity, in the light of confinement, it refers to more than merely the epistemic accessibility of the entity as a separate individual. Rather, it had widened its reference to a combination of some structural features without any assignment of individual observability to the entity.

Conceived this way, it became easier or even natural for physicists to acknowledge the reality of quarks and gluons. But then the meaning of reality has already been radically changed. Now the meaning of being real is more complicated than being observable individually. Thus, with the advent of QCD, physicists had convincingly revised metaphysical criterion of what counts as a real existence. An important point here is that metaphysics as reflections on physics, rather than a prescription for physics, cannot be detached from physics. With the advancements of physics, it has to move forward and revise itself so that it can accommodate to new situations.

In my view, Gell-Mann's painful experience in his personal struggle with the reality of quarks in fact was only a human expression of the difficult adjustment of metaphysics under the pressure of the advancement of physics. Gell-Mann's understanding of the metaphysical nature of quarks, real or not, was clearly not superior to other physicists' in the 1960s; but one can find no sense to claim that in the metaphysical assessment of quarks' nature in the 1960s, Gell-Mann was inferior to other physicists or philosophers. Some quark realists in the 1960s did claim the reality of quarks, but only in the sense that quarks would be sooner or later be observed in isolation in laboratory. But now we know this was only a wrong conviction.

What is the logic of the genesis of QCD?

QCD was constructed under the structural constraints of scaling, confinement and the existence of colour quantum number, which stemmed from structurally probing hadrons externally and internally, through current algebraic explorations and field theoretical construction. Since the evidential status of theoretical entities is encoded in theory, the reality of quarks and gluons (gauge bosons) cannot be separated from the validity of QCD, as Gell-Mann realized and we mentioned above, within which they are constructed. This point has resolutely deprived the constituent quark model as well as Nambu's wrongly conceived gauge theory of any constructive role in the genesis of QCD.

So we have good reasons for celebrating QCD's 40th anniversary now because it was created 40 years ago in 1972 by Murray Gell-Mann and Harald Fritzsch, on the basis of the works done by Stephen Adler, James Bjorken and others as I briefly surveyed here.

References

Abbreviations:
AP Annals of Physics (N.Y.)
NC IL Nuovo Cimento
PL Physics Letters
PR Physical Review
PRL Physical Review Letters
PTP Progress of Theoretical Physics

Adler, S. (1965a): "Consistency Conditions on the Strong Interactions Implied by a Partially Conserved Axial Vector Current, I and II" *PR*, **137**: B1022--B1033; *PR*, **139**: B1638--1643

Adler, S. (1965b): "Calculation of the Axial-Vector Coupling Constant Renormalization in Beta Decay" *PRL*, **14**: 1051--1055

Adler, S. (1965c): "Sum Rules for the Axial-Vector coupling Constant Renormalization in Beta Decay" *PR*, **140**: B736—B747

Adler, S. (1966): "Sum Rules Giving Tests of Local Current commutation Relations in High Energy Neutrino Reactions" *PR*, **143**: 1144—1155

Bell, J. (1967): "Current algebra and gauge invariance" *NC*, **50A**: 129-134.

Bjorken, J. (1966): "Inequality for Electron and Muon Scattering from Nucleons" *PRL*, **16**: 408

Bjorken, J. (1969): "Asymptotic sum rules at infinite momentum" *PR*, **179**: 1547-1553

Bjorken, J. (1997): "Deep-Inelastic Scattering: From Current algebra to Partons" The Rise of The Standard Model (eds. Hoddeson, L., Brown, L., Riordan, M, and Dressden, M; Cambridge University Press) 589-599

Cao, T.Y. (1997): *Conceptual Developments of 20th Century Field Theories* (Cambridge U. Press).

Cao, T.Y. (1999): *Conceptual Foundations of Quantum Field Theory* (editor; Cambridge U. Press).

Cao, T.Y. (2010): *From Current Algebra to Quantum Chromodynamics – A Case for Structural Realism* (Cambridge University Press).

Chew, G. F. (1967): "Closure, locality, and the bootstrap" *PRL*, **19**: 1492-1495

Crewther, R. (1972): "Nonperturbative Evaluation of the Anomalies in Low-Energy Theorems" *PRL*, **28**: 1421–1424

Dalitz (1966): "Quark models for the 'elementary particles'" *High Energy Physics* (eds. DeWitt, C. and Jacob, M., Gordon and Breach, New York) 251-324

Deser, S. (1999): Introduction to the panel discussion, p. 368 in Cao (1999)

Fritzsch, H., and Gell-Mann, M. (1971a): "Scale invariance and the light come" *Proceedings of the 1971 Coral Gables Conference on Fundamental Interactions at High Energy* (eds. Dal Cin, M, et al. Gorfon snf Breach) 1-53

Fritzsch, H., and Gell-Mann, M. (1971b): "Light cone current algebra,"*Proceedings of the International Conference on Duality and Symmetry in Hadron Physics* (ed. Gotsman, E.; Weizmann Science Press) 317-374

Fritzsch, H., and Gell-Mann, M. (1972): "Current algebra, quarks and what else?" *Proceedings of the XVI International Conference on High Energy Physics*, September 6-13, 1972 (eds. Jackson, J. D. and Roberts, A.; National Accelerator Laboratory, Batavia), vol. 2: 135-165

Fritzsch, H., Gell-Mann, M., and Leutwyler, H. (1973): "Advantages of the color octet gluon picture" *PL*, **47B**: 267-271

Fubini, S and Furlan, G. (1965). "Renormalization effects for partially conserved currents", *Physics*, **1**: 229-247.

Gell-Mann, M. (1962): "Symmetries of baryons and mesons" *PR*, **125**: 1067-1084

Gell-Mann, M. (1964a): "A schematic model of baryons and mesons" *PL*, **8**: 214-5

Gell-Mann, M. (1964b): "The symmetry group of vector and axial vector currents," *Physics*, **1**: 63-75.

Gell-Mann, M. (1972a): "Quarks: developments in the quark theory of hadrons" *Acta Phys. Austriaca*, **9**: 733-761

Gell-Mann, M. (1972b): "General Status: Summary and Outlook," *Proceedings of the XVI International Conference on High Energy Physics,* September 6-13, 1972 (eds. Jackson, J. D. and Roberts, A.; National Accelerator Laboratory, Batavia), **vol. 4**: 333-355

Gell-Mann, M. (1997): "Quarks, Color, and QCD" *The Rise of the Standard Model* (eds. Hoddeson, L., Brown, L., Riordan, M., and Dresden, M.; Cambridge University Press) 625-633

Gell-Mann, M. and Pais, A. (1954): "Theoretical views on the new particles," *Proceedings of the Glaskow International Conference on Nuclear Physics* (eds. Bellamy, E. H. and Moorhouse, R. G., Pergmon, Oxford), 342-350.
Greenberg, O. (1964): "Spin and Unitary-Spin Independence in a Paraquark Model of Baryons and Mesons" *PRL*, **13**: 598-602
Greenberg, O., and Zwanziger, D. (1966): "Saturation in Triplet Models of Hadrons" *PR*, **150**: 1177–1180
Gross, D. and Wilczek, F. (1973a): Ultraviolet behavior of non-abelian gauge Theories" *PRL*, **30**: 1343-1346.
Gross, D. and Wilczek, F. (1973b): "Asymptotic free gauge theories: I" *PR*, **D8**: 3633-3652.
Gross, D. and Wilczek, F. (1973c): "Asymptotic free gauge theories: II" *PR*, **D9**: 980-983
Gürsey, F. and Radicati, L. A. (1964): "Spin and unitary spin independence," *PRL*, **13**: 173-175.
Han, M. Y. and Nambu, Y. (1965): "Three-triplet model with double SU(3) Symmetry" *PR*, **139B**: 1006-1010.
Leutwyler, H. (1968): "Is the group SU(3)l an exact symmetry of one-particle-states?" Preprint (University of Bern)
Lipkin, H. (1997): "Quark models and quark phenomenology" *The Rise of the Standard* Model (eds. Hoddeson, L., Brown, L., Riordan, M., and Dresden, M.; Cambridge U. Press) 542-560
Nambu, Y. (1957): "Possible existence of a heavy neutral meson," *PR*, **106**: 1366-7
Nambu, Y. (1965a): "Dynamical symmetries and fundamental fields" in *Symmetry Principles at High Energies* (eds. Kursunoglu, B., Perlmutter, A. and Sakmar, A.; Freeman, San Francisco and London), 274-285.
Nambu, Y. (1965b): "A systematics of hadrons in sybnuclear physics" *Preludes in Theoretical Physics* (eds. De Shalit, A., Feshbach, H., and Van Hove, L.; North- Holland, 1966) 133-42
Pais, A. (1964): "Implications of spin-unitary spin independence" *PRL*, **13**: 175-7.
Pais, A. (1986): *Inward Bound* (Oxford University Press).
Pauli, W. (1953): Two letters to Abraham Pais, in *Wissenschaftlicher Briefwechsel*, Vol. IV, Part II (Springer-Verlag, edited by K. V. Meyenn). Letter [1614]: to A. Pais (July 25, 1953); letter [1682]: to A. Pais (December 6, 1953).
Pauli (1999); two seminar lectures, reported in Paolo Gulmanelli (1954)*: Su una Teoria dello Spin Isotropico* (Pubblicazioni della Sezione di Milano dell'istituto Nazionale di Fisica Nucleare, Casa Editrice Pleion, Milano).
Pickering, A. (1984): *Constructing Quarks* (Edinburgh University Press)
Sakata, S. (1956): "On the composite model for the new particles" *PTP*, **16**: 686-8.
Sakita, B. (1964): "Supermultiplets of elementay particles" *PR*, **B136**: 1756-1760.
Sakurai, J. J. (1960): "Theory of strong interactions" *AP*, **11**: 1-48.
Susskind, L (1968): "Models of self-induced strong interactions" *PR*, **165**: 1535-1546
Veltman, M. (1966): "Divergence conditions and sum rules" *PRL*, **17**: 553-556.
Weisberger, W. (1965): "Renormalization of the weak axial-vector coupling Constant" *PRL*, **14**: 1047-1055.
Wilson, K. G. (1969): "Non-Lagrangian models of current algebra" *PR*, **179**: 1499-1512.
Yang, C. N. and Mills, R. L. (1954): "Isotopic spin conservation and a generalized gauge invariance" *PR*, **95**: 631; "Conservation of isotopic spin and isotopic gauge invariance" *PR*, **96**: 191-195.
Zweig, G. (1964): "An SU3 model for strong interaction symmetry and its breaking: I and II", *CERN preprint TH-401* (17 Jan. 1964) and *TH-412* (21 Feb. 1964); "Fractionally charged particles and SU(6)" *Symmetries in Elementary Particle Physics* (ed. Zichichi, A.; Academic Press, New York) 192-243.

Hot Theoretical Topics

Perturbative Gravity from Gauge Theory

Z. Bern

Department of Physics and Astronomy
University of California at Los Angeles
Los Angeles, CA 90095
USA

Abstract

In this lecture, we describe a recently conjectured duality between color and kinematics for gauge-theory amplitudes. Whenever this duality is manifest, the integrands of loop-level gravity scattering amplitudes can be obtained from corresponding gauge-theory amplitudes via a double-copy relation. This duality has been used to enormously simplify a number of explicit multiloop supergravity calculations. The results of these computations is that supergravity theories have a surprisingly tame ultraviolet behavior, and in some cases may even be finite. As an example, we summarize a recent calculation showing that half-maximal $\mathcal{N} = 4$ supergravity in four space-time dimensions is ultraviolet finite at three loops, contrary to previous expectations.

1 Introduction

Recent years have seen remarkable progress in understanding scattering processes in gauge and gravity theories, both for phenomenological and theoretical purposes. In this talk, we will focus on recent theoretical progress in quantum gravity. In particular, we will describe a surprising relation between gauge and gravity theories. For nearly 30 years, physicists have been convinced that point-like theories of gravity along the lines of Einstein's theory are incompatible with quantum field theory in part because they lead to ultraviolet infinities. In a classic paper, 't Hooft and Veltman demonstrated that in four dimensions gravity coupled to matter generically diverges at the first quantum loop order [1]. Supersymmetry offers a mechanism for delaying the onset of divergences in gravity theories. However, supersymmetry alone apparently cannot eliminate the ultraviolet divergences in gravity theories because of the increasingly worse divergences at each loop order in gravity theories. The consensus opinion from the late 70's and early 80's was that all such theories would likely diverge by three loops. (See, for example, ref. [2] for a review of the situation in that era.)

Here we will discuss some new work challenging the understanding of the ultraviolet behavior of gravity. A new approach for doing so uses the recently conjectured all-loop relation between gravity and gauge-theory loop integrands [3, 4]. This greatly simplifies the construction of multiloop supergravity scattering amplitudes [5, 6, 7, 8], allowing us to directly probe their ultraviolet structure. Although gravity and gauge theories have obvious superficial similarities, their detailed dynamics appears to be rather different. Nevertheless, we know from the celebrated AdS/CFT correspondence [9] between gauge and gravity theories that there is an equivalence between weak coupling in one theory and strong coupling in the other. The connection we use here is a different showing that in a precise sense gravity at weak coupling is a double copy of gauge theory. We write this schematically as

$$\text{gravity} \sim (\text{gauge theory}) \times (\text{gauge theory}). \tag{1}$$

This was first understood at tree level over 25 years ago using string theory [10], but today we have a much simpler description [3], allowing for a straightforward extension to loop level [4].

Here we will describe how the new ideas relating gauge and gravity theories have led to significant progress on understanding the ultraviolet properties of supergravity theories, especially the cases of $\mathcal{N} = 8$ [11] and $\mathcal{N} = 4$ supergravity [12]. More generally, there have also been major advances in developing powerful tools for carrying out the explicit computations of scattering amplitudes. These include the unitarity method [13, 14] and its refinement known as the method of maximal cuts [15], used in the constructions described here. These tools have played a central role in uncovering the duality between color and kinematics [3, 4]. There has also been enormous progress in carrying out the required loop integration needed to determine the explicit values of potential divergences, from the constructed amplitudes (see, for example, ref. [16]). The obtain results has stimulated significant new progress in understanding the role of supersymmetry and duality symmetry in constraining potential counterterms in supergravity theories [17].

Even with the new progress, the question of whether supergravity theories diverge in perturbation theory is still an open one. (For a recent optimistic opinion in favor of ultraviolet finiteness of $\mathcal{N} = 8$ supergravity see ref. [18]. For a recent pessimistic opinion see ref. [19].) While conventional counterterm symmetry considerations suggest that all supergravity amplitudes will diverge at some loop order, studies of multiloop scattering amplitudes suggest that additional ultraviolet cancellations exist beyond these [20]. Surprisingly, even pure Einstein gravity at one loop exhibits remarkable cancellations as the number of external legs increases [21]. The double-copy structure also implies that gravity amplitudes are much more highly constrained than implied by standard symmetries.

The case of half-maximal $\mathcal{N} = 4$ supergravity [12] at three loops is especially interesting. A recent calculation [7], as well as string arguments [22], shows that the potential three-loop R^4 divergence [23] is absent, contrary to earlier symmetry expectations [24]. While the possible existence of a certain harmonic superspace may lead to an explanations [25], further studies are needed. Other more exotic potential explanations are based on the double-copy property of gravity [8] and on a hidden superconformal invariance [26]. A four-loop calculation, in particular, would shed important new light on this.

2 The unitarity method

In recent years, there has been a fundamental shift in how we view scattering amplitudes. In the traditional Feynman diagram approach one starts from an off-shell Lagrangian and constructs Feynman diagrams according to a set of rules. The diagrams encode algebraic expressions describing the scattering process. The diagrams depend on the gauge and field-variable choices. Gauge invariance is restored only at the end of a computation, when one puts all external states on shell and all pieces are added together. The lack of gauge invariance for the individual diagrams can lead to enormously complicated expressions, which simplify only after a nontrivial effort to combine terms.

In contrast, on-shell methods construct new amplitudes directly from simpler gauge-invariant on-shell amplitudes. Since the simpler amplitudes are already gauge invariant they can be greatly simplified before being used in the construction of more complex amplitudes. The two basic on-shell methods are the unitarity method [13, 15, 27] and on-shell recursion [28]. For studying multiloop gravity, the current method of choice is the unitarity method. This method was originally developed in the context of one-loop supersymmetric amplitudes [13], but further refinements [29, 15, 27, 30], make it offers powerful formalism for massless amplitudes at high loop orders, including non-planar contributions. This method has been reviewed numerous times (see for example ref. [31]), so here we give only a brief outline. Powerful methods for constructing multiloop integrands using on-shell recursion have also been devised [32].

Unitarity has been a basic principle in quantum field theory since its inception. For a description of unitarity during the 1960's see ref. [33]. However, a variety of difficulties prevented its widespread use as a means of constructing amplitudes, especially after the rise of gauge theories in the 1970's. These difficulties include non-convergence of dispersion relations and its inapplicability to massless particles. It was also unclear how one could fully reconstruct loop amplitudes beyond four points from their unitarity cuts. The modern unitarity method overcomes these difficulties, allowing for the complete construction of loop amplitudes at any loop order. It does so by avoiding dispersion relations, and instead using the existence of an underlying Feynman diagram representations to fully reconstruct amplitudes. By construction the obtained Feynman-like integrands have the correct analytic properties in all channels.

Although the unitarity method applies just as well to supersymmetric and non-supersymmetric theories, it is usually much simpler to deal with the supersymmetric cases because they have a simpler analytic structure. Indeed, the original application of the unitarity method was to construct one-loop supersymmetric amplitudes with arbitrary numbers of external legs [13].

3 Comparing Gravity to Gauge Theory

We start by comparing gravity to gauge theory using standard off-shell methods. Feynman rules are generated from the Yang-Mills and Einstein-Hilbert Lagrangians,

$$\mathcal{L}_{\rm YM} = -\frac{1}{4} F^a_{\mu\nu} F^{a\,\mu\nu}\,, \qquad\qquad \mathcal{L}_{\rm EH} = \frac{2}{\kappa^2}\sqrt{-g}R\,. \tag{2}$$

From the viewpoint of Feynman diagrams, these two Lagrangians have rather different properties. With standard gauge choices gauge theories have three- and four-point interactions, while gravity has an infinite number of contact interactions. Perhaps more striking than the infinite number of interactions is the complexity of these interactions.

To be more concrete, consider the three-gluon vertex in Feynman gauge,

$$V^{abc}_{3\mu,\nu,\sigma}(k_1,k_2,k_3) = g f^{abc}\Big[(k_1-k_2)_\sigma \eta_{\mu\nu} + \text{cyclic}\Big]\,, \tag{3}$$

where g is the coupling, f^{abc} the usual group theory structure constants, $\eta_{\mu\nu}$ the flat metric and the k_i the momenta of the vertex. This vertex is relatively simple. We may compare this to the three-graviton interaction in, for example, de Donder gauge,

$$G_{3\mu\alpha,\nu\beta,\sigma\gamma}(k_1,k_2,k_3) = i\frac{\kappa}{2}\Big[-\frac{1}{2}k_1\cdot k_2\eta_{\mu\alpha}\eta_{\nu\beta}\eta_{\sigma\gamma} - \frac{1}{2}k_{1\nu}k_{1\beta}\eta_{\mu\alpha}\eta_{\sigma\gamma} + \cdots\Big]\,, \tag{4}$$

where we have displayed two terms out of about 100. Here the gravitational coupling κ is related to Newton's constant by $\kappa^2 = 32\pi^2 G_N$. The precise form of the vertex depends on the gauge, but in any case the three vertex is a rather involved and unenlightening object. The complete expression can be found in refs. [1, 34].

Comparing the vertex in eq. (3) to the one in eq. (4), it certainly would appear that gravity is much more complicated than gauge theory. Moreover, there does not appear to be any simplicity or obvious relation between the gauge and gravity vertices. The former leads to complicated diagrams, but the latter appears hopelessly complicated. One can do somewhat better with special gauge choices and appropriate field redefinitions [35, 36], considerably simplifying the Feynman rules. Still, multiloop Feynman diagram calculations in (super) gravity are extremely difficult, and generally out of reach using even the most powerful supercomputers.

Now let us reconsider the same process but from the on-shell vantage point. If we take the three-graviton vertex in eq. (4) and dot the three legs with physical polarizations tensors satisfying the physical-state conditions, $k_i^2 = 0$, $\varepsilon_i^{\mu\nu}k_{i\mu} = \varepsilon_i^{\mu\nu}k_{i\nu} = \varepsilon^\mu{}_\mu = 0$, we obtain a greatly simplified vertex,

$$G_3(k_1,k_2,k_3) = -i\kappa\varepsilon_1^{\mu\alpha}\varepsilon_2^{\nu\beta}\varepsilon_3^{\sigma\gamma}\Big[(k_1)_\sigma\eta_{\mu\nu} + \text{cyclic}\Big]\Big[(k_1)_\gamma\eta_{\alpha\beta} + \text{cyclic}\Big]\,. \tag{5}$$

Remarkably, up to overall factors, this is just a double copy of the kinematic part of the on-shell Yang-Mills vertex,

$$V_3^{abc}(k_1,k_2,k_3) = 2\varepsilon_1^\mu\varepsilon_2^\nu\varepsilon_3^\sigma g f^{abc}\Big[(k_1)_\sigma\eta_{\mu\nu} + \text{cyclic}\Big]\,, \tag{6}$$

where the polarization vector satisfies $\varepsilon_i^\mu k_{i\mu} = 0$. To make the comparison, we identify the graviton polarization tensor as a product of gluon polarization vectors, $\varepsilon_i^{\mu\nu} = \varepsilon_i^\mu \times \varepsilon_i^\nu$. Similar considerations allow us to express all three-point vertices in supergravity as products of super-Yang-Mills vertices. Using on-shell recursion [28], these three vertices are sufficient to construct any tree-level gauge or gravity amplitude. The unitarity method then allows us to construct any loop amplitude.

Clearly, there is a rather striking relationship between gravity and gauge theory, but to make it visible we need to keep external states on shell. As we shall see below, the double-copy structure in eq. (5) is not accidental, but appears likely to extend to all

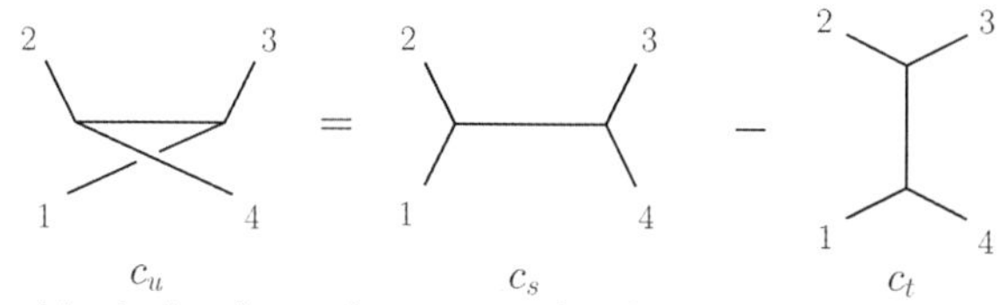

Figure 1: The Jacobi relation for at four points for the three channels labeled by s, t and u. These diagram represent either color factors or kinematic numerators.

loop orders. As such, it reflects a profound and important property of quantum gravity, pointing to unification of the two theories, perhaps along the lines of string theory.

The above relation between gravity and gauge theory is a special case of the recently discovered duality between color and kinematics [3, 4]. In general, we can write any m-point tree-level gauge-theory amplitude with all particles in the adjoint representation as,

$$\mathcal{A}^{\text{tree}}(1, 2, 3, \ldots, m) = g^{m-2} \sum_i \frac{n_i \, c_i}{\prod_{\alpha_i} p_{\alpha_i}^2}, \tag{7}$$

where the sum runs over the set of m-point L-loop diagrams with only cubic vertices. These include distinct permutations of external legs. The product in the denominator runs over all propagators of each cubic diagram. The c_i are the color factors obtained by dressing every three vertex with an $\tilde{f}^{abc} = i\sqrt{2} f^{abc}$ structure constant, the n_i are kinematic numerator factors depending on momenta, polarizations and spinors and g is the coupling constant. The form (7) can be obtained straightforwardly, for example, from Feynman diagrams, by representing all contact terms as inverse propagators in the kinematic numerators that cancel propagators. For supersymmetric amplitudes expressed in superspace, there are also Grassmann parameters in the numerators.

The color factors satisfy Jacobi identities. In fig. 1 the basic four-point identity is shown, where the diagrams represent color factors obtained by dressing each vertex with an $\tilde{f}^{abc}$. The duality requires there to exists a way of reorganizing the diagrams so that the numerators n_i satisfy equations in one-to-one correspondence with the Jacobi identities of the color factors. That is, for each triplet of color factors satisfying a Jacobi identity, the corresponding numerators satisfy a similar identity:

$$c_i = c_j - c_k \;\Rightarrow\; n_i = n_j - n_k \,. \tag{8}$$

This duality is conjectured to hold to all multiplicity at tree level in a large variety of theories, including supersymmetric extensions of Yang-Mills theory. In fig. 1 we display the Jacobi relation at four points. The duality conjecture states that there exists representations of the amplitude, such that the color factors and numerators of the diagrams satisfy the relations. As explained in some detail in refs. [37, 6], the numerator relations are functional equations, depending nontrivially on the momenta and external polarizations and spinors. For four-point tree amplitudes such relations were noticed long ago using Feynman diagrams [38]. Beyond the four-point tree level, the relations are highly nontrivial and hold only after appropriate rearrangements of the amplitudes.

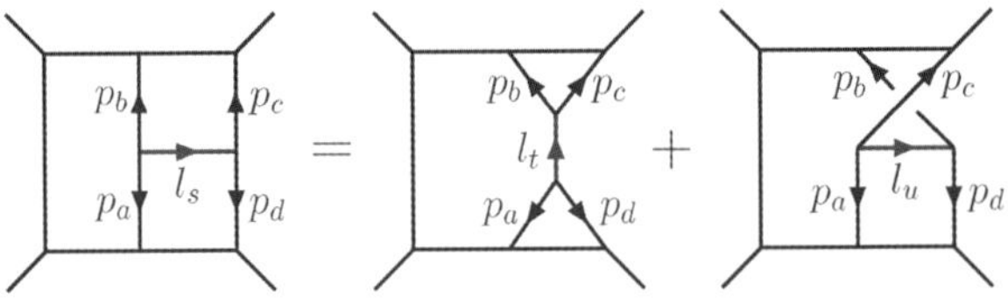

Figure 2: An example of a duality relation satisfied by numerators for diagrams of the three-loop four-point amplitude. Both color factors and numerator factors satisfy these relations.

At tree level, explicit forms of amplitudes satisfying the duality have been found for an arbitrary number of legs [39]. The duality has also been studied in string theory [40, 41]. An interesting consequence of the duality is that color-ordered partial tree amplitudes satisfy nontrivial relations [3], proven in gauge theory and in string theory [42]. Although we do not yet have a satisfactory Lagrangian understanding, some progress in this direction can be found in refs. [43, 44]. The duality (8) has also been expressed in terms of an alternative trace-based representation [45]. Progress has also been made in understanding the underlying infinite-dimensional Lie algebra [44, 46] responsible for the duality and in finding new representations that satisfy the duality [47]. The duality between color and kinematics also appears to hold in three-dimensional theories based on three algebras [48], as well as in some cases with higher-dimension operators [49].

Perhaps more surprising than the duality itself is a related conjecture that once the gauge-theory amplitudes are arranged into a form satisfying the duality (8), corresponding gravity amplitudes can be obtained simply by replacing the c_i color factor in eq. (7) with a second copy of a numerator factor $\tilde{n}_i$ [3, 4],

$$\mathcal{M}^{\text{tree}}(1,2,\ldots,m) = i\left(\frac{\kappa}{2}\right)\sum_i \frac{n_i\,\tilde{n}_i}{\prod_{\alpha_i} p^2_{\alpha_i}}\,, \tag{9}$$

where n_j and $\tilde{n}_j$ are kinematic numerator factors from gauge-theory amplitudes and κ is the gravitational coupling. The sum runs over the same set of diagrams with cubic vertices, as in eq. (7). This is expected to hold in a large class of gravity theories, including theories that are the low-energy limits of string theories. At tree level, this double-copy property encodes what are known as KLT relations between gravity and gauge-theory tree amplitudes [10]. The double-copy formula (9) has been proven via on-shell recursion [28] for pure gravity and for $\mathcal{N}=8$ supergravity tree amplitudes, when the duality (8) holds in the corresponding gauge theories [43]. It turns out that for the double-copy formula to hold, only one gauge-theory copy needs to satisfy the duality (8) because a generalized gauge invariance gives us the ability to rearrange the numerators of the second copy [4, 43].

More recently, the above relations have been conjectured to hold at loop level [4]. Any m-point L-loop gauge-theory amplitude with all particles in the adjoint representation can be written in the form,

$$\mathcal{A}_m^{L-\text{loop}} = i^L g^{m-2+2L} \sum_j \int \prod_{l=1}^{L} \frac{d^D p_l}{(2\pi)^D} \frac{1}{S_j} \frac{n_j c_j}{\prod_{\alpha_j} p_{\alpha_j}^2}, \tag{10}$$

where the sum labeled by j runs over the set of distinct m-point L-loop graphs with only cubic vertices, including distinct relabelings of external legs. The factor S_j is the symmetry factor of graph j that removes overcounts from internal symmetry. In this representation contact terms are absorbed into graphs with cubic vertices by multiplying and dividing by appropriate inverse propagators. The integrals are over L independent D-dimensional loop momenta. The product in the denominator runs over all Feynman propagators of graph j. As at tree level, the c_j are the color factors of the diagrams, and the n_j are kinematic numerators of graph j depending on momenta, polarizations and spinors. According to the duality conjecture of ref. [4], a representation of L-loop m-point amplitudes should exist where kinematic numerators satisfy the same algebraic properties as corresponding color factors (8).

The duality has been confirmed to hold up to four loops for the four-point amplitudes of $\mathcal{N} = 4$ super-Yang-Mills (sYM) theory [4, 6], and for the five-point one- and two-loop amplitudes of this theory [50]. Moreover, the infrared singularities do appear to be consistent with the duality to all loop orders [51]. The duality is also known to hold for the identical-helicity one- and two-loop four-point amplitudes of pure Yang-Mills theory [4]. It has also been tested in one and two-loop gravity examples in cases with fewer supersymmetries than the maximum [5] and for form factors [52]. There has also been progress in understanding more general one-loop amplitudes [53].

As a relatively simple multiloop example of the duality, consider the three-loop four-point amplitude of $\mathcal{N} = 4$ sYM theory. A form of the amplitude manifesting the duality is given by the 12 cubic diagrams in fig. 3 with the kinematic numerators given in table 1 [4]. For example, consider the numerator relation in fig. 2. Since the numerator of the first diagram on the right hand side vanishes, the numerator of diagram (e) in fig. 3 must equal the one of diagram (f). Indeed, this holds, as indicated in table 2. Similarly, all other numerator Jacobi relations hold.

As at tree level, gravity amplitudes can be obtained from gauge theory amplitudes simply by replacing the color factors with a second gauge-theory numerator factor, $c_j \to \tilde{n}_j$. That is, associated with the conjectured duality between color and kinematics is a double-copy formula for m-point L-loop gravity amplitudes [4]

$$\mathcal{M}_m^{\text{loop}} = i^{L+1} \left(\frac{\kappa}{2}\right)^{m-2+2L} \sum_j \int \prod_{l=1}^{L} \frac{d^D p_l}{(2\pi)^D} \frac{1}{S_j} \frac{n_j \tilde{n}_j}{\prod_{\alpha_j} p_{\alpha_j}^2}. \tag{11}$$

The particular gravity theory obtained by the double-copy formula (11) is dictated by the choice of gauge theories. As at tree level, in eq. (11) only one of the two copies needs to satisfy the duality (8) [4, 43]. The other gauge-theory amplitude can be in another convenient representation arranged into graphs with only cubic vertices.

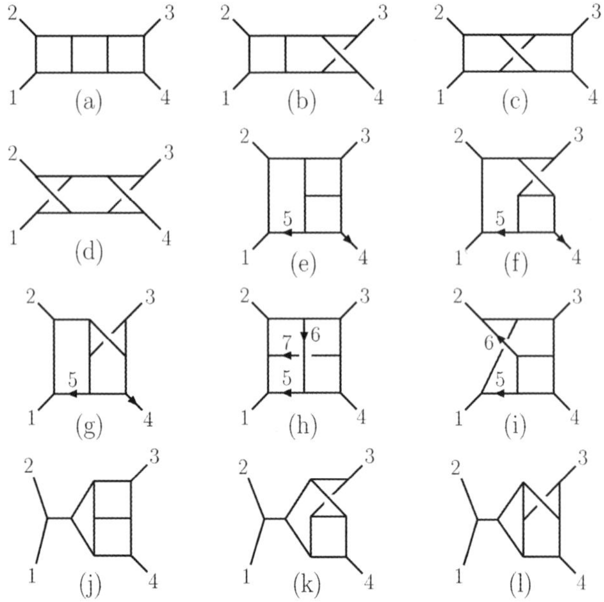

Figure 3: The twelve graphs of the $\mathcal{N}=4$ sYM three-loop four-point amplitude in a representation satisfying the duality between color and kinematics [4].

Integral $I^{(x)}$	$\mathcal{N}=4$ Super-Yang-Mills numerator
(a)–(d)	s^2
(e)–(g)	$\big(s\,(-\tau_{35}+\tau_{45}+t)-t\,(\tau_{25}+\tau_{45})+u\,(\tau_{25}+\tau_{35})-s^2\big)/3$
(h)	$\big(s\,(2\tau_{15}-\tau_{16}+2\tau_{26}-\tau_{27}+2\tau_{35}+\tau_{36}+\tau_{37}-u)$ $+t\,(\tau_{16}+\tau_{26}-\tau_{37}+2\tau_{36}-2\tau_{15}-2\tau_{27}-2\tau_{35}-3\tau_{17})+s^2\big)/3$
(i)	$\big(s\,(-\tau_{25}-\tau_{26}-\tau_{35}+\tau_{36}+\tau_{45}+2t)$ $+t\,(\tau_{26}+\tau_{35}+2\tau_{36}+2\tau_{45}+3\tau_{46})+u\,\tau_{25}+s^2\big)/3$
(j)-(l)	$s(t-u)/3$

Table 1: The numerator factors of the integrals $I^{(x)}$ in fig. 3. The first column labels the integral, the second column the numerator factor for $\mathcal{N}=4$ sYM theory. An overall factor of stA_4^{tree} has been removed, s,t,u are Mandelstam invariants corresponding to $(k_1+k_2)^2, (k_2+k_3)^2, (k_1+k_3)^2$ and $\tau_{ij}=2k_i\cdot l_j$, where k_i and l_j are momenta as labeled in fig. 3.

4 Half maximal supergravity at three loops

The case of $\mathcal{N}=4$ half-maximal supergravity is an especially interesting one, because of the expected appearance of a three-loop R^4 divergence based on symmetry consideration [24]. By constructing the three-loop four-point amplitude we can test this. To obtain pure supergravity amplitudes with no additional matter, we take one component gauge theory to be $\mathcal{N} = 4$ sYM theory and the second component to be nonsupersymmetric pure Yang-Mills theory [5, 12]. The $\mathcal{N} = 4$ sYM copy is the one discussed above [4], making the duality between color and kinematics manifest. What representation should we use for the nonsupersymmetric pure Yang-Mills amplitude? As explained in ref. [7], for the pure Yang-Mills copy, a convenient representation is ordinary Feynman-gauge Feynman diagrams. Any contact contributions are assigned to diagrams with only cubic vertices according to their color factor. Enormous simplifications follow from the fact that in this construction most Feynman graphs are irrelevant because they get multiplied by vanishing $\mathcal{N} = 4$ sYM graph numerators. In fact, only the twelve graphs displayed in fig. 3 have nonvanishing numerators on the $\mathcal{N} = 4$ sYM side. Using these observations the complete three-loop four-point integrand of $\mathcal{N} = 4$ supergravity was obtained in ref. [7] for all external states of the theory.

Once the integrand of the amplitude is obtained, systematic methods for extracting the ultraviolet divergences from the constructed integrand can be applied [54]. We expand in large loop momenta or equivalently in small external momenta. This gives a set of vacuum graphs containing both infrared and ultraviolet divergences. To separate out the infrared divergences, a uniform mass m is introduced following the procedure in ref. [54]. Although ultimately there are no one- and two-loop ultraviolet divergences in $\mathcal{N} = 4$ supergravity, individual integrals generally will contain subdivergences, so it is best to systematically subtract these to ensure that no regulator dependence remains. Extractions of ultraviolet divergences in higher-dimensional $\mathcal{N} = 8$ supergravity were discussed recently in refs. [27, 6].

The next task is to calculate the ultraviolet divergences of the vacuum integrals. To evaluate these integrals, Lorentz invariance is used to replace tensors composed of loop momenta in the numerators by linear combinations of products of metric tensors $\eta^{\mu\nu}$ and dot products of momenta amongst themselves. (See ref. [6] for a more detailed discussion of simplifying tensor vacuum integrals.) A reduction of the scalar integrals to a basis, is carried out using integration-by-parts identities, as implemented in the program FIRE [55]. The resulting basis scalar integrals have been evaluated in refs. [56]. Combining this give the ultraviolet divergence of the amplitude [7].

In table 2, the the derived divergences of the three-loop four-graviton amplitude for each graph in fig. 3 for external gravitons. The individual graphs are not gauge invariant; in particular, the vanishing the contribution (a)-(d) in table 2 is due to a particular gauge choice. The results shown in the table for each graph are summed over independent permutations and include symmetry factors. We have divided out a prefactor depending on the four-point color-ordered sYM tree amplitude, spinor inner products and the usual Mandelstam invariants s and t. The sum over the contributions of all graphs vanish, so the three-loop four-graviton amplitude is ultraviolet finite. In fact, this cancellation holds for all external states of the theory [7], demonstrating that there are no three-loop four-point divergences in pure $\mathcal{N} = 4$ supergravity.

A key question is whether this result can be understood in terms of supersymmetry and duality symmetry, or whether a new mechanism is involved. The most conservative potential explanation would be to conjecture that a harmonic superspace exists mani-

Graph	(divergence)$/(\langle 12\rangle^2[34]^2 st A^{\text{tree}}(\frac{\kappa}{2})^8)$
(a)-(d)	0
(e)	$\frac{263}{768}\frac{1}{\epsilon^3} + \frac{205}{27648}\frac{1}{\epsilon^2} + \left(-\frac{5551}{768}\zeta_3 + \frac{326317}{110592}\right)\frac{1}{\epsilon}$
(f)	$-\frac{175}{2304}\frac{1}{\epsilon^3} - \frac{1}{4}\frac{1}{\epsilon^2} + \left(\frac{593}{288}\zeta_3 - \frac{217571}{165888}\right)\frac{1}{\epsilon}$
(g)	$-\frac{11}{36}\frac{1}{\epsilon^3} + \frac{2057}{6912}\frac{1}{\epsilon^2} + \left(\frac{10769}{2304}\zeta_3 - \frac{226201}{165888}\right)\frac{1}{\epsilon}$
(h)	$-\frac{3}{32}\frac{1}{\epsilon^3} - \frac{41}{1536}\frac{1}{\epsilon^2} + \left(\frac{3227}{2304}\zeta_3 - \frac{3329}{18432}\right)\frac{1}{\epsilon}$
(i)	$\frac{17}{128}\frac{1}{\epsilon^3} - \frac{29}{1024}\frac{1}{\epsilon^2} + \left(-\frac{2087}{2304}\zeta_3 - \frac{10495}{110592}\right)\frac{1}{\epsilon}$
(j)	$-\frac{15}{32}\frac{1}{\epsilon^3} + \frac{9}{64}\frac{1}{\epsilon^2} + \left(\frac{101}{12}\zeta_3 - \frac{3227}{1152}\right)\frac{1}{\epsilon}$
(k)	$\frac{5}{64}\frac{1}{\epsilon^3} + \frac{89}{1152}\frac{1}{\epsilon^2} + \left(-\frac{377}{144}\zeta_3 + \frac{287}{432}\right)\frac{1}{\epsilon}$
(l)	$\frac{25}{64}\frac{1}{\epsilon^3} - \frac{251}{1152}\frac{1}{\epsilon^2} + \left(-\frac{835}{144}\zeta_3 + \frac{7385}{3456}\right)\frac{1}{\epsilon}$

Table 2: The divergences for each graph in the four-graviton amplitude with helicities $(1^-2^-3^+4^+)$, as computed in ref. [7]. Each expression includes a permutation sum over external legs, with the symmetry factor appropriate to the graph. The sum over all contributions vanishes.

festing the full supersymmetry [25]. If this were the explanation we would expect that the theory diverges at four loops; this would be important to test. Other potential explanations are based on a hidden superconformal invariance [26] or nontrivial links to cancellations appearing in subleading-color terms of gauge theories [8].

5 Conclusions and outlook

In this talk, we described a new duality between color and kinematics and how it offers a new way to study the structure of supergravity scattering amplitudes, including their ultraviolet properties. Although there has been good progress in understanding the underlying origin and structure of this duality [40, 43, 44, 46, 47], much more would be be needed to fully unravel it. Besides desiring a group-theoretic description of the duality, it would be important to find systematic constructive rules for building duality satisfying numerators at loop level. In the mean time, it offers a powerful means for studying the scattering amplitudes of gravity amplitudes. Indeed when combined with the unitarity method, it offers a practical means for directly computing the ultraviolet divergences at high loop orders.

As one particularly interesting recent example, we summarized the demonstration that coefficient of the potential three-loop four-point divergence of $\mathcal{N}=4$ supergravity [7] vanishes. This is concrete example of an ultraviolet cancellation in supergravity that has not yet been understood from more traditional symmetry considerations [24]. Some recent conjectures, point to possible explanations [26, 8, 25]. These conjectures emphasize the importance of carrying out further explicit computations to determine if any of these explanations are viable. Obvious next steps are to determine the ultraviolet properties of $\mathcal{N}=8$ supergravity at five loops and $\mathcal{N}=4$ supergravity at four loops and to find the mechanism underlying all uncovered ultraviolet cancellations.

In summary, the newly discovered duality between color and kinematics and the associated double-copy formula for gravity integrands gives us a powerful new way to study gauge and gravity theories. We can look forward to many new results in the coming years.

The work presented in this talk was carried out in various papers in collaborations with John Joseph Carrasco, Scott Davies, Tristan Dennen, Lance Dixon, Yu-tin Huang, Henrik Johansson, and Radu Roiban. This research was supported in part by the US Department of Energy under contract DE–FG03–91ER40662.

References

[1] G. 't Hooft and M. J. Veltman, *Annales Poincare Phys. Theor.* A **20**, 69 (1974);

[2] P. S. Howe and K. S. Stelle, Int. J. Mod. Phys. A **4**, 1871 (1989).

[3] Z. Bern, J. J. M. Carrasco and H. Johansson, Phys. Rev. D **78**, 085011 (2008) [arXiv:0805.3993 [hep-ph]].

[4] Z. Bern, J. J. M. Carrasco and H. Johansson, Phys. Rev. Lett. **105**, 061602 (2010) [arXiv:1004.0476 [hep-th]].

[5] Z. Bern, C. Boucher-Veronneau and H. Johansson, Phys. Rev. D **84**, 105035 (2011) [arXiv:1107.1935 [hep-th]]; C. Boucher-Veronneau and L. J. Dixon, JHEP **1112**, 046 (2011) [arXiv:1110.1132 [hep-th]]; J. J. M. Carrasco, M. Chiodaroli, M. Gunaydin and R. Roiban, arXiv:1212.1146 [hep-th].

[6] Z. Bern, J. J. M. Carrasco, L. J. Dixon, H. Johansson and R. Roiban, arXiv:1201.5366 [hep-th].

[7] Z. Bern, S. Davies, T. Dennen and Y.-t. Huang, Phys. Rev. Lett. **108**, 201301 (2012) [arXiv:1202.3423 [hep-th]].

[8] Z. Bern, S. Davies, T. Dennen and Y.-t. Huang, arXiv:1209.2472 [hep-th]; R. H. Boels and R. S. Isermann, arXiv:1212.3473 [hep-th].

[9] J. M. Maldacena, Adv. Thew. Math. Phys. **2**, 231 (1998) [Int. J. Theor. Phys. **38**, 1113 (1999)] [hep-th/9711200]; S. S. Gubser, I. R. Klebanov and A. M. Polyakov, Phys. Lett. B **428**, 105 (1998) [hep-th/9802109].

[10] H. Kawai, D. C. Lewellen and S.-H. H. Tye, Nucl. Phys. B **269**, 1 (1986); Z. Bern, Living Rev. Rel. **5**, 5 (2002) [gr-qc/0206071].

[11] E. Cremmer and B. Julia, Nucl. Phys. B **159**, 141 (1979).

[12] E. Cremmer, J. Scherk and S. Ferrara, Phys. Lett. B **74**, 61 (1978).

[13] Z. Bern, L. J. Dixon, D. C. Dunbar and D. A. Kosower, Nucl. Phys. B **425**, 217 (1994) [arXiv:hep-ph/9403226]; Nucl. Phys. B **435**, 59 (1995) [arXiv:hep-ph/9409265].

[14] Z. Bern, J. S. Rozowsky and B. Yan, Phys. Lett. B **401**, 273 (1997) [arXiv:hep-ph/9702424]; Z. Bern, L. J. Dixon, D. C. Dunbar, M. Perelstein and J. S. Rozowsky, Nucl. Phys. B **530**, 401 (1998) [arXiv:hep-th/9802162]; Z. Bern, J. J. Carrasco, L. J. Dixon, H. Johansson, D. A. Kosower and R. Roiban, Phys. Rev. Lett. **98**, 161303 (2007) [arXiv:hep-th/0702112]; Z. Bern, J. J. Carrasco, L. J. Dixon, H. Johansson and R. Roiban, Phys. Rev. Lett. **103**, 081301 (2009) [arXiv:0905.2326 [hep-th]].

[15] Z. Bern, J. J. M. Carrasco, H. Johansson and D. A. Kosower, Phys. Rev. D **76**, 125020 (2007) [arXiv:0705.1864 [hep-th]].

[16] V. A. Smirnov, *Feynman integral calculus,* Berlin, Germany: Springer (2006) 283 p

[17] M. B. Green, J. G. Russo and P. Vanhove, JHEP **1006**, 075 (2010)[arXiv:1002.3805 [hep-th]]; J. Bjornsson and M. B. Green, JHEP **1008**, 132 (2010) [arXiv:1004.2692 [hep-th]]. H. Elvang and M. Kiermaier, JHEP **1010**, 108 (2010) [arXiv:1007.4813 [hep-th]]; G. Bossard, P. S. Howe and K. S. Stelle, JHEP **1101**, 020 (2011) [arXiv:1009.0743 [hep-th]]; N. Beisert, H. Elvang, D. Z. Freedman, M. Kiermaier, A. Morales and S. Stieberger, Phys. Lett. B **694**, 265 (2010) [arXiv:1009.1643 [hep-th]].

[18] R. Kallosh, JHEP **1203**, 083 (2012) [arXiv:1103.4115 [hep-th]].

[19] T. Banks, arXiv:1205.5768 [hep-th].

[20] Z. Bern, L. J. Dixon and R. Roiban, Phys. Lett. B **644**, 265 (2007) [arXiv:hep-th/0611086].

[21] Z. Bern, J. J. Carrasco, D. Forde, H. Ita and H. Johansson, Phys. Rev. D **77**, 025010 (2008) [arXiv:0707.1035 [hep-th]].

[22] P. Tourkine and P. Vanhove, Class. Quant. Grav. **29**, 115006 (2012) [arXiv:1202.3692 [hep-th]].

[23] S. Deser, J. H. Kay and K. S. Stelle, Phys. Rev. Lett. **38**, 527 (1977);
S. Deser and J. H. Kay, Phys. Lett. B **76**, 400 (1978);
M. T. Grisaru and W. Siegel, Nucl. Phys. B **201**, 292 (1982) [Erratum-ibid. B **206**, 496 (1982)].

[24] G. Bossard, P. S. Howe, K. S. Stelle and P. Vanhove, Class. Quant. Grav. **28**, 215005 (2011) [arXiv:1105.6087 [hep-th]].

[25] G. Bossard, P. S. Howe and K. S. Stelle, arXiv:1212.0841 [hep-th].

[26] S. Ferrara, R. Kallosh and A. Van Proeyen, arXiv:1209.0418 [hep-th].

[27] Z. Bern, J. J. M. Carrasco, L. J. Dixon, H. Johansson and R. Roiban, Phys. Rev. D **82**, 125040 (2010) [arXiv:1008.3327 [hep-th]].

[28] R. Britto, F. Cachazo, B. Feng and E. Witten, Phys. Rev. Lett. **94**, 181602 (2005) [hep-th/0501052].

[29] Z. Bern and A. G. Morgan, Nucl. Phys. B **467**, 479 (1996) [hep-ph/9511336]; Z. Bern, L. J. Dixon, D. C. Dunbar and D. A. Kosower, Phys. Lett. B **394**, 105 (1997) [hep-th/9611127]; Z. Bern, L. J. Dixon and D. A. Kosower, JHEP **0001**, 027 (2000) [hep-ph/0001001]; Z. Bern, L. J. Dixon and D. A. Kosower, Nucl. Phys. B **513**, 3 (1998) [hep-ph/9708239]; Z. Bern, L. J. Dixon and D. A. Kosower, JHEP **0408**, 012 (2004) [hep-ph/0404293]; Z. Bern, V. Del Duca, L. J. Dixon and D. A. Kosower, Phys. Rev. D **71**, 045006 (2005) [hep-th/0410224]; R. Britto, F. Cachazo and B. Feng, Nucl. Phys. B **725**, 275 (2005) [hep-th/0412103].

[30] Z. Bern, J. J. Carrasco, T. Dennen, Y.t. Huang and H. Ita, Phys. Rev. D **83**, 085022 (2011) [arXiv:1010.0494 [hep-th]]; A. Brandhuber, D. Korres, D. Koschade and G. Travaglini, JHEP **1102**, 077 (2011) [arXiv:1010.1515 [hep-th]].

[31] Z. Bern, L. J. Dixon and D. A. Kosower, Ann. Rev. Nucl. Part. Sci. **46**, 109 (1996) [hep-ph/9602280]; Z. Bern, L. J. Dixon and D. A. Kosower, Annals Phys. **322**, 1587 (2007) [arXiv:0704.2798 [hep-ph]]; Z. Bern and Y.t. Huang, J. Phys. A **44**, 454003 (2011) [arXiv:1103.1869 [hep-th]].

[32] N. Arkani-Hamed, J. L. Bourjaily, F. Cachazo, S. Caron-Huot and J. Trnka, arXiv:1008.2958 [hep-th]; N. Arkani-Hamed, J. L. Bourjaily, F. Cachazo and J. Trnka, arXiv:1012.6032 [hep-th].

[33] R. J. Eden, P. V. Landshoff, D. I. Olive, J. C. Polkinghorne, *The Analytic S Matrix* (Cambridge University Press, 1966).

[34] B.S. DeWitt, Phys. Rev. 162:1239 (1967); M. Veltman, in *Les Houches 1975, Proceedings, Methods In Field Theory*, eds. R. Balian and J. Zinn-Justin (North-Holland, Amsterdam, 1976); S. Sannan, Phys. Rev. D34:1749 (1986).

[35] A. E. M. van de Ven, *Nucl. Phys.* B **378**, 309 (1992).

[36] Z. Bern and A. K. Grant, Phys. Lett. B **457**, 23 (1999) [hep-th/9904026].

[37] J. J. M. Carrasco and H. Johansson, J. Phys. A **44**, 454004 (2011) [arXiv:1103.3298 [hep-th]].

[38] D. Zhu, Phys. Rev. D **22**, 2266 (1980); C. J. Goebel, F. Halzen and J. P. Leveille, Phys. Rev. D **23**, 2682 (1981).

[39] N. E. J. Bjerrum-Bohr, P. H. Damgaard, T. Sondergaard and P. Vanhove, JHEP **1101**, 001 (2011) [1010.3933 [hep-th]]; C. R. Mafra, O. Schlotterer and S. Stieberger, 1104.5224 [hep-th].

[40] N. E. J. Bjerrum-Bohr, P. H. Damgaard, T. Sondergaard and P. Vanhove, arXiv:1003.2403 [hep-th];

[41] S.-H. H. Tye and Y. Zhang, JHEP **1006**, 071 (2010) [arXiv:1003.1732 [hep-th]].

[42] N. E. J. Bjerrum-Bohr, P. H. Damgaard and P. Vanhove, Phys. Rev. Lett. **103**, 161602 (2009) [0907.1425 [hep-th]]; S. Stieberger, 0907.2211 [hep-th]. B. Feng, R. Huang and Y. Jia, Phys. Lett. B **695**, 350 (2011) [arXiv:1004.3417 [hep-th]]. Y. X. Chen, Y. J. Du and B. Feng, JHEP **1102**, 112 (2011) [1101.0009 [hep-th]]; F. Cachazo, arXiv:1206.5970 [hep-th].

[43] Z. Bern, T. Dennen, Y.t. Huang and M. Kiermaier, Phys. Rev. D **82**, 065003 (2010) [arXiv:1004.0693 [hep-th]].

[44] R. Monteiro and D. O'Connell, JHEP **1107**, 007 (2011) [arXiv:1105.2565 [hep-th]].

[45] Z. Bern and T. Dennen, 1103.0312 [hep-th].

[46] N. E. J. Bjerrum-Bohr, P. H. Damgaard, R. Monteiro and D. O'Connell, JHEP **1206**, 061 (2012) [arXiv:1203.0944 [hep-th]].

[47] J. Broedel and J. J. M. Carrasco, Phys. Rev. D **84**, 085009 (2011) [arXiv:1107.4802 [hep-th]]; A. S. Vera, E. S. Campillo and M. A. Vazquez-Mozo, arXiv:1212.5103 [hep-th]; C. -H. Fu, Y. -J. Du and B. Feng, arXiv:1212.6168 [hep-th].

[48] T. Bargheer, S. He and T. McLoughlin, arXiv:1203.0562 [hep-th]; Y.-t. Huang and H. Johansson, arXiv:1210.2255 [hep-th].

[49] J. Broedel and L. J. Dixon, [arXiv:1208.0876 [hep-th]].

[50] J. J. Carrasco and H. Johansson, Phys. Rev. D **85**, 025006 (2012) [arXiv:1106.4711 [hep-th]].

[51] S. Oxburgh and C. D. White, arXiv:1210.1110 [hep-th].

[52] R. H. Boels, B. A. Kniehl, O. V. Tarasov and G. Yang, arXiv:1211.7028 [hep-th].

[53] N. E. J. Bjerrum-Bohr, P. H. Damgaard, H. Johansson and T. Sondergaard, JHEP **1105**, 039 (2011) [arXiv:1103.6190 [hep-th]]; R. H. Boels and R. S. Isermann, JHEP **1203**, 051 (2012) [arXiv:1110.4462 [hep-th]]; E. Y. Yuan, arXiv:1210.1816 [hep-th].

[54] A. A. Vladimirov, Theor. Math. Phys. **43**, 417 (1980) [Teor. Mat. Fiz. **43**, 210 (1980)]; N. Marcus and A. Sagnotti, Nuovo Cim. A **87**, 1 (1985).

[55] A. V. Smirnov, JHEP **0810**, 107 (2008) [arXiv:0807.3243 [hep-ph]].

[56] K. G. Chetyrkin, M. Misiak and M. Munz, Nucl. Phys. B **518**, 473 (1998) [hep-ph/9711266]; M. Czakon, Nucl. Phys. B **710**, 485 (2005) [hep-ph/0411261].

CHAIRMAN: Z. BERN

Scientific Secretaries: L. Alberte, A. Khmelnitskiy

DISCUSSION

- *Klopot:*

Can the approach you developed be applied to higher-order calculations in QCD?

- *Bern:*

Yes. I was very quick in the talk, but the unitarity method has been applied to the next to leading order QCD for LHC physic. In particular, it is being applied to multi-jet processes, to processes like $Z + 3$ or 4 jet production. In fact, it is the only way how to do these calculations today. Even calculations as complex as $W + 5$ jets, which people have considered to be extremely difficult if not impossible via Feynman diagrams are doable. The on-shell techniques really do help us to do QCD calculations directly relevant to collider physics.

- *Klopot:*

As far as I understand heavy computer algebra calculations are done to obtain the results. Did any other group check your results? Are you sure that everything is correct?

- *Bern:*

The way not to make a mistake is always to be extremely paranoid about mistakes. At the high loop orders we are now working other groups have not checked directly though various parts have been checked. Other groups have certainly checked the results where we say that there is a cancellation and something is finite. Then they go off and they find, or at least in the past, they have found symmetry explanations confirming our results. In addition there are examples, such as gluon-gluon scattering in QCD that has been done by on-shell techniques by us and it has also been done with Feynman diagram by Nigel Glover and his group; the results are in complete agreement. There are also extremely strong checks on many of the computations. For example, we know a lot about infrared singularities in four dimension thanks to the work of Magnea and Sterman and others. This allows us to check in many cases whether the infrared divegences of amplitudes come out correctly. More generally, we spend an enormous amount of time in dealing with exactly your question: how do we know that we actually did it correctly? In this game if you make a bad mistake once, people won't believe next time. So we try to be careful.

- *Knoops:*

This might be a naive question. If $N = 8$ supergravities turn out to be finite, you said that then there would be some new symmetries that we do not know whether there are there yet. Could these symmetries be exactly the ones that you started with, namely, the symmetry between the color and the kinematics? Could this be exactly the reason why it would be finite?

- *Bern:*

The answer is yes. I am convinced that the duality between color and kinematics plays a central role, but unfortunately as of today we do not know how to use that to construct a proof. But there is absolutely no question in my mind that this is a key property of gravity and arguments on counterterms which do not account for it are incomplete. No one has figured out how to include it in a good way in the Lagrangian formulation to make the double copy property is obvious. I do believe that the answer to your question will be yes at some time in the future, but right now we just do not know enough to make a definite statement.

- *Knoops:*

In the beginning of your talk you seem to be putting every particle on shell. Would this mean that this is really a physical thing to do? That there is no such thing as an off shell particle in the end in what you do?

- *Bern:*

Part of it is that the quantities we are looking at are the scattering amplitudes which by definition have the external legs on shell. So I guess you are asking about the inside of the interaction. The point is that the construction of this object is not the way you read about it in textbooks but you can construct it entirely using only on shell information. That is the precise claim. As far as the S-matrix is concerned there is no off shell information needed. Off shell information is completely irrelevant.

- *Sleptsov:*

As we know ordinary quantum gravity is a non-renormalizable theory. Can the gauge-gravity duality you described help us to solve this problem and to deal with this problem? Or the quantum gravity is just a toy model for this duality?

- *Bern:*

In nonsupersymmetric pure gravity none of this can help because people have done explicit calculations showing divergence and they have been carefully checked by other groups. But with the right matter contents, that is basically what supersymmetry is, we have been arguing that it might be finite. Whether this is true, it remains to be seen.

- *Sleptsov:*

Recently a nice AGT (Alday-Gaiotto-Tachikawa) relation has been discovered. Can this relation be incorporated in your duality, in your picture somehow?

- *Bern:*

I do not know. Right now we only have limited examples where the duality works. It works in pure supergravity theories, quite generally in D dimensions. It also works in Bagger-Lambert theory. But I don not know of other examples. I am pretty sure no one has yet tried to connect it to the AGT relations.

- *Csanàd:*

You calculated basically in this framework the three and four loop diagrams and they turned out to be finite. I am wondering how far one can go with that and what happens if at some loops you encounter some problems?

- *Bern:*

The way I see it is that the people who should be very excited about us doing these calculations are those who think that we will encounter problems proving their beliefs. To prove that there is a divergence you actually have to do the calculation. Power counting arguments can never prove that a theory diverges, leaving the question undecided. Of course, I do believe it will turn out finite, but much more important we now have in place the tools needed to settle the issue. In my talk I showed a concrete example in $N = 4$ supergravity where power counting based on known symmetries is misleading. Obviously, at some point you run out of the steam and cannot do the computations, obviously we are not going to be doing a hundred loops explicitly. The basic point of the explicit calculations is to push things far enough so that all the potential standard symmetry explanations that people can possibly come up run out. At that point you have the smoking gun that something important is going on. More important, with the explicit calculations we should be able to learn enough to be able to construct the proof of finiteness, assuming it is true. For example, the duality between color and kinematics it all came from doing these explicit calculations. That is certainly a concrete example of how explicit calculations can teach you about the fundamental structure of the theory.

- Twigg:

To derive your duality relation you looked at the flat space. Could you do it for another background, maybe de Sitter and anti de Sitter space?

- Bern:

We have tried to look at this question, but nothing useful has come out of it yet. There may be way of doing it but no one has succeeded.

- Zoller:

I have also a question about the duality between color and kinematics in QCD. You showed as an example the duality between structure constants and metric tensors. I assume this also applies if you include quarks so that you have traces of gamma matrices and so on.

- Bern:

Yes, fermions can be included, as appears in super-Yang-Mills theories. However, it is important for them to be in the adjoint representation. It certainly becomes more confusing if the fermions are in the fundamental representation.

- Zoller:

It becomes more confusing or it cannot be done?

- Bern:

It becomes more confusing. Certainly, the most naïve way of doing it does not work. We have not really tried it very hard.

- Artymowski:

Can you use the same machinery for curved background. Not for Minkowski but for de Sitter for example.

- Bern:

I wish we could. We have looked a little bit for it, but no one has succeeded in doing that. One of the problems is even in flat space it is not so easy to understand the duality between the kinematics at the level we would like. In gravity things become obscure quickly because of algebraic complexity. I do believe there should be a way of doing it, but our attempts to do it have not gone far enough to really say anything useful about curved space. But I think it will work, once we have a better understanding of the flat space case.

- Artymowski:

Another very short question. If you have matter fields which are non-minimally coupled to gravity. The question is if you could use your machinery to calculate corrections to running of coupling constants, unitarity limits and so on. For example there are theories in which you couple Higgs field to Ricci scalar and this could produce inflation so that you do not need any new physics to produce inflation. Is it possible to calculate such things?

- *Bern:*

The thing is the formalism right now is very limited because we only understand the duality between the color and the kinematics and the double copy in the flat space and in pure supergravity theories. There are also some examples when you couple in certain matter multiplets and also unpublished work on higher dimension operators. But this is not a formalism that is ready for use in general cases. I certainly think that is an extremely interesting issue and we have devoted at least some attention to this question how to go beyond the relatively straightforward theories we are dealing with. Most questions in gravity are not in flat space and unfortunately we do not yet have an understanding to go out of flat space or to theories that have extra Higgs or non-minimal couplings.

- *Tumanov:*

The point you are trying to make is that $N = 8$ supergravity has some chance to be an ultraviolet finite model but I wanted to know your opinion on another matter. As you know perturbation theory deals only with small fluctuations of fields and if fluctuations are large, then there may be problems in some theories like for example scalar field ϕ^3 theory, which is stable for small fluctuations but it is not defined for large fluctuations of the field. The same situation I believe is in gravitational theory. What is your opinion on these large fluctuations and if $N = 8$ theory is well defined for the large fluctuations.

- *Bern:*

The way I would think about what we are doing here is to understand the structure and the symmetries of the theory using the tool of perturbation theory. There are all sorts of questions you can ask about large fluctuations or high energy behavior which perturbation theory is not really the right tool. The part we are focusing on is cancellations strong enough to make it finite. There must be a symmetry in the theory or a deep reason for this to happen. Once you have a proper understanding of the symmetries then I think we will be able to start answering many questions along the lines of what you want. But the questions are beyond perturbation theory so we do not have much to say about it at present.

- *Karakostas:*

In your presentation this morning you mentioned that you are looking for the information in the near future for your theories. What did you mean by this?

- *Bern:*

I meant that we need more information---theoretical data which comes out from doing explicit computations. You do a calculation and you see what the result is. It is not as good as experimental data, but the theoretical data is very important for guiding the thinking and knowing what you should do next. It has proved crucial for settling theoretical arguments. That is all I was referring to.

- *Melia:*

I guess when you are calculating supergravity amplitudes because there is no color ordering like in QED there should be some permutation symmetry when you swap any two of the external particles. What I am interested in knowing is if the method you described gives you the answer whether this permutation symmetry is really manifest or it is one of those things nontrivially hidden.

- *Bern:*

The permutation sum is quite explicit as it sits outside amplitude: you are supposed to sum over all the different orderings of legs. The importance of the permutation sum depends on which calculation you are doing. For example in $N = 8$ supergravity at four loops the key cancellations are independent of permutation sums. But for $N = 4$ supergravity at the cancellations become apparent only after carrying out the permutation sum.

- *Melia:*

So you color-order and then sum over color-ordered objects.

- *Bern:*

Yes you could say it that way if you want. At the end when you write down the sets of diagrams and their contributions, they follow very closely to the way things are set up in Yang-Mills theory. The difference is in Yang-Mills theory there are color factors that prevent different terms from mixing, where one term has one color factor and another term has a different color factor. In gravity all of this mixes, allowing for nontrivial cancellations to take place, but the structure of the formulas is exactly the same.

- *`t Hooft:*

Is it fair to say that the beautiful cancellations you see are due to the fact that the denominators are massless or do you have a hope that theories with masses will have them as well. Like in spontaneously broken gauge theories?

- *Bern:*

Yes, one way you can put in masses considering higher dimensional theories and then doing dimensional reduction. Then you can construct gravity theories that have masses. I do believe that the ultraviolet cancellations we see will happen in those theories. It certainly sounds like an interesting problem to look at.

CERN Preprint 2012-274

Black holes and Supersymmetry

Laura Andrianopoli[1], Riccardo D'Auria[1] and Sergio Ferrara[2]

[1] *DISAT, Politecnico di Torino, Corso Duca degli Abruzzi 24, I-10129 Turin, Italy and INFN Sezione di Torino, Italy;*
E-mail: laura.andrianopoli@polito.it; riccardo.dauria@polito.it;
[2] *Physics Department, Theory Unit, CERN, 1211 Geneva 23, Switzerland and INFN LNF, Via E. Fermi 40 00044 Frascati, Italy;*
E-mail: sergio.ferrara@cern.ch

Abstract

Some aspects of black holes in supersymmetric theories of gravity are reviewed and some recent results outlined.

Contribution to the Proceedings of the International School of Subnuclear Physics, 50th Course: "What We Would Like LHC to Give Us"
Erice, 23 June - 2 July 2012, based on a talk given by Sergio Ferrara.

1 Introduction

Einstein's theory of General Relativity was the result of a deep intuition by A. Einstein on the implications of the equivalence principle, whilst trying to extend Special Relativity to gravitational phenomena which, in the Newtonian formulation, imply an instantaneous action-at-distance interaction among masses. The final assessment of the theory required the use of the mathematical formalism of Riemannian geometry, that was fully developed at the beginning of the XXth century by many famous mathematicians like D. Hilbert, T. Levi-Civita and G. Ricci. In this way, he could implement all the far reaching consequences of the equivalence principle in terms of a well established mathematical theory of gravitation. In this setting, the space-time geometry is determined by the energy and momentum content of the Universe. On of the most striking consequences of the theory is the prediction of the existence of black holes, very non intuitive objects surrounded by an event horizon from the interior of which no physical signal can escape.

From an experimental point of view, black holes are now observed in the sky as relics of collapsing stars. They seem to encompass many of the mysteries of the evolution of our Universe from its creation to its final destiny, the so-called Big Crunch, or its eternal existence, namely an endless expansion. Fundamental advances in the study of black holes have been given by Bekenstein and Hawking, who recognized that, because of the proportionality between Black-Hole area and Entropy, the laws governing the physics of black holes can be interpreted in terms of the laws of classical Thermodynamics. This strongly suggests the possibility of giving a statistical interpretation of the black-hole entropy in terms of a more fundamental microscopic theory underlying classical general Relativity. Furthermore, the strong gravitational field confined in finite regions such us the horizon surrounding of black holes, makes their study a useful arena for possible insights into a field theoretical description of Quantum Gravity. Actually, from a theoretical point of view, Superstring or M-Theory are possible theoretical frameworks to make precise predictions as to the existence and properties of black holes.

In general, astrophysical black holes have huge masses, typically of the order of magnitude of the solar mass scale, 2×10^{30} Kg, while Quantum Gravity black holes have tiny masses, of the order of the Planck mass scale, namely 2×10^{-8} Kg. Note that this is still much bigger than the typical mass of the atomic nuclei, that is from one to ten proton masses (the mass of a proton being 1.6×10^{-27} Kg).

In the following we shall be concerned with Supergravity black holes that is the black holes of the Superworld [1]. Supersymmetry requires that they are extremal, that is that they have vanishing temperature, are marginally stable but carry Entropy. Actually, the black-hole Entropy makes a bridge between classical gravity and Quantum Gravity. In fact, we recall that the macroscopic definition of the black-hole entropy (Bekenstein – Hawking Entropy) [2, 3] connects its value to the black-hole horizon area A_H:

$$S_{BH}^{macro} = \frac{k_B c^3}{G\hbar} \frac{1}{4} A_H \tag{1.1}$$

The microscopic definition of the black-hole entropy, instead, relates its value to the number N_{mic} of microstates of the quantum system underlying the black hole, namely:

$$S_{BH}^{micro} = k_B \ln(N_{\text{mic}}) \tag{1.2}$$

Remarkably these formulae, computed with appropriate approximations, give the same result in Superstring Theory [4].

From now on, we generally use Natural Units, where $c = \hbar = G = k_B = 1$.

2 Introducing the Superworld

In order to understand what Superworld is, one first has to introduce the notion of Superspace [6, 7]. This is a geometrical entity which extends the notion of Riemannian manifold to that of Supermanifold. A point on a D-dimensional Riemannian manifold $\mathcal{M}_D$, endowed with a Lorentz signature (*H. A. Lorentz, 1853-1928*), is identified by giving numerical coordinates x^μ, $(\mu = 1, \cdots, D)$. To identify a point in a Supermanifold we need, besides the coordinates x^μ, also a set of Grassmann (*H. Grassmann, 1809-1877*) anticommuting coordinates θ_α $(\alpha = 1, \cdots, 2^{[D/2]})$ with two basic properties:

1. $\theta_\alpha \theta_\beta = -\theta_\beta \theta_\alpha$, which implies nilpotency: $\theta_\alpha^2 = 0$;

2. They transform as spinors (*E. Cartan, 1869-1951, H. Weyl, 1885-1955*) under the action of the Lorentz group and their properties are related to modules of Clifford Algebras (*W. K. Clifford, 1845-1879*) and to the Spin Group, namely the universal covering group of the D-dimensional Lorentz group [5].

Superworld is the physical entity corresponding to the mathematical concept of supermanifold, whose environment is not ordinary space but superspace. The group of motion in Superspace is supersymmetry, as much as the group of motion in ordinary space-time is the Poincaré group (*H. Poincaré, 1854-1912*). An infinitesimal supersymmetry transformation with spinorial parameter ϵ_α acts on the coordinates of superspace as follows:

$$x^\mu \to x^\mu + \mathrm{i}\bar{\epsilon}^\alpha (\gamma^\mu)_\alpha{}^\beta \theta_\beta \quad \Leftrightarrow \quad \delta x^\mu = \mathrm{i}\bar{\epsilon}^\alpha (\gamma^\mu)_\alpha{}^\beta \theta_\beta \tag{2.1}$$

$$\theta_\alpha \to \theta_\alpha + \epsilon_\alpha \quad \Leftrightarrow \quad \delta\theta_\alpha = \epsilon_\alpha \tag{2.2}$$

where γ^μ is a matrix satisfying the Clifford Algebra and $\bar{\epsilon}$ denotes the Dirac conjugate spinor, namely $\bar{\epsilon} = \epsilon^\dagger \gamma^0$. Commuting twice the action of such transformations with parameters ϵ_1 and ϵ_2 respectively, one finds that x^μ undergoes an infinitesimal translation:

$$[\delta_1, \delta_2] x^\mu = 2\mathrm{i}\bar{\epsilon}_2^\alpha (\gamma^\mu)_\alpha{}^\beta \epsilon_{1\beta} \,. \tag{2.3}$$

The supersymmetry algebra is a graded Lie algebra [8] (*S. Lie, 1842-1899*) with basic anticommutator: [9, 10, 11]

$$\{Q_\alpha, \bar{Q}_\beta\} = 2(\gamma^\mu C)_{\alpha\beta} p_\mu \tag{2.4}$$

where the supersymmetry generators Q_α are Majorana spinors (*E. Majorana, 1906-1938*) and C denotes the charge-conjugation matrix.

The supergroup associated to the supersymmetry algebra acts on a supermanifold, denoted by $\mathcal{M}_{b,f} \equiv \mathcal{M}_{D,2^{[D/2]}}$, where b and f denote the number of bosonic and fermionic coordinates respectively. the total (graded) dimension of a supermanifold is $b + f$. As we will see in the following, the maximal possible number of coordinates of superspace is $b_{max} = 11$, $f_{max} = 32$.

Actually, one can extend the super Lie algebra by introducing N supersymmetry generators $Q_{\alpha I}$ $(I = 1, \cdots, N)$ acting on an N-extended superspace labeled by N Grassmannian spinor coordinates $\theta_{\alpha I}$. The basic anticommutators now become

$$\{Q_{\alpha I}, \bar{Q}^J_\beta\} = 2(\gamma^\mu C)_{\alpha\beta} p_\mu \delta_I^J \tag{2.5}$$

$$\{Q_{\alpha I}, Q_{\beta J}\} = \varepsilon_{\alpha\beta} Z_{IJ} \tag{2.6}$$

where Z_{IJ} are "central terms" which commute with all the rest of the superalgebra, including the Lorentz generators. It is precisely the presence of the central charges Z_{IJ} which makes it possible the existence of supersymmetric Black Holes, as will be shown in the next section.

The interaction in the superworld are described by supersymmetric theories. It is remarkable that such theories may encompass gauge interactions, in particular Yang–Mills theories [12, 6], as well as gravitational interactions. In the latter case, the corresponding theory is called supergravity [13, 14]. However these theories exist only for few values of the number N of supersymmetries and of the space-time dimension D [15, 16]. In particular, Super Yang–Mills theories in $D = 4$ require $1 \leq N \leq 4$ and at most they live in $D = 10$ [17]. On the other hand supergravity theories at $D = 4$ require $1 \leq N \leq 8$ and at most they live in $D = 11$ dimensions [18].

3 General Properties of Extremal Black Holes.

The celebrated Black-Hole solution of pure Einstein theory looks, in a chosen frame of spherical coordinates

$$ds^2_{\text{Schw}} = -\left(1 - \frac{2M}{r}\right) dt^2 + \left(1 - \frac{2M}{r}\right)^{-1} dr^2 + r^2 d\Omega^2 \tag{3.1}$$

where M denotes the ADM mass, that is the total energy of the black-hole configuration. The naked singularity at $r = 0$ is covered by the *event horizon* at $r = 2M$. By event horizon we mean a surface where the gravitational red-shift is infinite, that is where the time intervals undergo an infinite dilation with respect to a distant observer. This is also a singularity of the metric but it is only a *coordinate* singularity which can be removed with an appropriate choice of coordinates, while the singularity at $r = 0$ is a real singularity of the theory, that is independent of the reference frame.

The generalization of the Schwarzschild solution to an electrically charged black hole in the Einstein–Maxwell theory is given by the Reissner–Nordström black hole, whose metric reads:

$$ds^2_{\text{RN}} = -\left(1 - \frac{2M}{r} + \frac{Q^2}{r^2}\right) dt^2 + \left(1 - \frac{2M}{r} + \frac{Q^2}{r^2}\right)^{-1} dr^2 + r^2 d\Omega^2 \tag{3.2}$$

Here M denotes the ADM mass and Q the electric charge of the space-time configuration. Such configuration can be easily generalized for dyonic configurations where also a magnetic charge P is present by replacing in (3.2) Q^2 with $Q^2 + P^2$. This metric exhibits two horizons, at

$$r_{\pm} = M \pm \sqrt{M^2 - Q^2} = M \pm r_0 \tag{3.3}$$

where r_+ corresponds to the event horizon and r_- to the Cauchy horizon, together with the physical singularity at $r = 0$.

In cosmology a Cosmic Censorship Principle is postulated (see for example [19]. An event horizon should always cover the singularity at $r = 0$, so that the singularity be not accessible to an observer external to the event horizon of the black hole. This can be rephrased by saying that no "naked" singularities can exist. For the Reissner–Nordström solution the Cosmic Censorship principle requires $M \geq Q$.

As shown by Steven Hawking, black holes obey laws which are formally the same as the laws of thermodynamics, after an appropriate identification of the quantum numbers of the solution is given. In particular, the thermodynamical properties of the black holes relate the area of the event horizon to the Entropy through the Bekenstein–Hawking formula (see eq. (1.1) in Natural Units):

$$S_{BH} = \frac{1}{4} A_H = \frac{1}{4} \pi R_+^2 \, , \tag{3.4}$$

where R_+ is the event horizon r_+ for the Reissner–Nordström black hole, while it becomes an effective radius in the presence of other black-hole attributes such as angular momentum J and/or scalar charges Σ. For instance, in the presence of the latter $R_+^2 = r_+^2 - \Sigma^2 \leq r_+^2$.

The fact that a black hole continuously increases its horizon area can be interpreted as the second law of thermodynamics if we identify the black hole entropy as proportional to the horizon area, as pointed out by S. Hawking. Further support to this interpretation is given by the other laws of thermodynamics. in particular, the 0th law of thermodynamics states that for a system in equilibrium there is a quantity, the temperature, which is constant. An analogous constant quantity exists for a black hole at equilibrium, the so-called surface gravity that for the Reissner–Nordström black hole is

$$\kappa = \frac{c}{(r_+ + r_-)r_+ - Q^2} \tag{3.5}$$

where

$$c = \frac{1}{2}(r_+ - r_-) \tag{3.6}$$

It is then possible to identify the black-hole temperature T_{BH} as

$$T_{BH} = \frac{1}{2\pi}\kappa = \frac{c}{2S_{BH}} \tag{3.7}$$

The analogy is completed by rewriting the first law of thermodynamics:

$$dE = TdS + \text{work terms} \tag{3.8}$$

as

$$dM = T_{BH}dS_{BH} + \cdots = \frac{\kappa}{2\pi}\frac{A_H}{4} + \cdots \tag{3.9}$$

and observing that the third law of thermodynamics, stating that it is impossible to achieve $T = 0$ by a finite number of physical processes, can be rephrased as the impossibility to achieve $\kappa = 0$ by a finite number of physical processes.

The black hole which reaches the limit equilibrium temperature $\kappa = 0$ is called extremal. This corresponds to $c = 0$, that is to $r_+ = r_-$. For the Reissner–Nordström black hole, this happens when $M = |Q|$. A supergravity black hole is supersymmetric (BPS saturated) if its ADM mass M equals the highest skew-eigenvalue of the central charge matrix $Z_{IJ} = -Z_{JI}$ evaluated at asymptotic infinity. In the presence of scalar charges Σ, the extremality condition allows both for supersymmetric and non-supersymmetric black-hole configurations.

For stationary but non-static solutions, that is rotating black holes of angular momentum J (Kerr–Newman black holes), the horizon radii become

$$r_\pm = M \pm \sqrt{M^2 - Q^2 - P^2 - J^2/M^2} \tag{3.10}$$

so that for a neutral spinning black hole (Kerr black hole) we reach extremality when $M^2 = J$, that is when the extremality parameter $a^* \equiv J/(GM^2) = 1$. Kerr black holes have been observed in our galaxy, in particular **GRS 1915+105** is the heaviest of the stellar black holes so far known [20] in the Milky Way Galaxy, with 10 to 18 times the mass of the Sun and a value of spin $J \simeq 10^{78}\hbar$. It was discovered on 15 August 1992. It is a nearly extremal black hole since in this case the extremality parameter is $a^* = 0.98 \simeq 1$. It has been argued that such black hole has an exact Conformal Field Theory dual [21].

4 Black Holes in Supergravity and Groups of Type E_7

One of the main properties of supergravity is the presence of scalar fields not minimally coupled to vector fields. The typical form of the Lagrangian of a set of electromagnetic field strengths $F^\Lambda = dA^\Lambda$ (enumerated by capital Greek indices Λ, Σ) in supergravity is of the form:

$$\mathcal{L} \propto g_{\Lambda\Sigma}(\varphi) F^\Lambda_{\mu\nu} F^{\Sigma|\mu\nu} + \Theta_{\Lambda\Sigma}(\varphi) \frac{1}{2} F^\Lambda_{\mu\nu} F^\Sigma_{\rho\sigma} \epsilon_{\mu\nu\rho\sigma} \tag{4.1}$$

where the couplings $g_{\Lambda\Sigma}$ and $\Theta_{\Lambda\Sigma}$ depend on a set of scalar fields enumerated by an index s. This has the implication that the Maxwell–Einstein black hole solution gets a non-trivial modification. In particular, the metric flow of the black hole towards the horizon $r = r_H$ is accompanied by trajectories of scalar-fields evolution from asymptotic infinity to the horizon:

$$\begin{aligned} \lim_{r\to\infty} \varphi^s(r) &= \varphi^s_\infty \in \mathcal{M} \\ \lim_{r\to r_H} \varphi^s(r) &= \varphi^s_{crit} \in \mathcal{M} \end{aligned} \tag{4.2}$$

The resulting analysis exploits the *attractor mechanism* [22]. Indeed scalar fields behave as dynamical systems which, in their evolution towards the black-hole horizon of an extremal black hole, loose memory of their initial conditions (at $\varphi = \varphi_\infty$) approaching a critical point where the first derivative vanishes:

$$\begin{aligned} \lim_{r\to r_H} \varphi^s(r) &= \varphi^s_{crit}(Q) \\ \lim_{r\to\infty} \frac{d}{dr}\varphi^s(r) &= 0\,, \end{aligned} \tag{4.3}$$

and whose value only depends on the set of charges Q. Consistency of the solution implies that φ_{crit} is a critical point of an *effective black-hole potential* $V_{BH}(\varphi, Q)$:

$$\lim_{r\to\infty} \frac{\partial}{\partial\varphi^s} V_{BH}|_{\varphi=\varphi_{crit}} = 0\,. \tag{4.4}$$

Moreover, the Bekenstein–Hawking entropy-area formula becomes [24, 25]:

$$S_{BH} = \frac{1}{4} A_H = \pi V_{BH}(Q, \varphi_{crit}(Q))\,. \tag{4.5}$$

Note that the critical value of the scalar fields for extremal black holes, satisfying the attractor mechanism, is given only in terms of the electric and magnetic vector of charges Q, and this explains why the entropy only depends on Q and not on the initial values of the scalar fields φ_∞.

The attractor mechanism allows to reduce the dynamical black-hole flow to a *first-order* evolution both for supersymmetric and non-supersymmetric extremal black holes [26, 27, 28, 29]. Indeed, the black-hole potential V_{BH} may be always written for extremal black holes as

$$V_{BH} = W^2 + 2\partial_s W \partial^s W \tag{4.6}$$

where $W(\varphi, Q)$ is known as the *(fake) superpotential.* There are several properties of W that make it important. First of all, in terms of W, the second order equations of motion of the theory reduce to a set of first order equations. Moreover W is invariant under the electric-magnetic duality group. It has a clear meaning in the context of the Hamilton–Jacobi theory, since it allows the interpretation of the flow as an Hamiltonian flow whose Hamilton characteristic function is actually simply related to W [30, 31].

The attractor mechanism allows to classify black-hole solutions, that is critical points of the black-hole potential, through the electric-magnetic duality symmetry of the theory. For each orbit of the duality symmetry, the fake superpotential W has a different expression. In the particular case of supersymmetric black holes, one obtains $W = |Z|$ where $|Z|$ is the highest skew-eigenvalue of the central charge matrix Z_{IJ}. The duality orbits are modules of *groups of type* E_7, as requested by the Gaillard–Zumino analysis [23] combined with the attractor mechanism. The group E_7 appears in supergravity as the duality group of of the maximally extended $N = 8$ theory in four dimensions, in its symplectic 56 dimensional module relating 28 electric to 28 magnetic charges. The orbits of the **56** module classify the black-hole solutions preserving different fractions of the original $N = 8$ supersymmetry. Moreover, E_7 controls the ultraviolet divergences of perturbation theory since it is anomaly free, and its arithmetic subgroups $G \subset E_7(\mathbb{Z})$ may encode the non-perturbative quantum corrections. It happens that all the duality groups of four dimensional supergravity theories with a number of supersymmetries $N < 8$ are groups of type E_7, that is they have symplectic representations admitting a symmetric quartic invariant polynomial, but not a quadratic one [32]. As an example, we have presented in Table 1 the possible $N = 2$ choices of groups G of the $\frac{G}{H}$ symmetric spaces and their symplectic representations

	G	R module	Primitive symm. inv.
$J_3^{\mathbb{O}}$	$E_{7(-25)}$	56	I_4
$J_3^{\mathbb{H}}$	$SO^*(12)$	32	I_4
$J_3^{\mathbb{C}}$	$SU(3,3)$	20	I_4
$J_3^{\mathbb{R}}$	$Sp(6,\mathbb{R})$	$14'$	I_4
T^3	$SL(2,\mathbb{R})$	4	I_4
$J_{2,n}$	$SL(2,\mathbb{R}) \times SO(2,n)$	$(2,2+n)$	I_4
CP^n	$U(1,n)$	$(1+n)_C$	I_2

Table 1: Supergravity sequence for $N = 2$ symmetric spaces.

R [33, 34]. The first column identifies the scalar manifold whose isometry group is G is given. In particular, for the first four entries, they are named with the Jordan algebras J_3 over octonions, quaternions, complex and real numbers respectively, to which they are related. As we see in the last column, all the duality groups listed are of Type E_7 groups [35] except the last one, which has a primitive quadratic invariant polynomial. For $N \geqslant 3$ supergravity theories the analogous sequence is given in Table 2. Again, all the groups appearing in the sequence are of type E_7 except the first one, which admits a primitive symmetric invariant polynomial of order 2.

5 Future directions of research

We are just at the beginning of the exploration of the beautiful intricacy given by supergravity black-hole physics, its group-theoretical structure and quantum perspectives. It is clear that much work and effort has to be done to unveil all the physics behind their structure which are emerging from supergravity considerations. In particular, we may mention few possible future directions of research where recent advances have occurred:

- Extension of black-hole solutions to multi-center configurations [36, 37], the classification of their orbits and the study of their dynamics, regarding their splitting behavior and their relation to the underlying stringy microstate counting [38, 39].
- Further study of the role of the group E_7 as far as quantum corrections are concerned [40, 41, 42].

N	G	**R**
$N = 3$	$U(3, n)$	$(\mathbf{3} + \mathbf{n})$
$N = 4$	$SL(2, \mathbb{R}) \otimes SO(6, n)$	$(\mathbf{2}, \mathbf{6} + \mathbf{n})$
$N = 5$	$SU(1, 5)$	$\mathbf{20}$
$N = 6$	$SO^*(12)$	$\mathbf{32}$
$N = 8$	$E_{7(7)}$	$\mathbf{56}$

Table 2: The supergravity sequence for $N \geq 3$

- Inclusion of the Attractor Mechanism in the presence of higher derivative modifications of gravity, as suggested by superstring theory [43].
- The role of $N = 8$ black holes in a perturbatively finite theory of $N = 8$ supergravity [44, 45, 46].

Acknowledgements

The work of S.F. is supported by the ERC Advanced Grant n. 226455, "Supergravity, Quantum Gravity and Gauge Fields". This work was supported in part by the MIUR-PRIN contract 2009-KHZKRX.

References

[1] Andrianopoli L., D'Auria R., Ferrara S. and Trigiante M. (2008). Extremal black holes in supergravity. In Gasperini M. and Maharana J.

(Ed.s). *String Theory and Fundamental Interactions. Lect. Notes Phys.* **737** 661. [hep-th/0611345].

[2] Hawking S. W. (1971). *Gravitational radiation from colliding black holes. Phys. Rev. Lett.* **26**, 1344.

[3] Bekenstein J. D. (1973). *Black holes and entropy. Phys. Rev. D* **7** , 2333.

[4] Strominger A. and Vafa C. (1996). *Microscopic origin of the Bekenstein-Hawking entropy. Phys. Lett. B* **379**, 99. [hep-th/9601029].

[5] Deligne P. (1999). Notes on spinors. In Deligne P., Etingof P., Freed D. S., Jeffrey L. C., Kazhdan D., Morgan J. W., Morrison D. R. and Witten E. (Ed.s). *Quantum fields and strings: A course for mathematicians. American Mathematical Society.*

[6] Salam A. and Strathdee J. A. (1975). *On Superfields and Fermi-Bose Symmetry. Phys. Rev. D* **11**, 1521.

[7] Ferrara S., Wess J. and Zumino B. (1974). *Supergauge Multiplets and Superfields. Phys. Lett. B* **51**, 239.

[8] Corwin L., Ne'eman Y. and Sternberg S. (1975). *Graded Lie Algebras in Mathematics and Physics (Bose-Fermi Symmetry). Rev. Mod. Phys.* **47**, 573.

[9] Wess J. and Zumino B. (1974). *Supergauge Transformations in Four-Dimensions. Nucl. Phys. B* **70**, 39.

[10] Golfand Y. .A. and Likhtman E. P. On N=1 Supersymmetry Algebra And Simple Models. In Shifman, M. A. (Ed.). *The many faces of the superworld*, 54-79

[11] Akulov V. P. and Volkov D. V. (1974). *Goldstone fields with spin 1/2. Theor. Math. Phys.* **18**, 28.

[12] Ferrara S. and Zumino B. (1974). *Supergauge Invariant Yang-Mills Theories. Nucl. Phys. B* **79**, 413.

[13] Freedman D. Z., van Nieuwenhuizen P. and Ferrara S. (1976). *Progress Toward a Theory of Supergravity. Phys. Rev. D* **13**, 3214.

[14] Deser S. and Zumino B. (1976). *Consistent Supergravity. Phys. Lett. B* **62**, 335.

[15] Gell-Mann M., Ramond P. and Slansky R. (1978). *Color Embeddings, Charge Assignments, and Proton Stability in Unified Gauge Theories. Rev. Mod. Phys.* **50**, 721.

Gell-Mann M., Ramond P. and Slansky R. (1979). *Complex Spinors And Unified Theories. Conf. Proc. C* **790927** 315.

[16] Nahm W. (1978). *Supersymmetries and their Representations. Nucl. Phys. B* **135**, 149.

[17] Brink L., Schwarz J. H. and Scherk J. (1977). *Supersymmetric Yang-Mills Theories. Nucl. Phys. B* **121**, 77.

[18] Cremmer E., Julia B. and Scherk J. (1978). *Supergravity Theory in Eleven-Dimensions. Phys. Lett. B* **76**, 409.

[19] Kallosh R., Linde A. D., Ortin T., Peet A. W. and Van Proeyen A. (1992). *Supersymmetry as a cosmic censor. Phys. Rev. D* **46**, 5278. [hep-th/9205027].

[20] McClintock J. E., Shafee R., Narayan R., Remillard R. A., Davis S. W. and Li L. -X. (2006). *The Spin of the Near-Extreme Kerr Black Hole GRS 1915+105. Astrophys. J.* **652**, 518. [astro-ph/0606076].

[21] Guica M., Hartman T., Song W. and Strominger A. (2009). *The Kerr/CFT Correspondence. Phys. Rev. D* **80**, 124008. [arXiv:0809.4266 [hep-th]].

[22] Ferrara S., Kallosh R. and Strominger A. (1995). *N=2 extremal black holes. Phys. Rev. D* **52**, 5412. [hep-th/9508072].

[23] Gaillard M. K. and Zumino B. (1981). *Duality Rotations for Interacting Fields. Nucl. Phys. B* **193**, 221.

[24] Ferrara S. and Kallosh R. (1996). *Supersymmetry and attractors. Phys. Rev. D* **54**, 1514. [hep-th/9602136].

[25] Ceresole A., D'Auria R. and Ferrara S. (1996). *The Symplectic structure of N=2 supergravity and its central extension. Nucl. Phys. Proc. Suppl.* **46**, 67. [hep-th/9509160].

[26] Ceresole A. and Dall'Agata G. (2007). *Flow Equations for Non-BPS Extremal Black Holes. JHEP* **0703**, 110. [hep-th/0702088].

[27] Andrianopoli L., D'Auria R., Orazi E. and Trigiante M. (2007). *First order description of black holes in moduli space. JHEP* **0711**, 032. [arXiv:0706.0712 [hep-th]].

[28] Ceresole A., Dall'Agata G., Ferrara S. and Yeranyan A. (2010). *First order flows for N=2 extremal black holes and duality invariants. Nucl. Phys. B* **824**, 239. [arXiv:0908.1110 [hep-th]].

Ceresole A., Dall'Agata G., Ferrara S. and Yeranyan A. (2010). *Universality of the superpotential for d = 4 extremal black holes. Nucl. Phys. B* **832**, 358. [arXiv:0910.2697 [hep-th]].

[29] Bossard G., Michel Y. and Pioline B. (2010). *Extremal black holes, nilpotent orbits and the true fake superpotential. JHEP* **1001**, 038. [arXiv:0908.1742 [hep-th]].

[30] Andrianopoli L., D'Auria R., Orazi E. and Trigiante M. (2010). *First Order Description of D=4 static Black Holes and the Hamilton-Jacobi equation. Nucl. Phys. B* **833**, 1. [arXiv:0905.3938 [hep-th]].

[31] Andrianopoli L., D'Auria R., Ferrara S. and Trigiante M. (2010). *Fake Superpotential for Large and Small Extremal Black Holes. JHEP* **1008**, 126. [arXiv:1002.4340 [hep-th]].

[32] Ferrara S. and Kallosh R. *Creation of Matter in the Universe and Groups of Type E7. arXiv:1110.4048 [hep-th].* To appear in *JHEP.*

[33] Gunaydin M., Sierra G. and Townsend P. K. (1984). *The Geometry of N=2 Maxwell-Einstein Supergravity and Jordan Algebras. Nucl. Phys. B* **242**, 244. Gunaydin M., Sierra G. and Townsend P. K. (1983). *Exceptional Supergravity Theories and the MAGIC Square. Phys. Lett. B* **133**, 72.

[34] Cremmer E. and Van Proeyen A. (1985). *Classification Of Kahler Manifolds In N=2 Vector Multiplet Supergravity Couplings.* Class. Quant. Grav. **2**, 445.

[35] Brown R. B. (1969). *Groups of Type E_7. J. Reine Angew. Math. 236*, 79.

Garibaldi R. S. (2001). *Groups of Type E_7 over Arbitrary Fields.*", *Commun. in Algebra 29*, 2689.

Meyberg K. (1968). *Eine Theorie der Freudenthalschen Triplesysteme. I, II. Nederl. Akad. Wetensch. Proc. Ser. A71*, 162.

[36] F. Denef, "Supergravity flows and D-brane stability," JHEP **0008** (2000) 050 [hep-th/0005049].

[37] G. Bossard and C. Ruef, "Interacting non-BPS black holes," Gen. Rel. Grav. **44** (2012) 21 [arXiv:1106.5806 [hep-th]].

[38] A. Dabholkar, F. Denef, G. W. Moore and B. Pioline, "Precision counting of small black holes," JHEP **0510** (2005) 096 [hep-th/0507014].

[39] I. Mandal and A. Sen, "Black Hole Microstate Counting and its Macroscopic Counterpart," Nucl. Phys. Proc. Suppl. **216** (2011) 147 [Class. Quant. Grav. **27** (2010) 214003] [arXiv:1008.3801 [hep-th]].

[40] G. Bossard, C. Hillmann and H. Nicolai, "E7(7) symmetry in perturbatively quantised N=8 supergravity," JHEP **1012** (2010) 052 [arXiv:1007.5472 [hep-th]].

[41] R. Kallosh and T. Kugo, "The Footprint of E(7(7)) amplitudes of N=8 supergravity," JHEP **0901** (2009) 072 [arXiv:0811.3414 [hep-th]].

[42] N. Beisert, H. Elvang, D. Z. Freedman, M. Kiermaier, A. Morales and S. Stieberger, "E7(7) constraints on counterterms in N=8 supergravity," Phys. Lett. B **694** (2010) 265 [arXiv:1009.1643 [hep-th]].

[43] A. Sen, "Black hole entropy function and the attractor mechanism in higher derivative gravity," JHEP **0509** (2005) 038 [hep-th/0506177].

[44] Z. Bern, J. J. M. Carrasco and H. Johansson, "Progress on Ultraviolet Finiteness of Supergravity," arXiv:0902.3765 [hep-th].

[45] M. Bianchi, S. Ferrara and R. Kallosh, "Perturbative and Non-perturbative N =8 Supergravity," Phys. Lett. B **690** (2010) 328 [arXiv:0910.3674 [hep-th]].

[46] T. Banks, "Arguments Against a Finite N=8 Supergravity," arXiv:1205.5768 [hep-th].

CHAIRMAN: S. FERRARA

Scientific Secretaries: V. Slepukhin, Ks. Bulycheva

DISCUSSION

- *Sleptsov:*

From the morning session, I didn't understand the attractor mechanism. Could you please explain this mechanism, and clarify its role in relation to black holes and SUSY?

- *Ferrara:*

Actually, this is a subject I didn't touch in the morning for reasons of time. The attractor mechanism is a phenomenon which occurs when one studies black holes in a theory which does not contain only Einstein and Maxwell fields, but also scalar fields. This is the situation which occurs in supergravity, where you have a number of scalar fields which essentially parameterize a non-linear sigma-model, for example is N=8 supergravity.

- *Zichichi:*

I apologize, but I missed the question. What is the question?

- *Ferrara:*

The question was the explanation of the attractor mechanism. This has to do with resolving the following paradox: usually, the entropy depends only on the electric and magnetic charges of the black hole, if you have an extremal black hole. However, when you look at the black hole mass at infinity, this mass depends also on the scalar fields parameters. These are just arbitrary couplings, which correspond to the value of the fields at infinity. The apparent paradox is how the mass depends on such parameters while the entropy should not depend. For instance, in N=8 supergravity there are 28 Maxwell fields, so you have to generalize the Reissner-Nordström black hole with 28 electric and 28 magnetic charges, but you also have 70 scalar fields coupled to the electric, magnetic and gravitational fields. The Lagrangian contains the Maxwell term and theta term, whose couplings depend on scalar fields. These couplings are fixed by electric-magnetic duality. Studying the evolution of the black hole towards the horizon, you discover that the scalar fields at infinity can take arbitrary values on the possible manifold of scalar field configurations, which in N=8 supergravity is the coset of the exceptional group E7(7) by its maximal compact subgroup SU(8). The scalar fields flow towards the horizon, their value being fixed to a critical value which is actually a solution to the extremization problem, as a function of the electric and magnetic charges. This phenomenon is essentially the core of the attractor mechanism: independently of the value of the scalar fields at infinity, all the scalar fields flow to a common value. This value is an attractor in the sense of statistical mechanics, namely the scalar fields at the attractor point have zero velocity. This attractor value is computed by minimizing a scalar potential named the black hole potential. This phenomenon also explains why that in N=8 supergravity the entropy is given by the quartic invariant of 56-dimensional representation, essentially related to E7(7).The dynamical scalar system has underlying E7(7) symmetry.

- *Zichichi:*

What is the parameter tau?

- *Ferrara:*

tau is essentially the radial parameter. If you take a solution which is spherically symmetric, the black hole evolution is given in terms of the radial parameter.

- *Zichichi:*

But when tau is large, one actually has no more black hole.

- *Ferrara:*

No. "tau infinity" corresponds to the asymptotical infinity, where the black hole becomes the flat space. The horizon is the opposite of having tau large. The horizon of the black hole has some universal feature, for example concerning the metric. The space near the horizon has the geometry of the two-dimensional anti-de Sitter space times the two-dimensional sphere. The area of the horizon parameterizes the radius of the AdS_2\times S_2 geometry, which characterizes the black hole near its event horizon. The attractor mechanism essentially concerns the flow of the scalar fields, such that when you compute the entropy in terms of some general formula, you see that the entropy is invariant of the E7(7) algebra in N = 8 sugra. The extremization condition corresponds to the condition that these phi fields have zero velocity and take values which are completely determined in terms of electric and magnetic charges.

- *Twigg:*

I'm wondering if your E7(7) electric-magnetic duality is essentially the Hodge star. Could you clarify this point?

- *Ferrara:*

Roughly speaking it is, but in a general sense, because in presence of scalar fields the duality symmetry transforms not only electric and magnetic fields, but acts also on the scalar fields. Since there are 28 electric and 28 magnetic fields, the duality is implemented as a 56 \times 56 matrix. The entropy is given by a mathematical formula, namely the quartic invariant which can be constructed in terms of the charge vector.

- *Twigg:*

So, do you think it possible to do this for infinite-dimensional Lie groups, in particularly E_{11}, as well?

- *Ferrara:*

I don't know, the infinite-dimensional groups do not actually occur here.

- *Twigg:*

Yes, but you might possibly set a similar duality in the infinite-dimensional case.

- *Ferrara:*

I don't know what is the implication of this infinite-dimensional symmetry on the entropy.

- *Csanad:*

How can one possibly observe extremal black holes? Is this connected to the fact that non-classically they may have T=0?

- *Ferrara:*

Well, this is a question which always occurs when I'm giving this seminar. Extremal black holes are not really realized in nature, as far as we don't see this kind of black holes in our Universe. When string states are considered, this is the object which appears in the spectrum of the fundamental theory at the Planck scale. But what is the implication of such black holes as far as observations are concerned? I don't have any answer at the moment.

- *Zenkevich:*

I have two questions. The first one is about attractor mechanism. Does a small variation of scalar fields near the horizon produce a large variation on the infinity?

- *Ferrara:*

This is not possible because the first derivative of the scalar fields at the horizon vanishes. The value of scalar fields is pretty unique, depending on reasonable conditions of extremality and regularity of the near-horizon geometry.

- *Zenkevich:*

A second question, could you tell us a little bit more about black holes in string theory? What are the macrostates counted by entropy.

- *Ferrara:*

Supergravity is the low-energy limit of string theory, and the Bekenstein-Hawking formula no longer holds beyond such a limit. However, our current understanding of black holes in string theory is limited to the low-energy limit. Entropy essentially counts configurations of wrapped branes and momenta.

- *Guts:*

What observational properties can be helpful to find out whether a black hole is supersymmetric?

- *Ferrara:*

In classical gravity, black holes cannot be extremal, hence, they have a finite temperature, while for supersymmetric ones it is zero. Supersymmetry implies extremality, but not the other way around.

- *'t Hooft:*

You mentioned that extremal black holes exist in supergravity theory, and you also pointed out that the temperature of black hole is zero. However, experimentalists tell us that the zero temperature is unreachable. On the other hand, if we consider the black hole with charge it will lose its charge due to Hawking radiation and it will eventually turn to extremal black hole.

- *Ferrara:*

To reach the extremal point isn't the same as to be at the extremal point. In fact, at the extremal point (T=0) there is an enhanced symmetry, so for extremal BH vanishing temperature is special, and is not the limiting case of very small temperature. One could use the same argument for renormalizability in presence of supersymmetry, in contrast to renormalization in non-sypersymmetric theories. In this case, the relation between couplings is due to SUSY Ward identities.

- *'t Hooft:*

So, the extremal case is a non-limiting case? If the system is in 98% of extremality, why cannot it reach 100% ?

- *Ferrara:*

Because the number of quantum states change suddenly, so for T=0 a kind a phase transition occurs.

- *'t Hooft:*

You still cannot make a temperature exactly zero, or entropy will vanish.

- *Ferrara:*

Indeed, this is consistent with the statement that the extremal point cannot be reached through a limit procedure.

- *Zichichi:*

In my opinion, extremal black holes simply do not exist.

- *Ferrara:*

From the mathematical point of view, they do exist. Let me insist once more: T=0 is a phase transition, so the number of states changes. Extremal spinning black holes have been seen in the sky, in our galaxy, heavy as ten solar masses. Particularly interesting is the GRS1915+105, discovered ten years ago, whose extremality parameter is bigger than 0.98! (See McClintock et al arXiv:0710.3165[astro-ph]) .

.

Composite Weak Bosons at the LHC

H. Fritzsch
Department für Physik
Ludwig-Maximilians-Universität
München, Germany

Abstract

The weak bosons are assumed to be composite particles, consisting of a fermion and an antifermion. In this case besides the standard weak bosons an isoscalar weak boson and orbital excitations of the weak bosons must exist, in particular twelve p-wave excitations. The isoscalar p-wave excitation with spin zero might have been observed recently at the *LHC*. The decays of this particle are discussed in detail. Finally we discuss the substructure of the leptons and quarks.

In the standard model of the electroweak interactions the weak bosons are elementary gauge bosons. Their masses and the masses of the leptons and quarks are due to the spontaneous breaking of the gauge symmetry. In QCD the hadronic masses are due to the field energy of the confined gluons and quarks inside the hadrons. In the absence of the quark masses the masses of the baryons and mesons are proportional to the QCD mass scale Λ_c.

It is not yet known, whether the mechanism of spontaneous symmetry breaking is realized in nature. If the weak bosons are composite particles, their masses would also be due to a confinement of the constituents, as the masses of the ρ -mesons in QCD. Many years ago the ρ -mesons were viewed as elementary gauge bosons. Their masses were generated by a spontaneous symmetry breaking, involving a complex doublet of scalar fields. After the symmetry breaking three scalars are absorbed by the ρ -mesons, which become massive particles.

Today we know that this model is not correct. In the theory of quantum chromodynamics the ρ -mesons are quark-antiquark bound states. The masses of the ρ -mesons are due to the field energies of the quarks and gluons inside the mesons.

The decay constant of the ρ -meson has been measured - it is about 220 MeV. Both the mass of the ρ -meson and its decay constant are proportional to the QCD mass scale. Their ratio is about 0.29.

If the weak bosons are composite particles, their constituents must be bound by a new confining gauge interaction. The standard weak bosons consist of a lefthanded fermion and its antiparticle, which are denoted as "haplons". The haplons interact with each other through the exchange of massless gauge bosons. A theory of this type was proposed in 1981 (see ref.(1), also ref.(2,3,4,5,6)).

The new confining chiral gauge theory is denoted as QHD. The QHD mass scale is given by a mass parameter Λ_h, which determines the size of the weak bosons. The QHD mass scale must be at least about 1000 times larger than the QCD mass scale (refs.(7,8)).

The interaction of the weak bosons violates parity, thus the QHD gauge theory must be a chiral gauge theory. We assume that the gauge group of the standard electroweak model $SU(2) \times U(1)$ is a subgroup of the left-right symmetric model, given by the gauge group $SU(2,L) \times SU(2,R) \times U(1)$. The charge of the $U(1)$ part of this group is the charge (B-L), where B is the baryon number and L is the lepton number. The QCD gauge group and this $U(1)$ group can be a subgroup of the group $SU(4)$. The lepton number can be interpreted as the fourth color (ref.(9)).

The masses of the gauge bosons of the righthanded group must be much larger than the masses of the standard weak bosons. These bosons are also composite systems. The gauge group of the QHD gauge theory consists of a lefthanded and a righthanded group, e.g. the group $SU(n,L) \times SU(n,R)$. There are two QHD mass scales $\Lambda_h(L)$ and $\Lambda_h(R)$. The mass scale $\Lambda_h(R)$ must be much larger than $\Lambda_h(L)$.

Two types of lefthanded and righthanded haplons are needed as constituents of the lefthanded and righthanded weak bosons, denoted by α(L), β(L) and α(R) and β(R). Their electric charges in units of e are +1/2 and -1/2.

The weak bosons have the following internal structure:

$$\begin{aligned} W^+ &= \overline{\beta(L)}\alpha(L)\,, \\ W^- &= \overline{\alpha(L)}\beta(L)\,, \\ W^3 &= \frac{1}{\sqrt{2}}\left(\overline{\alpha(L)}\alpha(L) - \overline{\beta(L)}\beta(L)\right)\,. \end{aligned} \tag{1}$$

The heavy weak bosons, coupled to the righthanded fermions, are bound states of the righthanded haplons α(R) and β(R).

In the absence of electromagnetism the three standard weak bosons are degenerate in mass. If the electromagnetic interaction is introduced, the mass of the neutral boson increases due to the mixing with the photon (see refs.(7,8)). This mixing is determined by the decay constant of the weak boson, which is analogous to the decay constant of the ρ -meson. The decay constant of the standard weak boson is given by the electroweak mixing angle and the mass of the weak boson (see. ref. (7,8)) - it is about 125 GeV. Both the decay constant and the weak boson mass are proportional to the mass scale $\Lambda_h(L)$. Their ratio is about 1.54. It is about 5 times larger than the corresponding ratio of the ρ -mesons. Probably this is due to the fact that the QHD theory is a chiral gauge theory, unlike QCD.

Using this decay constant, we can estimate the mass scale $\Lambda_h(L)$. It depends on the QHD gauge group and should be in the range 0.15 TeV ... 1.5 TeV.

The neutral weak boson is an isotriplet. There must also exist an isoscalar neutral weak boson X, which has the internal structure:

$$X = \frac{1}{\sqrt{2}}\left(\overline{\alpha}\alpha + \overline{\beta}\beta\right) . \tag{2}$$

This boson is not present in the Standard Model. The present lower limit on the mass of the X-boson is about 0.4 TeV. As the Z-boson the X-boson will also mix with the photon, but due to its the high mass the mixing angle is very small. We shall neglect this mixing.

In strong interaction physics the mass of the ρ-meson and of the ω-meson are nearly the same. Thus one would expect that the mass of the X–boson is about the same as the mass of the Z-boson, but this is excluded by the experiments.

The fact that the X–boson, if it exists, is much heavier than the Z-boson, might be related to the QHD analogy of the gluonic anomaly of QCD. The latter implies that the mass of the η'-meson is different from zero in the chiral limit, while the masses of the π-mesons and of the η-meson vanish. In QHD the isospin singlet axial vector current also has an anomaly, and this might be the reason why the X-boson is very heavy. The anomalous divergence of the singlet axial vector current increases the mass of the X-boson. For our further discussion we shall assume a mass of 0.8 TeV for the X-boson.

The X-boson would couple to the leptons and quarks with about the same strength as the Z-boson. Thus it can be produced at the LHC by quark-antiquark-fusion. The cross section for the production of Z-bosons at the LHC is estimated to about 60 nb. If the X-boson has a mass of 0.8 TeV, the cross section for its production at the LHC should be about 0.8 nb.

An interesting decay mode of the X-boson is the decay into lepton pairs, e.g. into muon pairs. The partial width for this decay can be estimated by comparing it with the decay of a charged weak boson into a muon and the muon antineutrino, which has a partial width of about 220 MeV. Using this result, we can calculate the partial width for the leptonic decay of the X-boson with a mass of 0.8 TeV:

$$\Gamma(X \to \mu^+\mu^-) \approx 3.6\, GeV. \tag{3}$$

Prominent decays of the X - boson would be the decays into three or four weak bosons or photons. We consider especially decays of the X-boson into two weak bosons or photons. If we assume a branching ratio B for the decay into two charged weak bosons, the branching ratio for the decay into two Z-bosons would be given by 0.59 B. The branching ratio of the decay into a Z-boson and a photon would be 0.36 B. For the decay into two photons we find 0.05 B. The total width of the X-boson should be in the range between 100 GeV and 200 GeV.

The best way to observe the X - boson at the LHC is to find the decays into two weak bosons, e.g. into two Z-bosons. The branching ratio of the X-decay into muon pairs is expected to be very small, probably less than 0.001. Thus the X-boson cannot be discovered by its decay into lepton pairs.

The QHD mass scale is three orders of magnitude larger than the QCD mass scale. In strong interaction physics complexities arise at the energy of 1 GeV and above. Analogously there should be complexities due to the QHD dynamics at the energy of 1 TeV and above.

In strong interaction physics there exist excited states of the vector mesons.

Analogously we expect that there exist excited states of the weak bosons. These states can have the total angular momentum 0, 1, 2, 3 etc. The excited state of a charged weak boson can decay into a charged weak boson and a Z-boson or a photon. The excited state of a Z-boson will decay mainly into two weak bosons. Decays of the excited weak bosons into quark pairs or leptons pairs are expected to be suppressed.

The weak bosons consist of pairs of haplons, which are in an s-wave. The spins of the two haplons are aligned, as the spins of the quarks in a ρ-meson. The first excited states are those, in which the two haplons are in a p-wave. We describe the quantum numbers of these states by $I(J)$. The $SU(2)$-representation is denoted by I, J describes the total angular momentum. There are three $SU(2)$ singlets, which we denote by S(J):

$$\begin{aligned} S(0) &= [0(0)] \\ S(1) &= [0(1)] \\ S(2) &= [0(2)] \end{aligned} \tag{4}$$

The internal structure of the S-states is given by:

$$S = \frac{1}{\sqrt{2}}\left(\overline{\alpha}\alpha + \overline{\beta}\beta\right) \tag{5}$$

The three $SU(2)$ triplets are denoted by T:

$$\begin{aligned} T(0) &= [1(0)] \\ T(1) &= [1(1)] \\ T(2) &= [1(2)] \end{aligned} \tag{6}$$

They have the internal structure:

$$\begin{aligned} T^{+} &= \overline{\beta}\alpha \\ T^{-} &= \overline{\alpha}\beta \\ T^{3} &= \frac{1}{\sqrt{2}}\left(\overline{\alpha}\alpha - \overline{\beta}\beta\right) \end{aligned} \tag{7}$$

We compare these states with the low mass mesons in strong interaction physics, in which the quarks are in a p-wave. The meson with the lowest mass is the scalar σ-meson (mass $\sim 600 - 800$ MeV). This meson can easily mix with glue mesons, i.e. the σ-meson would be a superposition of a quark-antiquark-meson and a glue meson. This explains the relatively low mass of this meson.

The σ-meson is the QCD analogue of the state $S(0)$. The spin one meson with isospin zero is $h_1(1170)$, analogous to the state S(1). The corresponding spin two meson is $f_2(1270)$, which is analogous to the state S(2).

We expect that the scalar state $S(0)$ with the quantum numbers $[0(0^+)]$ mixes with the bound states of the QHD gauge bosons, analogous to the mixing of the σ-meson with glue mesons. Due to this mixing the mass of $S(0)$ is expected to be much lower than the masses of the other p-wave states.

Now we consider the isospin triplet mesons: the scalar meson $a_0(980)$, the vector meson $b_1(1235)$ and the spin two meson $a_2(1320)$. These mesons are analogous to the bosons T(0), T(1) and T(2) respectively.

What is the mass spectrum of the excited weak bosons, in particular of the S and T bosons? The boson $S(0)$ should have a relatively small mass, as discussed above. We identify the $S(0)$ boson with the particle, which might have been observed recently at CERN (refs. (10,11)). This particle has a mass of about 125 GeV and has been observed in the decay into weak bosons and into two photons. We shall assume that the mass of $S(0)$ is 125 GeV.

In analogy to QCD we expect that the masses of the other p-wave states are in the range 0.2 - 0.6 TeV. The mixing of the bosons $S(1)$ and $S(2)$ with the bound states of the QHD gauge bosons is not as strong as the corresponding mixing for the $S(0)$ boson. We estimate the mass of the $S(1)$ - boson to be just above 0.3 TeV. The mass of the $S(2)$ - boson should be in the range 0.4 - 0.5 TeV.

The $SU(2)$ - triplet bosons T cannot mix with the bound states of the QHD gauge bosons. For this reason their masses should be larger than the masses of the S - bosons. The mass of the $T(0)$ - boson should be about 0.3 TeV, the mass of the $T(1)$ - boson just above 0.4 TeV, and the mass of the $T(0)$ - boson should be in the range 0.5 - 0.6 TeV. But these mass estimates are very uncertain.

The $S(0)$ - boson will decay mainly into two charged weak bosons or into two Z-bosons (one of them virtual respectively). The Z-boson is the boson W^3, mixed with the photon. The mixing angle is the weak angle, measured to about 28.7 degrees. Using this angle, we can calculate the branching ratios for the various decays. The boson W^3 has a chance of 77 % to be a Z-boson and a chance of 23 % to be a photon. Thus we find the following branching ratios - the branching ratio for the decay into charged weak bosons is denoted by B:

$$S(0) \Longrightarrow ("W^{+}" + W^{-}) \qquad \text{BR} \simeq B,$$

$$S(0) \Longrightarrow ("W^{-}" + W^{+}) \qquad \text{BR} \simeq B,$$

$$S(0) \Longrightarrow ("Z" + Z) \qquad \text{BR} \simeq 0.59\ B,$$

$$S(0) \Longrightarrow (Z + \gamma) \qquad \text{BR} \simeq 0.36\ B,$$

$$S(0) \Longrightarrow (\gamma + \gamma) \qquad \text{BR} \simeq 0.05\ B$$

- "Z" means "virtual Z" etc.

We would not expect that the decay rates for the decays of $S(0)$ into letons and quarks are given by the mass of the fermion, as they are for the Higgs boson. The branching ratio for the decay into a lepton pair should be universal:

$$\text{BR}\ (S(0) \Longrightarrow (e^+ + e^-)\ ) \simeq \text{BR}\ (S(0) \Longrightarrow (\mu^+ + \mu^-)\ )$$

$$\simeq \text{BR}\ (S(0) \Longrightarrow (\tau^+ + \tau^-)\ ) \simeq \text{BR}\ (S(0) \Longrightarrow (\nu + \bar{\nu})\ )$$

(ν stands for the electron, muon and tau neutrino).

Due to the color of the quarks the branching ratios for the decays into quarks are three times larger than the branching ratios for the decays into leptons.

The $S(0)$ boson decays mainly into two weak bosons. Decays into lepton pairs or quark pairs are strongly suppressed. The branching ratio for the decay into a lepton pair can be less than 0.1 %. This explains why the $S(0)$ boson has not been observed at LEP in the collisions of electrons and positrons.

It is not certain, that the σ-meson is a quark-antiquark bound state. It might be a state consisting mainly of two quarks and two antiquarks. Analogously the $S(0)$ boson might be composed of two haplons and two antihaplons: S(0)=($\overline{\alpha}\overline{\beta}\alpha\beta$). In this case we expect that the decays of the $S(0)$ boson are more complicated than outlined above.

The masses of the bosons $S(1)$ and $S(2)$ are higher than the mass as the $S(0)$ - boson. These bosons will decay mainly into three or four weak bosons. Decays into into lepton pairs or quarks pairs would be suppressed.

The $SU(2)$ - triplet bosons $T(0)$, $T(1)$ and $T(2)$ will decay mainly into three or four weak bosons or photons. Again the decays into lepton or quark pairs would be suppressed.

If the weak bosons are bound states of haplons, these haplons should also be inside the leptons and quarks. The quarks and leptons must also be composite systems, as discussed in ref. (1). We consider the simplest model, in which the quarks and leptons are bound states of a spin 1/2 haplon and a spin 0 haplon. The four haplons α(L), β(L), α(R) and β(R) generate the electroweak group $SU(2,L) \times SU(2,R) \times U(1)$.

The group $SU(4)$, describing the three colors of the quarks and the lepton number, is generated by the four scalars r, g, b and l. The scalar l is inside the leptons, the scalars r, g, b are inside the colored quarks. The leptonic scalar has the electric charge $-1/2$, the electric charge of the colored scalars is $+1/6$. The sum of the electric charges of the scalar haplons vanishes.

Inside the known leptons and quarks the haplons have no orbital angular momentum. We expect that there exist excited leptons and quarks, in which the haplons have a certain orbital angular momentum. The states, in which the haplons are in a p-wave, would be the ones with the lowest mass. The p-wave states can have a total angular momentum 1/2 or 3/2. Thus we expect that there are excited leptons with spin 1/2 or 3/2, e.g. $e^*(1/2)$ or $\mu^*(3/2)$, and excited quarks with spin 1/2 or 3/2, e.g. $u^*(1/2)$ or $d^*(3/2)$. The masses of these excited leptons and quarks are expected to be in the range 0.5 TeV ... 1 TeV. They decay into the ground state by emitting a photon or a Z-boson, e.g. $\mu^* => \mu + \gamma$ or $\mu^* => \mu + Z$.

The excited quarks and leptons can be produced at the *LHC* in quark-antiquark collisions. The colliding quark-antiquark system produces a resonance, which produces an excited quark and an antiquark or vice versa. The excited quark decays into a quark and a photon or a Z-boson. This can be observed at the *LHC* - a quark jet and a photon or a Z-boson are emitted.

Likewise an excited lepton can be produced, which decays into a lepton and a photon or a Z-boson. For example, a muon pair and a photon can be observed. The invariant mass of the muon and the photon determines the mass of the excited muon.

References

[1] H. Fritzsch and G. Mandelbaum, Phys. Lett. B102 (1981) 319; Phys. Lett. B 109 (1982) 224.

[2] R. Barbieri, R. Mohapatra and A. Masiero, Phys. Lett. B 105 (1981) 369.

[3] H. Fritzsch. D. Schildknecht and R. Kogerler, Phys. Lett. B 114 (1982) 157.

[4] L. F. Abbott and E. Farhi, Phys. Lett. B 101, 69 (1981).

[5] T. Kugo, S. Uehara and T. Yanagida, Phys. Lett. B 147, 321 (1984).

[6] S. Uehara and T. Yanagida, Phys. Lett. B 165, 94 (1985).

[7] H. Fritzsch, Mod. Phys. Lett. A26, 2305 (2011)

[8] H. Fritzsch, Phys. Lett. B 712, 231 (2012)

[9] J. Pati and A. Salam, Phys. Rev. D10, 275 (1974)

[10] Atlas collaboration:
arXiv:1202.1414,arXiv:1202.1408, arXiv:1202.1415.

[11] CMS collaboration:
arXiv:1202.1488, arXiv:1202.1489, arXiv:1202.1487.

CHAIRMAN: H. FRITZSCH

Scientific Secretaries: I. Ivanchenko, I. Rudenok

DISCUSSION

- *Zoller:*

Unitarity in W-scattering is restored in the SM by the Higgs particle. How does this work in QHD without a Higgs?

- *Fritzsch:*

We have the same feature as in the strong interactions physics with ρ-mesons. Many new particles are being created and the unitary is not violated. It is a bound state picture. I don't need a Higgs particle - I see no problem here.

- *Knoops:*

Can you predict the parameter m, corresponding to the mixing angle?

- *Fritzsch:*

The parameter m in the SM is the Weinberg angle. But in my approach m is not a mixing angle. It is related to the decay constant of the weak boson and given by the bound state structure of weak boson, which is very similar to the ρ-meson. This parameter is unknown, but in the gauge theory of QHD this decay constant probably can be calculated in the future, just like now we can calculate the decay constant of the ρ-meson, which is proportional to the Λ_{QCD}.

- *Knoops:*

Are the new particles again composite? Where does it end?

- *Fritzsch:*

No, I don't think so. It's just before the end right now, probably. In the future we can calculate the masses of the leptons and quarks and of the weak bosons.

- *'t Hooft:*

I have to make a remark. In whatever model you have as a primordial model, you have a large number of free adjustable parameters, so why do you think that you can calculate everything?

- *Fritzsch:*

In principle I don't have any free parameters. In principle I can calculate everything in this gauge theory. I don't need any free parameter. I can calculate in principle the Weinberg angle, the mass of the electron, the mass of the top quark, etc.

- *Zichichi:*

Can you calculate the masses of α and β?

- *Fritzsch:*

α and β are supposed to be massless.

- *Karakostas:*

Are there any experimental observations of the composite weak bosons?

- *Fritzsch:*

No, we don't know if they are composite or not. We know that the muon must have a radius which is less than 10^{-17} cm. I'd like to mention, that my friend Richard Feynman was telling me, that he believed that electrons and muons are bound states of smaller things with a radius about 10^{-18} cm, which is consistent with my approach.

- *Karakostas:*

What would be the optimum conditions for an observation of the composite weak bosons?

- *Fritzsch:*

It is very hard to see the bound state structure directly. I think the first signal will be from the mentioned p-wave bosons or the X-boson. The X-boson is interesting candidate. It is a bit heavy, since it must have a mass larger than 400 GeV. The X-boson could be seen with the LHC. The other bosons are the p-wave bosons. In the future we must see the Z-gamma decay of the S(0). It must be in 5 times higher than the two-gamma decay.

- *Carlucci:*

Going deeper with the theoretical consistency of your model, can you discuss the eventual presence of anomalies?

- *Fritzsch:*

I don't worry about the anomalies. They are probably canceled if there are any. I don't worry about flavor changing neutral currents. They are related to the lepton and quark masses, to the mixing angles of neutrinos and of quarks, and what we have to do with the generation of masses of neutrino and quarks. At the present time, it's too early to worry about such details.

- *'t Hooft:*

The one thing to worry about is why some quarks and leptons become very light in this theory. If everything is composite, fine, but how do you make a very light particle which consists of very heavy particles?

- *Fritzsch:*

They consist of massless haplons. The light quarks get their masses due to the color and due to electromagnetism, the masses are radiative corrections.

- *'t Hooft:*

But, don't you want a handle that enables you to compute such things?

- *Fritzsch:*

In the future we can compute this. It is too early now. QHD is a theory without free parameters, and everything can be calculated. I don't know whether my theory is correct, we have to see. Maybe it is totally wrong. But I also don't think that the SM is the correct theory for the future, in particular the spontaneous symmetry breaking and the mass generation. Albert Einstein would not like this possibility, because in his theory the curvature of space-time is related to the mass, and he would prefer a theory in which the mass generation is one unique mechanism. In the SM we have the proton mass, generated by confinement, and we have the electron mass, generated by spontaneous symmetry breaking two different possibilities. I don't like this.

- *Klopot:*

How are the masses of SM particles generated? Is it a Higgs mechanism or something else?

- *Fritzsch:*

It is not a Higgs mechanism, the mass is coming from radiative corrections.

- *Inguglia:*

Can you predict the quark masses and compare them with the observed values?

- *Fritzsch:*

I think that the masses of quarks can be calculated, just like the lepton masses. But at this moment in the QHD theory we just know the bound state structure of the W-boson.

- *Close:*

Why is the X-boson mass much larger than the W-mass? The ω-meson mass is not much larger than the ρ-meson mass.

- *Fritzsch:*

I've worried about it. Probably is is connected with a similar phenomenon in QCD, related to the pseudo-scalar mesons. As you well know, the η-meson is relatively light and the η'-meson is very heavy. This is connected to the mixing with gluonic states. A similar mechanism might exist in QHD. Thus the isoscalar mass is much larger than the isovector mass.

- *Slepukhin:*

Why we should use haplons to compose weak boson, but cannot use usual quarks? For example unknown quarks? Why we should use a new particle?

- *Fritzsch:*

Because the quarks have color and they are confined at a distance of about 1 Fermi. We need constituents, which are confined at a distance of about 0.001 Fermi.

- *Slepukhin:*

Can the S(0) boson be a Higgs boson?

- *Fritzsch:*

No, in my theory there is no Higgs boson.

- *Mager:*

Why are you waiting for the results from CERN? Why don't you check your parameter-free model with the existing data, e.g. by calculating the parameters of the Standard Model?

- *Fritzsch:*

First we have to find out whether the weak bosons are composite systems. Then we can calculate the parameters of the Standard Model.

High energy scattering in QCD and in quantum gravity

L.N. Lipatov
Theoretical Physics Department,
Petersburg Nuclear Physics Institute,
Gatchina, 188300, St. Petersburg, Russia

Abstract

The theory of the high energy scattering in QCD is based on the BFKL equation for the Pomeron wave function and on its generalization for composite multi-gluon states in the crossing channel. At a large number of colors the equations for the gluon composite states have remarkable mathematical properties including their Möbius invariance, holomorphic separability, duality symmetry and integrability. High energy QCD interactions local in the particle rapidities are formulated in the form of the gauge invariant effective action. In the maximally extended $N = 4$ super-symmetry the Pomeron turns out to be dual to the reggeized graviton in the 10-dimensional anti-de-Sitter space. As a result, the Gribov calculus for the Pomeron interactions should be reformulated here as a generally covariant effective field theory for the reggeized gravitons. We construct the corresponding effective action, which gives a possibility to calculate their trajectory and couplings. The graviton trajectory in the leading order contains an ultraviolet divergency meaning the presence of the double-logarithmic (DL) terms. We sum the DL contributions in all orders of the perturbation theory in the Einstein-Hilbert gravity and in its super-symmetric generalizations. In the $N = 8$ super gravity the ratio of the scattering amplitude in the DL approximation to the Born expression tends to zero at large energies.

1 Pomeron in hadron interactions

For the hadron-hadron scattering amplitudes at high energies and a small momentum transfer

$$s = 4E^2 \gg (-t) = \vec{q}^2 \sim m^2 \tag{1}$$

the main contribution appears from the Pomeron pole in the crossing channel partial wave. This pole was invented to provide the fulfillment of the Pomeranchuck theorem

$$A_P(s,t) \approx is\,\gamma^2(t)\,s^{\omega(t)}\,,\ \omega(t) = \Delta + \alpha' t\,,\ \sigma_{pp} = \sigma_{p\bar{p}} \sim s^{\Delta}\,. \tag{2}$$

But the exchange of two or more Pomerons in the crossing channel leads to the Mandelstam cut contributions of the type

$$A_{Mand}(s,t) = -is \int \frac{d^2k}{(2\pi)^2} \Phi^2(k, q-k) s^{\omega(-k^2)} s^{\omega(-(q-k)^2)}\,. \tag{3}$$

These ω-plane singularities appear in the t-channel partial wave $f_\omega(t)$

$$A(s,t) = is \int_{a-i\infty}^{a+i\infty} \frac{d\omega}{2\pi i} s^{\omega} f_\omega(t)\,, \tag{4}$$

satisfying the multi-particle unitarity

$$\Im_t f_\omega(t) \sim \sum_n \int d\Omega_n |f_\omega^{(n)}|^2\,.$$

At high energies the hadrons are produced in clusters ordered in their rapidities

$$y_r = \frac{1}{2} \ln \frac{\sqrt{k^2 + m_r^2} + k}{\sqrt{k^2 + m_r^2} - k}\,,\ 1 \ll y_r - y_{r-1} \ll \ln s\,. \tag{5}$$

The Pomeron contribution corresponds to a non-relativistic form of the propagator for the t-channel partial wave

$$G_0 = \frac{1}{E + \Delta - \frac{k^2}{2m}}\,,\ E = -\omega\,,\ \alpha' = \frac{1}{2m}\,.$$

To find the contribution of all possible Pomeron interactions V.N. Gribov constructed the effective action

$$S = \int dy\, d^2\rho \left(\phi^*(\partial_y - \Delta)\phi + \frac{1}{2m} |\partial_\mu \phi|^2) + i\lambda(\phi^* \phi^2 + \phi^* \phi^2) + ... \right)\,, \tag{6}$$

where the Pomeron field ϕ depends on two-dimensional impact parameters $\vec{\rho}$ and the rapidity y playing the role of time. The action is local in rapidity.

2 High energy scattering amplitudes in QCD

QCD Born amplitude at high energies $s \gg t$ has the simple factorized form

$$M_{AB_{Born}}^{A'B'} = 2s\, g\, T^c_{A'A}\, \delta_{\lambda_{A'}\lambda_A}\, \frac{1}{t}\, g\, T^c_{B'B}\, \delta_{\lambda_{B'}\lambda_B}\,, \tag{7}$$

corresponding to the helicity conservaion for each of the colliding particles. In the leading logarithmic approximation (LLA) the gluon in the crossing channel is reggeized [1]

$$M(s,t) = M_{|Born}\, s^{\omega(t)}\,,\ \alpha_s \ln s \sim 1\,,\ \alpha_s = \frac{g^2}{4\pi} \ll 1\,. \tag{8}$$

The gluon trajectory in LLA is given below

$$\omega(-|q|^2) = -\frac{\alpha_s N_c}{4\pi^2}\int d^2k\, \frac{|q|^2}{|k|^2|q-k|^2} \approx -\frac{\alpha_s N_c}{2\pi}\ln\frac{|q^2|}{\lambda^2}\,. \tag{9}$$

The infrared divergency at $\lambda \to 0$ is canceled in the total cross section. The partial wave, containing the pole singularity in the t-channel partial wave satisfies the bootstrap equation

$$\omega f = 1 + \left(\omega(-|k|^2) + \omega(-|q-k|^2) + \hat{K}_8\right) f\,,\ f = \frac{1}{\omega - \omega(-|q|^2)}\,,$$

corresponding to the fact, that the reggeized gluon is a composite state of two reggeized gluons.

Similarly the amplitude for the production of n particles in the high energy kinematics $s \gg s_i \gg |t_i|$ has the factorized form [1]

$$M_{2\to 2+n}^{BFKL} \sim \frac{s_1^{\omega_1}}{|q_1|^2}\, gT^{d_1}_{c_2c_1}\, C(q_2,q_1)\, \frac{s_2^{\omega_2}}{|q_2|^2} ... gT^{d_n}_{c_{n+1}c_n}\, C(q_{n+1},q_n)\, \frac{s_{n+1}^{\omega_{n+1}}}{|q_{n+1}|^2}\,, \tag{10}$$

where ω_r are the trajectories of poles in the partial waves of t_i-channels and $C(q_r, q_{r-1})$ are the effective vertices for the emission of the gluon with a definite helicity

$$\omega_r = -\frac{\alpha_s N_c}{2\pi}\ln\frac{|q_r^2|}{\lambda^2}\,,\ C(q_2,q_1) = \frac{q_2\, q_1^*}{q_2^* - q_1^*}\,,\ \sigma_t = \sum_n \int d\Gamma_n\, |M_{2\to 2+n}|^2\,. \tag{11}$$

3 BFKL equation

Due to the factorized form of the production amplitude the total cross-section satisfies the Bethe-Salpeter-type equation. The equation for the corresponding elastic amplitude with the color singlet quantum numbers in the crossing channel was obtained by Balitsky, Fadin, Kuraev and Lipatov (BFKL) many years ago [1]. Next-to-leading corrections to it were calculated comparatively recently [2]. The homogeneous BFKL equation in LLA coincides with the Schrödinger equation for the wave function of the Pomeron constructed from two gluons

$$E\,\Psi(\vec{\rho}_1,\vec{\rho}_2) = H_{12}\,\Psi(\vec{\rho}_1,\vec{\rho}_2)\,,\ \sigma_t \sim s^{\Delta}\,,\ \Delta = -\frac{\alpha_s N_c}{2\pi}E_0\,. \tag{12}$$

Its hamiltonian can be written in the following operator form [3]

$$H_{12} = \frac{1}{p_1p_2^*}(\ln|\rho_{12}|^2)p_1p_2^* + \frac{1}{p_1^*p_2}(\ln|\rho_{12}|^2)p_1^*p_2 + \ln|p_1p_2|^2 - 4\psi(1)\,, \tag{13}$$

where we introduced the complex notations

$$\rho_{12} = \rho_1 - \rho_2\,,\ \rho_r = x_r + iy_r\,.$$

The operator H_{12} is invariant under the Möbius transformations [4]

$$\rho_k \to \frac{a\rho_k + b}{c\rho_k + d}\,,\ m = \gamma + n/2\,,\ \widetilde{m} = \gamma - n/2\,,\ \gamma = 1/2 + i\nu\,, \tag{14}$$

where γ is the anomalous dimension for the twist-2 operators and n is the integer conformal spin. The eigenvalues of H_{12} are given below

$$E = \epsilon_m + \epsilon_{\widetilde{m}}\,,\ \epsilon_m = \psi(m) + \psi(1-m) - 2\psi(1)\,, \tag{15}$$

which allows to calculate the Pomeron intercept

$$\Delta = \frac{g^2 N_c}{\pi^2} \ln 2 > 0\,. \tag{16}$$

As a result, in LLA one obtains the violation of the Froissart bound for the total cross section $\sigma_t < c \ln^2 s$.

To restore the s-channel unitarity lost in the LLA approximaton, we should take into account the contribution of diagrams with an arbitrary number of the gluons in the crossing channel. The equation for the colorless composite states of n reggeized gluons was obtained by Bartels, Kwiecinskii and Prascalowicz (BKP) [5]. In the large N_c limit the corresponding hamiltonian has the property of the holomorphic separability [6]

$$H = \frac{1}{2}(h + h^*)\,,\ h = \sum_{k=1}^{n} \left(\ln(p_k p_{k+1}) + \frac{1}{p_k} \ln\left(\rho_{k,k+1}\rho_{k-1,k}\right) p_k - 2\psi(1) \right)) \tag{17}$$

It leads to the holomorphic factorization of the eigenfunctions

$$\Psi(\vec{\rho}_1, \vec{\rho}_2, ..., \vec{\rho}_n) = \sum_{r,s} a_{r,s}\, \Psi_r(\rho_1, ..., \rho_n)\, \Psi_s(\rho_1^*, ..., \rho_n^*)\,. \tag{18}$$

The holomorphic factorization in the two-dimensional conformal field theories is a consequence of the infinitely dimensional Virasoro group [7]. The BKP equation has also the duality symmetry under the transformations [8]

$$\rho_r \to p_r \to \rho_{r+1}\,, \tag{19}$$

which is similar to the symmetry of the harmonic oscillator.

The holomorphic BKP equation is integrable [3, 9]. Indeed, one can introduce the monodromy matrix

$$t(u) = L_1 L_2 ... L_n = \begin{pmatrix} A(u) & B(u) \\ C(u) & D(u) \end{pmatrix}\,,\ L_k = \begin{pmatrix} u + \rho_k\, p_k & p_k \\ -\rho_k^2\, p_k & u - \rho_k\, p_k \end{pmatrix}\,, \tag{20}$$

which satisfies the Yang-Baxter equation [9, 10]

$$t_{r_1'}^{s_1}(u)\, t_{r_2'}^{s_2}(v)\, l_{r_1 r_2}^{r_1' r_2'}(v-u) = l_{s_1' s_2'}^{s_1 s_2}(v-u)\, t_{r_2}^{s_2'}(v)\, t_{r_1}^{s_1'}(u)\,,\ \hat{l} = u\,\hat{1} + i\,\hat{P}\,. \tag{21}$$

The trace of the monodromy matrix is the transfer matrix which is the generating function for integrals of motion. It turns out, that the BKP hamiltonian is the local hamiltonian of the integrable Heisenberg spin chain in which the spins coincide with the generators of the Möbius group [11].

4 High energy effective action in QCD

To calculate the trajectories and all possible couplings of the reggeized gluons one can use the effective action [12] describing the interaction of these gluons with usual quarks and gluons in the form local at the rapidity space

$$y = \frac{1}{2}\ln\frac{\epsilon_k + |k|}{\epsilon_k - |k|}\,,\;\; |y - y_0| < \eta\,,\;\; \eta << \ln s\,. \tag{22}$$

The usual and reggeized gluons are described by the fields

$$v_\mu(x) = -iT^a v_\mu^a(x)\,,\;\; A_\pm(x) = -iT^a A_\pm^a(x)\,,\;\; \delta A_\pm(x) = 0\,. \tag{23}$$

The gauge-invariant fields $A_\pm$ describe the production and annihilation of the reggeized gluons in the crossing channel. They satisfy the kinematical constraints

$$\partial_\mp A_\pm = 0\,. \tag{24}$$

The effective action is written in terms of these fields

$$S = \int d^4x \left(L_{QCD} + Tr(V_+\partial_\mu^2 A_- + V_-\partial_\mu^2 A_+)\right)\,, \tag{25}$$

where

$$V_+ = -\frac{1}{g}\partial_+ P\exp\left(-g\int_{-\infty}^{x^+} v_+(x')d(x')^+\right) = v_+ - gv_+\frac{1}{\partial_+}v_+ + \dots\,. \tag{26}$$

The Feynman rules for the corresponding effective diagrams are known [13].

As an example of the application of this effective action, we present below the expression for the high energy amplitude of two gluons production in the gluon-gluon collisions in $N=4$ SUSY. This expression was obtained in the next-to-leading logarithmic approximation for the amplitudes with the maximal helicity violation at large N_c. The corresponding remainder function R for the Bern-Dixon-Smirnov ansatz [14] has the following form [15]

$$R\,e^{i\pi\delta} = \cos\pi\omega_{ab} + i\frac{a}{2}\sum_{n=-\infty}^{\infty}(-1)^n e^{i\phi n}\int_{-\infty}^{\infty}\frac{|w|^{2i\nu}d\nu}{\nu^2+\frac{n^2}{4}}\,\Phi(\nu,n)\left(\frac{-1}{\sqrt{u_2u_3}}\right)^{\omega(\nu,n)}\,, \tag{27}$$

depending on three anharmonic ratios

$$u_1 = \frac{ss_2}{s_{012}s_{123}}\,,\;\; u_2 = \frac{s_1t_3}{s_{012}t_2}\,,\;\; u_2 = \frac{s_3t_1}{s_{123}t_2}\,,\;\; w = \sqrt{\frac{u_2}{u_3}}\,e^{i\phi}\,,\;\; \cos\phi = \frac{1-u_1-u_2-u_3}{2\sqrt{u_2u_3}}\,, \tag{28}$$

where we used the notations

$$\delta = \frac{\gamma_K}{8}\ln\frac{|w|^2}{|1+w|^4}\,,\;\; \omega_{ab} = \frac{\gamma_K}{8}\ln|w|^2\,,\;\; \Phi(\nu,n) = 1 - a\left(\frac{E_{\nu n}^2}{2} + \frac{3}{8}n^2/(\nu^2+\frac{n^2}{4})^2 + \zeta(2)\right)$$

and γ_K is the casp anomalous dimension. The eigenvalue of the BFKL equation in the adjoint representation are given below

$$\omega(\nu,n) = -aE_{\nu,n} - a^2(\epsilon_{\nu n}^{FL} + 3\zeta(3))\,,\;\; E_{\nu n} = -\frac{|n|/2}{\nu^2+\frac{n^2}{4}} + 2\Re\psi(1+i\nu+\frac{|n|}{2}) - 2\psi(1)$$

and the next-to-leading correction to it looks as follows [15]

$$\epsilon_{\nu n}^{FL} = -\frac{\Re}{2}\left(\psi''(1+i\nu+\frac{|n|}{2}) - \frac{2i\nu\psi'(1+i\nu+\frac{|n|}{2})}{\nu^2+\frac{n^2}{4}}\right) - \zeta(2)\,E_{\nu n} - \frac{1}{4}\frac{|n|\,\left(\nu^2-\frac{n^2}{4}\right)}{\left(\nu^2+\frac{n^2}{4}\right)^3}\,.$$

Similar next-to-leading corrections to the kernel of the BKP equation were calculated with the use of the effective action approach in Refs. [16].

5 Pomeron and graviton in N=4 SUSY

According to the AdS/CFT correspondence [17, 18, 19] the BFKL Pomeron in the conformal $N = 4$ supersymmetric gauge theory is related to the graviton in the dual anti de Sitter 10-dimensional space [20]. Eigenvalue of the BFKL kernel in a diffusion approximation

$$j = 2 - \Delta - \Delta\,\nu^2\,,\ \gamma = 1 + \frac{j-2}{2} + i\nu \tag{29}$$

satisfies the energy-momentum conservation $\gamma(j = 2) = 0$ and depends on the parameter Δ which can be calculated in the perturbation theory [2]. On another hand one can use the known correspondence [18, 19] to present the BFKL kernel eigenvalue in the diffusion approximation in the form of the linear graviton Regge trajectry in the 10-dimensional anti de Sitter space

$$j = 2 + \frac{\alpha'}{2}\,t\,,\ t = E^2/R^2\,,\ \alpha' = \frac{R^2}{2}\,\Delta\,. \tag{30}$$

Using the result of the Gubser, Klebanov and Polyakov for the asymptotic behavior of the anomalous dimensions of the twist-2 operators at the large coupling constant

$$\gamma = 1 - \sqrt{1 + (j-2)/\Delta} \tag{31}$$

we can obtain the leading correction to the intercept of graviton $j = 2$ at large coupling constant [22, 23]

$$\Delta = \frac{1}{\sqrt{\lambda}}\,,\ \lambda = g^2 N_c\,. \tag{32}$$

Note, that the slope of the anomalous dimension at $j = 2$ is closely related to the intercept of the Pomeron. It was calculated up to the fifth order of the perturbation theory [22, 24]. Recently the quantity $\gamma'(2)$ was derived in all orders [25]

$$\gamma'(2) = -\frac{\lambda}{24} + \frac{1}{2}\frac{\lambda^2}{24^2} - \frac{2}{5}\frac{\lambda^3}{24^2} + \frac{7}{20}\frac{\lambda^4}{24^4} - \frac{11}{35}\frac{\lambda^5}{24^5} + \ldots = -\frac{\sqrt{\lambda}}{4}\frac{I_3(\sqrt{\lambda})}{I_2(\sqrt{\lambda})} \tag{33}$$

This result was obtained with the use of the asymptotic Bethe ansatz for the anomalous dimensions in the N=4 SUSY [25]. Note, that the integrable Heisenberg spin chain for the anomalous dimensions of quasi-partonic operators [26] in LLA at N=4 SUSY was discovered firstly in Ref. [27]. The duality between the BFKL Pomeron and the graviton in the framework of the AdS/CFT correspondence in $N = 4$ SUSY means, that the Gribov Pomeron calculus should be reformulated here in terms of the general covariant theory for the reggeized gravitons. It is the reason, why we study below the high energy effective action local in the graviton rapidities.

6 High energy amplitudes in gravity

The multi-graviton production amplitude for the high energy graviton-graviton collisions was obtained in Ref. [28]

$$A_{2\to n} = -s^2\Gamma^{\mu'\nu'}_{\mu\nu}\frac{s_1^{\omega(q_1^2)}}{q_1^2}\Gamma_{\rho_1\sigma_1}\frac{s_2^{\omega(q_2^2)}}{q_2^2}\Gamma_{\rho_2\sigma_2}\ldots\Gamma^{\rho'\sigma'}_{\rho\sigma}\,. \tag{34}$$

The effective graviton-graviton-reggeized graviton vertex

$$\Gamma_{\mu\nu}^{\mu'\nu'} = \frac{\kappa}{4}\left(\Gamma_{\mu\mu'}\Gamma_{\nu\nu'} + \Gamma_{\mu\nu'}\Gamma_{\nu\mu'}\right) \tag{35}$$

is expressed in terms of the effective gluon-gluon-reggeized gluon vertex

$$\Gamma_{\mu\mu'} = -\delta_{\mu\mu'} + \frac{p^A_{\mu'}p^B_{\mu} + p^{A'}_{\mu}p^B_{\mu'}}{p^Ap^B} + \frac{q^2}{2}\frac{p^B_{\mu}p^B_{\mu'}}{(p^Ap^B)^2}\,. \tag{36}$$

Similarly the reggeon–reggeon-graviton vertex

$$\Gamma_{\rho\sigma} = \frac{\kappa}{4}\left(C_\rho C_\sigma - N_\rho N_\sigma\right)\,,\ N = \sqrt{q_1^2q_2^2}\left(\frac{p^A}{kp^A} - \frac{p^B}{kp^B}\right) \tag{37}$$

contains apart from the QED bremstrahlung vertex N also the QCD effective reggeon-reggeon-gluon vertex

$$C = -q_1^{\perp} - q_2^{\perp} + p^A\left(\frac{q_1^2}{kp^A} + \frac{kp^B}{p^Ap^B}\right) - p^B\left(\frac{q_2^2}{kp^B} + \frac{kp^A}{p^Ap^B}\right)\,. \tag{38}$$

The graviton trajectory is given below [28]

$$\omega(q^2) = \frac{\alpha}{\pi}\int\frac{q^2\,d^2k}{k^2(q-k)^2}\left((k,q-k)^2\left(\frac{1}{k^2} + \frac{1}{(q-k)^2}\right) - q^2 + \frac{N}{2}(k,q-k)\right)\,, \tag{39}$$

where κ is the Einstein constant, $\alpha = \kappa^2/8\pi^2$ and N is the number of gravitinos appearing in the N-extended super-gravity. The gravitino action is well known

$$S_{3/2} = -\frac{1}{2}\epsilon^{\mu\nu\rho\sigma}\int d^4x\sum_{r=1}^{N}\bar{\psi}^r_\mu\gamma_5\gamma_\nu\partial_\rho\psi^r_\sigma\,.$$

The graviton trajectory contains both the infrared and ultraviolet divergencies

$$\omega(q^2) = -\alpha|q|^2\left(\ln\frac{|q|^2}{\lambda^2} + \frac{N-4}{2}\ln\frac{|\Lambda|^2}{|q|^2}\right)\,.$$

The infrared divergency is universal, but the divergency of the trajectory in the ultraviolet region depends on N and is absent at the $N = 4$ super-gravity.

To calculate the graviton trajectory and graviton couplings in upper orders of the perturbation theory we can construct the effective action local in the rapidities of reggeized and usual gravitons

$$y = \frac{1}{2}\ln\frac{\epsilon_k + |k|}{\epsilon_k - |k|}\,,\ |y - y_0| < \eta\,,\ \eta << \ln s\,. \tag{40}$$

The fields A^{++} and A^{--} describing the production and annihilation of the reggeized gravitons are generally covariant and satisfy additional kinematical constraints

$$\delta A^{++}(x) = \delta A^{--}(x) = 0\,,\ \partial_+A^{++}(x) = \partial_-A^{--}(x) = 0\,. \tag{41}$$

In terms of these fields the effective action has the following form [29]

$$S = -\frac{1}{2\kappa}\int d^4x\left(\sqrt{-g}\,R + \frac{1}{2}\left(\partial_+j^-\,\partial^2_\mu A^{++} + \partial_-j^+\,\partial^2_\mu A^{--}\right)\right)\,. \tag{42}$$

The currents $j^{\pm} = 2x^{\pm} - \omega^{\pm}$ satisfy the Hamilton-Jacobi equation

$$g^{\mu\nu}\partial_\mu\omega^{\pm}\partial_\nu\omega^{\pm} = 0\,,\ \partial_{\pm}j^{\mp} = h_{\pm\pm} - \left(h_{\rho\pm} - \frac{1}{2}\frac{\partial_\rho}{\partial_\pm}h_{\pm\pm}\right)^2 + \ldots \tag{43}$$

which can be solved in the parturbation theory. One can calculate them explicitly in the shock wave gravitational field with the Aichelburg-Sexl metric [29].

7 Double-logarithmic asymptotics in gravity

For the graviton-graviton scattering amplitude in the double-logarithmic approximation it is convenient to use the Mellin transformation

$$A(s,t) = A_{Born}\, s^{-\alpha|q|^2 \ln \frac{|q|^2}{\lambda^2}} \int_{a-i\infty}^{a+i\infty} \frac{d\omega}{2\pi i\,\omega} \left(\frac{s}{|q|^2}\right)^{\omega} f_\omega\,, \quad \alpha = \frac{\kappa^2}{8\pi^2}\,. \tag{44}$$

For a general case of the N-extended super-gravity the t-channel partial wave satisfies the infrared evolution equation [30]

$$f_\omega = 1 + b\frac{d}{d\omega}\frac{f_\omega}{\omega} - b\frac{N-6}{2}\frac{f_\omega^2}{\omega^2}\,, \quad b = \alpha|q|^2 \tag{45}$$

similar to that obtained earlier in QED and QCD [31]. One can find its explicit solution in terms of the parabolic cylinder function

$$\frac{f_\omega^{(N)}}{\omega} = \frac{2}{6-N}\frac{1}{\sqrt{b}}\frac{d}{dx}\ln d^{(N)}(x)\,, \quad d^{(N)}(x) = e^{\frac{x^2}{4}} D_{\frac{6-N}{2}}(x)\,, \quad x = \frac{\omega}{\sqrt{b}}\,. \tag{46}$$

The perturbative expansion of the scattering amplitudes in the DL parameter $\xi = \alpha\,|t|\,\ln^2 \frac{s}{|q|^2}$ has the form

$$A(s,t) = A_{Born}\, s^{-\alpha|q|^2 \ln \frac{|q|^2}{\lambda^2}} \left(1 - \frac{N-4}{2}\frac{\xi}{2} + \frac{(N-4)(N-3)}{2}\frac{\xi^2}{4!} + \ldots\right), \tag{47}$$

being in an agreement with the exact two-loop results. It turns out, that for large ξ the amplitude asymptotics depends on the number of gravitinos in such a way, that at $N > 5$ the scattering amplitude monotonically decreases in the comparison with the Born contribution.

One can calculate also in the DL approximation the multi-graviton production amplitudes

$$s \gg s_i = (k_{i-1}+k_i)^2 \gg |k_r^\perp|^2\,, \quad \prod_{r=1}^{n+1} s_r = s \prod_{r=1}^{n} |k_r|^2\,.$$

For example, one graviton production amplitude in this approximation has the form

$$A_{2\to 3}^{(N)} = A_{2\to 3}^{Born}\, s_1^{-\alpha|q_1|^2 \ln \frac{|q_1^2|}{\mu^2}}\, s_2^{-\alpha|q_2|^2 \ln \frac{|q_2^2|}{\mu^2}}\, r^{(N)}(\rho_1,\rho_2)\,, \quad \rho_r = \sqrt{\alpha}\ln s_r/|q_r|^2\,, \tag{48}$$

where

$$r^{(N)} = \int_{-i\infty}^{+i\infty} \frac{dx_1 dx_2}{(2\pi i)^2}\, e^{x_1\rho_1 + x_2\rho_2}\, \frac{d^{(N)}\big(\sqrt{\frac{|q_2|}{|q_1|}}x_1\big)\, d^{(N)}\big(\sqrt{\frac{|q_1|}{|q_2|}}x_2\big)}{d^{(N)}(x_1)\, d^{(N)}(x_2)}\, \phi^{(N)}(x_1,x_2)\,, \tag{49}$$

and

$$\phi^{(N)}(x_1,x_2) = \frac{2}{6-N}\, \frac{\frac{\left(d^{(N)}\left(\sqrt{\frac{|q_2|}{|q_1|}}x_1\right)\right)'}{d^{(N)}\left(\sqrt{\frac{|q_2|}{|q_1|}}x_1\right)} - \frac{\left(d^{(N)}\left(\sqrt{\frac{|q_1|}{|q_2|}}x_2\right)\right)'}{d^{(N)}\left(\sqrt{\frac{|q_1|}{|q_2|}}x_2\right)}}{\sqrt{\frac{|q_1|}{|q_2|}}x_2 - \sqrt{\frac{|q_2|}{|q_1|}}x_1}\,. \tag{50}$$

8 Discussion

As we demonstrated above, the Pomeron in LLA appears as a composite state of two reggeized gluons. To construct all possible gluon vertices one can use the effective action approach. At large N_c the BKP equation for the colorless gluon composite states turns out to be integrable. In the $N = 4$ extended super-symmetric gauge model the Pomeron is dual to the graviton in the 10 dimensional space. As a result, the Gribov Pomeron calculus in this model should be based on the theory of the reggeized graviton interactions. The graviton trajectories and vertices in the perturbation theory can be calculated from the effective theory local in the particle rapidities. In the lowest order the graviton trajectory contains ultraviolet logarithmic divergency. Really it means, that in gravity near the point $j = 2$ we have rather complicated singularities generated by the sum of the double-logarithmic terms. It was shown by finding these terms in all orders of perturbation theory with the use of the infrared evolution equation, that in this approximation the supersymmetry leads to the the amplitudes decreasing at high energies in comparison with the Born expression.

References

[1] L. N. Lipatov, *Sov. J. Nucl. Phys.* **23** (1976) 338;
V. S. Fadin, E. A. Kuraev and L. N. Lipatov, *Phys. Lett.* **B60** (1975) 50;
E. A. Kuraev, L. N. Lipatov and V. S. Fadin, *Sov. Phys. JETP* **44** (1976) 443 ; **45** (1977) 199;
I. I. Balitsky and L. N. Lipatov *Sov. J. Nucl. Phys.* **28** (1978) 822.

[2] V. S. Fadin and L. N. Lipatov, *Phys. Lett.* **B429** (1998) 127;
G. Camici and M. Ciafaloni, *Phys. Lett.* **B430** (1998) 349.;
A. V. Kotikov and L. N. Lipatov, *Nucl. Phys.* **B582** (2000).

[3] L. N. Lipatov, *Phys. Lett.* **B309** (1993) 394.

[4] L. N. Lipatov, *Sov. Phys. JETP* **63** (1986) 904.

[5] J. Bartels, *Nucl. Phys.* **B175** (1980) 365;
J. Kwiecinskii and M. Praszalowicz, *Phys. Lett.* **B94** (1980) 413.

[6] L. N. Lipatov, *Phys. Lett.* B **251** (1990) 284; *Pomeron in QCD*, in "Perturbative QCD", ed. by A. N. Mueller, World Scientific, 1989.

[7] A. A. Belavin, A. B. Zamolodchikov and A. M. Polyakov, *Nucl. Phys.* **B241** (1984) 333.

[8] L. N. Lipatov, *Nucl. Phys.* **B548** (1999) 328.

[9] L. N. Lipatov, *High energy asymptotics of multi-colour QCD and exactly solvable lattice models*, Padova preprint DFPD/93/TH/70, hep-th/9311037, unpublished.

[10] R. J. Baxter, *Exactly Solved Models in Statistical Mechanics*, (Academic Press, New York, 1982).

[11] L. N. Lipatov, *JETP Lett.* **59** (1994) 596;
L. D. Faddeev and G. P. Korchemsky, *Phys. Lett.* **B342** (1995) 311.

[12] L. N. Lipatov, *Nucl. Phys.* **B452** (1995) 369; *Phys. Rept.* **286** (1997) 131.

[13] E. N. Antonov, L. N. Lipatov, E, A. Kuraev and I. O. Cherednikov, *Nucl. Phys.* **B721** (2005) 111.

[14] Z. Bern, L. J. Dixon and V. A. Smirnov, *Phys. Rev.* **D72** (2005) 085001; *Phys. Rev.* **D72** (2005) 085001.

[15] V. S. Fadin, L. N. Lipatov, *Phys. Lett* **B706** (2012) 170.

[16] J. Bartels, L. N. Lipatov, G. P. Vacca, hep-th/1205.2530; J. Bartels, V. S. Fadin, L. N. Lipatov, G. P. Vacca, hep-th/1210.0797.

[17] J. Maldacena, *Adv. Theor. Math. Phys.* **2** (1998) 231.

[18] S. S. Gubser, I. R. Klebanov and A. M. Polyakov, *Phys. Lett.* **B428** (1998) 105.

[19] E. Witten, *Adv. Theor. Math. Phys.* **2** (1998) 253.

[20] J. Polchinski and M. J. Strassler, hep-th/0209211.

[21] A. V. Kotikov and L. N. Lipatov, *Nucl. Phys.* **B661** (2003) 19.

[22] A. V. Kotikov, L. N. Lipatov, A. V. Onishchenko and V. N. Velizhanin, *Phys. Lett.* **B595** (2004) 521; [Erratum-ibid. **B632** (2006) 754].

[23] R. C. Brower, J. Polchinski, M. J. Strassler and C. I. Tan, *JHEP* **0712** (2007) 005.

[24] V. N. Velizhanin, hep-th/0912.1624

[25] B. Basso, hep-th/1109.3154

[26] A. P. Bukhvostov, G. V. Frolov, E. A. Kuraev and L. N. Lipatov, *Nucl. Phys.* **B 258** (1985) 601.

[27] L, N, Lipatov, talk at "Perspectives in Hadronics Physics", Proceedings of the ICTP conference (World Scientific, Singapore, 1997).

[28] L. N. Lipatov, *Phys. Lett.* **B116** (1982) 411; Sov. Phys. *JETP* **55** (1982) 582.

[29] L. N. Lipatov, hep-th/1105.3127.

[30] J. Bartels, L. N. Lipatov, A. Sabio Vera, hep-th/1208.3423.

[31] R. Kirschner and L. N. Lipatov, *Nucl. Phys.* **B 258** (1983) 122.

CHAIRMAN: L. LIPATOV

Scientific Secretaries: R. Lopes de Sá, L. Spodyneiko

DISCUSSION

- *Csanad:*

At the LHC there are now precision data available on proton-proton diffraction, for example, dσ/d|t| at the TOTEM experiment and these are compared to theories. I wonder if, based on your calculations, an advanced model could be built and compared to this data.

- *Lipatov:*

Of course, the experiments at the LHC are performed at higher Q^2, higher energies, and therefore the theory which I discussed today should be applicable to these experiments because, due to asymptotic freedom, the growth in Q^2 makes the predictions better and better. But there are additional complexities in the high energy limit, namely, at the small-x region the equations are rather complicated. We cannot restrict ourselves only to the use of the renormalization group approach when solving the BFKL equation. I believe that the BFKL equation can describe the behavior of scattering amplitudes at large energies in QCD but to find its solutions we must impose boundary conditions at small and large $k_\perp$. In the region of large $k_\perp$, asymptotic freedom allows one to derive the boundary conditions with good accuracy. If there is new physics beyond the Standard Model, such as SUSY, extra dimensions, Planck-scale unification, these boundary conditions will be different, which would modify the spectrum of singularities of the crossing channel partial waves in the angular momentum plane, as discussed. Another problem for the BFKL equation is the infra-red region because we do not understand confinement very well.

Professor t' Hooft said earlier that it would be very nice to have a consistent theory at large distances. In the framework of this theory we would be able to predict everything with the same accuracy as it was done in QED. Now such theory is absent and we have large uncertainties. Thus, to find a solution of the BFKL equation one needs infra-red boundary conditions that are not well known. Nevertheless, as it was described in our paper with Kowalski and Ross, in the framework of BFKL theory, reasonable assumption about the boundary conditions leads to a discrete spectrum for the Pomeron singularities, which is in an agreement with the HERA experimental data and therefore, using the same parameters (what is important), the LHC experimental data at larger values of Q^2 and smaller values of x can be described. We want to do this in the near future.

- *Csanad:*

I am asking because almost all models fail to describe the dσ/d|t| data, so...

- *Lipatov:*

That is another problem, the problem of elastic scattering. In principle, if the theory is correct, we know the spectrum of Pomerons and their residues, which is enough to describe the proton-proton elastic scattering using the factorization of the residues for the t-channel partial wave poles. Collinear and Regge factorization can be used together with other properties to predict the behavior of scattering amplitudes at the LHC. For this purpose it would be useful to have additional large invariants, like Q^2 in DIS. Such invariants exist in the Drell-Yan process and similar semi-hard inclusive reactions at the LHC.

- *Fazio:*

About the unification of the soft and the hard Pomerons: how is it possible to test this unification by a purely string-theoretical calculation?

- *Lipatov:*

In principle, in my talk, I considered the superstring model predictions. But we do not have a consistent string theory dual to QCD, and that is the problem. There are only super-symmetric string models in 10 dimensions. Therefore, one should find additional arguments for applying these string theory predictions to QCD. I do not know if it is possible or not to use the ideas of the AdS/CFT correspondence, introducing some phenomenological string parameters, to describe experimental data at the LHC at small scattering angles, like in the previous question. In N=4 SYM one can realize such construction, but we want to describe the real world. I do not know a string theory dual to QCD and not to N=4 SYM. Only with such theory we would be able to describe the large-distance interactions, not only their small-distance behavior obtained from perturbative QCD. But we do not have now the appropriate realistic string model.

- *Fazio:*

In the Standard Model, we know about the existence of an effective coupling between the Higgs and gluons when one resumes the internal quark loops. Could this be translated into a double-diffractive Higgs production by the fusion of two Pomerons? And is it possible to make predictions without having to have a model for the survival probability, that would make the prediction clearer?

- *Lipatov:*

Clearer prediction? Indeed, the prediction of the Durham group for the Higgs production in gluon-gluon collisions is not very clear and explicit. I do not know its accuracy: possibly it is 10-20%. Nevertheless, using analogous ideas these authors predicted the production of heavy quarks at Fermilab and these results are in agreement with experimental data. And therefore, I think, that their approach is reliable for exclusive production of Higgs particles at the LHC. In my opinion the different string-type mechanism for this process should not work because we do not have now a good string model dual to QCD.

- *Fazio:*

Do you mean to say that there is not a consistent triple vertex pomeron-pomeron-Higgs?

- *Lipatov:*

Of course, this vertex was calculated with a good accuracy in QCD. But I do not know how to find this vertex in string theory. Only in N=4 SYM, where the Pomeron is dual to the graviton, one could calculate this vertex at large couplings. But the coupling constant in QCD due to asymptotic freedom cannot be very large as in N=4 SYM.

- *Carrasco:*

As discussed by Zvi Bern, we see a weak-weak connection between S-matrix elements of gluons (YM) and gravitons (gravity) against a flat background, the ``double copy". In your talk, you have identified a strong-weak connection of the AdS graviton and the flat-space Pomerons. If you were inclined to relate these two ideas and results, which steps would you take? How would you start to draw the connection between what we see in flat-space graviton calculations and the AdS gravitons in terms of Pomerons?

- *Lipatov:*

In my talk I derived the effective action for the case of asymptotically flat space. Of course, the space is curved, because the action contains the gravity field. But I did not take into account the AdS property of the space at infinity. In my opinion, it is not difficult to do due to the general covariance of the result. One can increase also the number of space-time dimensions up to 10. The most difficult technical problem for me is to include supersymmetry in the effective action. I am not a specialist in the extended supersymmetry and have not yet generalized the action for the case of the extended supergravity. When this will be done, then one can indeed construct the effective action for Pomerons. But again, not for the real world, but for N=4 SYM. There is still a long road to the real world. But it is very interesting that the Gribov theory, which was considered earlier as a purely phenomenological model, in N=4 SYM contains this remarkable mathematical construction, the extended supergravity in AdS space. In my opinion, this is interesting: how a phenomenological model can turn into a beautiful theory during 50 years. Everything started in 60's with the Pomeron hypothesis which was developed into the Gribov reggeon calculus. G. Veneziano constructed the dual amplitude with the linear Regge trajectories. Later it was understood that the Veneziano amplitude corresponds to a string theory. Comparatively, recently J. Maldacena guessed that the Superstring Theory in the AdS space is equivalent to a conformal field model in the four-dimensional space. Now again this theory returned to the high energy scattering problems which were not solved 50 years ago. In my opinion it is important that the duality between Pomeron and graviton increased the theoretical status of the Gribov Pomeron calculus.

One-Parameter Model for the Superworld

Tianjun Li,[1,2] James A. Maxin,[2] Dimitri V. Nanopoulos,[2,3,4,*] and Joel W. Walker[5]

[1]*State Key Laboratory of Theoretical Physics and Kavli Institute for Theoretical Physics China (KITPC), Institute of Theoretical Physics, Chinese Academy of Sciences, Beijing 100190, P. R. China*

[2]*George P. and Cynthia W. Mitchell Institute for Fundamental Physics and Astronomy, Texas A&M University, College Station, TX 77843, USA*

[3]*Astroparticle Physics Group, Houston Advanced Research Center (HARC), Mitchell Campus, Woodlands, TX 77381, USA*

[4]*Academy of Athens, Division of Natural Sciences, 28 Panepistimiou Avenue, Athens 10679, Greece*

[5]*Department of Physics, Sam Houston State University, Huntsville, TX 77341, USA*

We review the No-Scale $\mathcal{F}$-$SU(5)$ model with extra TeV-scale vector-like flippon multiplets and its associated collider phenomenology in the search for supersymmetry at the LHC. The model framework possesses the rather unique capacity to provide a light CP-even Higgs boson mass in the favored 124–126 GeV window while simultaneously retaining a testably light SUSY spectrum that is consistent with emerging low-statistics excesses beyond the Standard Model expectation in the ATLAS and CMS multijet data.

I. NO-SCALE $\mathcal{F}$-$SU(5)$

A. Phenomenological Overview

We have demonstrated [1, 2] the unique phenomenological consistency and profound predictive capacity of a model dubbed No-Scale $\mathcal{F}$-$SU(5)$, resting essentially and in equal

*Contribution to the Proceedings of the International School of Subnuclear Physics - 50th, What we would like LHC to give us, Erice, Sicily, Italy, 23 June - 2 July 2012, based on a talk given by Dimitri V. Nanopoulos.

measure on the tripodal foundations of the $\mathcal{F}$-lipped $SU(5)$ Grand Unified Theory (GUT) [3–5], two pairs of hypothetical TeV scale vector-like supersymmetric multiplets with origins in $\mathcal{F}$-theory model building [6–10], and the dynamically established boundary conditions of No-Scale Supergravity (SUGRA) [11–15]. It appears that the No-Scale scenario, and most stringently the vanishing of the Higgs bilinear soft term B_μ, comes into its own only when applied at an elevated scale, approaching the Planck mass. $M_{\mathcal{F}}$, the point of the ultimate second stage $SU(5) \times U(1)_X$ unification, emerges in turn as a suitable candidate scale only when substantially decoupled from the penultimate GUT scale unification of $SU(3)_C \times SU(2)_L$ at $M_{32} \simeq 10^{16}$ GeV via the modification to the renormalization group equations (RGEs) from the extra vector-like multiplets.

We have systematically established the hyper-surface within the $\tan\beta$, top quark mass m_t, gaugino mass $M_{1/2}$, and vector-like particle mass M_V parameter volume which is compatible with the application of the simplest No-Scale SUGRA boundary conditions [11–15]. We have demonstrated that simultaneous adherence to all current experimental constraints, most importantly contributions to the muon anomalous magnetic moment $(g-2)_\mu$, the branching ratio limit on $(b \to s\gamma)$, and the 7-year WMAP relic density measurement [16], dramatically reduces the allowed solutions to a highly non-trivial "golden strip", tightly confining $\tan\beta$, m_t, $M_{1/2}$, and M_V, effectively eliminating all extraneously tunable model parameters. Finally, taking a fixed Z-boson mass, we have dynamically determined the universal gaugino mass $M_{1/2}$ and fixed $\tan\beta$ via the "Super No-Scale" mechanism [17], that being the secondary minimization, at a local *minimum minimorum*, of the minimum $V_{\rm min}$ of the Higgs potential for the electroweak symmetry breaking (EWSB) vacuum.

This model is moreover quite interesting from a phenomenological point of view [7, 8]. The predicted vector-like particles can be observed at the Large Hadron Collider (LHC), though possibly not during the initial run. The partial lifetime for proton decay in the leading $(e|\mu)^+\pi^0$ channels falls around 5×10^{34} years [9, 10], testable at the future Hyper-Kamiokande [18] and Deep Underground Science and Engineering Laboratory (DUSEL) [19] experiments [9, 10, 20]. The lightest CP-even Higgs boson mass can be increased [21], hybrid inflation can be naturally realized, and the correct cosmic primordial density fluctuations can be generated [22].

B. The $\mathcal{F}$-lipped SU(5) GUT

Gauge coupling unification strongly suggests the existence of a GUT. In minimal supersymmetric $SU(5)$ models there are problems with doublet-triplet splitting and dimension five proton decay by colored Higgsino exchange [5]. These difficulties can be elegantly overcome in Flipped $SU(5)$ GUT models [3–5] via the missing partner mechanism [5].

Written in full, the gauge group of Flipped $SU(5)$ is $SU(5) \times U(1)_X$, which can be embedded into $SO(10)$. The generator $U(1)_{Y'}$ is defined for fundamental five-plets as $-1/3$ for the triplet members, and $+1/2$ for the doublet. The hypercharge is given by $Q_Y = (Q_X - Q_{Y'})/5$. There are three families of Standard Model (SM) fermions, whose quantum numbers under the $SU(5) \times U(1)_X$ gauge group are

$$F_i = (\mathbf{10}, \mathbf{1}), \quad \bar{f}_i = (\bar{\mathbf{5}}, -\mathbf{3}), \quad \bar{l}_i = (\mathbf{1}, \mathbf{5}), \tag{1}$$

where $i = 1, 2, 3$. To break the GUT and electroweak gauge symmetries, we introduce two pairs of Higgs fields: a pair of ten-plet Higgs for breaking the GUT symmetry, and a pair of five-plet Higgs for electroweak symmetry breaking.

$$H = (\mathbf{10}, \mathbf{1}) \quad ; \quad \overline{H} = (\overline{\mathbf{10}}, -\mathbf{1}) \tag{2}$$

$$h = (\mathbf{5}, -\mathbf{2}) \quad ; \quad \overline{h} = (\bar{\mathbf{5}}, \mathbf{2}) \tag{3}$$

A most notable intrinsic feature of the Flipped $SU(5)$ GUT is the presence of dual unification scales, with the ultimate merger of $SU(5) \times U(1)_{\mathrm{X}}$ occurring subsequent in energy to the penultimate $SU(3)_{\mathrm{C}}$ and $SU(2)_{\mathrm{L}}$ mixing at M_{32}. In the more traditional Flipped $SU(5)$ formulations, the scale $M_{\mathcal{F}}$ has been only slightly elevated from M_{32}, larger by a factor of perhaps only two or three [23]. Our interest however, is in scenarios where the ratio $M_{\mathcal{F}}/M_{32}$ is considerably larger, on the order of 10 to 100.

Key motivations for this picture include the desire to address the monopole problem via hybrid inflation, and the opportunity for realizing true string scale gauge coupling unification in the free fermionic model building context [6, 24], or the decoupling scenario in F-theory models [7, 8]. We have previously also considered the favorable effect of such considerations on the decay rate of the proton [9, 10]. Our greatest present interest however, is the effortless manner in which the elevation of the $SU(5) \times U(1)_{\mathrm{X}}$ scale salvages the dynamically established boundary conditions of No-Scale Supergravity. Being highly predictive, these conditions are thus also intrinsically highly constrained, and notoriously difficult to realize generically.

C. $\mathcal{F}$-theory Vector-Like Multiplets

We have introduced additional vector-like particle multiplets derived within the $\mathcal{F}$-theory [6] model building context to address the "little hierarchy" problem, altering the β-coefficients of the renormalization group to dynamically elevate the secondary $SU(5)\times U(1)_X$ unification at $M_{\mathcal{F}}$ to near the Planck scale, while leaving the $SU(3)_C \times SU(2)_L$ unification at M_{32} close to the traditional GUT scale. In other words, one obtains true string-scale gauge coupling unification in free fermionic string models [6, 24] or the decoupling scenario in F-theory models [7, 8]. To avoid a Landau pole for the strong coupling constant, we are restricted around the TeV scale to one of the following two multiplet sets [6].

$$\begin{aligned} Z1: &\ \left(XF_{(\mathbf{10},\mathbf{1})} \equiv (XQ, XD^c, XN^c),\ \overline{XF}_{(\overline{\mathbf{10}},-\mathbf{1})}\right) \\ Z2: &\ \left(XF,\ \overline{XF}, Xl_{(\mathbf{1},-\mathbf{5})},\ \overline{Xl}_{(\mathbf{1},\mathbf{5})} \equiv XE^c\right) \end{aligned} \tag{4}$$

In the prior, XQ, XD^c, XE^c, XN^c have the same quantum numbers as the quark doublet, the right-handed down-type quark, charged lepton, and neutrino, respectively. We have argued [2] that the feasibly near-term detectability of these hypothetical fields in collider experiments, coupled with the distinctive flipped charge assignments within the multiplet structure, represents a smoking gun signature for Flipped $SU(5)$, and have thus coined the term *flippons* to collectively describe them. In this paper, we consider only the $Z2$ set, although discussion for the $Z1$ set, if supplemented by heavy threshold corrections, can be similar.

We emphasize that the specific representations of vector-like fields which we currently employ have been explicitly constructed within the local F-theory model building context [7, 8]. However, the mass of these fields, and even the fact of their existence, is not mandated by the F-theory, wherein it is also possible to realize models with only the traditional Flipped (or Standard) $SU(5)$ field content. We claim only an inherent consistency of their conceptual origin out of the F-theoretic construction, and take the manifest phenomenological benefits which accompany the elevation of $M_{\mathcal{F}}$ as justification for the greater esteem which we hold for this particular model above other alternatives.

D. No-Scale Supergravity

The Higgs boson, being a Lorentz scalar, is not stable in the SM against quadratic quantum mass corrections which drive it toward the dominant Planck scale, some seventeen orders of magnitude above the value required for consistent EWSB. Supersymmetry naturally solves this fine tuning problem by pairing the Higgs with a chiral spin-1/2 "Higgsino" partner field, and following suit with a corresponding bosonic (fermionic) superpartner for all fermionic (bosonic) SM fields, introducing the full set of quantum counter terms. Localizing the supersymmetry (SUSY) algebra, which includes the generator of spacetime translations (the momentum operator), induces general coordinate invariance, producing the supergravity (SUGRA) theories.

Since we do not observe mass degenerate superpartners for the known SM fields, SUSY must itself be broken around the TeV scale. In the traditional framework, supersymmetry is broken in the hidden sector, and the effect is mediated to the observable sector via gravity or gauge interactions. In GUTs with minimal gravity mediated supersymmetry breaking, called mSUGRA, one can fully characterize the supersymmetry breaking soft terms by four universal parameters (gaugino mass $M_{1/2}$, scalar mass M_0, trilinear coupling A, and the low energy ratio $\tan\beta$ of up- to down-type Higgs VEVs, plus the sign of the Higgs bilinear mass term μ. The μ term and its bilinear soft term B_μ are determined by the Z-boson mass M_Z and $\tan\beta$ after the electroweak (EW) symmetry breaking.

No-Scale Supergravity was proposed [11–15] to address the cosmological flatness problem, and defined as the subset of supergravity models which satisfy the following three constraints [11]: (i) The vacuum energy vanishes automatically due to the suitable Kähler potential; (ii) At the minimum of the scalar potential, there are flat directions which leave the gravitino mass $M_{3/2}$ undetermined; (iii) The super-trace quantity $\mathrm{Str}\mathcal{M}^2$ is zero at the minimum. Without this, the large one-loop corrections would force $M_{3/2}$ to be either zero or of Planck scale. The defining Kähler potential [14]

$$K = -3\ln(T + \overline{T} - \sum_i \overline{\Phi}_i \Phi_i) \ , \tag{5}$$

automatically satisfies the first two conditions, while the third is model dependent and can always be satisfied in principle [25].

In Eq. (5), T is a modulus field, while the Φ_i are N_C scalar matter fields which parameterize the coset space $SU(N_C+1,1)/(SU(N_C+1)\times U(1))$. The scalar potential is automatically positive semi-definite, and has a flat direction along the T field. The non-compact structure of the symmetry implies that the classical vacuum is not only constant but actually identical to zero. Moreover, the simplest No-Scale boundary conditions $M_0 = A = B_\mu = 0$ are dynamically established, while $M_{1/2} > 0$ is allowed, and indeed required for SUSY breaking. A one-parameter model of similar form has been much studied in the past (See [26] and references therein. For a review, see [27]). The CP violation problem and the flavor changing neutral current problems are automatically solved in turn. All low energy scales are dynamically generated by quantum corrections, *i.e.* running under the RGEs, to the classically flat potential.

E. The Stringy Super No-Scale Mechanism

The fiercely reductionist No-Scale picture inherits an associative weight of motivation from its robustly generic and natural appearance, for example, in the compactification of the weakly coupled heterotic string theory [28], compactification of M-theory on S^1/Z_2 at the leading order [29], and potentially also directly in F-theory models [30–33].

In the simplest stringy No-Scale SUGRA, the Kähler modulus T, a characteristic of the Calabi-Yau manifold, is the single relevant modulus field, the dilaton coupling being irrelevant. The F-term of T generates the gravitino mass $M_{3/2}$, which is proportionally equivalent to $M_{1/2}$. Exploiting the simplest No-Scale boundary condition at $M_{\mathcal{F}}$ and running from high energy to low energy under the RGEs, there can be a secondary minimization, or *minimum minimorum*, of the minimum of the Higgs potential $V_{\rm min}$ for the EWSB vacuum. Since $V_{\rm min}$ depends on $M_{1/2}$, the universal gaugino mass $M_{1/2}$ is consequently dynamically determined by the equation $dV_{\rm min}/dM_{1/2} = 0$, aptly referred to as the "Super No-Scale" mechanism; We have argued by the combined action of this mechanism, the transmutative role of the the RGEs, and the stabilizing counter-balance of supersymmetry, that No-Scale $\mathcal{F}$-$SU(5)$ addresses the various aspects of the gauge hierarchy problem [17].

The three parameters M_0, A, B_μ are once again identically zero at the boundary because of the defining Kähler potential, and are thus known at all other scales as well by the RGEs. The minimization of the Higgs scalar potential with respect to the neutral elements of both SUSY Higgs doublets gives two conditions, the first of which fixes the magnitude of μ.

The second condition, which would traditionally be used to fix B_μ, instead here enforces a consistency relationship on the remaining parameters, being that B_μ is already constrained.

In general, the $B_\mu = 0$ condition gives a hypersurface of solutions cut out from a very large parameter space. If we lock all but one parameter, it will give the final value. If we take a slice of two dimensional space, as has been described, it will give a relation between two parameters for all others fixed. In a three-dimensional view with B_μ on the vertical axis, this curve is the "flat direction" line along the bottom of the trench of $B_\mu = 0$ solutions. In general, we must vary at least two parameters rather than just one in isolation, in order that their mutual compensation may transport the solution along this curve. The most natural first choice is in some sense the pair of prominent unknown inputs $M_{1/2}$ and $\tan\beta$, as demonstrated in Ref. [17].

It must be emphasized that the $B_\mu = 0$ No-Scale boundary condition is the central agent affording this determination, as it is the extraction of the parameterized parabolic curve of solutions in the two compensating variables which allows for a localized, bound nadir point to be isolated by the Super No-Scale condition, dynamically determining *both* parameters. The background surface of $V_{\rm min}$ for the full parameter space outside the viable $B_\mu = 0$ subset is, in contrast, a steadily inclined and uninteresting function. In our prior study, the local *minimum minimorum* of $V_{\rm min}$ for selected inputs of M_V and m_t was taken to dynamically establish the values of $M_{1/2}$ and $\tan\beta$. Although $M_{1/2}$ and $\tan\beta$ have no *directly* established experimental values, they are severely indirectly constrained by phenomenology in the context of this model [1, 2]. It is highly non-trivial that there should be a strong accord between the top-down and bottom-up perspectives, but this is indeed precisely what has been observed [17].

II. $\mathcal{F}$-$SU(5)$ SUSY MULTIJETS AT THE LHC

The $\mathcal{F}$-$SU(5)$ model space is bounded primarily by a set of "bare-minimal" experimental constraints distinguished by a great longevity of relevance, as defined in Ref. [34]. These include the top quark mass 172.2 GeV $\leq m_t \leq$ 174.4 GeV, 7-year WMAP cold dark matter relic density $0.1088 \leq \Omega_{\rm CDM}h^2 \leq 0.1158$ [16], and precision LEP constraints on the SUSY mass content. We further append to this classification an adherence to the defining high-scale boundary conditions of the model. In light of recent developments, the favored parameter space may be further circumscribed by the demands of a 124–126 GeV Higgs boson mass. The surviving region is comprised of a narrow strip of space confined to $400 \leq M_{1/2} \leq 900$ GeV, $19.4 \leq \tan\beta \leq 23$, and $950 \leq M_V \leq 6000$ GeV. The border at the minimum $M_{1/2} = 400$ GeV is required by the LEP constraints, while the maximum boundary at $M_{1/2} = 900$ GeV prevents a charged stau LSP at around $\tan\beta \cong 23$. In the bulk of the model space the lightness of the stau is leveraged to facilitate an appropriate dark matter relic density via stau-neutralino coannihilation.

The convergence of the predicted $\mathcal{F}$-$SU(5)$ Higgs mass with the collider measured value is achieved through contributions to the lightest CP-even Higgs boson mass from the flippons, calculated from the RGE improved one-loop effective Higgs potential approach [35, 36]. The mechanism for the serendipitous mass shift is a pair of Yukawa interaction terms between the Higgs and vector-like flippons in the superpotential, resulting in a 3–4 GeV upward shift in the Higgs mass to the experimentally measured range [37]. Using the relevant shift in the Higgs mass-square as approximated in Refs. [21, 37], which implements a leading dependence of the flippon mass M_V, larger shifts correspond to lighter vector-like flippons. This flippon induced mechanism operates in synthesis with the top quark mass, whose elevation similarly raises the non-flippon contributed Higgs mass. The cumulative result is a very narrow strip of model space, with the lower strip boundary truncated by the upper top quark mass extremity, and the upper strip boundary situated at the minimum Higgs mass of 124 GeV, conveniently establishing a stable, thin band of experimentally viable points with which to explore new physics.

The same flippon induced perturbation to the RGE unification structure of $\mathcal{F}$-$SU(5)$ that was responsible for facilitating a consistent application of the No-Scale boundary conditions near the Planck mass also produces a key phenomenological signature. The flat RGE evolution of the $SU(3)_C$ gaugino mass M_3, which mirrors the flatness of the β-coefficient $b_3 = 0$, suppresses the standard logarithmic mass enhancement at low-energy and yields a SUSY spectrum $M(\widetilde{t}_1) < M(\widetilde{g}) < M(\widetilde{q})$ where the light stop $\widetilde{t}_1$ and gluino $\widetilde{g}$ are both less massive than all other squarks. This highly unusual hierarchy produces a distinct event topology initiated by the pair-production of heavy first or second generation squarks $\widetilde{q}$ and/or gluinos in the hard scattering process, with the heavy squark likely to yield a quark-gluino pair $\widetilde{q} \rightarrow q\widetilde{g}$. The gluino then has only two main channels available in the cascade decay, $\widetilde{g} \rightarrow \widetilde{t}_1\bar{t}$ or $\widetilde{g} \rightarrow q\bar{q}\widetilde{\chi}_1^0$, with $\widetilde{t}_1 \rightarrow t\widetilde{\chi}_1^0$ or $\widetilde{t}_1 \rightarrow b\widetilde{\chi}_1^{\pm}$. As $M_{1/2}$ increases, the stop-top channel be-

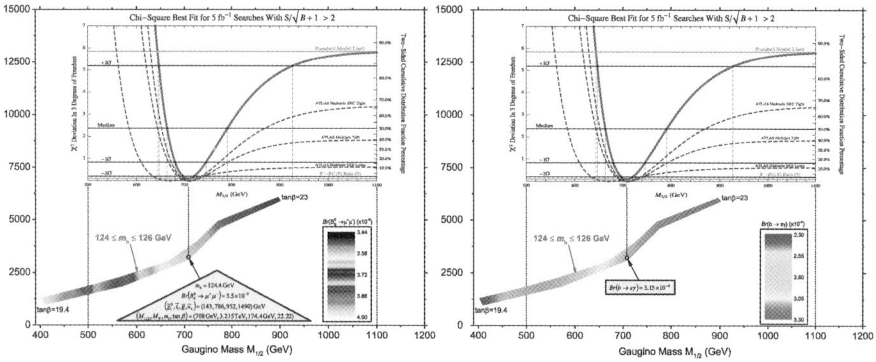

FIG. 1: We depict the experimentally viable parameter space of No-Scale $\mathcal{F}$-$SU(5)$ as a function of the gaugino mass $M_{1/2}$ and flippon mass M_V. The inset diagrams (with linked horizontal scale) are the multi-axis cumulative χ^2 fitting of Ref. [38].

comes dominant, ultimately reaching 100% for $M_{1/2} \geq 729$ GeV. For $M_{1/2} < 729$ GeV, both avenues have sufficient branching fractions to produce observable events at the LHC. Each gluino produces 2–6 hadronic jets, with the maximum of six jets realized in the gluino-mediated stop decay, so that a single gluino-gluino pair-production event can net 4–12 jets. After further fragmentation processes, the final event is characterized by a definitive SUSY signal of high-multiplicity jets.

The most robust test of any supersymmetric model is the prediction of a unique signature plainly accounting for observed anomalies in collider data. The exceptional mass ordering in No-Scale $\mathcal{F}$-$SU(5)$ provides a distinctive marker at the LHC, since multijet events are expected to dominate a probed $\mathcal{F}$-$SU(5)$ framework. We first suggested in March 2011 [39] that SUSY in an $\mathcal{F}$-$SU(5)$ universe would become manifest at the colliders in high-multiplicity jet events, extending this initial study in Refs. [38, 40–42]. The first ample accumulation of multijet data was released by the collaborations later in 2011 in Refs. [43–45], based upon 1 fb^{-1} of luminosity. Though the number of events remaining after the collaboration data cuts was less than ten, there did appear small but curious excesses beyond the SM estimates in these searches targeting multijet events. The most prominent examples came from ATLAS, where the 7j80 ($\geq$ 7 jets and jet $p_T > 80$ GeV) search of Ref. [45] and High Mass ($\geq$ 4 jets and jet $p_T > 80$ GeV) search of Ref. [44] displayed interesting event production over the data-driven background estimates. Employing the signal significance metric $S/\sqrt{B+1}$, we computed a value of 1.1 for 7j80 and 1.3 for the High Mass search. Despite the weak signal, reasonably attributable to statistical fluctuations, No-Scale $\mathcal{F}$-$SU(5)$ provided a neat and efficient explanation for the minor over-productions in these two searches. Despite the long odds at that time, those clean fits prompted us to extrapolate from the ATLAS published statistics of Refs. [44, 45] to predict signal strengths of $S/\sqrt{B+1} = 1.9$ for 7j80 and $S/\sqrt{B+1} = 3.0$ for the High Mass [41] search in the forthcoming 5 fb^{-1} data set at 7 TeV, assuming a legitimate physics origin for the intriguing over-production.

We provided a detailed analysis of the ATLAS and CMS 5 fb^{-1} observations at the 7 TeV LHC in Ref. [38], focused on those search strategies where the signal significance was strongest and the largest number of events had accumulated, imposing $S/\sqrt{B+1} > 2.0$ as a minimal boundary. Strikingly, the 7j80 [46] search and the composite successors to the High Mass search were the only 5 fb^{-1} strategies to surmount this significance hurdle. To elaborate, ATLAS essentially segregated the former High Mass $\geq$4 jet SUSY search of Ref. [44] into three separate searches of 4 jets, 5 jets, and 6 jets for the latter study, intended to isolate the $\widetilde{g}\widetilde{g}$, $\widetilde{q}\widetilde{g}$, and $\widetilde{q}\widetilde{q}$ 0-lepton channels via $\widetilde{q} \to q\widetilde{g}$ and $\widetilde{g} \to q\overline{q}\widetilde{\chi}_1^0$ [47]. In addition to the ATLAS 7j80 [46], these ATLAS 4-jet and 6-jet searches of Ref. [47], referred to as SRC Tight and SRE Loose, respectively, were the only other 5 fb^{-1} searches to achieve $S/\sqrt{B+1} > 2.0$ in all the ATLAS and CMS 5 fb^{-1} studies analyzed at that time. Granting that the 1 fb^{-1} data sample is a subset of the 5 fb^{-1} data, the signal strength nevertheless expanded in the precise proportionality expected. The final 5 fb^{-1} 7 TeV ATLAS observations computed signal significances of $S/\sqrt{B+1} = 2.1$ for 7j80 [46], $S/\sqrt{B+1} = 3.2$ for SRC Tight (4j) [47], and $S/\sqrt{B+1} = 2.6$ for SRE Loose (6j) [47], in line with our predictions and very consistent with the signal growth expected to be observed in an $\mathcal{F}$-$SU(5)$ universe.

This enlarged signal strength simultaneously presented a golden opportunity to derive a best fit SUSY mass to the 5 fb^{-1} data through a χ^2 fitting procedure. We demonstrated [38] clear internal consistency in the $\mathcal{F}$-$SU(5)$ mass scale favored by the various search windows, in addition to the described correlation across time in the signal growth. This analysis favored sparticle masses of $m_{\widetilde{\chi}_1^0} = 143$ GeV, $m_{\widetilde{t}_1} = 786$ GeV, $m_{\widetilde{g}} = 952$ GeV, and $m_{\widetilde{u}_L} = 1490$ GeV, complementing a Higgs mass of $m_h = 124.4$ GeV at the $M_{1/2} = 708$ GeV well of the 5 fb^{-1} multi-axis cumulative χ^2 curve, combining the 7j80, SRC Tight, and SRE Loose search channels. To exemplify this best fit at the χ^2 minimum, we chose an $M_{1/2} = 708$ GeV point as our standing favored benchmark [38]. The superimposed cumulative χ^2 curve of Ref. [38] visibly showcases how the ATLAS 7j80, SRC Tight, and SRE Loose over-productive search strategies illuminate the $\mathcal{F}$-$SU(5)$ model space as naturally conforming to the collider observations. By lowering the minimum threshold for signal significance to $S/\sqrt{B+1} > 1.0$, the CMS 5 fb^{-1} MT2 search strategy [48] was included into our 5 fb^{-1} multi-axis χ^2 fitting in Ref. [42] along with an additional ATLAS search, namely the 8j55 case from Ref. [46]. It was demonstrated in this manner that further non-trivial correlations exist between the mass scale favored by independently productive ATLAS and CMS SUSY searches, bolstering the case against attribution of the excesses to random statistical fluctuations.

The data observations for the ATLAS multijet searches discussed here have shown a very natural progression from 1 fb^{-1} to 5 fb^{-1}. In fact, the $S/\sqrt{B+1} \sim 3$ signal significance of the combined ATLAS 5 fb^{-1} multijet searches, which we can consider to be about 3σ, is near the same signal level as the Higgs boson after 5 fb^{-1} at 7 TeV. With the Higgs boson now at the discovery threshold of 5σ in the first 8 TeV data tranche, it would only be fitting if the ATLAS multijet SUSY searches continued to track the Higgs signal strength. Looking forward and preparing for potentially more significant SUSY production as we shift to forthcoming larger LHC beam collision energies and hence greater numbers of statistics, we transition here to a more appropriate metric for measuring signal strength in the presence of larger excess event production beyond expectations, $2 \times (\sqrt{S+B} - \sqrt{B})$. We employ the background statistics derived by the ATLAS Collaboration for 5 fb^{-1} at 7 TeV from Ref. [46], though to determine an estimate of the SM background for 8 TeV, we scale up these ATLAS statistics using the same factor observed in our Monte Carlo for $\mathcal{F}$-$SU(5)$ simulations. This estimator, while serving our limited scope here satisfactorily, can only be as reliable as the expectation of statistical, dynamic and procedural stability across the transition in energy, luminosity and model. We further assume here a static data cutting strategy between the ATLAS 7 and 8 TeV multijet searches. We indeed project that there should be a visible multijet signal strength sufficient for SUSY discovery in the isolated 15 fb^{-1} 8 TeV data, expected to be recorded in 2012 and processed in 2013, if the existing signal in the 5 fb^{-1} 7 TeV data is legitimately and wholly attributable to new physics. More precisely, assuming no important modifications to the background calibration procedures by ATLAS, we can project the 7j80 SUSY search tactics of Ref. [46] to yield a signal significance of $2 \times (\sqrt{S+B} - \sqrt{B}) \sim 6$ for 15 fb^{-1} at 8 TeV, and $2 \times (\sqrt{S+B} - \sqrt{B}) \sim$ 7–8 for the SRC-

Tight and SRE-Loose search strategies of Ref. [47]. Although potentially quite susceptible to large statistical fluctuation, these rather strong signal projections nonetheless indicate that a probing of the $\mathcal{F}$-$SU(5)$ framework at the LHC could indeed yield further tantalizing, and possibly convincing, evidence that nature herself is fundamentally supersymmetric. The summation of the 5 fb^{-1} of 7 TeV data to the 8 TeV data only improves the signal significance modestly. Moreover, the presence of excess events in the 15 fb^{-1} ATLAS multijet searches at 8 TeV will resoundingly indicate that random background anomalies are not the source of the 7 TeV multijet over-production. We find the predictable evolution of our SUSY exploration from the initial 1 fb^{-1} at 7 TeV to the 5 fb^{-1} at 7 TeV to warrant such positive speculation as we move forward to the 15 fb^{-1} at 8 TeV.

Acknowledgments

Dimitri Nanopoulos would like to thank Professor A. Zichichi for his warm hospitality at Erice. This research was supported in part by the DOE grant DE-FG03-95-Er-40917 (TL and DVN), by the Natural Science Foundation of China under grant numbers 10821504, 11075194, 11135003, and 11275246 (TL), and by the Mitchell-Heep Chair in High Energy Physics (JAM). We also thank Sam Houston State University for providing high performance computing resources.

[1] T. Li, J. A. Maxin, D. V. Nanopoulos, and J. W. Walker, "The Golden Point of No-Scale and No-Parameter $\mathcal{F}$-$SU(5)$," Phys. Rev. **D83**, 056015 (2011), 1007.5100.

[2] T. Li, J. A. Maxin, D. V. Nanopoulos, and J. W. Walker, "The Golden Strip of Correlated Top Quark, Gaugino, and Vectorlike Mass In No-Scale, No-Parameter F-SU(5)," Phys. Lett. **B699**, 164 (2011), 1009.2981.

[3] S. M. Barr, "A New Symmetry Breaking Pattern for $SO(10)$ and Proton Decay," Phys. Lett. **B112**, 219 (1982).

[4] J. P. Derendinger, J. E. Kim, and D. V. Nanopoulos, "Anti-$SU(5)$," Phys. Lett. **B139**, 170 (1984).

[5] I. Antoniadis, J. R. Ellis, J. S. Hagelin, and D. V. Nanopoulos, "Supersymmetric Flipped

$SU(5)$ Revitalized," Phys. Lett. **B194**, 231 (1987).

[6] J. Jiang, T. Li, and D. V. Nanopoulos, "Testable Flipped $SU(5) \times U(1)_X$ Models," Nucl. Phys. **B772**, 49 (2007), hep-ph/0610054.

[7] J. Jiang, T. Li, D. V. Nanopoulos, and D. Xie, "F-$SU(5)$," Phys. Lett. **B677**, 322 (2009).

[8] J. Jiang, T. Li, D. V. Nanopoulos, and D. Xie, "Flipped $SU(5) \times U(1)_X$ Models from F-Theory," Nucl. Phys. **B830**, 195 (2010), 0905.3394.

[9] T. Li, D. V. Nanopoulos, and J. W. Walker, "Elements of F-ast Proton Decay," Nucl. Phys. **B846**, 43 (2011), 1003.2570.

[10] T. Li, J. A. Maxin, D. V. Nanopoulos, and J. W. Walker, "Dark Matter, Proton Decay and Other Phenomenological Constraints in $\mathcal{F}$-SU(5)," Nucl.Phys. **B848**, 314 (2011), 1003.4186.

[11] E. Cremmer, S. Ferrara, C. Kounnas, and D. V. Nanopoulos, "Naturally Vanishing Cosmological Constant in $N = 1$ Supergravity," Phys. Lett. **B133**, 61 (1983).

[12] J. R. Ellis, A. B. Lahanas, D. V. Nanopoulos, and K. Tamvakis, "No-Scale Supersymmetric Standard Model," Phys. Lett. **B134**, 429 (1984).

[13] J. R. Ellis, C. Kounnas, and D. V. Nanopoulos, "Phenomenological $SU(1,1)$ Supergravity," Nucl. Phys. **B241**, 406 (1984).

[14] J. R. Ellis, C. Kounnas, and D. V. Nanopoulos, "No Scale Supersymmetric Guts," Nucl. Phys. **B247**, 373 (1984).

[15] A. B. Lahanas and D. V. Nanopoulos, "The Road to No Scale Supergravity," Phys. Rept. **145**, 1 (1987).

[16] E. Komatsu et al. (WMAP), "Seven-Year Wilkinson Microwave Anisotropy Probe (WMAP) Observations: Cosmological Interpretation," Astrophys.J.Suppl. **192**, 18 (2010), 1001.4538.

[17] T. Li, J. A. Maxin, D. V. Nanopoulos, and J. W. Walker, "Super No-Scale $\mathcal{F}$-$SU(5)$: Resolving the Gauge Hierarchy Problem by Dynamic Determination of $M_{1/2}$ and $\tan\beta$," Phys. Lett. B **703**, 469 (2011), 1010.4550.

[18] K. Nakamura, "Hyper-Kamiokande: A next generation water Cherenkov detector," Int. J. Mod. Phys. **A18**, 4053 (2003).

[19] S. Raby et al., "DUSEL Theory White Paper," (2008), 0810.4551.

[20] T. Li, D. V. Nanopoulos, and J. W. Walker, "Fast Proton Decay," Phys. Lett. **B693**, 580 (2010), 0910.0860.

[21] Y. Huo, T. Li, D. V. Nanopoulos, and C. Tong, "The Lightest CP-Even Higgs Boson Mass in

the Testable Flipped $SU(5) \times U(1)_X$ Models from F-Theory," (2011), 1109.2329.

[22] B. Kyae and Q. Shafi, "Flipped SU(5) predicts delta(T)/T," Phys. Lett. **B635**, 247 (2006), hep-ph/0510105.

[23] J. R. Ellis, D. V. Nanopoulos, and J. Walker, "Flipping SU(5) out of trouble," Phys. Lett. **B550**, 99 (2002), hep-ph/0205336.

[24] J. L. Lopez, D. V. Nanopoulos, and K.-j. Yuan, "The Search for a realistic flipped $SU(5)$ string model," Nucl. Phys. **B399**, 654 (1993), hep-th/9203025.

[25] S. Ferrara, C. Kounnas, and F. Zwirner, "Mass formulae and natural hierarchy in string effective supergravities," Nucl. Phys. **B429**, 589 (1994), hep-th/9405188.

[26] J. L. Lopez, D. V. Nanopoulos, and A. Zichichi, "A String no scale supergravity model and its experimental consequences," Phys.Rev. **D52**, 4178 (1995), hep-ph/9502414.

[27] J. L. Lopez, D. V. Nanopoulos, and A. Zichichi, *Searching for the Superworld: A Volume in Honor of Antonino Zichichi on the Occasion of the Sixth Centenary Celebrations of the University of Turin, Italy* (World Scientific Series in 20th Century Physics, 2007), vol. 39 Part B., pp. 226–516.

[28] E. Witten, "Dimensional Reduction of Superstring Models," Phys. Lett. **B155**, 151 (1985).

[29] T.-j. Li, J. L. Lopez, and D. V. Nanopoulos, "Compactifications of M theory and their phenomenological consequences," Phys.Rev. **D56**, 2602 (1997), hep-ph/9704247.

[30] C. Beasley, J. J. Heckman, and C. Vafa, "GUTs and Exceptional Branes in F-theory - I," JHEP **01**, 058 (2009), 0802.3391.

[31] C. Beasley, J. J. Heckman, and C. Vafa, "GUTs and Exceptional Branes in F-theory - II: Experimental Predictions," JHEP **01**, 059 (2009), 0806.0102.

[32] R. Donagi and M. Wijnholt, "Model Building with F-Theory," (2008), 0802.2969.

[33] R. Donagi and M. Wijnholt, "Breaking GUT Groups in F-Theory," (2008), 0808.2223.

[34] T. Li, J. A. Maxin, D. V. Nanopoulos, and J. W. Walker, "The Unification of Dynamical Determination and Bare Minimal Phenomenological Constraints in No-Scale F- SU(5)," Phys.Rev. **D85**, 056007 (2012), 1105.3988.

[35] K. Babu, I. Gogoladze, M. U. Rehman, and Q. Shafi, "Higgs Boson Mass, Sparticle Spectrum and Little Hierarchy Problem in Extended MSSM," Phys.Rev. **D78**, 055017 (2008), 0807.3055.

[36] S. P. Martin, "Extra vector-like matter and the lightest Higgs scalar boson mass in low-energy supersymmetry," Phys.Rev. **D81**, 035004 (2010), 0910.2732.

[37] T. Li, J. A. Maxin, D. V. Nanopoulos, and J. W. Walker, "A Higgs Mass Shift to 125 GeV and A Multi-Jet Supersymmetry Signal: Miracle of the Flippons at the $\sqrt{s} = 7$ TeV LHC," Phys.Lett. **B710**, 207 (2012), 1112.3024.

[38] T. Li, J. A. Maxin, D. V. Nanopoulos, and J. W. Walker, "Chanel $N^o5(\mathrm{fb}^{-1})$: The Sweet Fragrance of SUSY," (2012), 1205.3052.

[39] T. Li, J. A. Maxin, D. V. Nanopoulos, and J. W. Walker, "The Ultrahigh jet multiplicity signal of stringy no-scale $\mathcal{F}$-$SU(5)$ at the $\sqrt{s} = 7$ TeV LHC," Phys.Rev. **D84**, 076003 (2011), 1103.4160.

[40] T. Li, J. A. Maxin, D. V. Nanopoulos, and J. W. Walker, "Profumo di SUSY: Suggestive Correlations in the ATLAS and CMS High Jet Multiplicity Data," (2011), 1111.4204.

[41] T. Li, J. A. Maxin, D. V. Nanopoulos, and J. W. Walker, "A Multi-Axis Best Fit to the Collider Supersymmetry Search: The Aroma of Stops and Gluinos at the $\sqrt{s} = 7$ TeV LHC," (2012), 1203.1918.

[42] T. Li, J. A. Maxin, D. V. Nanopoulos, and J. W. Walker, "Non-trivial Supersymmetry Correlations between ATLAS and CMS Observations," (2012), 1206.0293.

[43] "Search for supersymmetry in all-hadronic events with α_{T}," (2011), CMS PAS SUS-11-003, URL `http://cdsweb.cern.ch`.

[44] G. Aad et al. (ATLAS Collaboration), "Search for squarks and gluinos using final states with jets and missing transverse momentum with the ATLAS detector in $\sqrt{s} = 7$ TeV proton-proton collisions," (2011), 1109.6572.

[45] G. Aad et al. (Atlas Collaboration), "Search for new phenomena in final states with large jet multiplicities and missing transverse momentum using $\sqrt{s} = 7$ TeV pp collisions with the ATLAS detector," JHEP **1111**, 099 (2011), 1110.2299.

[46] G. Aad et al. (ATLAS Collaboration), "Hunt for new phenomena using large jet multiplicities and missing transverse momentum with ATLAS in 4.7 fb^{-1} of $\sqrt{s} = 7$ TeV proton-proton collisions," JHEP **1207**, 167 (2012), 1206.1760.

[47] "Search for squarks and gluinos with the ATLAS detector using final states with jets and missing transverse momentum and 4.7 fb^{-1} of $\sqrt{s} = 7$ TeV proton-proton collision data," (2012), ATLAS-CONF-2012-033, 1208.0949, URL `http://cdsweb.cern.ch`.

[48] S. Chatrchyan et al. (CMS Collaboration), "Search for supersymmetry in hadronic final states using MT2 in pp collisions at $\sqrt{s} = 7$ TeV," (2012), 1207.1798.

Beyond Relativistic Quantum String Theory

or

Discreteness and Determinism in Superstrings

Gerard 't Hooft

Institute for Theoretical Physics
Utrecht University

and

Spinoza Institute
Postbox 80.195
3508 TD Utrecht, the Netherlands

e-mail: `g.thooft@uu.nl`
`internet: http://www.phys.uu.nl/~thooft`

Summary

The question "What lies beyond the Quantized String or Superstring Theory?" and the question "What lies beyond Quantum Mechanics itself?" might have one common answer: a discretized, classical version of string theory, which lives on a lattice in Minkowski space. The size a of the meshes on this lattice in Minkowski space is determined by the string slope parameter, α'.

1. Introduction

In spite of the beautiful mathematical underpinnings of string- and superstring theory[1][2], the physical foundations of these ideas are embarrassingly weak. Here, I am not referring to the lack of experimental evidence; the fact that this theory is almost impossible to subject to non-trivial tests or observations is an inevitable aspect of the domain of the physical phenomena it is designed to describe, and none of the competing approaches towards 'physics at the Planck scale' are faring any better than (super)string theory. Rather, the weakness of its foundations is the fact the we do not understand in a satisfactory way what the equations mean physically, as soon as one attempts to go beyond the level of perturbation theory. Term-by-term, the string perturbation theory appears to be finite and well-behaved, but what the theory means beyond this perturbation expansion is obscure. In particular, we have the fact that strings propagate in a space-time that itself gets curved because of the presence of strings, and the fact that the energy quanta of curved space-time, the gravitons, themselves must be strings again, to be precise, the lightest spin-two excitations of a closed string.

If string theory is finite perturbatively, what does this mean for the non-perturbative system, whatever that is? From its formulation, one would have thought that, at least in principle, strings should be able to form infinitely complex curves in arbitrarily small regions of space, and, when one includes such complex structures in a functional integral formalism, this should be expected to give rise to quite complex infinities. The most experienced practitioners of the theory usually respond to this that "the theory takes care of such infinities by itself", and indeed, this appears to be exactly what happens in any explicit calculation. Now, in these lectures, we arrive at an even simpler answer: everything is finite because the string actually is a deterministic system that lives in a lattice space-time; target space is a lattice, whose meshes have a length $2\pi\sqrt{\alpha'}$.

1.1. The Discrete Lorentz Group

Consider a theory with all its dynamical variables defined on a lattice: $(t, x, y, z) \in \mathbb{Z}^4$. Equations of motion could for instance take the form:

$$\phi(x, y, z, t+1) = F\Big(\phi(x \pm 1,\ y \pm 1,\ z \pm 1,\ t)\Big)\ . \tag{1.1}$$

Question: is there *any* such discrete theory for which an invariance, or covariance, property can be formulated under *discrete Lorentz transformations* (and indeed also under discrete Poincaré transformations)?

$$X^\mu \rightarrow L^\mu{}_\nu X^\nu\ , \tag{1.2}$$

(where all matrix elements $L^\mu{}_\nu$ of the $O(D-1,1)$ matrix L are integers:

$$L \in O(D-1, 1, \mathbb{Z})\ . \tag{1.3}$$

Examples of such matrices are: $L_1 = \begin{pmatrix} 2 & 1 & 1 & 1 \\ 1 & 0 & 1 & 1 \\ 1 & 1 & 0 & 1 \\ 1 & 1 & 1 & 0 \end{pmatrix}$ or $R_1 = \begin{pmatrix} 1 & 0 & 0 & 0 \\ 0 & 0 & 1 & 0 \\ 0 & 0 & 0 & 1 \\ 0 & 1 & 0 & 0 \end{pmatrix}$.

The discrete Lorentz group is *non compact.*

2. String Theory in a Nutshell

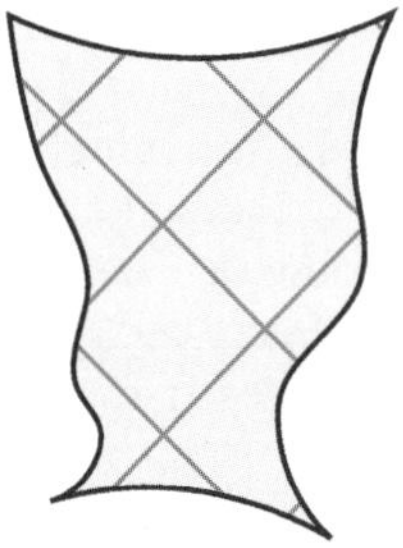

2.1. Bosonic String Theory

Anticipating the notation that will be useful later, points in a continuous D dimensional space-time will be indicated as q^μ, $\mu = 0, 1, \cdots, D-1$. The 2 dimensional world sheet is given two coordinates (x, t). A string is given by a function $q^\mu(x,t)$. Its equations of motion are defined by means of the string action. Defining light cone coordinates on the world sheet, $x^\pm = \frac{1}{\sqrt{2}}(t \pm x)$, we have

$$S = T \int \mathrm{d}x^+ \mathrm{d}x^- \sqrt{(\partial_+ q^\mu\, \partial_- q^\mu)^2 - (\partial_+ q^\mu)^2 (\partial_- q^\nu)^2}\ . \tag{2.4}$$

Here, $T = 1/(2\pi\alpha')$ is the string tension constant.

Since this action is invariant under all reparametrizations of the two coordinates x^+ and x^-, they can be chosen at will. Choosing them to follow light signals in the two directions on the string, implies that we demand

$$(\partial_+ q^\mu)^2 = (\partial_- q^\nu)^2 = 0\ ; \qquad \text{furthermore,} \quad t = \tfrac{1}{\sqrt{2}} q^+\ , \tag{2.5}$$

the latter implying that we also use light cone coordinates in target space. This simplifies the action:

$$S \ = \ T\,\partial_+ q^\mu \partial_- q^\mu \ = \ T(\tfrac{1}{2}(\partial_t q^\mu)^2 - \tfrac{1}{2}(\partial_x q^\mu)^2)\ . \tag{2.6}$$

Let us normalize our units such that

$$T \equiv 1\ . \tag{2.7}$$

This action is a bilinear one (the sign ambiguity in the square root can easily be resolved in the classical theory), and hence the equation of motion is linear:

$$\partial_+ \partial_-\, q^\mu(x^+, x^-) = 0\ ; \quad \partial_t^2 q^\mu - \partial_x^2 q^\mu = 0\ , \tag{2.8}$$

which is easily solved:

$$q^\mu(x,t) = q_L^\mu(x+t) + q_R^\mu(x-t)\ ; \tag{2.9}$$

this however must still must obey the constraints (2.5), which imply:

$$(\partial_x q_L^\mu)^2 = (\partial_x q_R^\mu)^2 = 0\ ; \qquad q_L^+ = x^+\ , \qquad q_R^+ = x^-\ . \tag{2.10}$$

Apart from these constraints, the action (2.6) is that of a massless, non-interacting bosonic field theory in 1+1 dimensions. As such it is easy to quantize. Note that the constraints do not directly affect the quantum dynamics of this system; they merely determine the functions $q_{L,R}^\pm$ when we know the $D-2$ transverse functions $q_{L,R}^a$.

2.2. Superstrings

Fermionic excitations are obtained if, besides the coordinate fields q^μ, we put fermionic fields on the string world sheet. Since the world sheet has only one space dimension, a fermionic field is a two component spinor, $\psi = (\psi_1, \psi_2)$, and if we assume that they are massless, they form Majorana fermions, and the world sheet Dirac equation also gives rise to left movers and right movers:

$$\psi_1(x,t) = \psi_L(x^+) , \qquad \psi_2(x,t) = \psi_R(x^-) . \tag{2.11}$$

If these spinors can furthermore be labeled with a Lorentz index μ just like the 'bosonic' fields q^μ, they together form $\mathcal{N} = 1$ supermultiplets. We thus obtain the superstring. Supersymmetry then demands that the 'bosonic' constraints (2.10) are replaced by:

$$(\partial_+ q^m_L)^2 + \tfrac{1}{2}\psi^\mu_L\,\hat{\kappa}\,\psi^\mu_L \;=\; 0\,, \quad (\partial_- q^\mu_R)^2 + \tfrac{1}{2}\psi^\mu_R\,\hat{\kappa}\,\psi^\mu_R \;=\; 0\,, \qquad \partial_+ q^+_L = \partial_- q^+_R = 1\,, \tag{2.12}$$

and that there are similar 'fermionic' constraints:

$$(\partial_+ q^\mu_L)\,\psi^\mu_L \;=\; 0\,, \qquad (\partial_- q^\mu_R)\,\psi^\mu_R \;=\; 0\,, \qquad \psi^+_{L,R} = 0\,. \tag{2.13}$$

Both bosonic strings and superstrings may be subject to different types of boundary conditions. In the absence of D-branes, one has only periodic boundary conditions, that is, strings form closed loops. Effectively, this reduces the independent degrees of freedom in the superstring to $D-2$ bosonic variables, the transverse coordinate fields q^a, and $D-2$ transverse fermionic fields ψ^a, moving to the left or to the right.

Interactions can be introduced by allowing strings to 'exchange their legs' when they coincide at a space-time point. This interaction is then associated with a constant, the string-interaction constant g_s.

3. Strings on a lattice

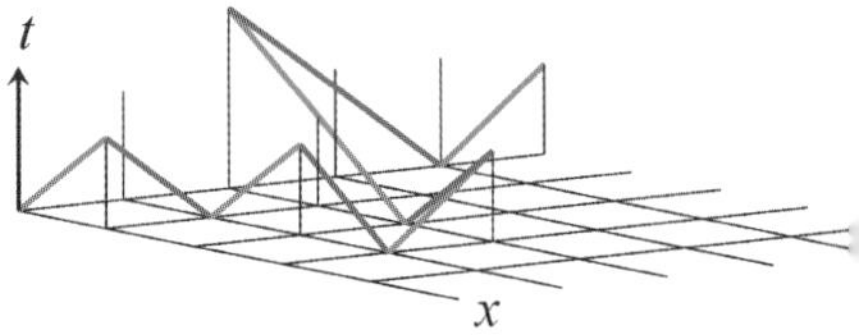

3.1. Bosonic String Theory on a Lattice

For this discussion, it is useful to replace the world sheet coordinates $(x,\,t)$ by a lattice[3]: $x \in \mathbb{Z}$, $t \in \mathbb{Z}$. We may or may not decide later to replace this lattice by a continuum[1]. The discretized equations of motion (2.8) turn into

$$q^\mu(x,\,t+1) = q^\mu(x-1,\,t) + q^\mu(x+1,\,t) - q^\mu(x,\,t-1)\,. \tag{3.1}$$

Again, we find left-movers and right-movers:

$$q^\mu(x,t) = q^\mu_L(x+t) + q^\mu_R(x-t)\,. \tag{3.2}$$

[1]Physically, this lattice might be not so meaningful. The world sheet coordinates are in principle arbitrary, so that this 'discretization' of the string produces string 'bits' [3] of unspecified length. One will have to send these lengths to zero, eventually.

To define the discretized constraint, we may define

$$\begin{aligned} a_L^\mu(x) &= q_L^\mu(x+1) - q_L^\mu(x-1)\ , \quad (a_L^\mu)^2 = 0\ ; \\ a_R^\mu(x) &= q_R^\mu(x-1) - q_R^\mu(x+1)\ , \quad (a_R^\mu)^2 = 0\ . \end{aligned} \tag{3.3}$$

A genuine lattice string theory emerges if we also put target space on a lattice:

$$q^\mu \to Q^\mu \in \mathbb{Z}^D\ , \quad a^\mu \to A^\mu \in \mathbb{Z}^D\ . \tag{3.4}$$

The discrete analogue of the constraints (2.10) would be

$$A_L^+ = 1\ , \quad A_L^- = \tfrac{1}{2}(A_L^{\rm tr})^2\ , \quad \text{and} \quad L \leftrightarrow R\ , \tag{3.5}$$

but there could be a problem with this:

$$A_L^+ = \tfrac{1}{\sqrt{2}}(A_L^0 + A_L^{D-1})\ , \quad A_L^- = \tfrac{1}{\sqrt{2}}(A_L^0 - A_L^{D-1})\ , \tag{3.6}$$

would modify the lattice in Minkowski space (this could be remedied by modifying the constraints with factors $\sqrt{2}$), but a more fundamental problem would be that the world sheet lattice itself already implies a gauge choice, so adding the constraints (3.6) may over determine the gauge. A Lorentz transformation may multiply A^+ with a factor and divide A^- by the same factor, which would knock us out of the Minkowski lattice by generating fractions. It seems that the entire procedure of fixing the gauge on the world sheet may have to be reconsidered when employing this lattice. Later, it will be argued that the constraints should be used in an altogether different fashion. As yet, we observe that the constraints will produce some sort of lattice in Minkowski space, as described in Subsection 1.1. See the figure at the top of this Section, describing a string at a given world sheet time t.

so if $A_L^0 + A_L^{D-1} > 1$ after a Lorentz transformation, imposing this gauge will require that x^+ have a gap of length > 1. This issue will not be pursued here.

3.2. String interactions

Note that, our description of strings on the Minkowski space lattice is entirely classical, obeying the deterministic equations (3.1). It is tempting now to demand that also the string interactions are deterministic. This can be done in exactly one way, see the figure here. In a deterministic string theory living on a space lattice, we can demand that two strings exchange their legs when they meet at one lattice point.

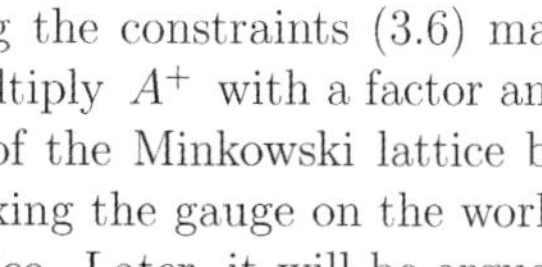

Originally, our theory only contained strings of infinite length. With this interaction, closed strings with finite lengths will emerge. Demanding that the interaction be deterministic has two other consequences. In quantum mechanics, superpositions are allowed. Here, we must choose *how* the legs are exchanged. This is unambiguous only if the strings have an *orientation*. So non-oriented strings are disallowed in this theory.

Secondly, the *strength* of the interaction is no longer adjustable. When two strings meet at one point, the interaction must be certain, so, in units yet to be identified, the string coupling constant, g_s, must be one.

Quantization will now be realized in a very unconventional manner. Conventionally, the quantum theory is quite different from the classical one; one could say that the classical theory merely serves as an inspiration, an educated guess, for producing an altogether different system, which however does have the property that in the limit $\hbar \to 0$ the classical theory is reproduced. In our second lecture (see Sections 4—8), it will be explained that, under special conditions, lattice string theory is *identical* to a fully quantized string theory in the continuum, and it will be either a bosonic string or a superstring. The field equations do *not* change. A different way of saying this is that the *states* of the lattice system are promoted to being the elements of a basis in Hilbert space. While we are describing the fully quantized theory, these particular basis elements continue to evolve according to the classical equations of motion.

3.3. Fermions

The classical counter parts of fermions are very simple. Besides the bosonic variables q^μ, the superstring has fermions (2.11) living on the world sheet. There are $D-2$ mutually independent fermion fields $\psi^a_A(x,t)$, $A = 1,2$, or equivalently, $A = L, R$. We saw in Subsection 2.2 that the solution on the world sheet behaves as

$$\psi^\mu_A(x,t) = \begin{pmatrix} \psi^\mu_L(x+t) \\ \psi^\mu_R(x-t) \end{pmatrix} , \tag{3.7}$$

so, the spinors also have left-movers and right-movers.

The deterministic lattice theory, instead, has Boolean degrees of freedom, $\sigma^a(x,t) = \pm 1$, $a = 1, \cdots, D-2$, obeying equations similar to Eqs. (3.1):

$$\sigma^a(x,\, t+1) = \sigma^a(x-1,\, t)\; \sigma^a(x-1,\, t)\; \sigma^a(x,\, t-1)\ . \tag{3.8}$$

This also splits up into left- and right-movers:

$$\sigma^a(x,t) = \sigma^a_L(x+t)\; \sigma^a_R(x-t)\ . \tag{3.9}$$

Our first impression would be that the superstring constraints (2.13) now have to be modified so as to determine the Boolean degrees of freedom σ^{D-1} and σ^0, or equivalently, $\sigma^\pm$, on the world sheet, but we will do this differently later.

4. Real numbers and integers

Imagine that, in contrast to appearances, the real world, at its most fundamental level, were *not* based on real numbers at all. We here consider systems where *only* integers describe what happens at a deeper level. Can one understand *why* our world *appears* to be based on real numbers? A mathematical tool is needed[4]. We shall first formulate it entirely in terms of quantum mechanical states. Consider two kinds of quantum systems.

System (D) is a quantum theory whose basis elements are entirely specified by a set of N pairs of integers, to be called Q_i and P_i, where $i = 1, \cdots, N$. So the Hilbert space is spanned by the products of N states of the form $|Q_i, P_i\rangle$, with $(P_i, Q_i) \in \mathbb{Z}^{2N}$.

System (R) is also a quantum system, but its Hilbert space is spanned by all wave functions $\psi(q_i)$, or equivalently, by all states $|q_i\rangle$, where again $i = 1, \cdots, N$. Here $q_i \in \mathbb{R}^N$. We also have operators p_i in this Hilbert space. Here, we define

$$p_i = -\tfrac{i}{2\pi}\partial/\partial q_i \ . \tag{4.1}$$

Our mathematical tool is a *mapping* between the states of system (D) and system (R). The mapping is a standard application of Fourier transforms. Mappings of this sort may become very important for instance when black holes are considered: it is generally believed that black holes come in a denumerable set of quantum states, whereas the particles we usually consider sit in spaces with continuous coordinates.

4.1. Notation

Define a useful symbol:

$$\epsilon \equiv e^{2\pi} \approx 535.49165\cdots \ ; \qquad \epsilon^{ipx} = e^{2\pi ipx} \ ; \qquad \epsilon^{iZ} = 1 \quad \text{if} \quad Z \in \mathbb{Z} \ . \tag{4.2}$$

Writing complex exponentials this way makes all periods 1 instead of 2π, which comes out handy when functions are normalized. In systems of the type (R), we have (suppressing the index i):

$$\begin{aligned} 2\pi[q,\, p] = i \ , \qquad & \langle q|\epsilon^{ipa}|p\rangle = \langle q+a\,|\,p\rangle \ . \\ \langle q|p\rangle = \epsilon^{ipq} \ , \qquad & \langle q_1|q_2\rangle = \delta(q_1 - q_2) \ , \\ \langle p_1|p_2\rangle = \delta(p_1 - p_2) \ . \qquad & \text{If } N > 1 \text{ insert } \delta_{ij} \ . \end{aligned} \tag{4.3}$$

Note that this means that Planck's constant h, not $\hbar$, is normalized to one. Let now in general

$$\begin{array}{lcl} \text{cap. Latin letters,} & N,\ P,\ Q,\ X,\ \cdots, & \text{indicate integers,} \\ \text{l.c. Latin letters,} & p,\ q,\ x,\ \cdots, & \text{indicate real numbers,} \\ \text{l.c. Greek letters,} & \alpha,\ \eta,\ \xi,\ \lambda,\ \cdots, & \text{are numbers } \textit{modulo } 1; \end{array} \tag{4.4}$$

the latter will also be called 'fractional numbers'. Typically, $-\frac{1}{2} < \eta \leq \frac{1}{2}$.

In system (D), we define the operators η_Q and η_P, with eigenvalues η_Q and η_P, such that

$$\begin{aligned} [P, Q] = 0 \ , \quad \langle Q|\,\eta_Q\rangle = \epsilon^{iQ\,\eta_Q} \ , \quad & \langle P|\,\eta_P\rangle = \epsilon^{iP\,\eta_P} \ , \\ \langle Q_1|\,Q_2\rangle = \delta_{Q_1Q_2} \ , \quad & \langle \eta_Q^1|\,\eta_Q^2\rangle = \delta(\eta_Q^1 - \eta_Q^2) \ . \end{aligned} \tag{4.5}$$

4.2. Q and P states

Until here, we have only considered quantum systems. These are systems whose states are described by Hilbert space, and where one may expect the evolution law te be described either by a hamiltonian, or more directly by some finite-time evolution operator $U(t_1-t_2)$. However, particularly systems of the type (D) are also suitable to describe *deterministic cellular automata*. A (deterministic) cellular automaton is here defined to be a discrete system where the evolution operator U takes a very special form. At regularly chosen time intervals, the evolution operator is assumed to be a pure permutation operator acting on the (discrete) basis elements that we picked out. Thus, in this basis of Hilbert space, at these time intervals, the evolution operator is required to be a unitary matrix containing only ones and zeros. This condition implies that cellular automata are simply a subset of all quantum systems. However, being described by simple permutations, also means that these systems are fundamentally classical. They are as deterministic, and as classical, as an ordinary laptop. The only difference is that a laptop has a delete button. Unitarity forbids the erasure of information; this makes cellular automata even more restrictive than laptops.

The essential theme of the idea displayed in these lectures is, that we can switch between the quantum notation and the entirely classical notation where a cellular automaton simply permutes its states $|Q\rangle$. We stress that, physically, we are then always talking about the same thing. If a classical automaton is described using a quantum state $|\psi\rangle$, this means nothing more than that the probabilities that different states Q are realized, are $|\langle Q|\psi\rangle|^2$. At the end of a calculation we will exclusively be interested in amplitudes ${}_{\rm out}\langle Q_i|Q_j\rangle_{\rm in}$. The classical and quantum probabilities will be the same.

In spite of the obvious correctness of such a procedure, the reader must be warned that most of the practitioners of the foundations of quantum mechanics have not yet relinquished their skepticism against it.

4.3. Operators

There is nothing wrong with treating discrete systems, including cellular automata, using the full machinery of quantum mechanics. Therefore, we simply introduce the operators η_Q and η_P. These will turn out to be handy for instance if we wish to identify some operator H such that

$$U(t)=\epsilon^{-iHt}\,. \tag{4.6}$$

The definition (4.5) defines the operators $\eta_{Q,P}$ uniquely if we also demand their eigenvalues to obey

$$-\tfrac{1}{2}<\eta\le\tfrac{1}{2}\,. \tag{4.7}$$

To find their matrix elements in the Q,P basis, one merely has to Fourier transform,

$$\eta=\sum_N\epsilon^{iN\eta}\int_{-\frac{1}{2}}^{\frac{1}{2}}\eta\,{\rm d}\eta\,\epsilon^{-iN\eta}\;=\;\sum_{N\ne0}\frac{i(-1)^N}{2\pi N}\epsilon^{iN\eta}\,, \tag{4.8}$$

and since

$$\epsilon^{iN\eta_Q}|Q\rangle = |Q-N\rangle \ , \tag{4.9}$$

we find the matrix elements:

$$\langle Q_1|\eta_Q|Q_2\rangle = \frac{i}{2\pi}(1-\delta_{Q_1Q_2})\frac{(-1)^{Q_1-Q_2}}{Q_1-Q_2} \ , \tag{4.10}$$

and similarly for η_P. Because the region (4.7) is symmetric, Eq. (4.8) vanishes ate $N=0$, and therefore the matrix elements (4.10) vanish when $Q_1 = Q_2$.

The commutator $[\eta_Q,\, Q]$ has the matrix elements

$$\langle Q_1|[\eta_Q,\, Q]|Q_2\rangle = \frac{i}{2\pi}\left(\delta_{Q_1Q_2} - (-1)^{Q_2-Q_1}\right) = \frac{i}{2\pi}(\mathbb{I} - |\psi_Q\rangle\langle\psi_Q|) \ , \tag{4.11}$$

where the state $|\psi_Q\rangle$ is what we will call an *edge state* :

$$\langle Q|\psi_Q\rangle = (-1)^Q \ ; \qquad \langle\eta|\psi\rangle = \delta(\eta - \tfrac{1}{2}) \ . \tag{4.12}$$

So, one finds that the commutator $[\eta_Q,\, Q] = i/(2\pi)$ only when it acts on a state that is orthogonal to the edge state.

4.4. The mapping

Of course, the operators Q and η_P commute, so they can be diagonalized simultaneously. Now, take any real number q. We can split it into its closest integer Q and the remainder, which is a fractional number, call that η_P:

$$q \equiv Q + \eta_P \ . \tag{4.13}$$

This split is unique and so it is a one-to-one mapping. This is the simplest way we can map (D) onto (R):

$$\langle Q,\, \eta_P|\psi\rangle \ = \ \langle q|\psi\rangle \ . \tag{4.14}$$

To find the operator p, all we have to do is calculate its eigen states[2],

$$\langle q|p\rangle = \epsilon^{ipq} \ . \tag{4.15}$$

All matrix elements can now be computed using Eqs. (4.3), (4.5) and (4.10).

---

[2]It is in equations such as (4.14) and (4.15), that our notation using ϵ as in (4.2) is useful: the normalizations do not change.

5. Orthogonality and Convergence

5.1. Wave functions

The wave functions of the states $|Q,\,P\rangle$ in the q representation are easy to find. The state $|0,\,0\rangle$ is

$$\langle q|0,\,0\rangle = \theta(\,\tfrac{1}{2} - |q|\,)\;, \tag{5.1}$$

see Fig. 1 *a*. Note, that it is properly normalized. The other states are

$$\langle q|Q,\,P\rangle = e^{iPq}\,\theta(\,\tfrac{1}{2} - |q-Q|\,)\;. \tag{5.2}$$

The wave functions in momentum space are obtained by acting with the matrix (4.15):

$$\langle p|Q,P\rangle = \int \mathrm{d}q\,\langle p|q\rangle\langle q|Q,P\rangle = \frac{\sin\pi(p-P)}{\pi(p-P)}e^{-ipQ}\;. \tag{5.3}$$

The wave function $\langle p|0,0\rangle$ is displayed in Fig. 1 *b*.

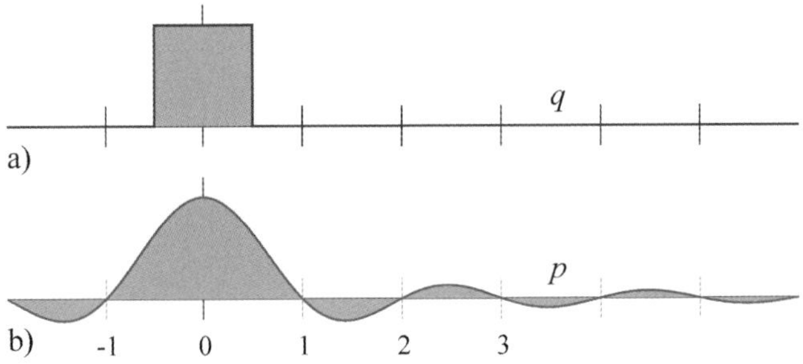

Figure 1: *a*) The state $|0,\,0\rangle$ in position space. *b*) in momentum space.

For our purposes, we are not quite content with the asymmetry between the states and their Fourier transforms, Eqs. (5.2) and (5.3), or, Figs. 1 *a* and *b*. We can do this better by first defining the operators q and p in the basis spanned by η_Q and η_P. If we take them to be periodic both in η_Q and in η_P, then a discontinuity arises at $\eta_P = \pm\frac{1}{2}$, but not at $\eta_Q = \pm\frac{1}{2}$.

To remedy this, we rotate over a phase angle $\phi(\eta_Q,\,\eta_P)$. It obeys

$$\phi(\eta,\xi+1) = \phi(\eta,\xi)+\eta; \qquad \phi(\eta+1,\xi) = \phi(\eta,\xi)\;. \tag{5.4}$$

It can also be forced to obey

$$\phi(\eta,\xi) = \eta\,\xi - \phi(\xi,\eta)\;. \tag{5.5}$$

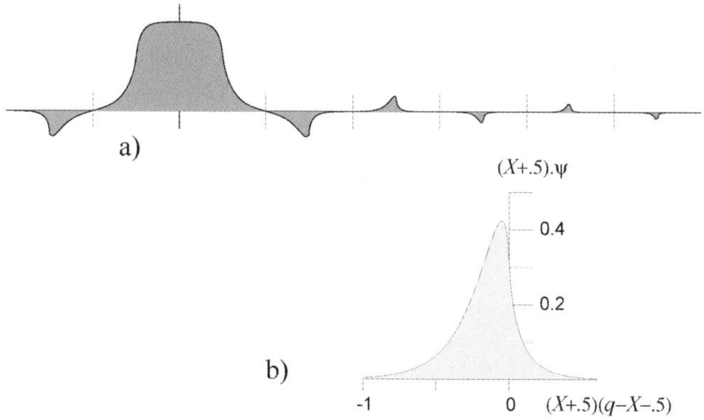

Figure 2: *a*) The modified wave function in q space, using Eqs. (5.6). It is equal to its own Fourier transform, so in p space, it is the same. *b*) The asymptotic form of the little peaks; they shrink rapidly.

Using some mathematics, this phase angle can be obtained from an elliptic integral[4]. One finds:

$$\begin{aligned} q &= \frac{-i}{2\pi}\frac{\partial}{\partial\eta_Q} + \left(\frac{\partial}{\partial\eta_Q}\phi(\eta_P,\eta_Q)\right) &= Q + \frac{\partial}{\partial\eta_Q}\phi(\eta_P,\eta_Q)\ ; \\ p &= \frac{-i}{2\pi}\frac{\partial}{\partial\eta_P} - \left(\frac{\partial}{\partial\eta_P}\phi(\eta_Q,\eta_P)\right) &= P - \frac{\partial}{\partial\eta_P}\phi(\eta_Q,\eta_P)\ . \end{aligned} \tag{5.6}$$

Eq. (5.5) implies that this obeys

$$[q,p] = i/2\pi\ . \tag{5.7}$$

The form of the state $|0,\,0\rangle$ in this q space is pictured in Fig, 2.

An explicit calculation of the matrix elements of q and p in Hilbert space spanned by the $|Q,\,P\rangle$ states, one finds:

$$\begin{aligned} q = Q + a_Q \quad &, \quad p = P + a_P\ , \\ \langle Q_1,P_1|a_Q|Q_2,P_2\rangle &= \frac{(-1)^{P+Q+1}\,iP}{2\pi(P^2+Q^2)} \\ \langle Q_1,P_1|a_P|Q_2,P_2\rangle &= \frac{(-1)^{P+Q}\,iQ}{2\pi(P^2+Q^2)}\ . \end{aligned} \tag{5.8}$$

From these, the commutator follows:

$$\begin{aligned} \langle Q_1,P_1|\,[q,p]\,|Q_2,P_2\rangle &= \tfrac{i}{2\pi}\,(-1)^{Q_1-Q_2+P_1-P_2}(\delta_{Q_1\,Q_2}\delta_{P_1\,P_2}\ -\ 1)\ . \\ = \tfrac{i}{2\pi}(\mathbb{I} - |\psi_{\rm edge}\rangle\langle\psi_{\rm edge}|), &\qquad \text{with} \qquad \langle Q,P|\psi_{\rm edge}\rangle = (-1)^{Q+P}\ . \end{aligned} \tag{5.9}$$

This contradicts Eq. (5.7), a contradiction that comes about from the fact that the phase function $\phi(\eta_Q, \eta_P)$ is singular at the point $\eta_Q = \pm\eta_P = \pm\frac{1}{2}$. This edge state cannot be removed, because the phase has a vortex there. The vortex is a necessary consequence of the twisted boundary condition (5.4). This is why, at at least one point on the (η_Q, η_P) torus, Eq. (5.7) fails to hold.

6. A harmonic oscillator

As harmonic oscillator hamiltonian we take[3]

$$H = \pi(p^2 + q^2) \ . \tag{6.1}$$

What are the matrix elements $\langle Q_1, P_1|H|Q_2, P_2\rangle$? We have

$$\langle Q_1, P_1|q^2|Q_2, P_2\rangle = \int_{-\infty}^{\infty} \mathrm{d}q \, \langle Q_1, P_1|q\rangle q^2 \langle q|Q_2, P_2\rangle \ . \tag{6.2}$$

Inspecting Fig. 2, one finds that this has a logarithmic divergence proportional to

$$(-1)^{Q_1-Q_2+P_1-P_2} \quad \propto \quad \langle Q_1, P_1|\psi_{\text{edge}}\rangle\langle\psi_{\text{edge}}|Q_2, P_2\rangle \ . \tag{6.3}$$

Only this edge state has a divergent hamiltonian. All other elements of the hamiltonian converge rapidly. Therefore, we introduce a constraint,

$$\langle\psi_{\text{edge}}|\psi\rangle = 0 \ . \tag{6.4}$$

We can write

$$|\psi(t)\rangle = \epsilon^{-iHt}|\psi(0)\rangle \ ; \qquad H = a^\dagger a = N \ ; \qquad |\psi(T)\rangle = |\psi(0)\rangle \ , \tag{6.5}$$

where N and T are integers (the period of this oscillator is 1). Now we know that, after a quarter phase rotation, the momentum and position operator of a harmonic oscillator interchange, Therefore, at steps $t_n = \frac{1}{4}n$, where $n = 1, 2, 3, 4$, we have $q \rightarrow p \rightarrow -q \rightarrow -p \rightarrow q$, and, because of our complete Q, P symmetry:

$$(Q, P) \rightarrow (P, -Q) \rightarrow (-Q, -P) \rightarrow (-P, Q) \rightarrow (Q, P) \ . \tag{6.6}$$

Therefore, at steps of one quarter of a full phase rotation, this system is a *"deterministic"* system. We call such a system an "automaton". Needless to say, this automaton is a very trivial one, but it will illustrate the way we argue:
This discrete, deterministic automaton is described by the hamiltonian (6.1) when transformed to Q, P space.
Of course, the states only perform a perfect permutation at time steps that are multiples of $1/4$. The importance of this finding is, that it is often far from easy to find such a hamiltonian that is properly bounded from below.

[3]The normalization is such that, with our commutation rules, Eqs. (6.5) will hold.

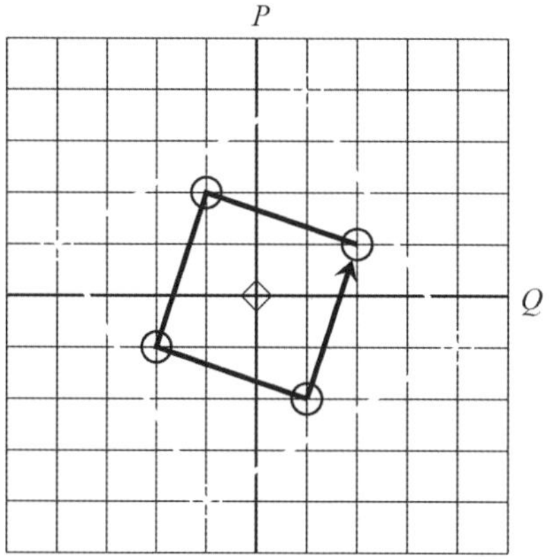

Figure 3: The trajectories of states of the harmonic oscillator in Q, P space.

7. QFT in 1+1 dimensions, and the 1+1 dimensional CA

7.1. Bosons

The model of the previous section was still quite trivial. To get a more interesting system, we need many oscillators. A beautiful example is a bosonic massless scalar quantum field theory in one space- and one time dimension. The quantum system would be defined by fields $\phi(x,t)$ and their canonical momentum fields $p(x,t)$, obeying commutation rules

$$[\phi(x,t),\, p(y,t)] = \frac{i}{2\pi}\delta(x-y)\ ; \qquad [\phi(x,t),\, \phi(y,t)] = [p(x,t),\, p(y,t)] = 0\ . \tag{7.7}$$

The hamiltonian is taken to be

$$H = \pi \int \mathrm{d}x (p(x)^2 + (\partial_x\phi)^2)\ . \tag{7.8}$$

Let us now, for simplicity, take space-time to be a lattice; in particular, x is discrete: $x \in \mathbb{Z}$. According to the field equations,

$$\phi(x, t+a) + \phi(x, t-a) = \phi(x-a, t) + \phi(x+a, t)\ . \tag{7.9}$$

We would like to *map* this model one-to-one on the cellular automaton:

$$Q(x, t+a) + Q(x, t-a) = Q(x-a, t) + Q(x+a, t)\ , \tag{7.10}$$

where Q are integers. a is the mesh size of the space-time lattice, which will be taken to be 1.

Applying our earlier methods directly does not work: adding real numbers does not imply exactly that their *integer parts* add up, since there will be rounding errors: our procedure is non-linear! It does work if real numbers are merely *interchanged.* This is exactly what happens if one considers *left movers* and *right movers* :

$$\phi(x,t)=\phi^L(x+t)+\phi^R(x-t)\ ;\ p(x,t)=\tfrac{1}{2}a^L(x+t)+\tfrac{1}{2}a^R(x-t)\ . \tag{7.11}$$

$$a^L=p+\partial_x\phi\ ;\qquad a^R=p-\partial_x\phi\ . \tag{7.12}$$

The hamiltonian obeys

$$H\ =\ \pi(p^2+(\partial_x\phi)^2)\ =\ \tfrac{1}{2}\pi({a^L}^2+{a^R}^2)\ , \tag{7.13}$$

$$[a^L,\,a^R]=0\ ;\qquad [a^L(x),\,a^L(y)]=\tfrac{i}{\pi}\partial_x\delta(x-y)\quad ; \tag{7.14}$$

If we replace space-time by a lattice: $(x,t)\in\mathbb{Z}$. the commutators are replaced by[4]

$$[\phi(x),\,p(y)]=\tfrac{i}{2\pi}\delta_{x,y} \tag{7.15}$$

$$[a^L(x),\,a^L(y)]=\pm\tfrac{i}{2\pi}\ \text{ if }\ y=x\pm 1\ ,\quad 0\ \text{otherwise.} \tag{7.16}$$

Now, we replace the hamiltonian of the continuum by a lattice hamiltonian in such a way that its action over one time unit is a pure replacement. Write $H=H^L+H^R$. In *momentum space* :

$$H^L=\tfrac{1}{2}\int_0^{1/2}\mathrm{d}k\,a^L(k)a^L(-k)\,M(k)\quad ;\quad M(\kappa)=\frac{\pi\kappa}{\sin(2\pi\kappa)}\ . \tag{7.17}$$

This hamiltonian turns $a^L(x)$ into a pure left-mover, and $a^R(x)$ into a right-mover. $M(\kappa)$ shows a κ dependence in such a way that the equations of motion for $a^L(\kappa)$ are no longer modified by lattice artifacts. This is important.

Now, we can try to express $a^L(x)$ in terms of the left-movers of the cellular automaton: $A^L(x+t)=Q(x,t+1)-Q(x-1,t)$

Demanding $[a^L(x),\,a^L(y)]=\pm\frac{i}{2\pi}$ if $y=x\pm 1$, and temporarily disregarding the effects of an edge state, we find that one can write, as one can easily check,

$$a^L(x)=A^L(x)+\eta^L_A(x-1)\ . \tag{7.18}$$

A better procedure, in η space, minimizes the effect of the edge state, as explained earlier:

$$a^L(x)=A^L(x)+\frac{\partial}{\partial\eta^L_A(x)}\Big(\phi(\eta^L_A(x+1),\eta^L_A(x))-\phi(\eta^L_A(x-1),\eta^L_A(x))\Big) \tag{7.19}$$

We can now substitute this identity in all operators of the discrete system to see how it maps to the states of the quantum field theory, and *vice versa.* The edge states in this case are limited to those corners where two consecutive $\eta^L_A(x)$ are $\pm\frac{1}{2}$

[4]The lattice artifacts could be chosen differently, but by redefinitions the commutators can always be recast in the form given here.

7.2. Fermions

Suppose we have left- and right moving boolean variables in our discrete deterministic theory:

$$\sigma_L(x+t) = \pm 1 \quad \text{and} \quad \sigma_R(x-t) = \pm 1 \ . \tag{7.20}$$

To map this system onto a system of quantized fermions, introduce operators

$$\sigma_3(x,t) = \sigma(x,t) = \begin{pmatrix} 1 & 0 \\ 0 & -1 \end{pmatrix} \ ; \tag{7.21}$$

then we can also define the operators

$$\sigma_1(x,t) = \begin{pmatrix} 0 & 1 \\ 1 & 0 \end{pmatrix} \quad ; \quad \sigma_2(x,t) = \begin{pmatrix} 0 & -i \\ i & 0 \end{pmatrix} \ , \tag{7.22}$$

which allows us to define the fermionic operators by means of the Jordan-Wigner[7] transformation,

$$\psi(x,t) = \sigma_1(x,t) \prod_{y<x} \sigma_3(y,t) \ ; \qquad \{\psi(x,t),\, \psi(y,t)\} = \delta_{x,y} \tag{7.23}$$

In this case, we also have $\psi(x,t) = \psi^\dagger(x,t)$, which makes this fermion into a Majorana field.

8. Conclusion

Superstring theory, as described in the first parts of these notes, is obtained if we regard the coordinates x and t to be the usual σ, τ coordinates of the string world sheet. To arrive at the model described here, a slight modification of superstring theory was brought about: the world sheet, normally described by continuous world sheet coordinates σ and τ, were replaced by a discrete lattice. At first sight this may seem to be a big departure from the original theory, but it must be remembered that the world sheet coordinates are arbitrary, so that the physical dimensions of this lattice are ill-defined. We may imagine taking the continuum limit immediately. This seems to be straightforward: all one has to do is limit the world sheet momentum values κ to be confined by the Brillouin zone, $|\kappa| \le \frac{1}{2}$. In practice however, this is not so easy, and more work must be done to understand exactly how to produce this continuum limit. The problem is strongly related to the question of the world sheet gauge constraints, essential for a proper formulation of the theory.

Be this as it may, we did arrive at a model closely related to superstring theory, and managed to replace its quantum degrees of freedom by sets of commuting integer-valued fields P^μ and Q^μ, and these describe the evolution of the string on the world sheet as if we were dealing with something as simple as a classical automaton. It is classical because all its independent variables, the integers $Q^a_{L,R}$, where a is the transverse index running from 1 to $D-2$, basically commute.

The fermions of the superstring could also be cast in a deterministic form, describing Boolean variables $\sigma^\mu = \pm 1$ on the world sheet of the string that evolve without turning into non-commuting observables. And finally, it seems that the superstring interactions, the string exchange operation, is also a totally classical, deterministic process, provided that the string coupling constant, g_s has a fixed value. What all this means is that, if a string of this sort starts out its life at $t = -\infty$ in one of the eigen states of these operators, it will continue to be in such an eigenstate. Such a model can be called an automaton. It is as deterministic as your laptop, and all of its typical quantum features then seem to have disappeared. Does this mean that we identified a deterministic theory underlying quantum mechanics?

It must be mentioned that theories of the nature just described are being hotly debated, and often dismissed, on the Internet. There are good reasons for many investigators to think that a quantum system can never be mimicked by a local deterministic model. Therefore, the question arises: what would be wrong about the model described here?

One answer often given is that we have been dealing with quantum amplitudes, and therefore, rather than pure probabilities, our theory is based on quantum amplitudes. Why however should we exclude the wave function of the universe that puts the superstring in one of these fundamental eigenstates (the 'ontological basis elements')? In fact, much of the mysteries of quantum mechanics, such as the 'collapse of the wave function', would disappear if we could assume this.

It is safe to say that, having a single string, or even a set of several strings that each evolve deterministically, is not the same as having a complete deterministic model of the Universe, where string theory should be second-quntized, so, as yet, one can indeed dismiss this model as representing a counter example of the famous theorems that tell that one can't have deterministic models of Quantum Mechanics. However, the model suggests that the truth may be more subtle. Perhaps our world is described by a small, non-commuting subset of observables of a system that itself is deterministic. Maybe the arguments that determinism can never be saved in Quantum Mechanics have to be revised.

We are tempted to ignore this question and search for more features of a deterministic realization of superstring theory.

In summary, the central theme of this work is that a unitary transformation can be written down that relates the familiar operators q_i and p_i with commutator $[q_i, p_j] = \frac{i}{2\pi}\delta_{ij}$, to pairs of integer valued operators Q_i and P_i that commute at all times. The real-valued operators q_i and p_i obey much more symmetry relations such as translation and rotation symmetries, as well as Lorentz symmetries, so that imposing such symmetries in physics automatically leads to non-commuting operators. Applying such unitary transformations on massless field theories in 1+1 dimensional space-time turns these theories into models that look conspicuously similar to ordinary classical automaton models. The trick is to apply the transformations directly to the left-movers and the right-movers of both systems.

If applied to quantized string theory, we find that the bosonic field variables, being described only by integers now, are coordinates living on a lattice space-time, the lattice length being exactly equal to $2\pi\sqrt{\alpha'}$. This may perhaps be regarded as much progress: according to the holographic principle, the amount of information in a given region of space-time should be limited very strongly at the Planck scale. The lattice is ideally suited for realizing such a constraint.

It is important for superstring theory to maintain Lorentz invariance. To understand this, one has to define the longitudinal degrees of freedom, both for the coordinates and for the fermions. They are normally dictated by the gauge constraints of the superstring, which will produce the longitudinal degrees of freedom as quantum operators. This turns the Lorentz transformation into a rather delicate quantum transformation, which is what should be demanded for the Lorentz invariance of the quantum theory.

In the lecture it was assumed that the lattice in $D-2$ dimensional transverse space should also imply that there is a complete lattice in all of Minkowski space-time. It would generate an immediate invariance under the subgroup $O(D-1,1,\mathbb{Z})$ of the Lorentz group. Alternatively, however, one could guess that the time component on the lattice remains strictly continuous. It remains to be seen what the best way is to describe most concisely the model we found.

Like Lorentz invariance, also supersymmetry would be an emergent symmetry of the quantum system, not yet obviously visible at the classical level.

More detailed accounts and discussions are to be found in Refs. [4]—[6].

References

[1] M.B. Green, J.H. Schwarz and E. Witten, *Superstring Theory*, ISBN-10: 0521357527 — ISBN-13: 978-0521357524, Cambridge Univ. Press, 1987.

[2] J. Polchinski, "String Theory", Volume I, An Introduction to the Bosonic String, ISBN 0-521-63303-6; Volume II, Superstring Theory and Beyond, ISBN 0-521-63304-4, Cambridge University Press, 1998.

[3] I. Klebanov and L. Susskind, "Continuum strings from discrete field theories", *Nucl. Physics* **B 309** (1988) 175.

[4] G. 't Hooft, "Relating the quantum mechanics of discrete systems to standard canonical quantum mechanics", arXiv:1204.4926.

[5] G. 't Hooft, "Duality between a deterministic cellular automaton and a bosonic quantum field theory in 1+1 dimensions", arXiv:1205.4107.

[6] G. 't Hooft, "Discreteness and Determinism in Superstrings", arXiv:1207.3612 [hep-th].

[7] P. Jordan and E. Wigner, Z. Phys. **47** (1928) 631.

CHAIRMAN: G. 't HOOFT

Scientific Secretaries: M. V. Carlucci, R. Knoops

DISCUSSION

- *Sleptsov:*

Superstring amplitudes have some properties and symmetries, for example under modular transformations. Are there some analogues or generalizations in your lattice model, or do they completely disappear?

- *'t Hooft:*

To the contrary, the suspicion, which we have not yet been able to prove, is that the lattice theory is mathematically identical to the continuum theory. In that case, of course, all the modular symmetries you can think of should also hold for this particular theory. That is so far a claim, and I will try to explain in my next lecture what I have about this and what is still to be done.

- *Alberte:*

When you were talking about the quantization of the bosonic string, you mentioned that only the states, not the equations of motion, have to be quantized. Do you mean that it is like in the usual light-cone quantization, when we first classically resolve the constraints, coming from gauge redundancy, and then we quantize only the physical states?

- *'t Hooft:*

In principle, as well in string theory, there is only one way to quantize it. So you can use light-cone coordinates, in principle you can use other coordinates, although light-cone is so much more convenient in string theory, that that is usually what you choose. I have to distinguish between the bulk of the string and the end points, or the boundary conditions, because they bring considerable complexity in the way you have to manage string theory. I have not yet even considered those boundary conditions, apart from the fact that when two strings collide they interchange; that is not really a boundary condition but it has an effect on the boundary conditions that you are going to use.

Boundaries are really complex, and I have not yet really investigated them in so much detail, but for the bulk properties in principle it should not matter how you determine your coordinates, the choice of gauge, it should not matter how you quantize the theory. There is only one correct answer and many wrong ones. In string theory, a few new constraints appear, when an anomaly is required to cancel. An example is the intercept of the string theory, which usually is taken to be a variable for which you find only one particular value that works, and other parameters like the number of dimensions; they must have particular values. All the other values simply fail to give a good quantum theory. This theory shouldn't be affected very much by the approach suggested in my lecture; the belief is that, for instance regarding the number of space-time dimensions, we eventually end up observing the same constraints. Maybe, just maybe, some new theory is to be found, because the compactification of a lattice may be a more drastic effect; possibly a different choice of lattice allows for generalizations that go even further than the normal string theories that we already have.

Note added in proof: I make an important distinction between quantizing the states (which is nothing more than a fancy way to describe correlations), and quantizing the equations of motion, which would be a replacement of a theory by a totally different theory. My claim is that in string theory only the states have to be quantized.

- *Twigg:*

Something like a theological question. Basing your theory on the integers you will at some point deal with those theorems that you have something which is true but you cannot prove. Do you regard this as a problem?

- *'t Hooft*:

Well, those integers are coordinates in space and time, so the only Gödel-like problem I will encounter is that the universe may be unbounded, so that the number of integers will tend to infinity. If the universe we live in were very small, we would just have periodic boundary conditions for the lattice in every direction; in that case, there would not be any Gödel problem because everything is finite. However, just because our universe is very large, you want to simplify things. Suppose I am looking at some scattering process right here, then I am not interested in what happens at the Andromeda nebula, or anywhere else in the universe, which should somehow decouple. This is called locality, and then you are already asking for more complex mathematics where one must be careful about interchanging limits. This is typically where you encounter Gödel-like problems.

- *Carlucci*:

This morning I asked you what is the mass scale of your theory, if maybe it is linked to the spacing of the lattice, and if the lattice of your theory is a physical object or a mathematical tool. You answered that the lattice should be physical, and that the spacing of the lattice is related to the string tension, but what are the typical values for that?

- *'t Hooft:*

People who do string theory are often extremely ambitious and they hope and expect a theory that actually even solves the problem of finding a Grand Unified Theory, the Theory of Everything and so on. The string tension, describing the interactions between strings, is basically in the Planck domain, so the string is just one or two orders of magnitude away from the Planck length; consequently, in terms of Planck units, the lattice length is just one or two orders of magnitude only. The strings are just very near the Planck scale and that means that also my lattice lengths are near the Plank scale. Other reasons to expect this are arguments concerning black holes, the observation that the amount of information that you can store in a certain region of space time depends on the volume, or the area, and the typical constant of proportionality there is again the Planck length, or the Planck length squared. So all the finiteness of nature seems to be situated in the Planck domain. Now the one caveat here that you have to remember is that if I indeed have extra dimensions, and if some of those are large, then information is spread over these extra dimensions, in such a way that the Planck scale will come down. It will be much lower in energy, and at relatively larger distance scales, though probably still out of reach for the LHC. The Planck length just comes a little bit closer if we have many dimensions. At present we simply don't know how the extra dimensions are compactified, how narrow they are, or how big they are, and this means we cannot tell exactly what the lattice size will be.

- *Aschieri:*

If you have a lattice, there exist preferred directions, namely those along the directions of the lattice. How do you think of them? Are these directions dynamical? Are they constant over all space-time and can they be detected?

- *'t Hooft:*

This is exactly the question I asked myself about a month ago. If indeed space and time are described by a lattice, there should be a way to figure out in which way the lattice is going. Because of this, there should, somehow, be a breakdown of rotational invariance. Today I am not so sure of this anymore. Even though nature is described by a lattice, the procedure that I will explain in the closing lecture, replaces the lattice again by a continuum. This is effectuated by a mathematical transformation. Once we have a continuum, the preferred directions disappear.

There is however still one indirect way in which the preferred directions can still manifest themselves. That is whenever you look at quantum mechanics being fundamental and 'decidable' by doing a Bell-like experiment. In such an experiment you measure the polarization of photons coming out of a certain particle process. Then you rotate the devices, and you get into contradictions as soon as you want to believe that this was a classical procedure. This is the reason why quantum mechanics is usually believed to be fundamental, and this is the reason why many researchers think that the classical theory can not possibly be right. Simply because the Bell inequalities disprove it.

The Bell inequalities assume that you can do an experiment while freely choosing the orientation of a device, and any orientation is to be treated the same way. If reality is described by a lattice, it is not true that you can rotate your experiment by 45 degrees and think that physically you have done nothing of importance. You have used rotational symmetry. This is no longer a true symmetry. A phenomenon could become a quantum superposition when you rotate it. This means that the interpretation of quantum mechanics requires a preferred direction in space, whereas the final mathematical scheme allows us to make superpositions where all preference for any direction in space-time seems to disappear.

This is an indirect answer which explains to you that you cannot come up with an experiment that allows you to determine in which direction the lattice is orientated. But perhaps reality is different. My first hunch was that it is impossible, but perhaps the experimentalists will nevertheless find a way to measure what I thought to be impossible. You should never underestimate the cleverness of the experimentalists, but as yet, my answer to your question is negative.

I have long advocated experimentalists to search for preferred directions in space-time. One way to do that, would be to see whether conservation of angular momentum could be violated at the tiniest distance scale. Angular momentum should be conserved when there is rotational invariance. If rotational invariance is violated, the angular momentum would no longer be conserved and rotating objects should precess around one of the preferred axes of a lattice. For this we need a high internal spin and the objects need to be as small as possible. I'm thinking about atomic nuclei with high spin, of which there are many examples. So I was hoping that nuclear physicists one day could do an experiment on a high spin nucleus and see whether there is any precession, even when one sets all the external fields to zero. This is technically very difficult, but this way one could notice an explicit breakdown of rotational invariance. I don't have much hope in finding something, but I would like to know what the practical limits are.

- Apresyan:

Is there a hope that the string theory will acquire a predictive power in the experimentally achievable domain? Are the experiments not looking at the right things?

- 't Hooft:

You are asking a very important question and this is also a very sensitive issue. First, string theorists have tried very hard to see whether there is any testable prediction. One of the predictions they came up with is supersymmetry. Supersymmetry is a very fundamental element in string theory and this might be observable at the LHC. The problem is that the ability to predict supersymmetry is not very good. Maybe we have to wait until we are able to increase the energy by a factor 100 and try again. If we don't find it there, maybe we should wait until we can increase the energy be another factor 100. This would probably be the response of superstring theorists.

I don't like this situation at all, but it is maybe a fact of reality that we have no way of attaining these energies. Fact is that string theory refers to the Plankian domain of space-time, which is very far away from any ordinary experiment. For instance, the experiments such as the ones that attempt to detect proton decay are very interesting because these can say something about grand unified theories which are not far away from the Plank scale. With this proton decay we can gain knowlegde about the very small distance scales of the universe, but even that are not really string theory predictions. We want to do better.

One reason for pursuing the present theory and try to improve it is to get rid of this nasty situation that we have not yet been able to make any really good prediction in superstring theory. We hope for instance that we can predict (or postdict) the Standard Model. We would love to do that, because the Standard Model has so many important pieces of information, like the Cabibbo–Kobayashi–Maskawa matrix, all the mass spectra, about 32 adjustable parameters,... We hope that string theory will be able to predict, or postdict, at least some of those fundamental important numbers. This is, for now, the best thing we can hope for.

On the other hand, string theory is a very powerful construction. Not only mathematical, but also in its predictions, even though these are not very useful to experimentalists today. String theory does say a lot about space-time and matter, it does say that if you go down in energy and up in distance scales, you do get quantum field theories. We are not able to identify which quantized field theory this is, it looks like the Standard Model, but it could be something else.

It also explains to us why there is gravity. It is not fair to claim that string theory predicts gravity, because there were theories of gravity long before string theory, but it gives us some insight of where gravity can come from. In this respect, string theory is a very important discovery of the 1980's, but more is needed to come to experimentally verifiable predictions.

- *Dorosti Hasankiadeh:*

Along the quantization procedure, it was mentioned that the transverse coordinate has to split in an odd and an even number. Is there any other way around this procedure such that the dimension of space-time could be anything else than 10 or 26 dimensions?

- *'t Hooft:*

It is solid mathematics that the bosonic theory has 26 dimensions and that the superstring theory has 10 dimensions. On the other hand, I am not happy with string theory's present formulation, and this is exactly the reason for my talk's title being "beyond string theory". I don't want to say it is wrong, but there may be much better ways of formulating the theory.

- *Mager:*

In set-theory we know that N=Q and that Q is dense in R. What is the real benefit of using N instead of R?

- *'t Hooft:*

The main clue is the finiteness of the number of states in the theory, but I will tell more about this in my closing lecture.

Hot Experimental Topics

Borexino latest results

New insights in the neutrino and Sun Physics

Gianpaolo Bellini
Istituto Nazionale di Fisica Nucleare- Sezione di Milano
Via Celoria 16- 20133 Milano

Abstract

Borexino is a unique detector, able to measure neutrino interactions with a lower energy threshold down to ~200 keV. This performance is allowed by an unprecedented radiopurity reached by means of a proper technology, developed in almost five year of R&D. Borexino succeeded to measure until now, and for the first time, the ^{7}Be and pep solar neutrino fluxes, the ^{8}B flux with a threshold down to 3. MeV, and reached a strict upper limit for the CNO neutrinos. In addition the day/night effect has been observed to be null for the ^{7}Be flux. These achievements allowed to check for the first time the oscillation MSW-LMA model, presently adopted, in Vacuum and in the transition region, and offered the possibility to measure the ratio between the ν_e survival probabilities in vacuum and in matter. The results obtained by Borexino are compared to the previsions of the Solar Standard Model. The Borexino further goal is to study also the pp neutrino flux and to reduce the total error on the pep and ^{8}B measurements, and therefore to allow a check of the hypothesis of Non Standard Neutrino Interactions. Borexino reached also evidence of the geo-neutrinos with 4.2 σ confidence level.

1. **Introduction**

Solar neutrinos are a useful tool in studying the neutrino oscillation and especially the matter effects. They are an energetically broadband, they cross quantities of matter not available to the terrestrial experiments, and their travel between production and detection is very long . In the meantime the study of solar neutrinos offers the opportunity to check the Solar Standard Model (SSM) predictions on the various solar neutrino fluxes and on the solar metallicity [1].

1.1 *The neutrino oscillation.* As it is known, the neutrino oscillation can be interpreted by means of the hypothesis of three neutrino mass eigenstates ν_i, which are connected to the known flavor eigenstates ν_α via the mixing matrix $U_{\alpha,i}$: $|\nu_\alpha\rangle = \sum_i U_{\alpha,i}|\nu_i\rangle$. The oscillation phenomenon can take place in vacuum and in matter and in both cases the transition probability P depends on the mixing angle θ and on the difference of mass squared Δm^2 between the mass eigenstates connected with the oscillating flavors [2].

In the two neutrino approach the transition probability P is the same, from a formal point of view, in vacuum and in matter, but Δm_M^2 and θ_M in matter are different from the same parameters in vacuum (Δm^2 and θ). They depend on Δm^2 and θ in vacuum and on a term X, which is a function of the Fermi coupling constant for the weak interactions G_F, of the electron density of the crossed matter n_e, of the neutrino energy, and finally of Δm^2 in vacuum: $X = \frac{2\sqrt{2}G_F n_e E}{\Delta m^2}$. In {1} and {2}, the transition probability in Vacuum and in matter are shown: they are written for the transition $\nu_e \rightarrow \nu_\mu$; for transitions between other couples of neutrinos the equations are very similar.

$$P(\nu_e \rightarrow \nu_\mu) = \sin^2 2\theta \cdot \sin^2\left(\Delta m^2 \frac{L}{4E}\right) \qquad \{1\}$$

$$P(\nu_e \rightarrow \nu_\mu) = \sin^2 2\theta_M \cdot \sin^2\left(\Delta m_M^2 \frac{L}{4E}\right) \qquad \{2\}$$

where:

$$\sin^2 2\theta_M = \frac{\sin^2 2\theta}{\sin^2 2\theta + (\cos 2\theta - X)^2} \quad and \quad \Delta m_{M,12}^2 = \Delta m_{12}^2 \sqrt{\sin^2 2\theta + (\cos 2\theta - X)^2}$$

If $\cos 2\theta < X$ the oscillation is matter enhanced; if, on the other hand, $X << \cos 2\theta$ the vacuum regime is dominating. The oscillation in matter is due to the different behavior of ν_e, which interacts in matter via charged and neutral current, and of $\nu_{\mu,\tau}$ which interact only via neutral current (this model is known as MSW, from the names of its authors [3]).

The two neutrino approach is a good approximation because the mass difference between ν_1 and ν_2 ($\Delta m_{solar}^2 \cong 7.5 \cdot 10^{-5} eV^2$) is almost negligible with respect the mass difference to ν_3 ($\Delta m_{atm}^2 \cong 2.5 \cdot 10^{-3} eV^2$).

In the three neutrino approach the parameters are five: the two mass differences mentioned before, and three mixing angles: θ_{12}, measured with the solar neutrinos; θ_{23}, measured with the atmospheric neutrinos; θ_{13}, very small, recently measured by T2K, Reno, Daya Bay [4].

The MSW model is presently adopted in studying the neutrino oscillation phenomenon, but, as explained later, several checks have to be done to validate it.

1.2 *The solar neutrinos.* Solar neutrinos are produced in the chains of nuclear reactions which take place within the Sun. The most important is the *pp* chain, which in our star is creating more than 99% of the total energy; the overall product of the reactions is summarized as follow:

$$4p \rightarrow {}^4He + 2e^+ + 2\nu_e + 26.73 MeV \qquad \{3\}$$

where the Hydrogen nuclei are transformed into Helium, releasing neutrinos and energy. In fig. 1 the *pp* chain is shown: the neutrinos, all ν_e, are emitted in the *pp*, *pep*, ^7Be, ^8B reactions. In addition a reaction *hep* is present, but with a very low rate [5].

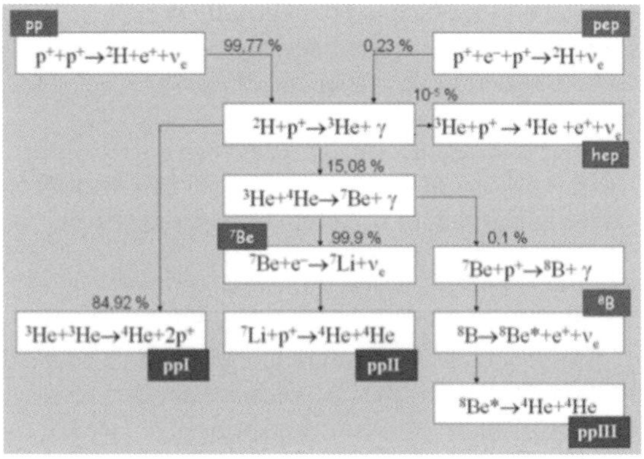

Fig.1

The energy spectrum is shown in fig. 2. We can observe that the *pp* neutrino rate is by far the highest one; the ^{7}Be and *pep* neutrinos are monochromatic being produced in two body reactions; the ^{8}B spectrum, with the lowest rate, extends up to ~ 16 MeV (the *hep* reaction is not taken into account here due to its very low flux).

The Cherenkov experiments (SNO and Superkamiokande) have explored <0.1% of the total spectrum studying the ^{8}B spectrum with a lower threshold >4-5 MeV, neutrino energy. The main reason of a so high threshold is the natural radioactivity, which produces a rate of interactions much higher than the solar neutrinos.

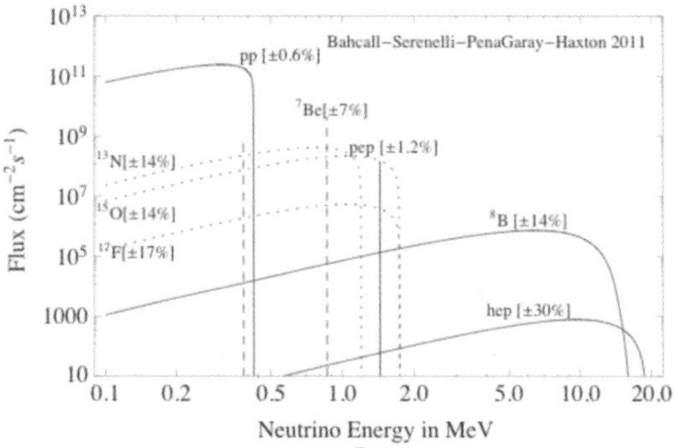

Fig.2

2. Borexino

2.1 The detector. The Borexino experiment has been conceived as a detector able to study the solar neutrinos at low energy, hopefully with a lower threshold down to 250 keV. Below this value the spectrum is dominated by the ^{14}C, a β emitter with a continuous spectrum up to ~150 keV.

The main problems in reaching the experiment goals are the external and the internal radiations. The external are mainly neutrons and gammas emitted by the rocks and the environment, and is due also to the radioactivity of the air, mostly due to ^{222}Rn, a gas β emitter. The external radiations can be strongly reduced and almost suppressed by means of a proper shielding with water and other liquids.

The internal background is due to the shielding and detector construction materials. Therefore all materials present in the detector have to be selected very carefully, liquid and gases must be purified of radioactive contaminants and special methods have to be adopted in the construction phase.

In fig.3 the detector design and layout are shown. Starting from outside a water tank (WT) contains a stainless steel sphere (SSS) surrounded by 2100 m^3 of highly purified water. The SSS supports 2212 Photomultipliers (PMT), coupled to optical concentrations (30% of optical coverage), and contains a nylon vessel (IV), 300 m^3 of volume, surrounded by 1050 m^3 of pseudocumene (buffer liquid), an aromatic chosen also as diluent of the two component liquid scintillator inserted within the IV. The nylon wall of the IV is 125 μm thick and, as a consequence, the buoyancy on it has to be very small: so the density of the external shielding liquid is approximately the same as the one of the internal scintillator solvent. A second nylon balloon functions as a barrier against the ^{222}Rn emitted mostly by the PMTs.

The scintillator is a two component liquid: pseudocumene (1,2,4-trimethylbenzene) plus ~1.5 g/l of PPO (2,5-diphenyloxazole). In the buffer the Pseudocumene is added of 5g/l of a quencher (DMP) to withdraw the scintillating power of the pseudocumene alone. The light emitted by the scintillator is detected by 2212 PMTs (Inner Detector-ID).

Finally between the SSS and the WT, 208 PMTs assure the cosmic muon identification by exploiting the Cherenkov light produced in the water (Outer Detector-OD).

The detector is installed at the underground Gran Sasso Laboratory in central Italy with an overburden of ~3800 m.w.e.; the cosmic muon flux which crosses the rock shielding is still significant: 1.2 muon $m^{-2}h^{-1}$.

Almost five years of R&D have been needed to develop techniques able to radio-clean Pseudocumene and PPO. The main tools are: water extraction, low temperature distillation, ultrafine filtration, stripping by means Nitrogen highly purified of ^{222}Rn, ^{39}Ar, ^{85}Kr. The crude oil for the Pseudocumene preparation has been procured from old layers to keep as low as possible the presence of ^{14}C. All stainless steel lines, walls, fittings, valves have been electropolished and the vessel nylon has been selected and extruded in clean room. Any operation has been carried out in clean rooms or in N_2/Ar atmosphere and the SSS itself has been equipped to obtain a clean room of class 10000 . The PMTs have been developed in collaboration with a British firm: very low radioactive glass and ceramics have been selected..

The final result has been a great success: the final contamination levels are better than the design requirements. In Table I the radioactive contaminants and other source of background are mentioned, and their typical concentration is compared to the Borexino final achievements. The reached results are world records, as it is the case for ^{14}C, ^{238}U and ^{232}Th equivalent, ^{222}Rn, and of almost all radioisotopes mentioned in the Table I [6].

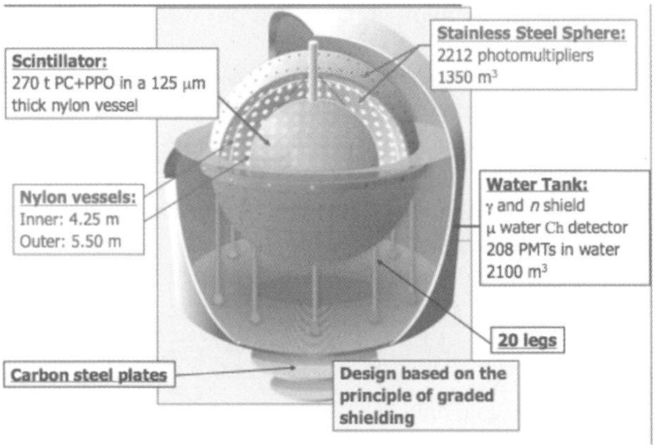

Fig. 3

Radio-Isotope		Concentration or Flux		Strategy for Reduction		Final
Name	*Source*	*Typical*	*Required*	*Hardware*	*Software*	*Achieved*
μ	cosmic	~ 200 s^{-1} m^{-2} @ sea level	<10^{-10} s^{-1} m^{-2}	underground water detector	Cherenkov PS analysis	<10^{-16} eff.>0.9992
γ	rock			water	fid. vol.	negligible
γ	PMTs, SSS			buffer	fid. vol.	negligible
^{14}C	intrinsic PC	~ 10^{-12} g/g	~ 10^{-18} g/g	selection	threshold	~ $2x10^{-18}$ g/g
^{238}U ^{232}Th	dust, metallic	10^{-5} – 10^{-6} g/g	<10^{-16} g/g	distillation, W.E., filtration, mat. selection, cleanliness	tagging, α/β	(5.0±0.9)10^{-18} g/g (3.0±1.0)10^{-18} g/g
^{7}Be	cosmogenic	~ $3x10^{-2}$ Bq/t	<10^{-6} Bq/t	distillation		not observed
^{40}K	dust, PPO	~ $2x10^{-6}$ g/g (dust)	<10^{-18} g/g	distillation, W.E.		not observed
^{210}Po	surface cont. from ^{222}Rn		<1 c/d/t	distillation, W.E. filtration, cleanliness	fit	May'07:70c/d/t Jan '10: ~1c/d/t
^{222}Rn	emanation from materials, rock	10Bq/l air, water 100-1000 Bq rock	<10 cpd/100t	N_2 stripping cleanliness	tagging α/β	<1 cpd 100 t
^{39}Ar	air, cosmogenic	17mBq/m^3 (air)	<1 cpd/100 t	N_2 stripping	fit	<<^{85}Kr
^{85}Kr	air, nuclear weapons	~ 1 Bq/m^3 (air)	<1 cpd/100 t	N_2 stripping	fit	30±5 cpd/100 t

Table I

2.2 *The measured quantities.* Borexino measures νs via the elastic scattering: $\nu + e \rightarrow \nu + e$, with a threshold at~ 60 keV (hardware) and ~200 keV (software) electron recoiled energy. The $\bar{\nu}$ are detected by means of the inverse beta decay: $\bar{\nu} + p \rightarrow e^{+} + n$, followed by the e^{+} annihilation prompt signal, with the emission of 2γ (1.1 MeV total energy), and a delayed (~240 μs) signal when the thermalized neutron is captured by a proton with the emission of a 2.2 MeV γ.

The detector measures the total energy released in the interactions via the photoelectron number, and the position via the PMT timing. In addition, Borexino is able to discriminate between α and β/γ [7], exploiting an important property of the scintillator to show different decay times for particles with high and low ionization.

The muons are identified if they cross the OD and/or the ID.

In order to further reduce the IV nylon and of the IV cap background, a fiducial volume (FV) is defined with a radius of 1m from the IV center and ± 1.67 cut in the vertical axis *Z*. The volume of the FV is ~88 m^3, which contains ~75.7 tons of scintillator.

3. **Some open problems in the neutrino and Sun physics before Borexino.**

We discuss here some problems of the Neutrino and Sun physics, still open before the Borexino data.

The first concerns the allowed oscillation regions in the Δm^2 and θ oscillation parameter plane. In fig. 4 the results of a global fit on all solar experiment data before Borexino (right plot) and of solar plus Kamland data (left plot) are shown. As it is known, Kamland is an experiment studying the reactor $\bar{\nu}$ s. The left plot shows only one region isolated at high mixing angle: it is called Large Mixing Angle region (LMA). On the other hand the fit on only solar data (right plot) isolates two regions: the LMA and, at lower Δm^2, a LOW region. The χ^2 for the LOW region is 11.83.

Therefore in order to isolate only the LMA solution we have to assume that the behaviors of ν and $\bar{\nu}$ oscillations are the same, i.e. that CPT is conserved.

The best fit values are quoted within the plots.

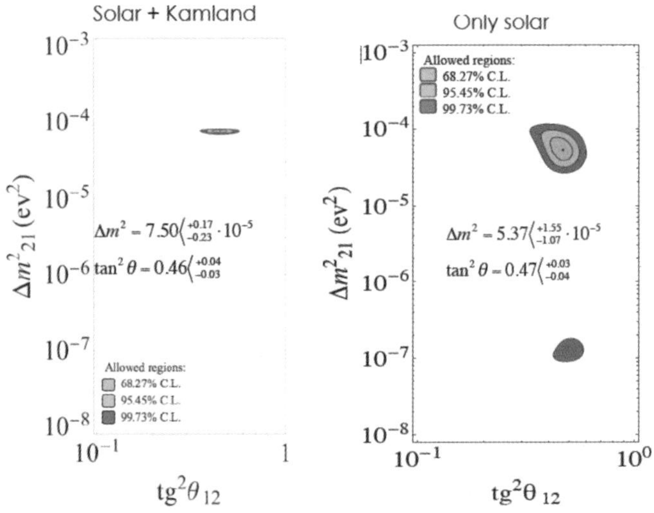

Fig. 4

The second problem concerns the experimental validation of the MSW-LMA model. In fig. 5 the ν_e survival probability P_{ee} is shown as a function of the neutrino energy. The grey band is the MSW-LMA prevision: it is a band and not a line due to the uncertainties of the SSM model and of the oscillation best fit parameters (all solar plus Kamland). The curve shows two plateaus corresponding to two different oscillation regimes: the left in *vacuum*, the right in *matter*; in between the *transition region* connects the two regime zones. P_{ee} goes as $\cong 1-\frac{1}{2}\sin^2 2\theta$ in the vacuum region and as $\cong \sin^2 2\theta$ in the matter one.

This model has been checked experimentally before Borexino only in the matter region by the Cherenkov experiment measurements of the ^{8}B neutrino flux [8] (black point in fig.5). The other two points have been obtained from the radiochemical experiment integrated fluxes once subtracted the ^{8}B neutrinos. Because in this calculation the unknown ^{7}Be flux has been missed, the errors are very large and these points cannot be used as a check of the model.

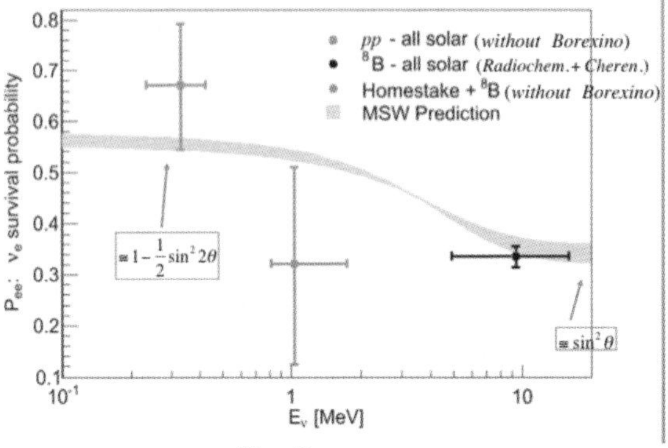

Fig. 5

The third problem concerns the so called Sun "metallicity". The Standard Solar Model in general is robust, but, few years ago, problems have arisen on the evaluation of the abundances of Carbon, Oxygen, Nitrogen, Neon and Argon in the solar surface. A 3D hydrodynamical model (AGSS09) [9] of the near-surface solar convection has changed the evaluation of these abundances, even reproducing perfectly the observed solar atmospheric line profile and asymmetries. These new calculations bring to reduced abundances (the so called "low metallicity") with respect to the results of an older calculation in 1D ("high metallicity"- GS98) [10]. But, while the 1D approach shows a good agreement with the helioseismology observations, the new 3D approach is in clear disagreement. This puzzle is known as the *solar abundance problem.* At present there is not satisfactory solutions for this controversy. Solar neutrino experimental measurements could help in solving it, because the previsions of the two approaches are different for many of them(see the first 4 columns of Table II).

We can note that only the CNO fluxes show significant differences compared to the model uncertainties. But it is not excluded that experimental determinations of other fluxes can give some indication.

4. **The Borexino data**.

In fig. 6 the energy spectrum of the Borexino data is shown. The black line is the row data; the blue one shows the spectrum after the FV cut and the removal of the muons and the muon induced events; the same events after the α subtraction give the red line distribution. In addition to the signal events, which will be discussed later, there are residues of radioactive contaminants.

ν flux	GS98	AGS09	$cm^{-2}s^{-1}$	Experimental results
pep	1.44±0.012	1.47±0.012	x 10^8	1.6±0.3 Borexino
^{7}Be	5.00±0.07	4.56±0.07	x 10^9	4.87±0.24 Borexino
^{8}B	5.58±0.14	4.59±0.14	x 10^6	5.2±0.3 SNO+SK+Borexino+Kamland 5.25±0.16+0.011-0.013 SNO-LETA
^{13}N **^{15}O** **^{17}F**	2.96±0.14 2.23±0.15 5.52±0.17	2.17±0.14 1.56±0.15 3.40±0.16	x 10^8	< 7.4 Borexino (total CNO)

Table II

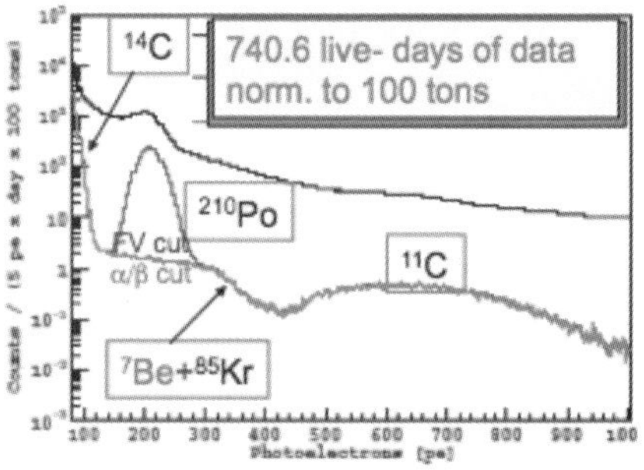

Fig. 6

Starting from low energy we can observe that: the steep tail of ^{14}C, a β emitter with a spectrum end point at 156 keV; ^{210}Po, with an average energy at ~ 200 photoelectrons (an α emitter with 200 days lifetime), which can be rejected via the $\alpha/\beta,\gamma$ discrimination; the residue ^{85}Kr, present in air and in nitrogen, which can be detected directly via the well tagged sequence $^{85}Kr \rightarrow {}^{85}Rb^* \rightarrow {}^{85}Rb$ with emission of a 173 keV β, followed, after ~1.464 μs, by a 514 keV γ (with a B.R. of 0.46%); the cosmogenic ^{11}C, around ~ 450 photoelectrons (β^+ emitter with a 29.4 minutes of lifetimes) which is produced by the 1.2 μ/m^2 h reaching the underground Gran Sasso Laboratory.

The light yield is ~500 photoelectrons/MeV; the energy resolution is 10% at 200 keV, 8% at 400 keV, 6% at 1 MeV; the spatial resolution, 35 cm at 200 keV, 16 cm at 500 keV. All these numbers are tuned with calibration campaigns via 13 artificial sources positioned at many different points within the detector volume.

The interference between these residual backgrounds and the neutrino signals can be better understood from the plots of fig. 7*a* and *b*, which show the energy spectrum up to about 1600 keV. The spectra have been fitted with Monte Carlo to obtain the ^{7}Be neutrino flux at 862 keV (plot *a* where αs are only partially subtracted with a procedure which does not reject any βs) and with an analytical algorithm (plot *b* with all αs rejected). The main parameters left free in the fits are: ^7Be rate and ^{85}Kr, ^{210}Bi, ^{11}C contributions. The results of the two fits are fully consistent. The data have been collected in 740.6 live days.

It is useful to recall that ^7Be neutrinos are monochromatic and therefore the spectrum of the elastic scattering on electron shows the typical Compton edge, clearly visible in the plots.

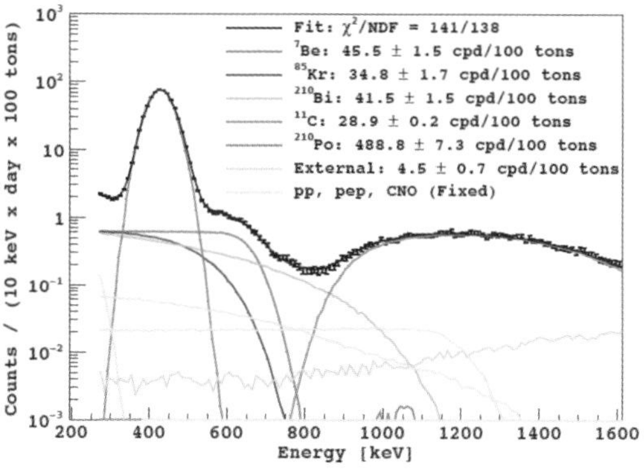

Fig. 7a

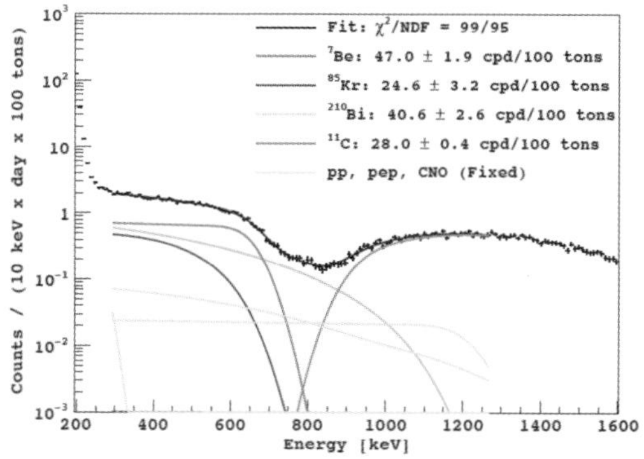

Fig. 7b

The ^7Be rate is $46 \pm 1.5(stat.) \pm^{+1.6}_{-1.5}(syst) cpd/10tons$. Assuming the present MSW best fit values of the oscillation parameters, the flux is: $\Phi(^7Be)$=(4.87±0.24) 10^9 cm^{-2} s^{-1}. The ratio to the SSM prediction is: f_{Be}=0.97±0.05±0.07 [11].

For the ^7Be ν_e also the day-night asymmetry has been studied. This effect can be produced by ν_μ and ν_τ, which can be transformed again to the original ν_e travelling through the Earth. The data for this study have been collected during 757.8 live days. Day and night are defined according to the Nadir angle θ_z between the vertical axis Z of the detector and the vector going from the Sun to the detector. Two approaches have been used. In the first one the rates in the day and in the night have been measured separately with the procedure used for the ^7Be total rate measurement. A second more sensitive approach consists in subtracting each other the day and night spectra, and to look on the subtracted plot for a possible residual component, having the shape of the electron recoiled spectrum typical of the ^7Be neutrino elastic scattering: if the day-night effect is null this plot must to be exactly flat. The day and night spectra are normalized to the same live time and the effects due to the seasonal variation of the Earth-Sun distance are corrected. The result is: $A_{ND} = 2\frac{R_N - R_D}{R_N + R_D} = \frac{R_{diff}}{R_{tot}} = 0.001 \pm 0.012(stat.) \pm 0.007(syst.)$ [12].

The measurement of the *pep* neutrino flux is more complicate because large part of the recoil electron energies falls in the region of ^{11}C decay products; in addition some contribution of the external background is present. *pep* neutrinos are monochromatic at 1442 keV and therefore the recoiled electron spectrum shows a shoulder due to the Compton edge. In fig. 8 the recoil electron spectrum from ~ 0.65 up to ~1.7 MeV is shown. The blue dashed line is the ^{11}C distribution; the continuous red, the pep spectrum; the continuous green, the ^{210}Bi background.

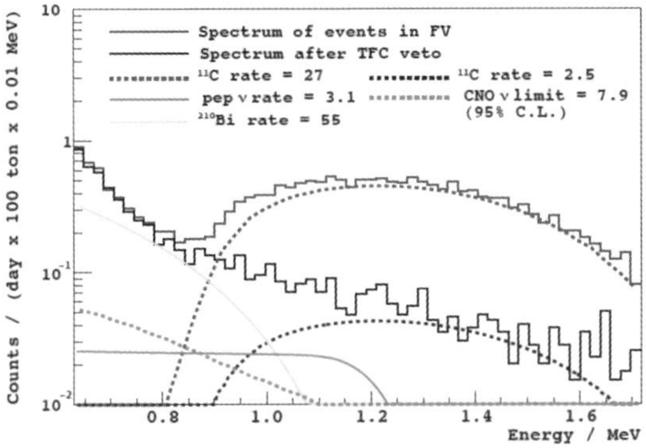

Fig.8

The first problem is to identify and reject ^{11}C. This can be done taking into account the incident muon and ^{11}C decay products. A muon strikes a ^{12}C nucleus, producing ^{11}C, and in more than 95% of the cases one or more neutrons are also emitted. The neutron is thermalized in ~ 236 μs, is captured by an hydrogen nucleus to form a deuteron with the emission of a 2.2 MeV γ. ^{11}C nuclides decay into $^{11}B + \nu_e + e^+$ with a lifetime of 29.4 minutes and then the positron annihilates with the emission of two 2.2 γs. The coincidence implies a cylindrical veto of 80 cm of radius around the muon track and a further spherical one around the position where the neutron is found, with 1m of radius. The vetoes are prolonged until 2 h after the detection time of the incident muon (see fig. 9).

This software threefold coincidence (TFC) is very effective and can suppress the ^{11}C contribution at ~99%, but it reduces also the fiducial exposure because part of the detector is frozen, due to the space and time vetoes. Therefore an optimum compromise between the ^{11}C rejection and the preservation of the fiducial volume has been reached leaving the residual ^{11}C rate at 9.1% (black dashed line in fig.8) and the residual exposure at 49.5%.

But 9.1% of ^{11}C is till too much with respect to the *pep* rate. A further selection can be obtained analyzing the pulse shape, which shows differences between the ^{11}C β^+ and the recoiled electron from the ν-e elastic scattering. The time distribution of β^+ decay is prolonged , with respect of the e^- one, due to the formation, in 50% of the cases, of ortho-positronium between the positron and one electron of the scintillator: the lifetime is 140 ns, reduced in the scintillator to

~3ns. In addition the two gammas, emitted in the positron annihilation, produce a more distributed multi-side topology than the electron.

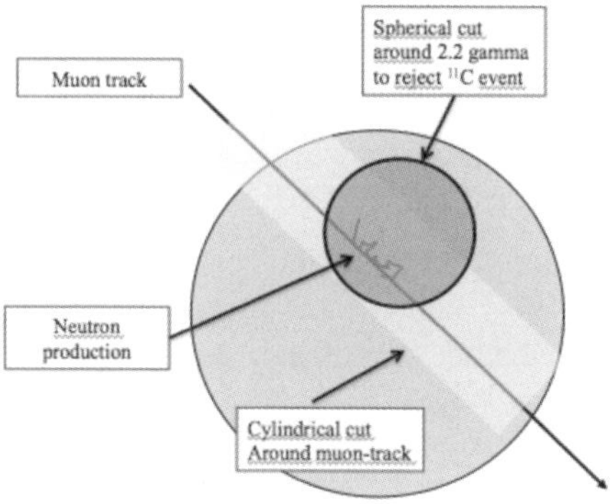

Fig. 9

Therefore a pulse shape parameter able to discriminate between e^+ and e^- has been developed using a boosted-decision-tree algorithm (BTD), which has been trained with the TFC selected ^{11}C, emitting an e^+, and ^{214}Bi emitting e^-. The ^{214}Bi are identified, as usually, using the ^{214}Bi-^{214}Po decay coincidence. Via the TFC analysis a further 4-5% of the ^{11}C, survived after the TFC section, is identified and rejected.

The external background is due mostly to the external ^{208}Tl and contributes with a small tail.

The fit is carried out on the energy spectrum once αs subtracted . All species, whose rate has been estimated to be ≥ 5% of the predicted rate for *pep* neutrinos, in the energy region of interest, are included in the fit. Therefore the internal background due to ^{210}Bi, ^{11}C, ^{10}C, ^{6}He, ^{40}K, ^{85}Kr, ^{234m}Pa (from ^{238}U decay chain) are taken into account together with the electron recoil signals from ^{7}*Be*, *pep* and *CNO* neutrinos. In addition the external γ-rays from ^{208}Tl, ^{214}Bi and ^{40}K are also considered. *pp* flux is fixed at the prediction of the Solar Standard Model (high metallicity) properly reduced for the oscillation, while for ^{8}B the mean value of the experimental results achieved at that time is assumed. Finally ^{214}Pb rate is experimentally determined following the ^{214}Bi-^{214}Po coincidence rate .

The fitting strategy makes use of a binned maximum likelihood multidimensional fit on the events surviving the TFC selection. The fit is performed simultaneously on the energy spectrum, on the radial distribution and on the pulse shape BDT parameter, with different fitting ranges. In addition the fit is performed in the same time on the events surviving the TFC vetoes and on the energy spectrum of the rejected events, constraining the non-cosmogenic species to be the same, because uncorrelated with the vetoes.

The best fit for the pep rate is :3.13±0.55(stat.)±0.23 (syst.) cpd/100 tons and the flux, calculated assuming the current MSW-LMA parameters, is : $\Phi(pep) = 1.6 \pm 0.3 \cdot 10^8 cm^{-2} s^{-1}$. The ratio to the SSM expectations with high metallicity is: f_{pep}=1.1±0.2 [13]. The fitted ^{7}*Be* rate is fully consistent with the previous analysis[11].

But an important check on the monochromatic *pep* neutrino fit is the observation of the Compton edge, which has to be shown by the *pep* energy spectrum as in the case of ^{7}Be. Fig. 10 shows the energy spectrum obtained after subtracted the rates of all species, except e^- recoils from *pep* and *CNO* neutrinos: the Compton edge is clearly visible. A likelihood test shows that the hypothesis of absence of *pep* neutrinos signal is rejected at 97% C.L.

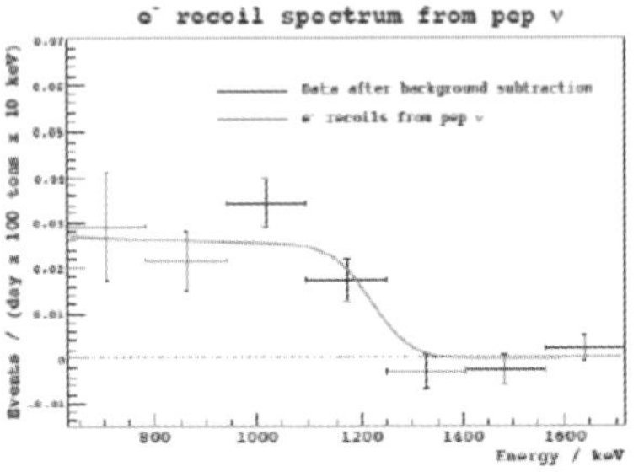

fig.10

The study of the *CNO* rate is more difficult due to the similarity of the shape of its spectrum to the ^{210}Bi one (same range and no clear tag- see fig.8). Therefore the fit on *CNO* flux has been carried out fixing the *pep* rate at the SSM prediction: this procedure leads to a *CNO* rate <7.6 cpd/100 tons. Assuming the MSW-LMA best fit parameters $\Phi(CNO) < 7.4 \cdot 10^{8} cm^{-2} s^{-1}$ and the ratio to the SSS prediction with high metallicity is $f_{CNO}<1.4$.

An effort has been done to measure the ^{8}B neutrinos with a lower threshold down to 3. MeV (3.2 MeV neutrino energy). Over 3. MeV additional backgrounds are present: the most important are: ^{208}Tl (Q=5. MeV) and ^{214}Bi emanated by the IV nylon; other contaminants are the cosmogenic isotopes with decay energy >3. MeV. ^{214}Bi-^{214}Po coincidences are a good tool to reject ^{214}Bi, while ^{208}Tl is subtracted exploiting the connection of ^{212}Bi-^{210}Po decay (τ=431 ns) with 64% B.R. to ^{218}Tl (τ=4.47 min) with 36% B.R..

The dangerous cosmogenic background is due to ^{6}He, ^{10}C, ^{7}Be. They are rejected either via a coincidence between the incident muon and their decay products, or subtracting them through Monte Carlo simulations.

The result gives for ^{8}B rate: 0.22±0.04±0.01 cpd/100 tons. The calculated flux, assuming the MSW-LMA model is $\Phi(^{8}B) = 2.4 \pm 0.4 \pm 0.1 \cdot 10^{6} cm^{-2} s^{-1}$ $\Phi_{^{8}B} = 2.4 \pm 0.4 \pm 0.1 \cdot 10^{6} cm^{-2} s^{-1}$ and the ratio to the high multiplicity SSM prediction, $f_{^{8}B} = 0.88 \pm 0.19$ [14]. In fig. 11 the ^{8}B measurements from various experiments are compared each other: the agreement of the Borexino results with the other experiments is quite good [15].

5. **Impact of the Borexino results on the neutrino and Sun physics.**

The Borexino results had and are having impact on the open problems considered in §3.

Starting with the oscillation allowed region of fig. 4, if we include in the global fit on only all solar the Borexino data, and specifically the day-night results, the LOW region now is ruled out at 8.5 σ C.L., allowing to isolate the LMA region without the Kamland $\bar{\nu}$ data and so without assuming the CPT symmetry (fig.12). The best fit of the oscillation parameters does not change in significant way (Table III) ; the different values between only solar and solar+Kamland global fit (see fig.4) [16] can be explained with a θ_{13} different from zero. The global fit gives θ_{13} values (Table III) in good agreement with the experimental results obtained by Daya Bay (0.023) and Reno (0.029) [17].

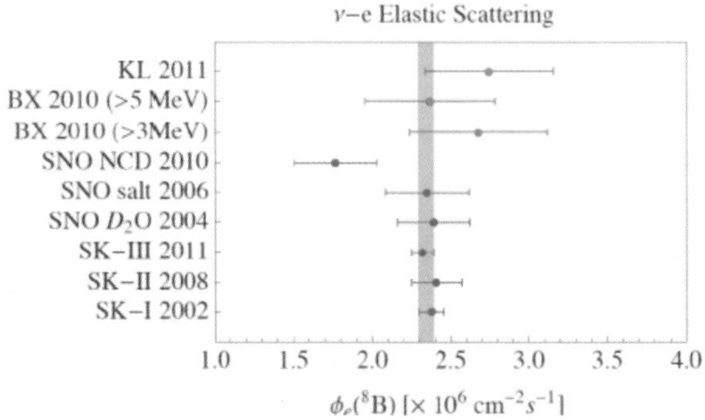

fig. 11

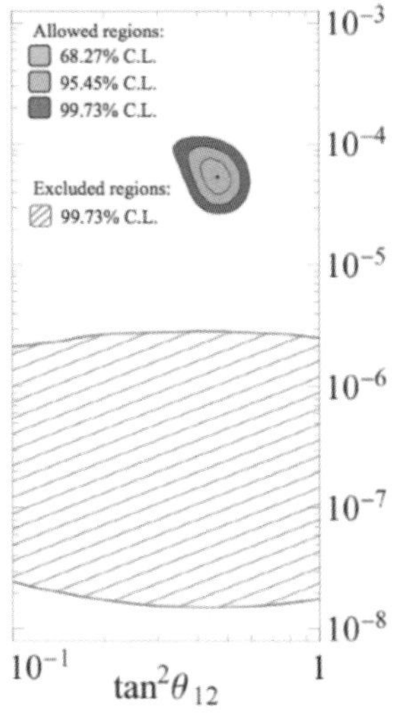

Fig. 12

EXPERIMENTAL DATA SET	Δm^2_{21} $[\times 10^{-5}\ eV^2]$	$\tan^2\theta_{12}$	$\sin^2\theta_{13}$
Solar without Borexino	$5.4^{+1.7}_{-1.1}$	$0.479^{+0.035}_{-0.042}$	< 0.029
Solar without Borexino + KL	$7.50^{+0.18}_{-0.22}$	$0.462^{+0.037}_{-0.033}$	$0.023^{+0.015}_{-0.018}$
Solar with Borexino	$5.4^{+1.7}_{-1.1}$	$0.468^{+0.031}_{-0.044}$	< 0.030
Solar with Borexino + KL	$7.50^{+0.18}_{-0.21}$	$0.457^{+0.038}_{-0.025}$	$0.023^{+0.014}_{-0.018}$

Table III

The ν_e survival probability versus neutrino energy, shown in fig.5, has been upgraded by the Borexino experimental results. In fig.13 the same plot with the Borexino data included, is shown. Starting from the low energy, the violet point is the *pp* determination: it is obtained from the radiochemical experiments once subtracted ^8B and ^7Be. The important reduction of the errors is due mostly to the ^7Be Borexino measurement.

pp falls in the vacuum oscillation regime and shows a good agreement with the MSW previsions in this region. The blue point is the Borexino ^7Be measured flux; it is close to the border vacuum-transition region and, within the errors, validates the model expectations. The *pep* red point has, for the moment, large errors; therefore it is premature to reach conclusions. Finally the Borexino ^8B measurement extends the data range down to 3. MeV (yellow).

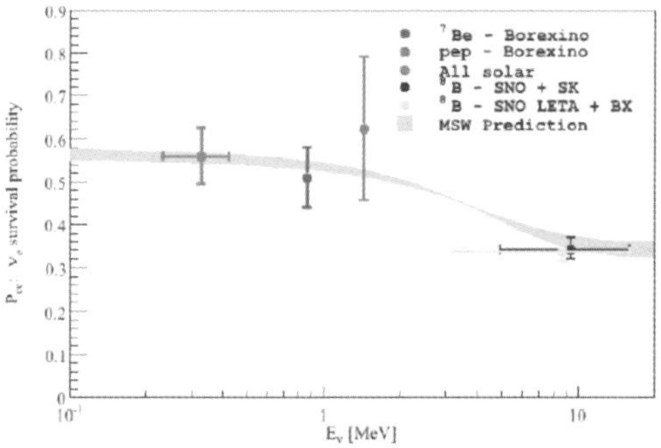

Fig. 13

The Borexino measurements allowed for the first time to determine the ratio between the ν_e survival probability in vacuum and in matter:

$$\frac{P_{ee}^{Vac}}{P_{ee}^{Matter}} = 1.63 \pm 0.26$$

The study of the shape of the transition region is an important check of the Non Standard neutrino Interactions (NSI) hypothesis and can allow to constraint the parameter ε (in the case of NSI the coupling constant $g = \varepsilon G_F$ [18], where G_F is the Fermi constant). Therefore the main effort is to reduce the errors on ^7Be, *pep* and ^8B at low energy (3.-5. MeV). This matter will be discussed again later.

An important open problem in the Sun physics is the metallicity puzzle, as mentioned in § 3 and Table II. Unfortunately the measured fluxes do not help until now in disentangling one or another solution due the uncertainties of the model and of the experimental values (fifth column of Table II). In the case of ^7Be and ^8B the experimental values just fall in between the SSM expectations for high and low metallicity and agree with both the approaches within the errors. The situation is different for the *pep* flux, but in that case the experimental uncertainty is still too large.

6. **The geoneutrinos**

The geological prevision of the Earth global heat loss ranges between 40 and 47 TW, but a global power of 30 TW is not excluded. Part of this caloric energy is due to radioactive decays, mostly of the natural ^{238}U and ^{232}Th families, and to ^{40}K. In these decays the ^{238}U chain emits 6 $\bar{\nu}$ s and 51.7 MeV of heat; the ^{232}Th chain,4 $\bar{\nu}$ s and 42.8 MeV; ^{40}K, 1 $\bar{\nu}$ and 1.32 MeV.

The drill-holes carried out by the geologists give information only on the Earth crust and cannot reach the mantle. The radioactivity of the crust can be measured, and it is expected that radioactive decays should be present also in the mantle, but probably not in the core, due to chemical affinity problems.

The canonical geological model, the Bulk Silicate Earth (BSE) predicts a contribution of 19 TW to the total Earth power from the radioactive decays, but also a contribution equal to the total terrestrial heat cannot be excluded .The only way to obtain a reliable information is to the detect the emitted $\bar{\nu}$ s (which are called *geo-neutrinos*).

In Borexino the $\bar{\nu}$ are well detected because their interactions are tagged via the inverse beta decay $\bar{\nu} + p \rightarrow e^+ + n$. The e^+ annihilates giving prompt signal with a 2γ production (total energy:1.2 MeV); the neutron, travelling in the scintillator along ~ 1m (Δt~260 μs), thermalizes and is captured by a ^{1}H nucleus to produce a deuteron and a 2.2 MeV γ. The threshold of this reaction is 1.8 MeV and this excludes the detection of ^{40}K decay, because the ^{40}K maximum $\bar{\nu}$ energy does not reach this threshold.

The background sources for this study are the reactor $\bar{\nu}$ and the internal radioactivity. The flux of reactor $\bar{\nu}$ at Gran Sasso is calculated considering 194 stations in Europe and 245 in the rest of the world (which contribute for 2.5% only). The calculation takes into account for each nuclear reactor: the thermal power versus time, the power fraction of isotopes, the average distance to the detector, the $\bar{\nu}$ survival probability. The data of the reactors have been obtained from EDF and IAEA.

The data analyzed for the geoneutrino study include 537.2 live days: along them the reactor $\bar{\nu}$ flux at Gran Sasso is 9.4±0.6 events [19]. Due to the Borexino very high radiopurity the contribution of the internal background is negligible.

In fig. 14 the data are shown: the white area corresponds to the reactor $\bar{\nu}$ interactions; the blue, to the internal background; the dashed, to the geoneutrino events. The geoneutrino rate is $9.9^{+4.1}_{-3.4}$, while the reactor $\bar{\nu}$ produce 6±2.5 events in agreement with the expected flux. The hypothesis of no geoneutrinos is disfavored at 4.2 σ [19].

The Kamland experiment reached an evidence similar to Borexino. A joint analysis Borexino-Kamland gives $20^{8.8}_{-8.6}$ TW for the radiogenic heat [20].

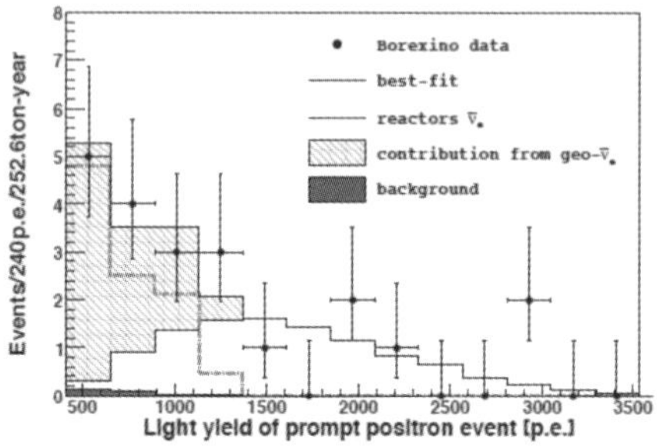

Fig. 14

7. Conclusions and perspectives

The phase 1 of the Borexino experiment is ended in 2010. The results achieved can be summarized as follows:

A) first measurement of the solar neutrino flux from ^{7}Be and *pep*, and first limit for *CNO*;

B) the flux from ^{8}B has been also measured with a lower threshold down to 3.2 MeV, neutrino energy. The measured integrated flux is in good agreement with the results obtained by the Cherenkov experiments;

C) the MSW-LMA model is now validated at low energy. The transition region needs more data to reduce the errors;

D) first measurement of the ratio vacuum/matter ν_e survival probabilities;

E) the LMA solution for the oscillation parameters is now isolated by solar data only, without assuming the CPT conservation;

F) a very robust evidence of geoneutrinos.

Since summer 2010 a campaign of further purification has been carried out in order to achieve an even better radiopurity. The campaign has been successful and has been completed in the fall 2011.The presence of the residual contaminants has further decreased. The ^{85}Kr has been reduced to a level close to zero, the ^{210}Bi is now lower than 20 cpd/100 tons, ^{230}U equivalent is <9.7 10^{-19} g/g and ^{232}Th equivalent <2.9 10^{-18} g/g.

Since fall 2011 the Borexino phase 2 is started. The main goals are now:

A') to reduce the uncertainties in the measurements of the ^{7}Be and pep flux and to reach a more stringent limit for *CNO*;

B') to accumulate more statistics for the ^{8}B flux in the range 3.-5. MeV;

C') first measurement of the neutrino flux from *pp*;

D') the goals A') and B') are especially devoted to check possible effects on the transition region due to the hypothesis of Non Standard neutrino Interactions;

E') A more precise measurement of the Earth heat from radiogenic sources with the geoneutrinos.

References

[1] See for instance J.N. Bahcall, Phys. Rev. C 65(2002) 025801

[2] See for instance M Fukugita and T. Yanagida, Physics of Neutrinos and Applications to Astrophysics, Springer-Verlag, 2003

[3] S.P. Mikheyev and A.Yu. Smirnov, Sov. Phys.-JEPT,64 (1986) 4; L.Wolfenstein, Phys. Rev. D 20 (1979) 2634.

[4] F. An et al., Phys. Rev. Lett. 108 (2012) 171803; J. Ahn et al., Phys. Rev. Lett. 108 (2012) 191802; K.Abe et al., Phys. Rev. Lett., 107 (2011)141801

[5] Bahcall, Neutrino Astrophysics, Cambridge Press, 1989; J.N. Bahcall, Phys. Rev. C 65 (2002) 025801

[6] For more information see G. Alimonti et al., N.I.M.a 609 (2009) 568; C.Arpesella et al., Astrop.Phys. 18 (2002) 1.

[7] E. Gatti et al., A linear method of discrimination between elementary particles in liquid scintillator, IAEA Wien (1965)

[8] K. Abe at al, Phys. Rev. D 83 (2011) 052010; B. Aharmim et al. arXiv1109.0763

[9] M. Asplund et al., ARA&A, 47 (2009) 481

[10] N. Grevesse et al., Space Science Reviews, 85 (1998) 161

[11] G.Bellini et al., Phys. Rev. Letters 107 (2011) 141362

[12] G. Bellini et al., Phys. Letters B707 (2012) 22

[13] G. Bellini et al., Phys. Rev. Letters 108 (2012) 051302

[14] G. Bellini at al., Phys. Rev. D82 033006

[15] S.Abe at al., arXiv:1106.1861v1 [hep-ex]

[16] G. Bellini, A. Ianni, G. Ranucci, Rivista Nuovo Cimento, vol.35, n. 9, 2012

[17] A.C. Re, PhD Thesis, february 2012.

[18] Y. Minakata and C. Pena-Garay, accepted in JHEP (2012)

[19] G. Bellini et al., Phys. Lett. B687 920120 299

[20] A. Gando et al., Nature GeoScience, 4 (2011) 647.

Highlights from LHC

Philippe Bloch
CERN, Geneva, Switzerland

Abstract

The Large Hadron Collider started operation in November 2009 at CERN. We present the current performance of the accelerator, some highlights from the initial physics results and prospects for the medium and long term evolution of hadron colliders at CERN.

1 Introduction

The Large Hadron Collider (LHC) project at CERN was launched in the mid-nineties with a double aim : understand the mechanism of electroweak symmetry breaking which is needed to complete the Standard Model, try to find signs of new physics beyond the Standard Model. To reach this goal, it was necessary to implement three components with unprecedented performance and complexity:

- a 27km long superconducting accelerator ring capable of accelerating beams of protons at an energy of 7TeV, with enough intensity to obtain the luminosity of $10^{34}\text{cm}^{-2}\text{s}^{-1}$ needed to study very rare processes, with cross sections of the order of 1 fb,
- large, high resolution experiments able to select and record the rare interesting events among a huge background,
- a world wide computing infrastructure with enough capacity to store and analyse Petabytes of data produced each year.

The LHC started operation in November 2009. In this lecture, we will present what has been achieved in less than 3 years and give an insight into the future plans. Section 2 is devoted to the accelerator performance, section 3 to a brief overview of the physics results already obtained with the experiments. In section 4, we present the current ideas about the future of the project and possible long term prospects.

2 The LHC accelerator performance

The LHC has reached in March 2010 half of its nominal energy (i.e. 3.5 TeV per beam or $\sqrt{s} = 7$ TeV in the center of mass). While the luminosity integrated in 2010 remained moderate (40 pb^{-1}), a large increase of performance in 2011 allowed the experiments to record an integrated luminosity of 5.3 fb^{-1}. The luminosity L of the accelerator at a collision point is given by equation 1 :

$$L = \frac{N^2 k_b f \gamma}{4\pi \epsilon_n \beta^*} F, \tag{1}$$

where k_b is the number of bunches in the ring, N the number of protons per bunch, ϵ_n the normalised emittance, f the revolution frequency, $\gamma = E/m$ the Lorentz factor, β^* the beta function at the intersection point and F a geometrical factor (< 1) which depends on the beams crossing angle.
The parameters f (11245Hz) and γ are fixed and given by the size of the ring and the beam energy, respectively. N, k_b and ϵ_n are given by the injectors chain properties and are strongly correlated. Finally, β^* is limited by the aperture of the focusing triplets.
Figure 1 illustrates the large progress achieved during 2011, by improving on the various parameters. At the end of 2011, the peak luminosity reached a maximum of $3.2 \times 10^{33}\text{cm}^{-2}\text{s}^{-1}$ using the parameters set listed in the second column of Table 1. It should be noted that several parameters (linked to the injectors performance) are already beyond their nominal value. This was however obtained with a reduced number of bunches (an interval of 50 ns between bunches). At the nominal bunch spacing of 25ns, a loss of performance was observed due to the "electrons cloud" effect: threshold effects lead to a build up of electrons inside the vacuum chamber of both the SPS and the LHC, resulting in heat load (in the cold sections), vacuum pressure rise and beam instability. The known solution is to condition the surface by electron bombardment, the so-called "scrubbing" method, which was found to be indeed effective but was discarded until now as it takes significant amounts of dedicated beam time.

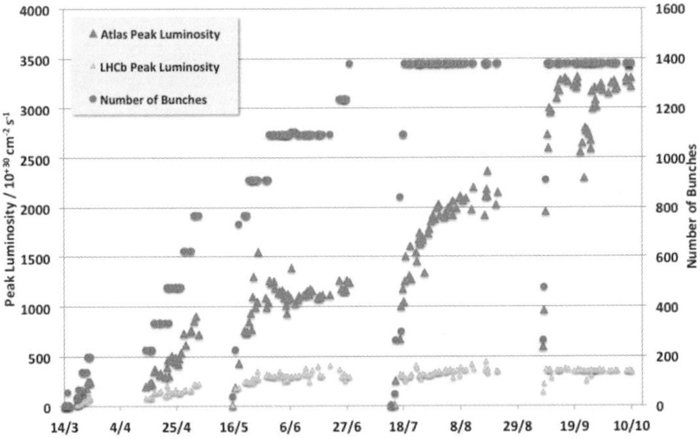

Figure 1: Evolution of the peak luminosity and the number of bunches during 2011.

In 2012, the center of mass energy was raised to 8 TeV. Furthermore, the operation in 2011 showed that, thanks to the control of the machine parameters and the quality of the alignment, the available aperture in the triplets is higher than expected, allowing one to further reduce β^* to 0.6m. As a result, the peak luminosity now routinely reaches $6.5 \times 10^{33}\text{cm}^{-2}\text{s}^{-1}$ (see Fig. 2) and the integrated luminosity has already surpassed the

Table 1: Luminosity parameters: nominal LHC, end of 2011 and mid 2012.

	Nominal	2011	mid-2012
$\sqrt{s}$(TeV)	14	7	8
k_b	2808	1331	1331
bunch spacing (ns)	25	50	50
N	1.2×10^{11}	1.7×10^{11}	1.6×10^{11}
$\epsilon_n(\mu m)$	3.7	≈ 2.0	≈ 2.0
β^* (m)	0.55	1.0	0.6

2011 total value by mid-June. It is expected to accumulate 15 to 20 fb^{-1} until the end of this year.

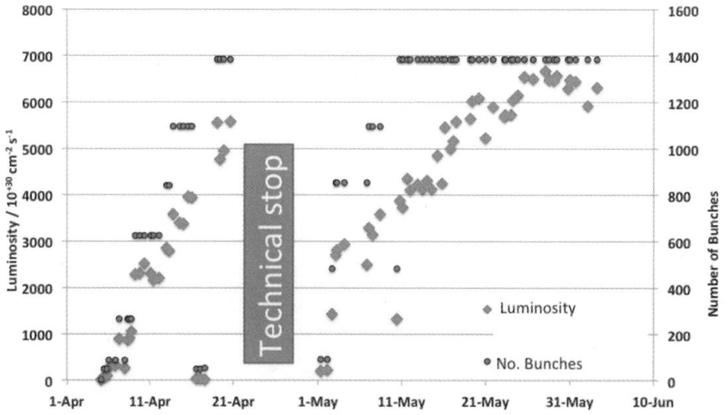

Figure 2: Peak Luminosity and number of bunches during the first half of 2012.

The drawback of the operation with 50ns bunch spacing is a large pileup in the experiments, with more than 35 simultaneous pp collisions for a single bunch crossing at the start of the fill, while the nominal value was ~ 20. The experiments have developed sophisticated subtraction methods, using in particular tracks assignments to the individual vertices, to cope with this increased pileup. Particularly important is to keep a constant trigger efficiency as a function of the number of simultaneous collisions, as the pileup conditions vary strongly during the long (sometimes more than 15 hours) fill.

3 Highlights of Physics results

In this section, we will present a short overview of the physics results obtained during the last months. This is a personal, non exhaustive choice, taking into account the fact that specific lectures are foreseen during the school on ATLAS, CMS and ALICE results.

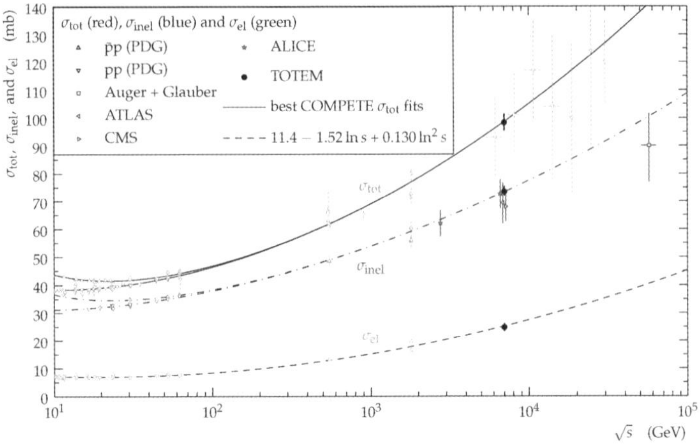

Figure 3: Compilation of total (σ_{tot}), inelastic (σ_{inel}) and elastic (σ_{el})cross-section measurements

3.1 Understanding the proton as a whole

An important measurement to be performed is the total pp cross section at LHC energy. This parameter is essential for measuring accurately the absolute luminosity of the accelerator and is also an important input for the extrapolation at higher energies, needed in particular for Ultra High Energy Cosmic Rays experiments. An accurate measurement has been obtained by the TOTEM experiment, using 3 different methods [1]:

1. measuring the rate of elastic events at zero momentum transfer t=0 and using the optical theorem, following equation 2:

$$\sigma_{tot}^2 = \frac{16\pi}{1+\rho^2}\frac{1}{L}\frac{dN_{el}}{dt}\bigg|_{t=0}, \tag{2}$$

   where ρ is the ratio of the real to imaginary part of the forward nuclear elastic amplitude. This requires the knowledge of ρ (predicted to be $0.14^{+0.01}_{-0.08}$) and of the luminosity, determined through Van der Meer scans,

2. measuring the rate of elastic and inelastic events , following equation 3:

$$\sigma_{tot} = \frac{N_{el} + N_{inel}}{L}, \tag{3}$$

   requiring again a separate determination of the luminosity L,

3. via a luminosity independent measurement following equation 4:

$$\sigma_{tot} = \frac{16\pi}{1+\rho^2}\frac{dN_{el}/dt|_{t=0}}{N_{el}+N_{inel}}, \tag{4}$$

The three measurements are in perfect agreement, yielding a value of $\sigma_{tot} = (98.3 \pm 0.2^{stat} \pm 2.8^{syst})$. A compilation of measurements of σ_{tot}, σ_{inel} and σ_{el} as a function of $\sqrt{s}$ is shown in Fig. 3.

3.2 Precise tests of the Standard Model (SM)

Precise tests of the Standard Model are a prerequisite to "qualify" the experiments before being able to claim any discovery. Furthermore, these SM processes are the source of background for the searches of new phenomena. All experiments have therefore precisely measured electroweak processes with increasingly smaller cross sections. This is best illustrated by Fig. 4 which shows a comparison of the measurements of CMS with theoretical predictions for single and double electroweak bosons production (γ, W and Z).

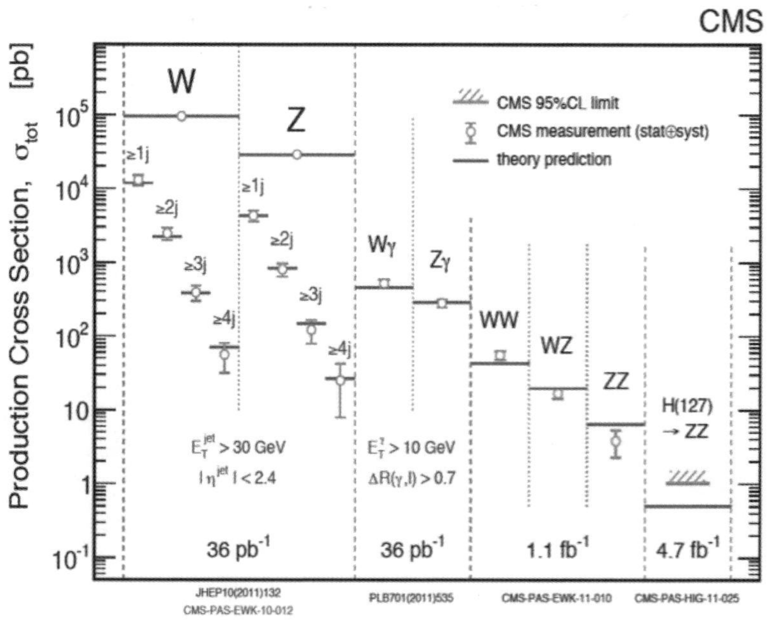

Figure 4: electroweak bosons cross-section measurements by the CMS experiment [2].

The agreement is really impressive. Similar results have been obtained by ATLAS. It should be noted that the measurements of W and Z cross sections as a function of rapidity produce already constraints on the proton structures functions (PDFs) determined previously with deep inelastic scattering.
The ultimate hope is of course to find deviations from the SM which would be an indication of new physics. Flavor physics is a particularly sensitive domain for these indirect searches. By selecting rare processes or processes involving small CP violation phases, one can hope to evidence new physics which would enhance these effects through lower order corrections. This can be illustrated by two recent measurements linked to B mesons physics:

- The $B_s \to \mu\mu$ decay is strongly suppressed in the SM, with a predicted branching ratio of $(3.2 \pm 0.2) \times 10^{-9}$. LHCb, ATLAS and CMS have looked for this decay channel. The present combined limit [3] for the branching ratio is 4.2×10^{-9} at 95% CL, i.e. only 35% above the SM prediction (Fig. 5 left). As shown in [4], this result provides strong constraints on SuperSymmetry (SUSY) models at high values of tanβ.

- CP violation is expected in B_s mixing and can be measured using the $B_s \to J/\psi X$ decays, where X represents the ϕ or the $\pi^+\pi^-$ final state. Unlike in the $B_d \to J/\psi K^0_s$ case, extensively measured at the B factories, the CP violation phase for the B_s case is expected to be small in the SM: $\phi_s = -0.036 \pm 0.002$ rad. Some tension with the SM was observed in previous measurements of CDF and D0. The more precise LHCb result $\phi_s = -0.002 \pm 0.083 \pm 0.027$ rad [5] is (unfortunately !) in excellent agreement with the SM prediction (Fig. 5 right).

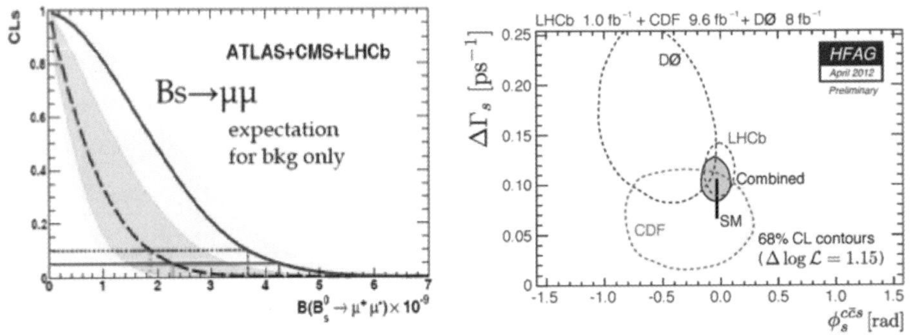

Figure 5: Left: Combined limit for the $B_s \to \mu\mu$ branching ratio. Right: 68% contours in the $\phi_s - \Delta\Gamma_s$ plane from $B_s \to J/\psi X$ decays for CDF, D0 and LHCb experiments and their combination.

The combination of $B_s \to \mu\mu$ and ϕ_s measurements is also restricting various SUSY models [6].
A surprise may have been found in the recent study of CP violation in charm decay by LHCb. The experiment has measured a difference ΔA_{CP} between the CP asymmetry observed in the $D^0 \to K^+K^-$ decay and the $D^0 \to \pi^+\pi^-$ decay:

$$\Delta A_{CP} = (-0.82 \pm 0.21 \pm 0.11)\% \quad [7].$$

Though it requires confirmation with higher statistics, this result has triggered considerable interest, as it was believed that direct CP violation in charm decay would signal new physics [8] . Theorists are now working hard to reevaluate whether the SM could account for the measured value [9].

3.3 Higgs Boson searches

The Higgs boson mechanism was proposed almost 50 years ago as a mechanism for the electroweak symmetry breaking. The search for the Higgs Boson is a top priority of the

LHC physics program[1].
Using the full dataset up to the end of 2011, ATLAS and CMS have put strong limits on the Higgs boson mass (Figs.6 and 7):

- CMS has excluded at 95% CL the mass interval [127-600] GeV/c^2 [10].
- ATLAS has excluded at 95% CL the mass intervals [111.4-116.6], [119.4-122.1] and [129.2-541] GeV/c^2 [11],

Both observe an excess in the region of 125GeV/c^2, but this is not significant enough to claim a discovery. Using the integrated luminosity of 6.5 fb^{-1} recorded during the first half of 2012, taking into account the higher cross section (by $\sim$ 27%) of the SM Higgs at 8TeV and also the analysis improvements obtained during the last 6 months which can be retroactively applied to the 2011 data, large progress in the Higgs search is expected in the next months.

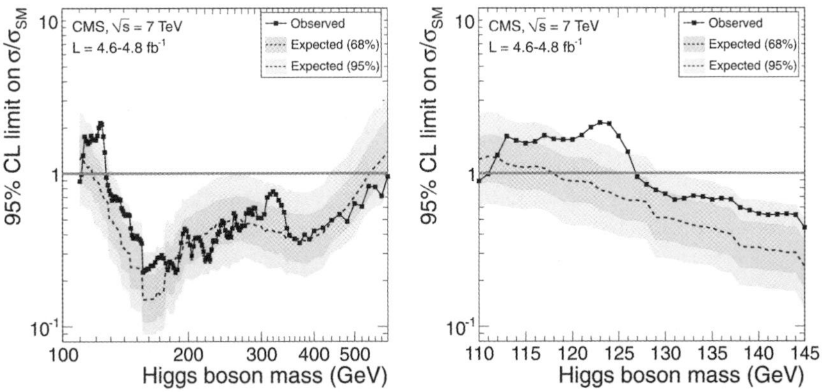

Figure 6: 95% Confidence level limits on the Higgs cross section normalised to the expected SM Higgs cross section for CMS. The right panel is a zoom in the low mass region. The observed values as a function of mass are shown by the solid line. The dashed line indicates the expected median of results for the background-only hypothesis, while the green (dark) and yellow (light) bands indicate the ranges that are expected to contain 68% and 95% of all observed excursions from the median, respectively.

3.4 Direct searches for Physics Beyond the Standard Model

Even if the Higgs Boson mechanism gets confirmed in the near future, we already know that the Standard Model can not provide the complete picture. It does not solve the hierarchy problem (why is the weak scale so high?), nor the complexity of the neutrino sector. It does not include the unification of gravity with the three other (weak, electromagnetic and strong) forces. Furthermore, cosmological data shows that ordinary matter

[1]At the time of writing these proceedings, the situation regarding the Higgs boson has dramatically evolved with the observation of a new particle compatible with the Higgs boson at $\sim$ 125GeV/c^2 reported by both ATLAS and CMS on July 4^{th}2012. The proceedings correspond to the status at the time of the school.

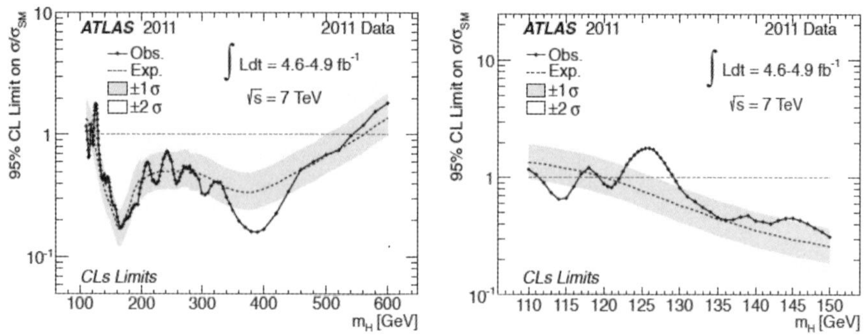

Figure 7: Same as Fig. 6, for the ATLAS experiment.

accounts for less than 5% of the mass of the universe, and the SM does not provide explanation for dark matter, without even mentioning dark energy. Many different theories have been proposed to solve some of these issues, the most popular being SuperSymmetry (SUSY) and Technicolor. More recently, the existence of extra dimensions has also been proposed. These theoretical models usually predict the existence of new particles or resonances which could be produced at the LHC. The search for these new particles is therefore a crucial part of the LHC physics program and actually accounts for more than 50% of the recent publications.

All the searches have so far been negative. Figures 8 and 9 show the status of SUSY searches by CMS [2] using 2011 data. In Fig. 8, limits obtained through various channels in the framework of the MUGRA/CMSSM model are plotted as a function of the common scalar mass (m_0) and the common gaugino mass ($\mathrm{m}_{1/2}$) for a fixed, not yet ruled out value of the 3 other parameters of the model, $\mathrm{A}_0 = 0$, $\mu > 0$ and $\tan\beta = 10$. As one can see, squarks and gluinos with a mass lower than $\sim$ 1TeV are excluded in this model.

There are many different SUSY models, and the experiments have tried to perform generic searches assuming the pair production of SUSY particles with various decay cascade configurations. Figure 9 displays some results of this approach. Limits from 400GeV to 1TeV are obtained, depending on the mass of the lightest supersymmetric particle $\widetilde{\chi}^0$.

3.5 Study of Quark Gluon Plasma

One month per year, the LHC accelerates lead ions to study Pb-Pb collisions. The QCD theory is predicting the existence of a new state of matter, the Quark Gluon Plasma (QGP), which must have existed after the Big-Bang, before quarks and gluons form hadronic bound states. Observations at the RHIC ions collider have shown evidence for this dense state of matter, and LHC data has beautifully confirmed it.
LHC ions collisions will allow us to actually study the QGP characteristics in detail. For example, measurements of the elliptic flow show that this dense medium behaves like an almost perfect fluid [12]. Experiments are studying energy loss in the medium through jet quenching [13], suppression of the excited states of Υ or J/ψ [14] and more generally

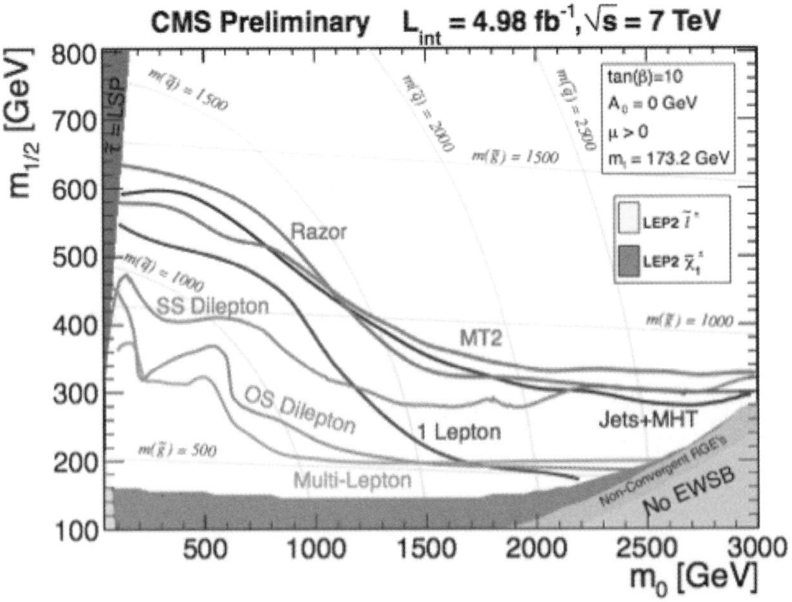

Figure 8: 95%CL limits in the m_0-$m_{1/2}$ plane in the framework of the MSUGRA/CMSSM theoretical model by the CMS experiment (see text) [2]. Lines of constant gluino $\widetilde{g}$ and squark $\widetilde{q}$ masses are shown by the faint grey lines.

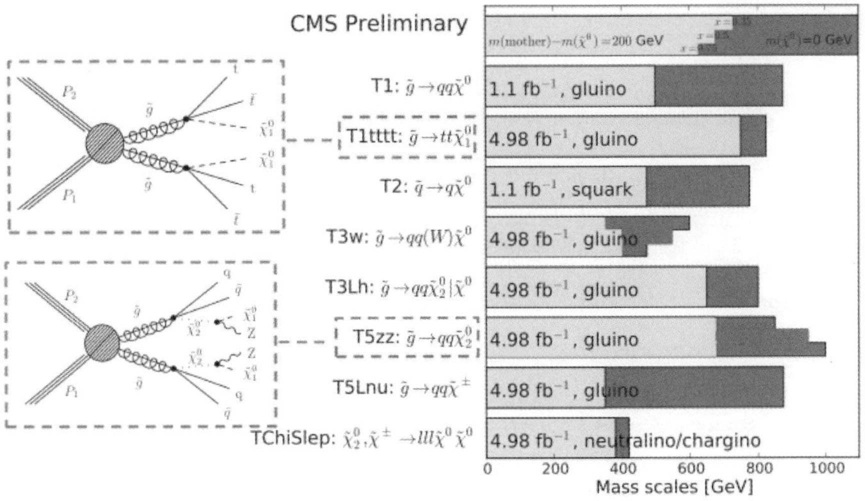

Figure 9: 95%CL limits on mass scales in simplified models by the CMS experiment [2].

by comparing particle production in Pb-Pb and pp collisions.

4 The LHC future

4.1 The foreseeable future

At the end of 2012 [2], the LHC will be shutdown for almost two years (LS1) to fix the magnet-connection problem which prevents running at full energy. It is likely that, due to the need of retraining some of the magnets, the energy in 2015 will correspond to $\sqrt{s} \sim 13.5$ TeV. Table 2 summarizes three possible scenarios for the 2015 configuration. The final outcome depends on mastering the electrons cloud effect, the beam emittance growth and also on the radiation resistance of the control electronics. The experiments strongly favor a 25ns bunch spacing to keep the pileup under acceptable limits. If this is not possible, a leveling of the luminosity during the first hours of the fill could be considered.

Table 2: Possible configurations in 2015.

	25 ns nominal	50 ns	25ns low emittance (to be tested)
$\sqrt{s}$(TeV)	13.5	13.5	13.5
$N \times 10^{11}$	1.15	1.7	1.15
$\epsilon_n(\mu m)$	2.8	2.1	1.4
bunch spacing (ns)	25	50	50
β^* (m)	0.5	0.5	0.5
peak luminosity $10^{34} cm^{-2} s^{-1}$	1.2	1.7	2.2
-		0.9 leveled	
pileup at start	28	76	46
-		40 leveled	
Int. luminosity/year fb^{-1}	30	40-50	55-60

In 2018, a second shutdown (LS2) is planned to upgrade the injectors and the LHC to reach the so-called ultimate luminosity of 2.2×10^{34}cm^{-2}s^{-1}, allowing one to accumulate about 100 fb^{-1} per year.
In $\sim$ 2022, a third long shutdown (LS3) is foreseen to consolidate the LHC and its injectors for another decade of running (in particular, a change of the focussing magnets at the intersection regions will be mandatory) and possibly improve its performance (High Luminosity or HL-LHC) using either a new optics or crab cavities to reduce β^* down to 10-20 cm. The aim is to accumulate about 3000 fb^{-1} before the stop of the facility.

The experiments are also planning for an intense program of upgrades, following the same timeline. Table 3 summarizes the main upgrades currently foreseen. Proposals or Letters of Intents have been submitted to the LHC Committee for the LS2 upgrades. To profit from the very high luminosity after LS3, ATLAS and CMS would require substantial modifications, in particular new inner Trackers, while LHCb and ALICE would continue

[2] After the school, it was announced that the stop of the LHC has been postponed to early March 2013.

Table 3: Main upgrades planned for the experiments.

Shutdown	Experiment	item
LS1	All	infrastructure consolidation
	ATLAS	reduced diameter beam pipe, new inner pixel layer
	CMS	reduced diameter beam pipe, 4th forward muon station
	ALICE	completion of TRD, EMCAL and DCAL
LS2	ATLAS	small Muons wheels, Forward Physics detectors
	ATLAS	FTK (Track) trigger, Trigger Upgrade
	CMS	new pixel detector, new photosensors for HCAL
	CMS	large rapidity Muon chambers, Trigger upgrade
	LHCb	replacement of Front-End electronics for 40 MHz readout, allowing a factor ×10 in luminosity
	ALICE	new Internal Tracker (ITS), faster TPC and DAQ
LS3	ATLAS	new central Tracker + ...
	CMS	new central Tracker + ...

with approximately the LS2 configuration to accumulate respectively $50 fb^{-1}$ of proton-proton and $50 nb^{-1}$ of Pb-Pb collisions.

4.2 LHeC ?

Recently, the Conceptual Design Report of an electron-proton facility using the LHC has been submitted to CERN [15]. It describes the physics program and the design of a new machine in which electrons of 60 up to 140 GeV would collide with protons of the LHC. The LHeC is designed to exceed the integrated luminosity collected at HERA by two orders of magnitude and the kinematic range in Q^2 by a factor of twenty. The electron and proton beams would collide in a single intersection, probably located at IR2 where ALICE is currently installed. The proponents aim at running the LHeC simultaneously with the LHC, for example during the last decade of LHC operation.
Two options have been proposed:

- the Ring-Ring option, in which one would add a new electron ring inside the LHC tunnel. The main difficulties are the spatial interference with the current LHC infrastructure and the need for two long bypasses around IR1 (ATLAS) and IR5 (CMS).
- the Linac-Ring option, where a new linac with energy recovery would be constructed tangentially to the LHC. This option would be more ambitious in terms of R&D and probably would provide a lower electron (and much lower positron) intensity, but would induce less nuisance to the LHC proton-proton program.

The LHeC would be able to address precision studies of QCD and the electroweak interaction, searches for phenomena beyond the standard model and physics at high parton densities.

Precision studies of QCD include a measurement of the inclusive proton structure function to per-mille accuracy and equally precise measurements of jet and heavy quark cross sections; these measurements would allow a determination of parton distributions and of the strong coupling constant to unprecedented precision. In electroweak physics, new measurements of the couplings of individual light quark flavors and a unique measurement of the running of the weak mixing angle up to the TeV scale in a single experiment could be envisaged. The LHeC would also provide sensitivity to new physics scenarios especially in the lepton sector (excited leptons, first generation leptoquarks, anomalous couplings). Finally, it would be able to study the high-energy limit of QCD in electron-proton and electron-nucleus interactions, thereby probing the previously inaccessible regime of ultra-high parton densities.

The future and priority of this project will have to be discussed during the European Strategy Update process in 2012/2013.

4.3 The longer term future: HE-LHC ?

Another idea for the long term future is to re-use the LHC tunnel for a higher energy hadron collider, the High Energy LHC (HE-LHC).

A working group started in April 2010, aiming at designing an accelerator able to exceed the LHC energy by a factor of 2 to 2.5. This would require very high field dipole magnets, with a magnetic field of 16 to 20 Teslas. It would also require a new injector, replacing the SPS, with an injection energy of about 1.3 TeV. Fairly conservative design parameters have been used so far, leading to a peak luminosity of $2 \times 10^{34}\text{cm}^{-2}\text{s}^{-1}$ with bunches of the same intensity as in LHC and a 50 ns bunch spacing. An interesting point to note is that at a beam energy of 16.5 TeV, synchrotron radiation emission becomes important, with a power of about 65 KW per beam. This would induce synchrotron radiation damping (like in e^+e^- machines !), improving the beam stability compared to the LHC case and possibly leading to an increase of luminosity during the fill.

Producing 20T magnets represents a non trivial step and requires a vigorous R&D. Neither the NbTi conductor of the LHC magnets, nor the Nb_3Sn conductor proposed for the intersection focussing magnets of HL-LHC would be suitable for such an application. High Temperature Superconductors (HTS) would therefore be needed. Today, they exist only in the form of very short tapes and their cost is more than one order of magnitude above the NbTi cost. Most probably, an hybrid coil using all available technologies would be required to keep the cost of the project at an affordable level. Figure 10 presents the sketch of such a magnet, fitting in the existing tunnel.

Clearly, it is the physics (and in particular the potential discoveries of the LHC after 2015) which will tell us whether a higher energy proton collider is useful or not for the future of High Energy Physics. Other options such as an electron-positron collider either at medium (ILC, 0.5 TeV) or high energy (CLIC, up to 3 TeV) are also being looked at. But it is clear that R&D should already be strongly pursued now, if one wants to keep the option of installing an HE-LHC soon after the LHC's stop, around 2030.

5 Conclusions

After only a few years of operation, the LHC project is fulfilling all its promises. The performance and reliability of the accelerator and of the detectors are truly impressive. Important physics results have already been obtained, showing the great potential of a project which has put Europe at the forefront of the Energy Frontier and is expected to continue and develop further for at least two decades.

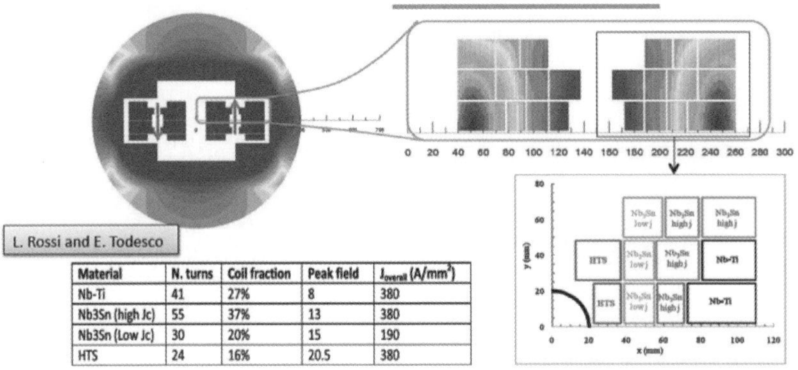

Material	N. turns	Coil fraction	Peak field	$J_{overall}$ (A/mm^2)
Nb-Ti	41	27%	8	380
Nb3Sn (high Jc)	55	37%	13	380
Nb3Sn (Low Jc)	30	20%	15	190
HTS	24	16%	20.5	380

Figure 10: A possible magnet for the High Energy LHC.

References

[1] First measurement of the total proton-proton cross section at the LHC energy of 7TeV, TOTEM Collaboration, EPL 96 (2011) 21002.

[2] CMS status report, Y.Sirois on behalf of the CMS collaboration, 110th LHCC meeting, June 13th 2012 (http://indico.cern.ch/conferenceDisplay.py?confId=194183).

[3] Search for the rare decays B_s and B_d to dimuons at the LHC with the ATLAS, CMS and LHCb experiments, CMS-PAS-BPH-12-009.

[4] Direct and indirect searches for New Physics. N. Mahmoudi, Rencontres de Moriond QCD and High Energy Interactions, La Thuile, March 10-17, 2012.

[5] Measurement of the CP-violating phase ϕ_s in $B_s \to J/\psi\pi^+\pi^-$ decays, LHCb collaboration, arXiv:1204.5675; Measurement of the CP-violating phase ϕ_s in $B_s \to J/\psi\phi$ decays, LHCb collaboration, Phys. Rev. Lett. 108, 101803 (2012).

[6] New physics searches in flavour physics. D.M. Straub, arXiv:1107.0266.

[7] Evidence for CP violation in time-integrated $D^0 \to h^-h^+$ decay rates, LHCb collaboration, Phys. Rev. Lett. 108 (2012) 111602.

[8] New Physics and CP Violation in Singly Cabibbo Suppressed D Decays, Y Grossman et al., Phys.Rev.D75:036008,2007.

[9] On the size of direct CP violation in singly Cabibbo-suppressed D decays, J. Brod et al., arXiv:1111.5000.

[10] Combined results of searches for the standard model Higgs boson in pp collisions at $\sqrt{s} =$ 7TeV, CMS Collaboration, Phys. Lett. B 710 (2012) 26-48 .

[11] Combined search for the Standard Model Higgs boson in pp collisions at $\sqrt{s} = 7$TeV with the ATLAS detector, arXiv:1207.0319.

[12] Elliptic flow of charged particles in Pb-Pb collisions at $\sqrt{s_{NN}} = 2.76$TeV / ALICE Collaboration, Phys. Rev. Lett. 105 (2010) 252302 ;
Measurement of the pseudorapidity and transverse momentum dependence of the elliptic flow of charged particles in lead-lead collisions at $\sqrt{s_{NN}} = 2.76$TeV with the ATLAS detector / ATLAS Collaboration, Phys. Lett. B 707 (2012) 330-348;
Measurement of the elliptic anisotropy of charged particles produced in PbPb collisions at nucleon-nucleon center-of-mass energy = 2.76TeV / CMS Collaboration arXiv:1204.1409.

[13] Observation of a Centrality-Dependent Dijet Asymmetry in Lead-Lead Collisions at $\sqrt{s_{NN}} = 2.76$TeV with the ATLAS Detector at the LHC / ATLAS Collaboration, Phys. Rev. Lett. 105 (2010) 252303;
Observation and studies of jet quenching in PbPb collisions at $\sqrt{s_{NN}} = 2.76$TeV / CMS Collaboration, Phys. Rev. C 84 (2011) 024906.

[14] J/ψ production at low transverse momentum in Pb-Pb collisions at $\sqrt{s_{NN}} = 2.76$TeV / ALICE Collaboration, arXiv:1202.1383;
Suppression of non-prompt J/ψ, prompt J/ψ, and (1S) in PbPb collisions at $\sqrt{s_{NN}} = 2.76$TeV / CMS collaboration,J. High Energy Phys. 05 (2012) 063;
Measurement of the centrality dependence of J/ψ yields and observation of Z production in lead-lead collisions with the ATLAS detector at the LHC / ATLAS Collaboration, Phys. Lett. B 697 (2011) 294-312

[15] Conceptual Design Report of the LHeC, arXiv:1206.2913.

HIGHLIGHTS FROM ATLAS

Peter Jenni
CERN, Switzerland

SUMMARY

For the past three years, experiments at the Large Hadron Collider (LHC) have started exploring physics at the high-energy frontier. Thanks to the superb turn-on of the LHC already a rich harvest of initial physics results has been obtained by the general-purpose experiment ATLAS, as well as its sister-experiment CMS. The initial data has allowed a test, at the highest collision energies ever reached in a laboratory, of the Standard Model (SM) of elementary particles, and to make many searches Beyond the Standard Model (BSM). Important results have already been obtained in the search for the Higgs Boson, which would establish the postulated electro-weak symmetry breaking mechanism in the SM, as well as for BSM physics like Supersymmetry (SUSY), heavy new particles, quark compositeness, and others. The important, and successful, SM physics measurements are giving confidence that the experiment is in good shape for their journey into the uncharted territory of new physics anticipated at the LHC.

Key index words
LHC, ATLAS experiment, experimental Standard Model measurements and experimental searches for physics Beyond the Standard Model.

1. INTRODUCTION

The first high-energy proton-proton collisions (3.5 + 3.5 TeV) at the LHC were registered on 30th March 2010, and since then the machine has operated in a superb way, providing the two general-purpose experiments ATLAS and CMS with data samples corresponding to an integrated luminosity of close to 5 fb^{-1} during the pp running periods in 2010 and 2011, and another 6 fb^{-1} at the slightly higher energy of 4 + 4 TeV in 2012 until this school. The excellent performance of the LHC has been a challenge for the experiment in many aspects, with typically 20 overlapping minimum bias events (pile-up) in the same bunch crossing as the events of interest. Nevertheless, ATLAS has recorded collision data in a very effective way, reaching a data taking efficiency of about 94% for the luminosity delivered by LHC in stable conditions. Thanks to a very careful and rather complete commissioning of the experiment over several years with cosmic ray data, and with the lower energy LHC collision data accumulated at the end of 2009 during the first LHC operation, ATLAS was able to quickly produce a rich harvest of first physics results. In fact, ATLAS has published more than 160 papers in scientific journals up to this school.

It would be impossible to review all these results; necessarily a very restrictive selection had to be made, in the spirit of giving illustrative examples. The results will be presented, roughly speaking, following a pattern of decreasing cross-sections. This will naturally first lead to measurements of the known Standard Model (SM) particles, of which the top quark is the heaviest known, with the smallest cross-section. All SM measurements, already with considerable accuracies and details, agree so far with the most sophisticated theoretical expectations. Next will be discussed the status of the search for the still missing element of the SM, the Higgs boson as the messenger of the electro-weak symmetry breaking mechanism. Finally, several examples of searches and resulting new limits on various physics processes Beyond the Standard Model (BSM) will be reported, mostly involving heavy new particles with small production cross-sections. For the searches the exploration at the LHC has only just begun, as much larger data samples are anticipated for the future, once the integrated luminosity will have reached the expected fourfold increase at the end of 2012, and further on also with the full LHC collision energy of 7 + 7 TeV for the years 2015 and beyond.

ATLAS is the result of an enormous collaborative effort that already lasts two decades including all its phases from conception, design, prototyping, construction, pre-assembling, installation in the underground cavern and commissioning, to finally starting up operation with recording the first collisions in November 2009. There is no way that any justice can be given here to this huge scientific, technological and human undertaking; the reader is referred to [1,2] and references therein. Just as an indication of the complexity and how well the apparatus is functioning, one may note that for all different detector layers, totaling about 100 Million electronic read-out channels, typically 98% of channels are working within design specification. Figure 1 shows as an example the detector status in February 2007 during the peak period of the installation work.

Figure 1. Photograph of one end of the ATLAS detector barrel with the calorimeter end-cap still retracted before its insertion into the barrel toroid magnet structure (February 2007 during the installation phase).

The ATLAS Collaboration currently brings together about 3000 scientists in a world-wide common effort from almost 200 Universities and Laboratories located in 38 countries. About one-third of the scientists are students working on their PhD with data accumulated with the ATLAS experiment.

The central role of computing in these successful first years cannot be stressed enough. The worldwide LHC Computing Grid (wLCG) has performed to high satisfaction, and it will be presented in a dedicated lecture at this school by J. Knobloch.

2. GENERAL EVENT PROPERTIES

The experiment has collected large samples of so-called minimum bias events (ordinary collision events without, or at most very minimal, selection criteria) in order to study general event properties. These properties are interesting in their own right as the physics of soft hadronic interactions (soft QCD), and an understanding of them is a crucial input to the modeling of background events for any measurements and searches of SM and BSM physics processes. The minimum bias events also allowed the experiments to verify in great detail that the average detector responses are well described in the Monte Carlo (MC) simulations, and that the detector elements are well aligned and calibrated, most convincingly demonstrated by the reconstruction and measurement of many well-known resonances, yielding the expected mass values and resolutions.

Charged particle production properties measured by ATLAS [3] over the central region in 7 TeV centre-of-mass proton-proton collisions are shown as an example in Fig. 2. The central region is expressed as $|\eta| < 2.5$ where η is the pseudorapidity defined in terms of the polar angle θ w.r.t. the beam axis as $\eta = -\ln\tan(\theta/2)$. Figure 2a shows the number of charged particles (multiplicities) per unit η, and Fig. 2b displays the transverse momentum p_T distribution w.r.t. the beam axis. Both measurements are compared with various Monte Carlo (MC) model simulations before tuning of the latter, and as can be seen, in particular from the MC over data ratio plots, the model descriptions require adjustments to better represent the measurement.

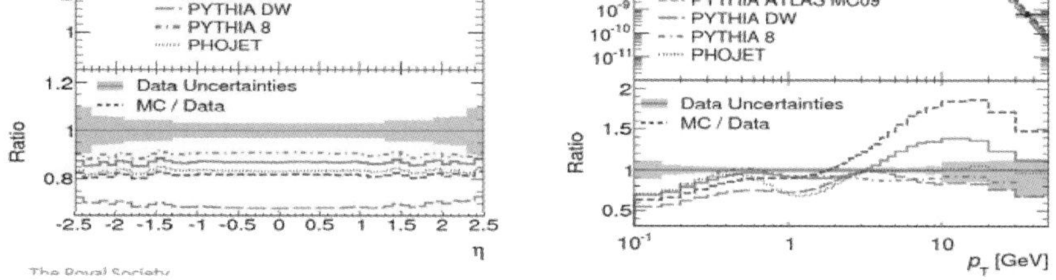

Figure 2. (a, left) Charged particle multiplicity per event and per unit η, (b, right) charged particle transverse momentum p_T distribution. The data (dots) are compared to various MC model simulations before tuning of the latter.

3. KNOWN STANDARD MODEL PHYSICS

Observing, and measuring accurately at the new collision energies the known particles from the SM, can be considered to be a necessary stepping stone towards exploring the full potential of the LHC with its many promises of possible new physics discoveries. The SM processes are often called 'standard candles' for the experiments. However there is much more value to measuring the SM processes than this: never before could the SM physics be studied at a hadron collider with such sophisticated and highly accurate detectors, ultimately allowing a test of detailed predictions of the SM with unprecedented precision and minimal instrumental systematic errors, as already published for several ATLAS and CMS QCD results.

The charged and neutral Intermediate Vector Bosons (IVB) W and Z are the major benchmark measurements at the LHC [4] for demonstrating the excellent detector performance, as well as for testing model predictions to a high degree of accuracy. The Z decays into electron and muon pairs can be extracted almost free of any backgrounds, as shown in Fig. 3.

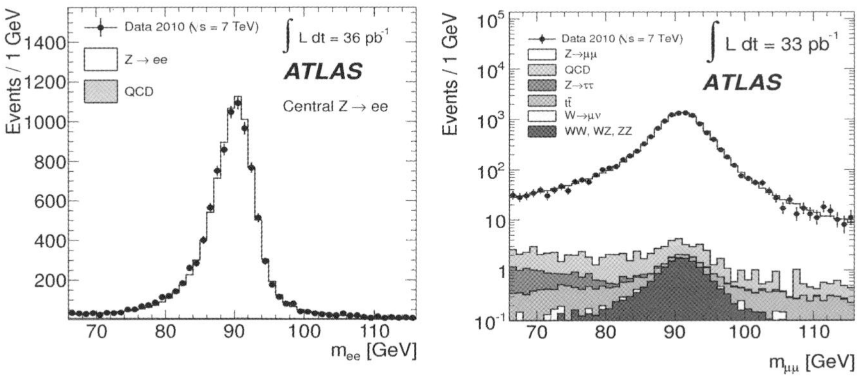

Figure 3. The electron-pair mass distribution on a linear (left) and the muon-pair mass distribution on a logarithmic (right) vertical scale, in the Z mass region. The estimated small background contributions are indicated, as well as the expected signal shape from MC simulations.

The classical W decay signatures into an electron or muon and the associated neutrino are an excellent test for the missing transverse energy (ETmiss) performance of the detector due to the undetected neutrino. ETmiss is inferred from the measured energy imbalance in the transverse projection of all observed signals w.r.t. the beam axis. The ETmiss spectrum for events with a well-identified muon candidate is shown in Fig. 4a, and shows a clear W signal over the expected background sources. After applying a selection of events with ETmiss > 25 GeV only a small residual background remains present under the W signal, as indicated in the transverse mass distribution (defined in [4]) given in Fig. 4b.

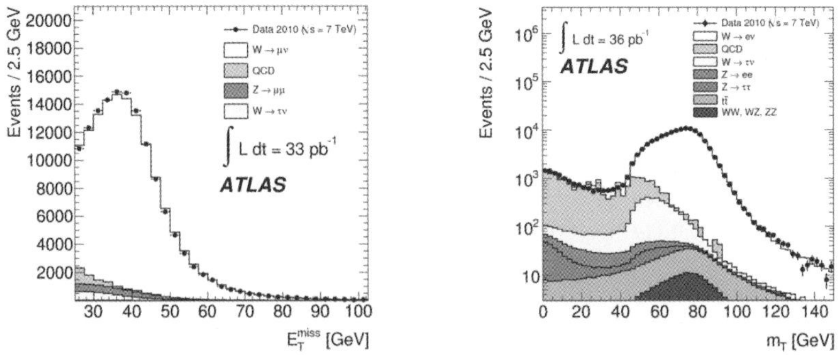

Figure 4. (a, left) Missing transverse energy distribution for events with a muon candidate. (b, right) Transverse mass distribution for W to electron decays before a cut on the transverse mass (see text). The expected background contributions are indicated as well.

The good agreement between the measured and expected cross-sections times the leptonic decay branching ratios (which is the expected rate for W bosons to be produced and then decay to leptons) is illustrated in Fig. 5. With the present data samples the experimental uncertainties still dominate, but with the addition of the 2011 data, the measurements will already constrain the theoretical model parameters. Figure 5a shows the W cross-section measurements and predictions as a function of the collision energy, whereas in (b) the W and Z cross-section results are displayed in a 2-dimensional plot including their correlated error ellipse, and compared to predictions with various parton distribution functions (describing the quark and gluon momentum distributions inside the protons). Detailed measurements of properties for IVB production and decay at the LHC have been published already, including for example the lepton charge asymmetry measurements for W decays [4] which were an important signature of the electro-weak nature of the W at the time of their discovery some 30 years ago.

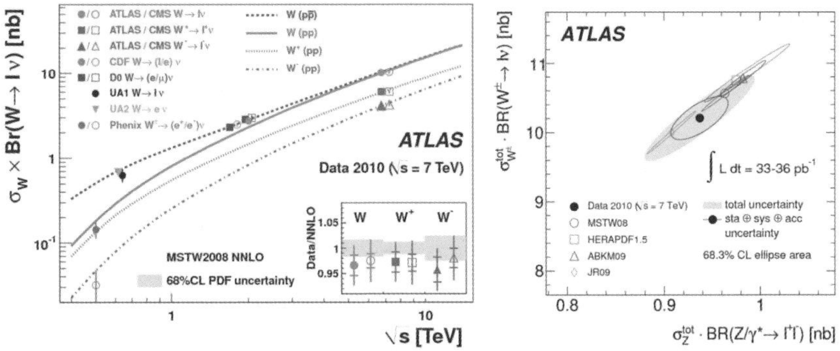

Figure 5. (a, left) W production cross-sections times leptonic branching ratio as a function of the collision energy, showing also previous measurements at lower energy colliders, (b, right) correlation of the measured (solid dot) leptonic W and Z cross-section as compared to theoretical expectations with various choices for the parton distribution functions (open symbols).

Hard collisions (characterized by having final state particles with significant transverse energy) at the LHC are dominated by the production of high transverse momentum jets, which are the collimated sprays of particles from the hadronization of the initially scattered partons (quarks, gluons) in the colliding protons. At work is the strong interaction described by Quantum Chromo Dynamics (QCD). Most commonly two jets emerge at opposite azimuth with balanced transverse momenta, from an initial lowest order parton-parton scattering process. However, higher order QCD corrections alter this picture significantly, and detailed measurements of multi-jet configurations are very important to constrain the QCD descriptions of hadronic processes.

The most impressive results at this stage are the inclusive jet and the di-jet cross-section measurements [5]; an example for them is shown in Fig.6a. These measurements cover unprecedented kinematical ranges spanning typically over jet transverse momenta from 20 GeV to 1.5 TeV, in many angular bins up to $|\eta| < 4.4$ (i.e. very close to the beam axis). The cross-sections vary over these ranges by up to 12 orders of magnitude. In general the agreement with perturbative QCD calculations including next to leading order (NLO) corrections is well within the systematic uncertainties. This cannot be seen in Fig. 6a directly, only in ratio plots measurement/theory for a given η-interval. The systematic uncertainties in the ratios are typically only 30%, which is a great achievement compared to previous such measurements. The systematic uncertainties on the measurements are dominated by the jet energy scale uncertainty (calibration of the detectors for the energy of jets), which, thanks to a considerable effort, has been determined to typically better than 3% [6]. Another example of precise QCD results is displayed in Fig. 6b where the inclusive single isolated prompt photon cross-section [7] is compared to one of the most advanced QCD calculation.

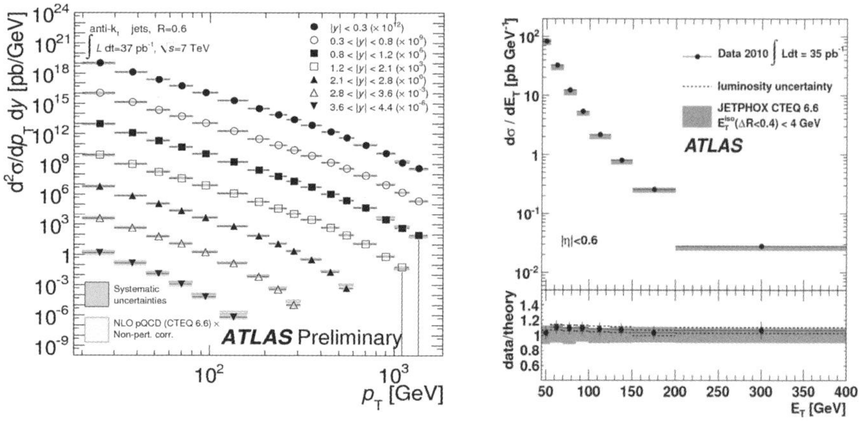

Figure 6. Inclusive jet (a, left) and single isolated prompt photon (b, right) cross-sections, compared to NLO perturbative QCD predictions.

Jets can also be produced together with W and Z bosons, so-called QCD corrections to the Intermediate Vector Boson production. First results of these processes have been published [8]. A good understanding of them is particularly important as they are, in many cases, a dominant source of backgrounds to the search for new particles, as well as to the measurements of top quark production discussed next.

The heaviest known particle in the SM is the top quark with its roughly 175 GeV mass. It decays almost exclusively into a W and a bottom quark. The measurement of top quark pair production typically requests that at least one of the W decays leptonically (also needed to trigger the events), and therefore the final states require one or two leptons (electrons or muons), ETmiss, and jets, some of which, coming from the b-quarks, can be tagged by the displaced secondary vertices due to the finite life times of b-hadrons. Whilst it is beyond the scope of this report to describe the sophisticated analyses employed, the message is that there are clear top pair signals in ATLAS [9], both in the single and two-lepton channels, when considering the correct jet topologies. The resulting cross-sections are shown in Fig. 7 which also illustrates the expected large rise of the cross-section with the collision energy increase from 2 TeV at the TeVatron to 7 TeV at the LHC. Good agreement with NLO QCD calculations is seen within the present 6% measurement errors. It can be mentioned that ATLAS has also reported first single top observations (events with just one top quark) at a rate in good agreement with QCD expectations.

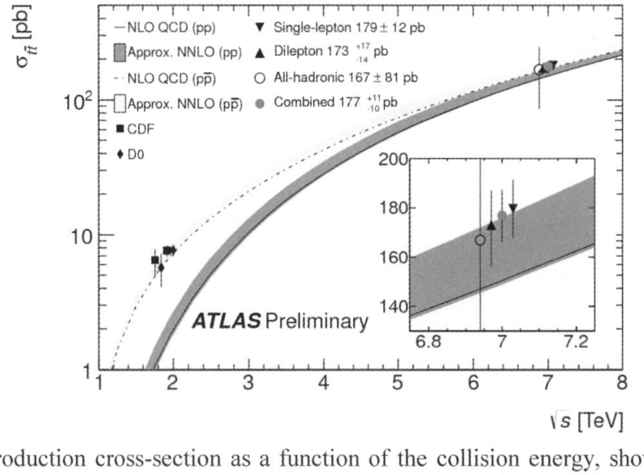

Figure 7. Top pair production cross-section as a function of the collision energy, showing the TeVatron and ATLAS LHC measurements.

A summary of many measurements of interesting Standard Model production cross-sections is given in Fig.8, which includes also a single top production measurement as well as the low cross-section processes of IVB pair productions, in some cases already with the most recent initial 8 TeV data from this year's collider run.

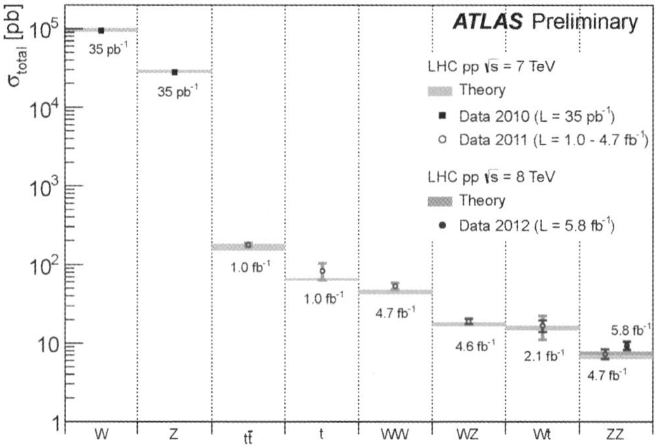

Figure 8. Summary of various Standard Model production cross-sections.

4. THE HUNT FOR THE HIGGS BOSON

The search for the Higgs boson H, as the decisive manifestation of the Brout, Englert, and Higgs mechanism for electro-weak symmetry breaking, postulated in 1964, was one of the major motivations for initiating the LHC project already more than 25 years ago. The ability to detect it unambiguously over the full possible mass range from its lower experimental limit (set finally at 114 GeV at the LEP collider) up to one TeV, with very different favoured final states (decay modes) at different masses, was the major benchmark in the conception of the ATLAS and CMS detector designs.

The most stringent limits at hadron colliders were set until a year ago, in spring 2011, by the combined Higgs search results from the TeVatron experiments CDF and D0, excluding at 95% confidence level (CL) the mass range 157 to 173 GeV. This was achieved by combining searches for an excess of events over the SM backgrounds in several Higgs decay channels, but dominated in this mass range by $H \rightarrow W^+W^-$ decays, with the Ws decaying in turn leptonically (electron, or muon plus neutrino channel). ATLAS and CMS have then presented their searches in many channels for the summer 2011 conferences, extending the H exclusion limits over a significantly larger mass range. Then first hints for a possible signal showed up at the end of 2011.

At the time of this School the public ATLAS Higgs search status [10] corresponded to the results presented at the March 2012 Moriond Conferences. Two examples are given in Fig. 9. The first (Fig. 9a) shows the relatively straightforward search for a mass peak from the process of the H decaying into two Z's (one might be virtual), which in turn decay into charged lepton pairs (electrons or muons in this figure). The second example (Fig. 9b) displays the search for the H decaying into WW, and each W decaying leptonically into an electron or muon and its associated neutrino. Because of the ETmiss from the neutrinos no mass peak can be reconstructed, only a broad enhancement in the transverse mass of the leptons and ETmiss can be expected. In both cases no significant excess is yet observed over the background distributions within the present data samples. The figures also illustrate the expected contributions from a Standard Model H boson.

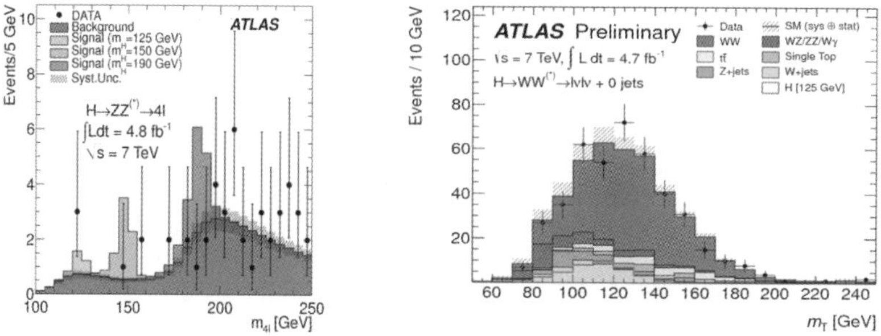

Figure 9. Examples of H search results, (a, left) in the four charged leptons channel and (b, right) in the two charged leptons plus two neutrinos channel (see text).

At this stage, the absence of any significant signal over the backgrounds in the analysis of many channels can be expressed in terms of 95% confidence level (CL) exclusion limits. For a graphical representation this is done in terms of a ratio between the limit cross-sections over the expected Standard Model H cross-sections, as shown in detail for several decay channels in Fig. 10a, and for their combination in Fig. 10b. The mass range for which this ratio is smaller than one is then excluded at the 95% CL. Combining all analysis channels, and taking into account also possible correlations, leads to exclude at 95% CL the SM H boson in the mass ranges 110-122.5 (except 117.5-118.5) and 129-539 GeV [10].

Note, however, that much progress in the H search can be expected on the basis of already accumulated new data, as well as the anticipated date for the rest of 2012. A definite statement about the existence or not of a SM H might well be in reach for the end of this year.

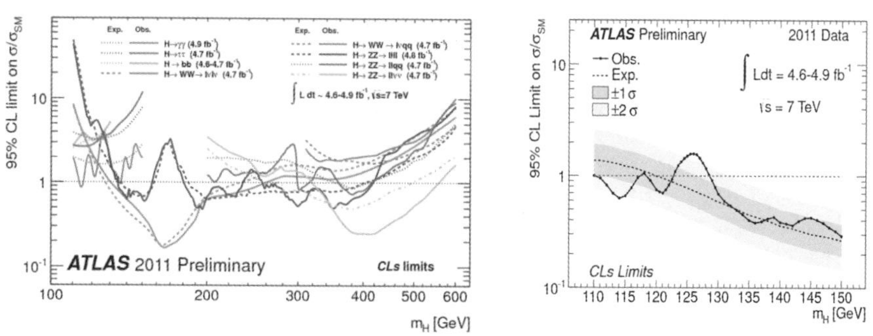

Figure 10. A summary of 95% CL limits (a, left), for various Standard Model Higgs search channels separately, as a function of the H mass (see text for explanations), and combined exclusion region around 126 GeV (b, right) where a hint of an excess (un-excluded region) is visible.

5. SEARCHES FOR PHYSICS BEYOND THE STANDARD MODEL

Besides the quest to elucidate the mechanism of the electro-weak symmetry breaking by searching for the Higgs boson, the major excitement for LHC comes from the great potential to explore uncharted territory of physics Beyond the Standard Model (BSM), thanks to its highest collision energy ever available in a laboratory so far. Since the beginning of the project, the search for Supersymmetry (SUSY) was a strong motivation, and besides the H boson it has been the other benchmark physics guiding the detector designs. However, many other hypothetical new processes can be searched for, and indeed ATLAS and CMS have already reported in many publications a very broad spectrum of searches for signatures of such processes (mass peaks for new particles or kinematical distributions with deviations from the expectations of known physics) for BSM physics. No such effect has yet been found, but all of these searches result in new exclusion limits, often well beyond the one TeV scale already. Only a few examples are shown in the following.

The most popular searches concern SUSY, which predicts additional fundamental particles. The search for SUSY is motivated in part by the prospect that the lightest stable neutral SUSY particle (LSP) could be an excellent candidate for explaining the Dark Matter (DM) in the Universe. The mysterious existence of DM was postulated by Zwicky, and rather convincingly evidenced by Rubin, both astronomers, in the 1930s and 1970s respectively. The SUSY searches at LHC are very complex as they must be sensitive to many (model-dependent) decay chains, implying a large variety of possible final state topologies. A common feature for most of them is the existence of significant missing transverse energy, ETmiss, due to the escaping LSPs (an experimental signature similar to that of the neutrinos in the W decays). Furthermore the SUSY signatures include high transverse momentum jets, some tagged as b-jets for third-generation squarks, and leptons. The expected topologies depend not only on the model parameters, but also on the mass relations between squarks and gluinos (the SUSY partners of the SM quarks and the gluons).

As an illustration of the typical data in the SUSY searches [11], Fig. 11 shows the effective mass (meff) distributions, defined as scalar sum of ETmiss and all jet transverse momenta, for a search for SUSY in events with at least two jets, ETmiss and no lepton. The distributions are shown in (a) for a background control region which is enhanced with QCD multi-jet events having only minimal contributions from a possible signal, and in (b) for a kinematical region where the signal would be strongest. In both distributions an expected signal for certain SUSY parameters is indicated by the dashed histogram, over the various SM backgrounds. The figure shows that the data are perfectly described at this stage with the background distributions only.

The data can then be used to extract 95% CL upper limits for the number of observable events and cross-sections independently of new physics models, in the background-only hypothesis. These cross-section limits in turn can be translated into lower limits for SUSY particle masses in SUSY models for a given set of model parameters.

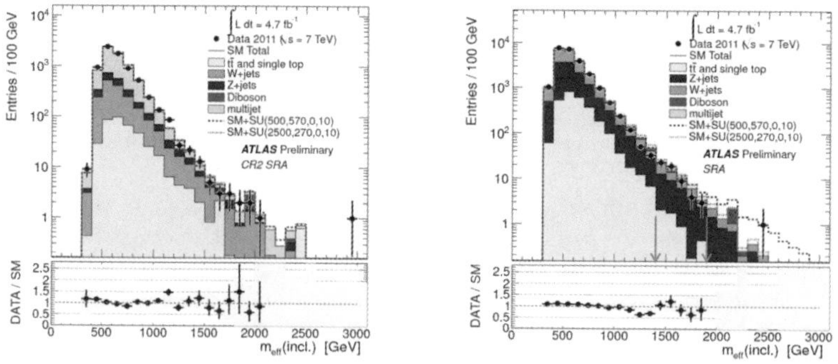

Figure 11. Typical meff distributions for (a, left) background enhanced and (b, right) signal enhanced SUSY search regions. A possible SUSY signal is indicated by the dashed histogramme (see text).

It is customary to represent the limits in a parameter plane correlating scalar (m_0) masses vs. gaugino ($m_{1/2}$) masses of SUSY representations [11]. The results for a simplified model are shown in Fig. 12 that assume only first- and second-generation squarks, gluinos, and massless neutralinos. Mass values smaller than indicated by the experimental limit are excluded by the data at 95% CL. It can be noted that the LHC results already extend significantly the search range from previous experiments. In this particular example, and assuming equal squark and gluino masses, their lower mass bound is 1.4 TeV at 95% CL. It is expected that the mass range will be quickly extended within more LHC running, and ultimately the LHC will be able to discover such particles up to about 2.5 TeV mass.

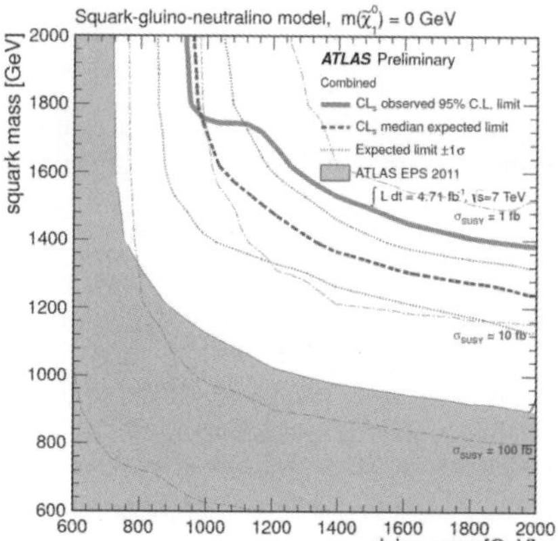

Figure 12. Example of an exclusion contour in a SUSY parameter plane of gluino mass vs. squark mass in the simplified SUSY model mentioned in the text. Lower mass values than indicated by the solid red line are excluded at 95% CL. Results from previous experiments are also indicated at lower masses.

Classical searches for BSM physics include the exploration of IVB-like signatures in the high mass range made kinematically accessible by the high collision energy of the LHC. The searches for hypothetical heavy W' and Z' particles is rather straightforward in the lepton-pair and lepton-neutrino decay channels, extending in mass coverage the SM IVB analyses mentioned before. Such heavy resonances arise in a variety of models mostly related to hypothetical new forces, either with the same properties as the SM IVBs, in that case called sequential (and labeled as SSM), or in other models as referenced in the ATLAS publications [12]. As examples of such searches, Fig. 13 shows the Z' and W' data distributions with the background expectations, which has also superimposed the expected signals for various IVB' masses. The absence of a signal in the data can be translated into 95% CL cross-section limits that are confronted with the expected theoretical cross-sections as a function of the mass of the new particles in order to establish lower mass limits. At this stage this typically leads to 95% CL lower limits for Z' and W' of 2.2 and 2.5 TeV, respectively. Ultimately, with the LHC running for several years at a collision energy of 14 TeV, the experiments will explore masses up to the order of 5 TeV for such particles.

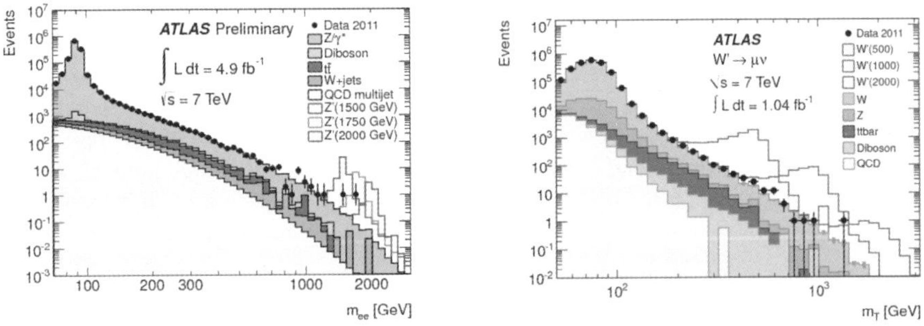

Figure 13. Examples of the di-electron mass distribution (a, left) and of the muon-neutrino transverse mass distribution (b, right) for the searches of heavy Z' and W' particles. The data are compatible with the SM backgrounds, the expected signals from hypothetical SSM signals are also indicated (see text).

Excited heavy quarks (q*), axigluons (massive gluons) or other objects are predicted in various models that would decay into a pair of partons, manifesting themselves as resonances in the final state di-jet mass distributions above the SM QCD spectrum. Searches for such di-jet resonances have been published in [13], and Fig. 14 shows an example of how well the data are actually described by the smooth QCD spectrum, allowing ATLAS to exclude at 95% CL typically q* with masses below 3.35 TeV.

Of great interest at this new energy frontier are also searches for a possible substructure of the quarks related to a new super-strong interaction. Such a substructure, called compositeness at a scale Λ, could manifest itself by deviations from QCD in the angular distribution of high-momentum transfer, large-angle scatterings, in analogy to the famous Rutherford scattering experiment some 100 years ago. No deviations have been observed so far [13], setting a 95% CL lower limits on Λ at 7.8 TeV, significantly higher than in previous experiments. Another possibility would be the formation of new bound states containing both quark and lepton properties that would decay into quarks (resulting in jets) and leptons, so-called lepto-quarks. New limits on their existence have been established in dedicated searches.

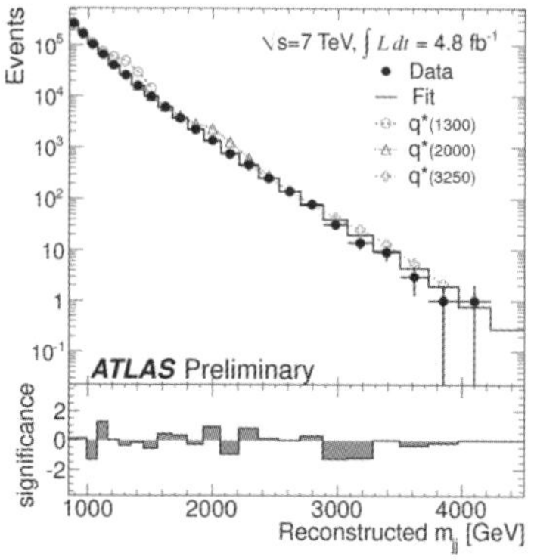

Figure 14. Example of a search for a resonance mass peak over the smooth QCD di-jet mass spectrum. Mass distributions for three hypothetical q* masses are indicated.

Many other searches aimed at exploring BSM physics have been conducted, all with finding no excess of event rates over the expected backgrounds from the SM. But in essentially all cases more stringent constraints could be established compared to previous experiments at lower collision energy. A non-exhaustive summary of such 95% CL limits, up-to-date until July 2012 including some initial 8 TeV results, is displayed in Fig. 15.

6. CONCLUSIONS

After only less than three years of high-energy operation of the LHC, ATLAS (as well as CMS, LHCb and ALICE) has already produced a very rich harvest of physics results. This should in the first place be seen as a success of the overall LHC project with its three major components having worked in an extraordinary coherent way so early on: the collider, the experiments and the novel computing grid structure wLCG.

ATLAS (as well as CMS) has already amply demonstrated that it masters the analyses of Standard Model physics, and has brought many SM measurements into domains of details and a degree of sophistication unprecedented at lower energy machines, setting a challenge for accurate theoretical descriptions. The early SM results also demonstrate that the experiment is well understood and operates at full potential to attack the main goals of LHC, namely to make the conclusive investigations of the electro-weak symmetry breaking mechanism, i.e. establish the existence, or otherwise, of the Higgs boson, and explore the physics Beyond the Standard Model. The searches for BSM physics are already exploring unchartered territory, thanks to the superb performance of the collider. One should remain aware, however, that the full physics potential will only become available in future years with the machine reaching its full design luminosity and energy. But there is no doubt: particle physics has exciting times ahead with the LHC project now operational.

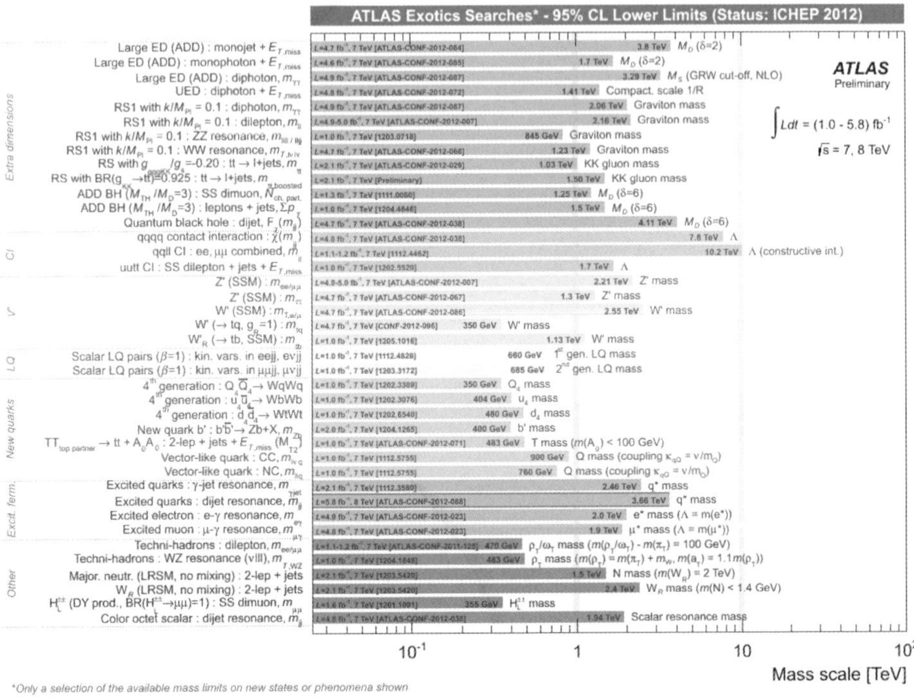

Figure 15. A non-exhaustive summary of 95% CL lower limits established so far on a variety of searches for BSM physics.

ACKNOWLEDGEMENT

I thank Professor Antonino Zichichi, Director of the Ettore Majorana Foundation and Centre for Scientific Culture, for having invited me to this stimulating school, and for the warm hospitality in the beautiful town of Erice.

REFERENCES

[1] The Large Hadron Collider: a Marvel of Technology, ed. L. Evans, EPFL Press, 2009, 182-199.
[2] ATLAS Collaboration, The ATLAS Experiment at the CERN Large Hadron Collider, JINST 3 (2008) S08003.
[3] ATLAS Collaboration, Charged-particle multiplicities in pp interactions measured with the ATLAS detector at the LHC, New J. Phys. 13 (2010) 053033.
[4] ATLAS Collaboration, Measurement of the inclusive W and Z cross-sections in the electron and muon decay channels in pp collisions at 7 TeV with the ATLAS detector, Phys. Rev. D85 (2012) 072004.
[5] ATLAS Collaboration, Measurement of inclusive jet and dijet production in pp collisions at 7 TeV using the ATLAS detector, Phys. Rev. D86 (2012) 014022.
[6] ATLAS Collaboration, Jet energy measurement with the ATLAS detector in proton-proton collisions at 7 TeV, arXiv:1112.6426[hep-ex], submitted to Eur. Phys. J. C.
[7] ATLAS Collaboration, Measurement of the inclusive isolated prompt photon cross-section in pp collisions at 7 TeV using 35 pb-1 of ATLAS data, Phys. Lett. B706 (2011) 150.
[8] ATLAS Collaboration, Measurement of the production cross-section for W-bosons in association with jets in pp collisions at 7 TeV with the ATLAS detector, Phys. Lett. B698 (2011) 325; ATLAS Collaboration, Measurement of the production cross-section for Z in association with jets in pp collisions at 7 TeV with the ATLAS detector, Phys. Rev. D85 (2012) 032009.
[9] ATLAS Collaboration, Measurement of the top quark-pair production cross-section with ATLAS in pp collisions at 7 TeV, Eur. Phys. J. C71 (2011) 1577; ATLAS Collaboration, Measurement of the top quark-pair production cross-section in pp collisions at 7 TeV in dilepton final states with ATLAS, Phys. Lett. B707 (2012) 469, and combination of all results in ATLAS-CONF-2012-024.
[10] ATLAS Collaboration, An update to the combined search for the Standard Model Higgs boson with the ATLAS detector at the LHC using up to 4.9 fb-1 of pp collisions data taken at 7 TeV, ATLAS-CONF-2012-019.
[11] ATLAS Collaboration, Search for squarks and gluinos using final states with jets and missing transverse momentum with the ATLAS detector at 7 TeV proton-proton collisions, ATLAS-CONF-2012-033, 037 and 041.
[12] ATLAS Collaboration, Search for high-mass dilepton resonances with 5 fb-1of pp collisions at 7 TeV with the ATLAS experiment, ATLAS-CONF-2012-007; ATLAS Collaboration, Search for a heavy gauge boson decaying to a charged lepton and a neutrino in 1 fb-1 of pp collisions at 7 TeV using the ATLAS detector, Phys. Lett. B705 (2011) 28.
[13] ATLAS Collaboration, Search for new phenomena in dijet mass and angular distributions using 4.8 fb-1 of pp collisions at 7 TeV collected by the ATLAS detector, ATLAS-CONF-2012-038.

CHAIRMEN: P. BLOCH, P. JENNI

Scientific Secretaries: T. Blake, K. Kornieieva, B.Vormwald, I. Yeletskikh

DISCUSSION

- Karakostas:

About ATLAS, since the design phase of ATLAS was a long time before ATLAS was installed in the cavern, how does the ATLAS management feel about their design choices?

- Jenni:

As you said, the detector was conceived in the early 90s, so a long time ago. As you can see from our results by large the ATLAS collaboration is pleased with the choices. I also think it was a good idea for CERN to have orthogonal magnetic field configurations for the General Purpose Detectors. In my view a toroidal configuration makes a lot of sense at a hadron collider because a lot of the physics is happening at large-p_T and of course a toroid gives you essentially a constant p_T-momentum resolution also at small angles (compared to a solenoid). So, we are very happy with these basic choices. We have a few things that we want to upgrade already in the early phase. First of all we will put an additional layer of pixels already next year. This will certainly improve and maintain the very satisfactory b-tagging we have now. We have seen this is much more important than we thought in the very beginning. So we will put a fourth layer to ensure that we can work in the phase from 2015-2022, when we will replace the full inner detector. We will also replace some of the very forward region for muons where the fluxes are highest, particularly in the inner most stations. We call this the small wheel, even though this wheel is 11m in diameter; it's not so small. We will put there a new technology to cope well with the rate we expect from 2018 onwards. So, to your question, I think by large we are satisfied. Of course we can improve by changes in the trigger electronics and things like that. This follows naturally the luminosity to come.

- Csanad:

I would like to ask to Prof. Bloch: You showed this nice figure of the pseudorapidity densities, compiled from the different experiments, the dN/dη. Basically, if you take into account the Bjoken's estimate, or other similar but advanced estimates, and take the multiplicity distribution you could estimate the initial energy density in proton-proton collisions and that may or may not be above 1 GeV/fm^3. So, are there any plans to exploit that?

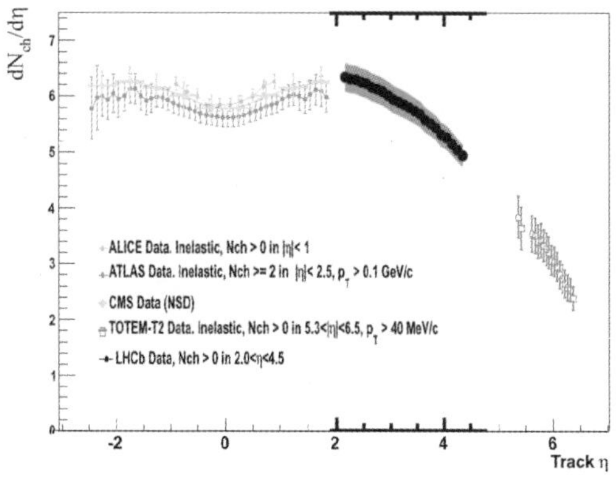

- *Bloch:*

So, I cannot really answer because I don't know. This is a very valid comment. This plot is pretty new and in fact so far the experiments have not yet collaborated to homogenize the measurements. These are currently independent measurements. As you know, we have organized some common working groups in the framework of the LHC Physics Center at CERN. Multiplicity and minimum bias are one of the working groups. I can assume in the future that the working group will try to collect these data to see if they fit together. There are a few percent differences between them. Even by eye you can see that. I look forward exploiting as much as possible this data.

- *Jenni:*

May I add something, of course what Philippe showed were separate measurements coming from ATLAS and CMS that were overlapping. What would be of course interesting would be to take single events where you can make full correlations over the rapidity space. That is of course not so easily achievable. There is maybe another thing to be said, for this type of physics, at the moment this is extremely difficult to do because of the (high) luminosity and the very large pile-up. For this type of investigation one may have to consider in the future, going back to a configuration of the machine with lower luminosity where you have only one event per crossing in order to make correlation studies.

- *Csanad:*

To Professor Jenni: In ATLAS, you have a quite good coverage in terms of angles. There was this suggestion in QCD that you might have structures similar to glue balls called QCD Buckey balls, which are made of Baryon junctions. They would decay to an equal number of baryons and anti-baryons, like 8 baryons and 8 anti-baryons. This would need extremely good identification possibilities. Is there a prospect for this in ATLAS?

- *Jenni:*

There is some rudimentary particle identification by dE/dx in particular in the pixel detector and also partially in the straw tracker part of the transition radiation detectors, but this is certainly not the optimal experiment for that. This particle identification I would rather see in an experiment like ALICE.

- *Bloch:*

Or LHCb in the restricted forward region, that's not so restricted, 2 to 4.5. There they have excellent particle identification. I don't think they have looked for that, but I know they have looked at multiple J/psi and this might be a good suggestion for them.

- *Apresyan:*

What are the reasons for LAr data quality efficiency being the lowest among all subdetectors, approximately 95%? Is there any relation to instantaneous luminosity?

ATLAS 2012 p-p run

Inner Tracker			Calorimeters		Muon Spectrometer				Magnets	
Pixel	SCT	TRT	LAr	Tile	MDT	RPC	CSC	TGC	Solenoid	Toroid
100	99.4	100	95.0	98.7	100	99.2	100	99.9	100	100

Luminosity weighted relative detector uptime and good quality data delivery during 2012 stable beams in pp collisions at $\sqrt{s}=8$ TeV between April 4th and May 31st (in %) – corresponding to 3.5 fb^{-1} of recorded data. The inefficiencies in the LAr calorimeter will partially be recovered in the future.

- *Jenni:*

There are 200000 LAr channels in the detector. The part of the channels which, are known to be dead, is small, in the order of 0.5% or less. Now, for some analysis the problem which we have is that in some modules we have short noise pulses or high voltage breakdown and then we have to rerun automatically the high voltage up again. This turns out to be not really very strongly luminosity dependent. So for analysis, where you don't really require all the channels to be alive, the efficiency number is higher, and will be further improved by re-processing the data. It's a kind of a global figure.

- *Apresyan:*

In a hypothetical scenario that all what LHC finds by ~1000 fb^{-1} is a SM-like Higgs boson, what would the CERN strategy be? Continue with HI-LHC or HL-LHC, or maybe invest in a Higgs factory?

- *Bloch:*

What you are describing is a bit of a nightmare. It is true that, if the only thing we find is the Higgs that completes the standard model and we don't have any sign of anything else, it is really an issue to define the future of our field. This is the reason why CERN has several projects open with R&D for different possibilities. This morning I only presented ideas about LHC and, clearly, if you wouldn't find anything, probably one of the key questions would be to see if we are able to increase energy, to see if the threshold for new physics was a bit higher. That's why many people believe in the high energy LHC. But we are keeping also open projects like e^+e^--collider which, in my opinion, would make sense either as a Higgs factory or if we find a sign of TeV-particles, and if we're not, we also try to see whether we should embark in the neutrino physics, which depends also on the situation in the United States where the neutrino physics is an important project. So we are in a quite puzzling situation: if anything we find is the Higgs, it will be not easy to decide in which direction we should aim.

- *Jenni:*

Well, first of all we don't make a decision on a time scale of the European strategy update which is in the first half of next year. What is absolutely top priority is to get the LHC running at its design energy. We should not only talk about the Higgs, all the discovery potential of cause will be largely dominated by what we can do when we finally reach 13-14 TeV. Then, there will be an upgrade of LHC in terms of luminosity, which will bring us from few 100/fb to 3000/fb, which are needed in any of the scenario if there is a Higgs or if there is no Higgs. If there is a Higgs, of course, one needs to study the couplings. One needs a large number of events of Higgs, produced together with another boson, e.g. Z and then look in an unbiased way at the Higgs decays. Now in terms of the strategy: People are making as realistic as possible simulations of precisions which one could get for the different Higgs couplings with the luminosity upgraded LHC. This is to be compared with the possibilities of a dedicated e^+e^--Higgs-factory, a linear collider of ILC type or CLIC. But the guidance will not only come from the Higgs, it will come also from what LHC will find in the coming years at 14 TeV in other domains of the physics.

- *Apresyan:*

The plot on page 49 shows comparison of measured cross-sections of various SM electroweak processes with prediction. Starting from ttbar, WW and ZZ, there is disagreement between data and prediction. Is this understood?

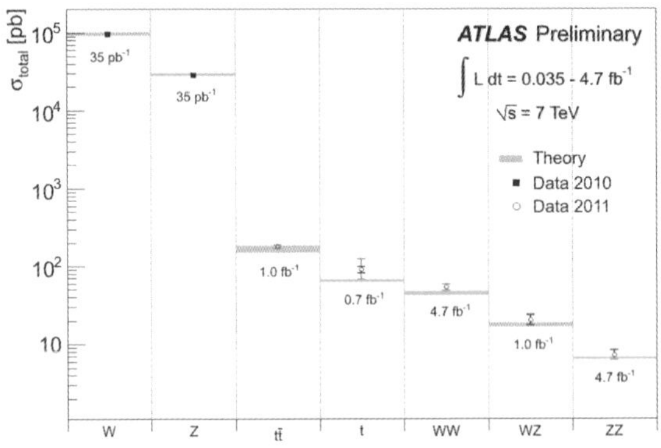

- *Jenni:*

I actually don't think there was a systematical disagreement. I showed in some cases more details. The cross section measurements, of course, have one common ingredient, which is determination of the luminosity. Big effort has been made to achieve a precision of the order of 4% on the luminosity normalization. That could be an error, which systematically affects all these measurements. But otherwise, I think, we conclude at this stage that there is a very good agreement with the standard model prediction.

- *Inguglia:*

The CMS Collaboration is using a tool for analysis of supersymmetric signatures, which is called RAZOR. This allows to analyze events with two missing particles and to measure their mass. Is the ATLAS collaboration thinking about using this, too?

- *Jenni:*

ATLAS is actually using it. We compare the results with other methods. There is in some cases some gain, in some cases not. But indeed, what CMS calls the RAZOR approach is been used. A similar method has actually been proposed by Lisa Randall. In the literature you can find some more details about both approaches (arXiv:1006.2727, arXiv:0806.1049).

- *Inguglia:*

Could the RAZOR approach also be used for looking for the Higgs boson in some channels like H→WW?

- *Jenni:*

I have never thought about that. Maybe some people have done this already, but I am not aware of.

- *Reece:*

In the RAZOR paper from 2010 there is a section for the process H→WW. It is shown, that it works.

- *Gazis:*

Thank you both for your nice comprehensive and essential talks given this morning. I keep the part of Prof. Bloch's talk among other interesting subsets, where the future plans of LHC presented, encouraging the young scientists for the great future for the particle physics by the planning of the high energy LHC operation. My question is if the realization of the HE-LHC program with 13TeV per beam can be made in a coherent mode in parallel with the CLIC project?

- *Bloch:*

I think we will have to make a choice at some moment, I don't think we can do everything at CERN. We are going to do more and more global projects. I believe that, at some moment, when physics will tell us which way to go, we will have to take a decision in Europe which way to go. The cooperation is very important with our colleagues in the United States and elsewhere, because we cannot do everything at CERN. I mean doing two machines at the same time.

- *Jenni:*

Both these projects are not in a stage that one could decide. We need to continue R&D on the CLIC technology in order to have the option in 2018/2019 to go that direction if the physics really demands that. This is even more valid for high-field magnets in order to reach e.g. 33 TeV collision energy in hardron-hardron collisions. Both cases need to be further developed. Nobody would ever think to have both these options at CERN or in Europe.

- *Bloch:*

Some people at CERN are studying, whether we could set a 2-2.5 times larger tunnel which would allow to go either to higher energy or to lower field, because the cost of these 20T high-temperature superconductors are one order of magnitude higher than those we are using for LHC now.

- *Jenni:*

Prof. Zichichi made comments that by far it's not guaranteed that such projects can be financed – we must have very serious arguments from physics and defend in a coherent way our plans.

- *Zoller:*

The systematic errors on the top mass determination are rather large at LHC. What are the chances to reduce these errors and get more precise values for it?

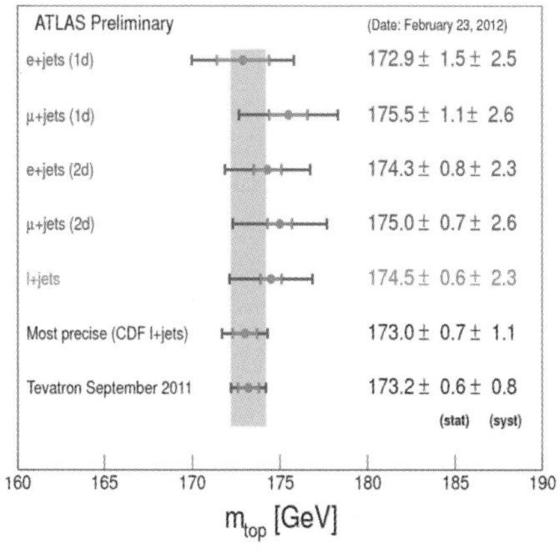

- *Jenni:*

Within reasonable time the systematic errors will go down to the level of what Tevatron has achieved, but to go much beyond that will be extremely difficult. For the precision tests you would like to measure for example the threshold of ttbar production. This can hardly be achieved by a hadron collider. Although we have been positively surprised in many cases, it is not obvious at the moment how to get to the precision of an e^+e^- collider for this particular measurement.

- *Zoller:*

What can future programs, like LHeC, do for precision physics?

- *Bloch:*

I don't know the answer. For further details you can check the LHeC CDR (arXiv:1206.2913).

- *Zoller:*

What is the quantity on the y-axis shown in the usual Higgs exclusion plot from LHC? How do you compute it and how do you extract it from data?

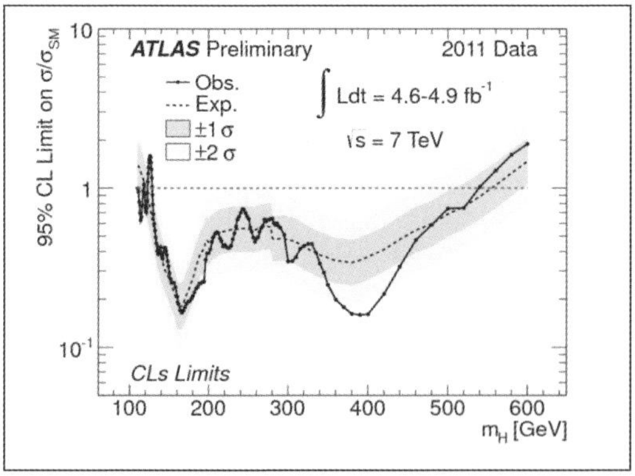

- *Jenni:*

For each channel, you scan over a whole masse range of different Higgs mass assumptions and you establish from the data a 95% confidence level limit on possible background fluctuation up to that. Then you divide that by the standard model expectation (using some defined set of structure functions. All channels are then combined with their statistical weight. This is the black curve.
What you also see in this plot is a smooth dashed curve, which is the expected exclusion in the absence of a Higgs signal. This is calculated by making many Monte Carlo simulations for each Higgs mass value, such that the data fluctuations of a single experiment disappear. The bands show the 1σ or 2σ fluctuations on this expectation.

- *Leontsinis:*

Is it planned to extent the ATLAS inner detector to $|\eta| = 2.7$ in order to do vertexing with muons of front region?

- *Jenni:*

There is no plan for the extension of inner detector coverage. It will stay at 2.5.

- Leontsinis:

How are your feelings and expectations to build the New Small Wheel with two new technologies (sTGC, MicroMegas) in such short time (until 2018)?

- Jenni:

Well, I would call it a "one-and-a-half" new technology. One of these new technologies are MicroMegas chambers which are really a new technology and they have a very strict schedule with milestones that are needed to be fulfilled.The half new technology is in a way the extrapolation of Thin Gap Chambers which exist already, but which will be more sophisticated, faster, etc. To have these ready for installation in 2018 there is a time scale and there are milestones. Both the ATLAS collaboration and the LHC-committee monitor this very closely.

- Bloch:

At CERN we have a collaboration (RD51) which studies detectors like MicroMegas. I think we are not starting from scratch today, for 2018. CERN has bought actually the equipments to build chambers of the size needed for both ATLAS and CMS for the upgrades. I know they have produced already chambers which are not far from what is needed in terms of dimensions. This upgrade is therefore not jumping into unknown – this is a small extrapolation of something that is already quite mature.So, we could be optimistic about the fact that this will be ready by 2018.

- Ieletskikh:

For the background estimate in the Higgs→gamma+gamma search the smooth function fit is used instead of the Monte Carlo simulation. What is the systematic uncertainty of this fit and how could it affected potentially observed signal significance?

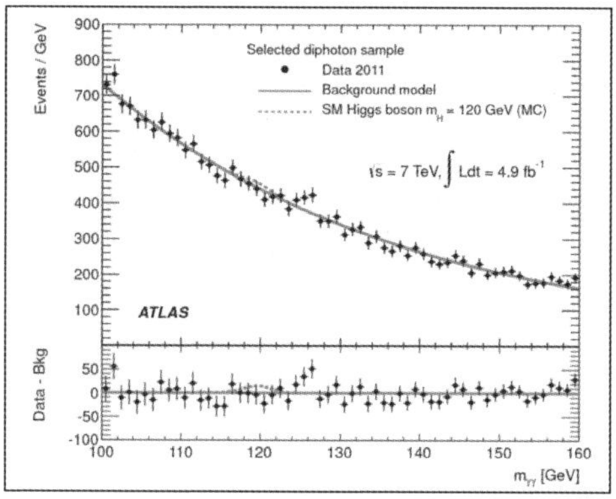

- Jenni:

This is a smooth function to the background that I showed, that's true. It's not a Monte-Carlo. Of cause, people have tried many different functions and looked what are the possible variations you can have on this smooth function. So, mathematically, I think, that is taking care of precisions and that's all I can tell you. As an experimentalist I am actually much more confident that such a fit like in the Higgs→gamma-gamma, where you look for a peak, over a nice smooth background behavior, that you use the data and not the Monte-Carlo to describe the background. I think this is the best thing we could do, because the data is always more reliable than the Monte-Carlo.

ORIGIN AND STATUS OF LUNA AT GRAN SASSO

Carlo Broggini
INFN – Padua, LUNA Collaboration

Abstract

The ultimate goal of nuclear astrophysic is to provide a comprehensive picture of the nuclear reactions which power the stars and, in doing so, synthesize the chemical elements. Deep underground in the Gran Sasso Laboratory the key reactions of the proton-proton chain and of the carbon-nitrogen-oxygen cycle have been studied down to the energies of astrophysical interest. The main results obtained in the past 20 years are reviewed and their influence on our understanding of the properties of the neutrino, the Sun, and the Universe itself is discussed. Finally, the evolution of underground nuclear astrophysics towards the study of helium burning is outlined.

Introduction

Only hydrogen, helium and lithium are synthesized in the first minutes after the big-bang. The other elements of the periodic table are produced in the thermonuclear reactions taking place inside the stars. Nuclear astrophysics studies these reactions which provide also the power that allows the stars to shine over their lifetimes. In particular, the knowledge of the reaction cross-section at the stellar energies is the heart of nuclear astrophysics.

The reaction rate in the hot plasma of a star, with temperatures in the range of tens to hundreds of millions degrees, is obtained by weighting the reaction cross section σ(E) with the energy distribution of the colliding nuclei: a Maxwell-Boltzmann Φ(E) peaked at energies of 1-10 keV. The product between Φ(E) and σ(E) identifies the energy window where the reaction occurs in the star: the Gamow peak. At lower energies the cross section is too small, whereas at higher energies the nuclei in the tail of the Maxwell-Boltzmann are too few.

Inside the Gamow peak, which is far below the Coulomb energy arising from the repulsion between nuclei, the reaction can take place only due to the quantum mechanical tunnel effect and, as a consequence, the reaction cross section σ(E) drops nearly exponentially with decreasing energy:

$$\sigma(E) = \frac{S(E)}{E} e^{-2\pi\eta(E)}$$

where S(E) is the astrophysical factor (which contains the nuclear physics information) and η(E) is the Sommerfeld parameter, given by $2\pi\eta = 31.29 \cdot Z_1 \cdot Z_2 \cdot (\mu/E)^{0.5}$. Z_1 and Z_2 are the charges of the interacting nuclei, μ is the reduced mass (in units of amu), and E is the center of mass energy (in units of keV). At low energies the cross sections are extremely small, because of the small probability to go through the Coulomb barrier. Such smallness makes the star life-time of the length we observe, but it also makes impossible the direct measurement in the laboratory.

The rate of the reactions, characterized by a typical energy release of a few MeV, is too low, down to a few events per year, in order to stand out from the laboratory background. Instead, the observed energy dependence of the cross-section at high energies is extrapolated to the low energy region, leading to substantial uncertainties. LUNA, Laboratory for Underground Nuclear Astrophysics, started twenty years ago to run nuclear physics experiments in an extremely low-background environment, the Gran Sasso Laboratory (LNGS), to reproduce in the laboratory what Nature makes inside the stars. As a matter of fact, the dolomite rock of Gran Sasso provides a natural shielding equivalent to at least 3800 meters of water which reduces the muon and neutron fluxes by a factor 10^6 and 10^3, respectively.

LUNA at Gran Sasso

Two electrostatic accelerators able to deliver hydrogen or helium beam have been installed underground: first a compact 50 kV "home made" machine [1] and then a commercial 400 kV one [2]. Common features of the two accelerators are the high beam current, the long term stability and the precise beam energy determination. The first feature maximizes the reaction rate, the second is demanded by the long time typically needed for a cross section measurement, and the third is important because of the exponential energy dependency of the cross section.
In particular, the 400 kV accelerator is embedded in a tank, a cylinder of 0.9 m diameter and 2.8 m long, filled with an insulating mixture of N_2/CO_2 gas at 20 bar. The high voltage is generated by an inline Cockcroft-Walton power supply located inside the tank. The radio frequency ion source directly mounted on the accelerator tube can provide beams of 1 mA hydrogen and 500 μA He^+ over a continuous operating time of 40 days. The ions can be sent into one of two different, parallel beam lines (Fig.1), allowing the installation of two different target setups. In the energy range between 150 and 400 keV, the accelerator can provide up to 500 μA of hydrogen and 250 μA of helium at the target stations, with 0.3 keV accuracy on the beam energy, 100 eV energy spread, and 5 eV h^{-1} long-term stability.

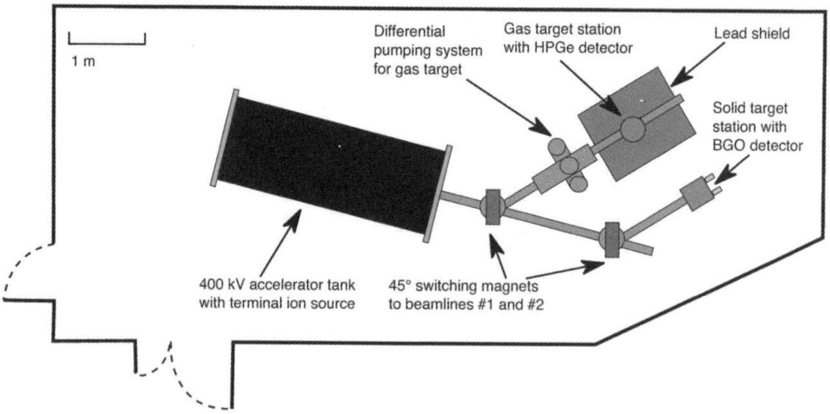

Figure 1: The LUNA set-up with the 400 kV accelerator.

Hydrogen Burning in the Sun

As the Sun mass contracted from an initially large gas cloud, half of the gravitational energy released was radiated into the space and half was converted into the kinetic energy of hydrogen and helium nuclei, thereby increasing the temperature of the system. At the central temperature of approximately 10 million degrees the kinetic energy of the hydrogen nuclei was high enough to penetrate the Coulomb barrier with significant probability and to switch on the hydrogen burning: $4^1H \rightarrow {}^4He + 2e^+ + 2\nu_e$.
Since the mass of the helium nucleus is lower than four times the proton mass, approximately 0.7% of the hydrogen rest mass is converted into energy in each of the transmutations. Hydrogen fusion supplies the necessary energy to halt the contraction, it provides all the energy required for the long life of the Sun, and it produces neutrinos detectable on Earth.
The Sun is a middle-aged main-sequence star which began to burn hydrogen approximately 4.5 billion years ago. In another 5 billion years it will begin burning helium and will eventually turn into a celestial body consisting mainly of carbon and oxygen. In the central region of the Sun, at the temperature of 15 million degrees and at the density of 150 g cm^{-3}, hydrogen burning does not take place in one step but rather

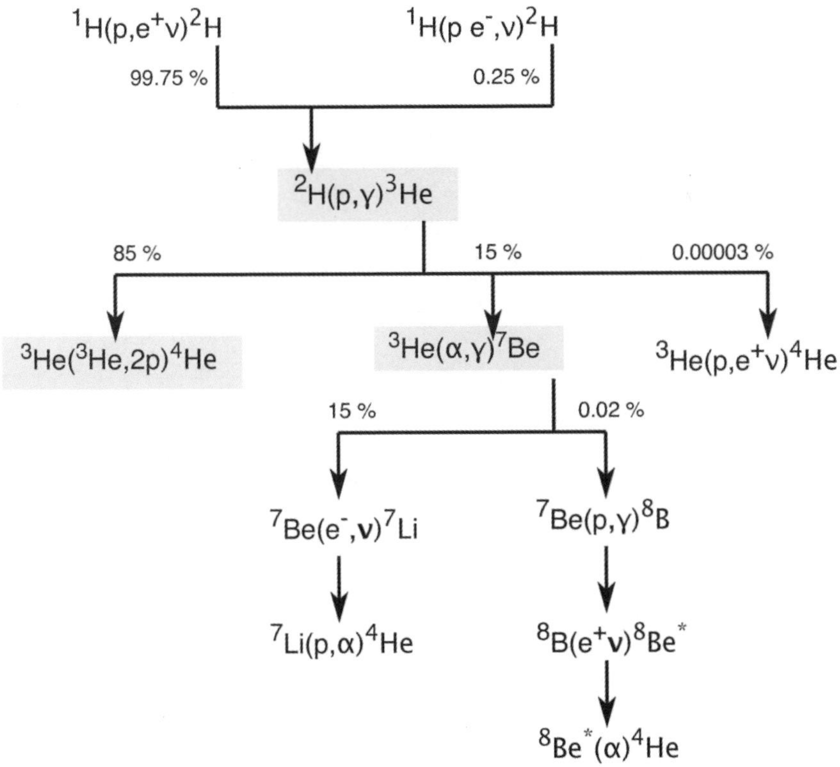

Figure 2: The proton-proton chain (pp) with the different branching ratios. The reactions studied by LUNA are highlighted.

proceeds through a sequence of two-body reactions: the proton-proton chain (Fig.2) and the CNO cycle. The relative importance of the different reactions is determined by the abundance of the nuclear species fusing together and by their fusion cross section at the Gamow peak energy. In particular, the pp chain is responsible for more than 99% of the luminosity of the Sun, whereas the CNO cycle accounts for less than 1% of nuclear energy production.
In 1964 both J.N. Bahcall and R. Davis proposed the detection of solar neutrinos in order to directly verify the hypothesis of nuclear-energy generation in stars and to study the inner region of the Sun.

The Solar Neutrino Problem and the Narrow Resonance

The initial activity of LUNA has been focused on the ^{3}He(^{3}He,2p)^{4}He cross section measurement within the solar Gamow peak (15-27 keV). Such a reaction has a Q-value of 12.86 MeV and it is a key one of the pp chain. As a matter of fact, it is the most probable fate of ^{3}He in the Sun (85% of the fusions).
The hypothesis of a narrow (~ keV) resonance in the cross section of ^{3}He(^{3}He,2p)^{4}He at low energy was advanced at the beginning of the seventies as a solution to the solar neutrino problem. Such a resonance would have explained the observed ^{8}B solar neutrino flux, about a factor 3 lower than the predicted one. Fewer than 1 neutrino in 10000 comes from the ^{8}B decay. However, their relatively high end-point energy (about 15 MeV) makes their detection much less difficult; unsurprisingly, ^{8}B neutrinos are the best studied neutrinos from the Sun. The resonance would have further enhanced the relative contribution of ^{3}He(^{3}He,2p)^{4}He, and the flux of the ^{8}B neutrinos, generated in the branch starting with ^{3}He(α,γ)^{7}Be, would have been suppressed.
The final set-up was made of eight 1 mm thick silicon detectors of 5x5 cm^2 area placed around the beam inside the windowless target chamber filled with ^{3}He at the pressure of 0.5 mbar. The simultaneous detection of two protons has been the signature which unambiguously identified a ^{3}He(^{3}He,2p)^{4}He fusion reaction.
Figure 3 shows the results from LUNA [3] together with the ones from a recent higher energy measurements. For the first time a nuclear reaction has been measured in the laboratory at the energy occurring in a star. Its cross section varies by more than two orders of magnitude in the measured energy range. In particular, at the lowest energy of 16.5 keV it has the value of 0.02 pbarn, which corresponds to a rate of about 2 events/month, rather low even for the "silent" experiments of underground physics.
No narrow resonance has been found within the solar Gamow peak and the astrophysical solution to the solar neutrino problem based on its existence has been definitely ruled out. We now know that the solution to the problem lies in the nature of the neutrino itself: it oscillates. Produced inside the Sun as an electron neutrino it may become a muon or tau neutrino by the time it reaches the Earth.
^{3}He(α,γ)^{7}Be (Q-value: 1.586 MeV), the competing reaction for ^{3}He burning, has been also studied in LUNA. The cross section of this reaction can be obtained by using two different experimental approaches: the measurement of the prompt γ rays and the detection of the delayed γ rays from ^{7}Be decay. Both techniques have been simultaneously employed underground, obtaining results in excellent agreement with each other. Thanks to the LUNA measurement the uncertainty on the ^{7}Be solar neutrino

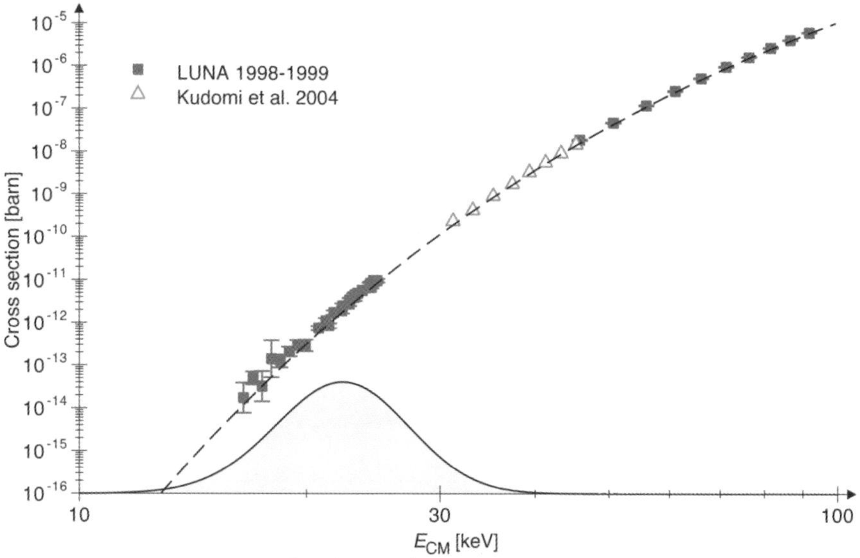

Figure 3: The cross section of $^3He(^3He,2p)^4He$ as function of the centre of mass energy.

flux went from 9.4% down to 5.5% [4,5]. The 0.86 MeV ^7Be neutrinos are the second largest component of the spectrum, amounting to 7% of the total and they have been very recently measured by Borexino with 5% error. The dominant component of the total neutrino flux (92%) are the neutrinos coming from $^1H(p,e^+\nu)^2H$. However, their continuous spectrum has an end-point energy of 0.42 MeV, which makes their detection extremely difficult and only possible, till now, in the radiochemical experiments with Gallium (Gallex/GNO and Sage).

The Metallicity of the Sun and the Age of the Universe

In the CNO cycle the conversion of hydrogen into helium is achieved with the aid of carbon and nitrogen previously synthesized in older stars (Fig.4). Carbon and nitrogen work as catalysts and they are not destroyed. $^{14}N(p,\gamma)^{15}O$ (Q-value: 7.297 MeV) is the bottleneck reaction of the cycle and it rules its energy production rate. In particular, it is the key reaction to know the ^{13}N and ^{15}O solar neutrino flux, which depends almost linearly on its cross section. The CNO neutrinos have end-point energies of 1.20 MeV (^{13}N) and 1.73 MeV (^{15}O) and they are a fraction of the total neutrino flux equal to the relative contribution of the CNO cycle to the luminosity of the Sun (less than 1%).
In the first phase of the LUNA study, data have been obtained down to 119 keV energy with solid targets of TiN and a germanium detector. This way, the five different radiative capture transitions which contribute to the cross section of $^{14}N(p,\gamma)^{15}O$ at low energy were measured. The total cross section was then studied down to very low energy in the second phase of the experiment by using the 4π BGO summing detector placed around a windowless gas target filled with nitrogen at 1 mbar pressure.

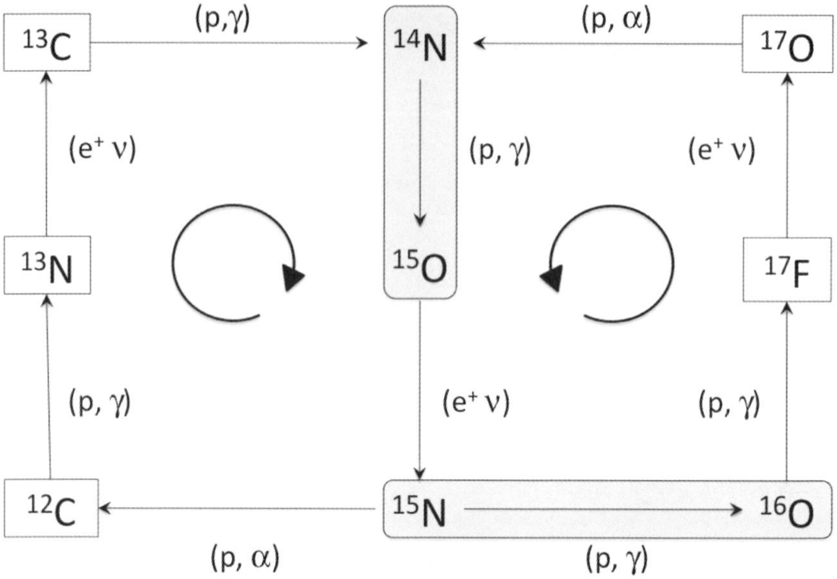

Figure 4: The first and second CNO cycles. Only the first cycle is important for the Sun because $^{15}N(p,\gamma)^{16}O$ has a cross section which is a factor 2000 lower than $^{15}N(p,\alpha)^{12}C$. The reactions studied by LUNA are highlighted.

At the lowest center of mass energy of 70 keV a cross section of 0.24 pbarn was measured, with an event rate of 11 counts/day from the reaction. The results obtained both with the germanium detector [6] and with the BGO set-up [7] were about a factor two lower than the existing extrapolation [8] from previous data at very low energy (Fig.4). As a consequence, the CNO neutrino yield in the Sun is decreased by about a factor two.

In order to provide more precise data for the ground state capture, the most difficult one to be measured because of the summing problem, we performed a third phase of the $^{14}N(p,\gamma)^{15}O$ study with a composite germanium detector. This way the total error on the S-factor has been reduced to 8%: $S_{1,14}(0)=1.57\pm0.13$ keV barn [9]. This is significant because, finally solved the solar neutrino problem, we are now facing the solar composition problem: the conflict between helioseismology and the new metal abundances (i.e. the amount of elements different from hydrogen and helium) that emerged from improved modeling of the photosphere [10].

Thanks to the relatively small error on $S_{1,14}$, it will be possible in the near future to measure the carbon and nitrogen content of the Sun core by comparing the predicted CNO neutrino flux with the measured one. As a matter of fact, the CNO neutrino flux, which almost linearly depends on the carbon and nitrogen content of the solar core where fusion reactions take place, is decreased by about 30% in going from the high to the new low metallicity scenario. This way it will be possible to test whether the early Sun was chemically homogeneous [11], a key assumption of the standard solar model.

$^{14}N(p,\gamma)^{15}O$ plays also a key role in determining the age of the globular clusters, which are the oldest stellar populations of a galaxy. The age is obtained from the luminosity of the turn off point in the Hertzsprung-Russell diagram of the cluster, which in turn depends on the $^{14}N(p,\gamma)^{15}O$ cross section (a star at the turn off is burning hydrogen through the CNO cycle). The age of the clusters gives then a lower limit to the age of the Universe. In particular, this limit is increased by 0.7-1 billion years [12] up to 14 billion year as a consequence of the factor 2 reduction measured by LUNA on the $^{14}N(p,\gamma)^{15}O$ cross section.

The lower cross section also affects stars that are much more evolved than the Sun. In particular, the dredge up of carbon to the surface of asymptotic giant branch stars is much more efficient [13].

$^{14}N(p,\gamma)^{15}O$ is an excellent example to explain why physicists have been studying for years these reactions which have such extremely small cross sections. From the nuclear point of view they are often of modest interest. It is the application to astrophysics, and to so many different astrophysical scenarios, which makes them so fascinating.

Hydrogen Burning at High Temperature

The solar phase of LUNA has reached the end. A new and rich program of nuclear astrophysics mainly devoted to CNO, Mg-Al and Ne-Na cycles has already started with the measurement of $^{15}N(p,\gamma)^{16}O$ [14] and $^{25}Mg(p,\gamma)^{26}Al$ [15]. Due to the higher Coulomb barrier of the reactions involved, these cycles become important at temperature higher than the one of our Sun: hydrogen burning in the shell of massive stars and Novae explosions. Relatively unimportant for energy generation, these cycles are essential for

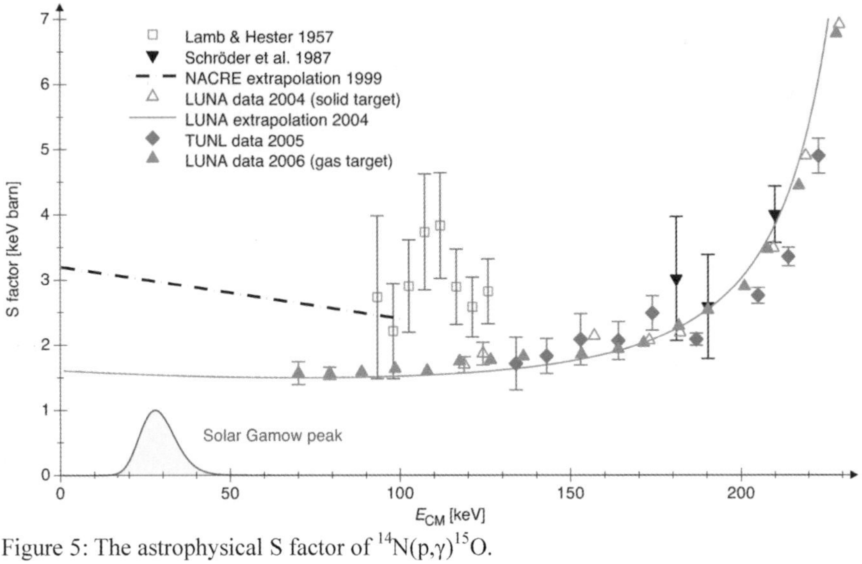

Figure 5: The astrophysical S factor of $^{14}N(p,\gamma)^{15}O$.

the synthesis of the different isotopes. In particular, LUNA is now measuring $^{17}O(p,)^{18}F$, the bridge reaction connecting the second to the third CNO cycle, and $^{2}H(\alpha,\gamma)^{6}$ Li, the key reaction of big-bang nucleosynthesis which determines the amount of primordial ^{6}Li in the Universe.

Helium Burning

LUNA has shown the advantages of the low background environment for the study of the hydrogen burning processes at the stellar energies. Natural evolution is the exploitation of the underground environment to study the next step in the fusion chain towards ^{56}Fe (the element with the highest binding energy per nucleon): the helium burning.

In particular, $^{12}C((\alpha,\gamma)^{16}O$ is the most important reaction of nuclear astrophysics and it determines the abundance ratio between carbon and oxygen, the two key elements to the development of life. This abundance ratio shapes the nucleosynthesis in massive stars up to the iron peak and the properties of supernovae. Of great significance are also $^{13}C(\alpha,n)^{16}O$ and $^{22}Ne(\alpha,n)^{25}Mg$, the stellar sources of the neutrons which synthesize most of the trans-iron elements through the S-process: neutron captures followed by β decays.

After several years of studies and discussions, this ambitious program is now becoming real: at the end of June 2012 a 3.5 MV accelerator has been financed by the Italian Authorities. Figure 6 shows the place where the new LUNA accelerator will be installed to study the key processes of stellar helium burning below the Gran Sasso peak.

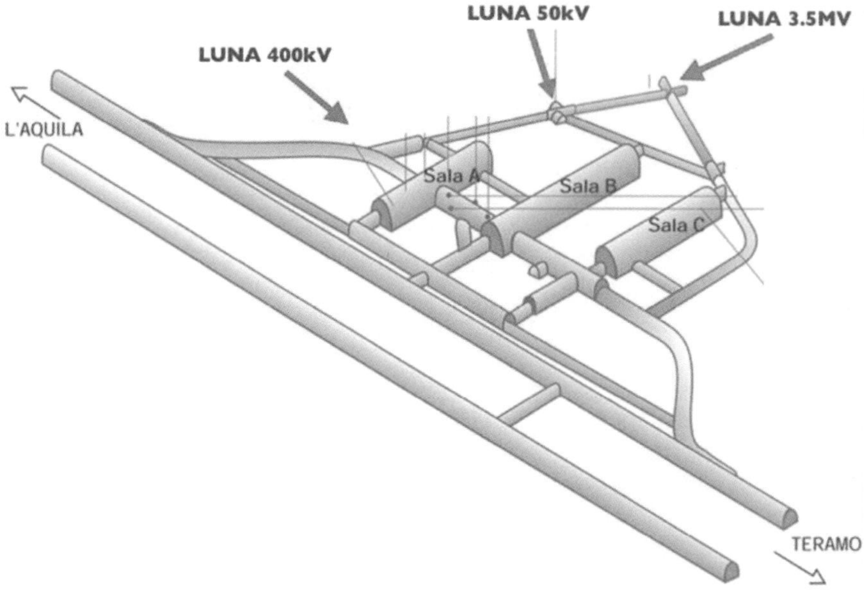

Figure 6: The map of the Gran Sasso underground laboratory with the position of the past, present and future LUNA accelerators.

Conclusions

Twenty years ago, LUNA started underground nuclear astrophysics in the core of Gran Sasso and it still remains the only facility of this kind in the world. The extremely low background has allowed experiments with count rates as low as a few events per year. For the first time, the important reactions which are responsible for the hydrogen burning in the Sun have been studied down to the relevant stellar energies. As a consequence, 50 years after the first pioneering cross section measurements, nuclear physics is no longer the dominant error source of the solar model. In particular, solar neutrinos can now be exploited to probe the deep interior of the Sun. When applied to the study of globular clusters, our results also increase the limit on the age of the Universe up to 14 billion years. Since a few years LUNA is not focused anymore on the Sun and it is studying the hydrogen burning reactions which are responsible for the abundance of the light elements up to aluminium. The study of stellar helium burning with the new 3.5 MV accelerator will be the key topic of LUNA for the next two decades.

References

[1] Greife, U., et al. (1994). Laboratory for Underground Nuclear Astrophysics (LUNA). *Nucl. Instrum. Methods A 350.* 327-337.
[2] Formicola, A., et al. (2003). The LUNA II 400 kV accelerator. *Nucl. Instrum. Methods A 507.* 609-616.
[3] Bonetti, R., et al. (1999). First measurement of the $^3He(^3He,2p)^4He$ cross section down to the lower edge of the solar Gamow peak. *Phys. Rev. Lett. 82.* 5205-5208.
[4] Bemmerer, D., et al. (2006). Activation Measurement of the $^3He(\alpha,\gamma)^7Be$ Cross Section at Low Energy. *Phys. Rev. Lett. 97* . 122502-1-5.
[5] Confortola, F., et al. (2007). Astrophysical S factor of the $^3He(\alpha,\gamma)^7Be$ reaction measured at low energy via detection of prompt and delayed γ rays. *Phys. Rev. C 75.* 065803-1-4.
[6] Formicola, A., et al. (2004). Astrophysical S-factor of $^{14}N(p,\gamma)^{15}O$. *Phys. Lett. B 591.* 61-68.
[7] Lemut, A., et al. (2006). First measurement of the $^{14}N(p,\gamma)^{15}O$ cross section down to 70 keV. *Phys. Lett. B 634.* 483-487.
[8] Angulo, C., et al. (1999). A compilation of charged-particle induced thermonuclear reaction rates. *Nucl. Phys. A* 656. 3-183.
[9] Marta, M., et al. (2008). Precision study of ground state capture in the $^{14}N(p,\gamma)^{15}O$ reaction. *Phys. Rev. C 78.* 022802(R)-1-4.
[10] Serenelli, A. M. (2009). New solar composition: the problem with solar models revisited. *Astrophys. J. Lett. 705.* L123-L127.

[11] Haxton , W. C., & Serenelli , A. M. (2008). CN cycle solar neutrinos and the Sun's primordial core metallicity. *Astrophys. J. 687*. 678-691.
[12] Imbriani, G. et al. (2004). The bottleneck of CNO burning and the age of Globular Clusters. *Astron. & Astrophys. 420.* 625-629.
[13] Herwig, F., & Austin, S. M. (2004). Nuclear reaction rates and carbon star formation. *Astrophys. J . 613.* L73-76.
[14] Caciolli, A., et al. (2011). Revision of the $^{15}N(p,\gamma)^{16}O$ rate and oxygen abundance in H-burning zones. *Astron. & Astrophys. 533.* A66-71.
[15] Strieder, F., et al. (2012). The $^{25}Mg(p,\gamma)^{26}Al$ reaction at low astrophysical energies. *Phys. Lett. B 707.* 60-65.

CHAIRMAN: C. BROGGINI

Scientific Secretaries: S. Kabana, T. Metcalf

DISCUSSION

- *Inguglia:*

Where does ^{12}C come from in the process you are discussing, i.e. the CNO cycle?

- *Broggini:*

In this process you have Hydrogen burning. Carbon is produced during Helium burning. Helium burning is a two step process, with first ^{4}He + ^{4}He giving ^{8}Be, and then, thanks to the Hoyle resonance (corresponding to an excited state at 7.56 MeV in ^{12}C), you have ^{4}He + ^{8}Be giving ^{12}C. Helium burning takes place at much higher density than the one of the Sun, around 10^{5}g/cm^{3}, and more than that, at a temperature of 100 million degrees, much higher than the 15 million degree temperature of the central region of the Sun. The CNO cycle cannot take place in first generation stars but only in stars where the carbon is already there, having been produced in a previous star that has exploded. Thus the origin of ^{12}C is that it has already been produced in another star.

- *Hemmer:*

To draw conclusions from the rate of CNO neutrinos on the Sun metallicity it is very important to know the cross section of the cycle. Why is the ^{14}N reaction so important?

- *Broggini:*

The point is that the cross section which is the important one in the cycle to give the rate of the cycle is the smallest one, which means the slowest reaction. This is by far ^{14}N+p, the bottleneck reaction of the CNO cycle.

Highlights from ALICE

Paolo Giubellino
INFN Torino and CERN Geneve
E-mail: `paolo.giubellino@CERN.ch`

Abstract

The LHC has delivered for the first time collisions of Nuclei in November 2010, at an energy of 2.76 TeV per nucleon pair, which represents a jump of more than an order of magnitude over the highest energy nuclear collisions ever studied before. The high energy, the quality of the state-of-the art detectors, and the readiness of the experimental collaborations at the LHC have allowed a rich harvest of important scientific results in a very short time. In this lecture a short overview will be given of how the results from the LHC, and in particular from the ALICE experiment, have provided new insight on the properties of matter under extreme conditions of temperature and pressure, analogous to the conditions present in the early phases of the evolution of the Universe.

1.Why high-energy Heavy-Ion Collisions

Collisions between very high-energy nuclei, normally referred to as Heavy-Ions since they are atoms stripped of their electrons to allow acceleration, produce each a huge number of particles, over ten thousand in a single event at the LHC. A typical event is shown in Fig. 1. The study of such extremely complex events poses a formidable challenge both to the detections systems and to the software and analysis methods to be used to treat and eventually understand the data. So, it is legitimate to ask first of all the question of why experimenters are willing to face such extreme problems, what are the underlying questions which we aim to answer.

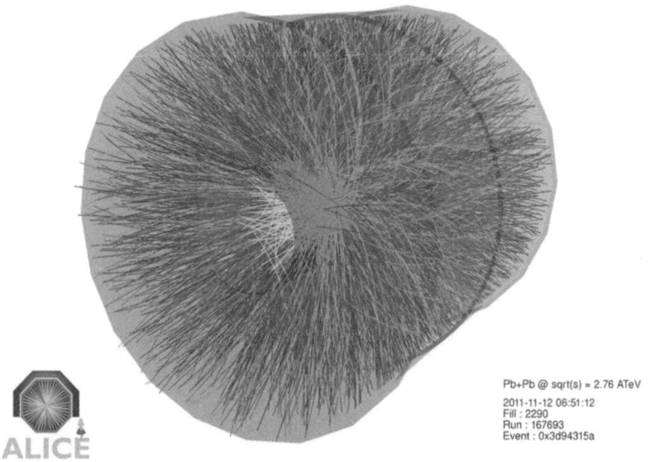

Fig. 1 A collision of two Lead Nuclei as seen by the ALICE detector at the LHC. (CERN-EX-1111290 04)

The strong force, which holds together the nucleons, protons and neutrons, in nuclei and the quarks within the nucleons, is described to great precision by a very successful theory, called Quantum Chromo Dynamics (QCD), in which the quarks possess a color charge and interact exchanging force carriers, the gluons, which are also colored. A fundamental feature of the interaction is the fact that it grows with the distance between the quarks, much like a spring connecting two balls. Therefore, any attempt to isolate a quark are in vain, since the energy stored in the spring will eventually be more than the mass of a quart-antiquark pair: the spring will break generating a pair, and the quarks will not be isolated. This property is called "confinement" and has major consequences. In particular, most of the mass of the hadrons is to be attributed to this feature: more technically, one can say that the mass is acquired dynamically because of the interaction between the quarks, whose Higgs mass would account for just a small fraction, about 1%, of the mass of the light hadrons as shown in Fig. 2. Therefore, since the mass of the matter around us is essentially the mass of the nuclei of the atoms, it can be said that, while the Higgs mechanism is at the base of the masses of the quarks, strong interaction is at the origin of most of the mass of ordinary matter.

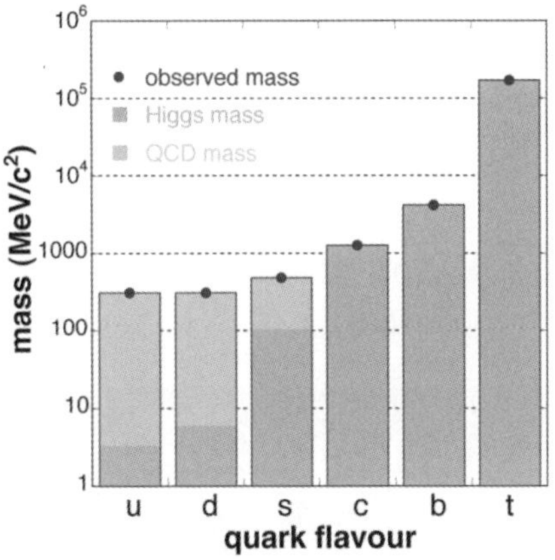

Fig. 2 QCD and Higgs mass components for the light and heay quarks from Ref [1]

The methodof colliding high-energy nuclei to heat and compress hadronic matter dates from the early eighties, but the path had been indicated already several years before. It was in 1975 that T.D. Lee pronounced the famous sentence "it would be interesting to explore new phenomena by distributing a high amount of energy or high nuclear density over a relatively large volume", and the idea of deconfinement had been around for some time. Already in 1965 Hagedorn had observed that the mass spectrum of hadronic states led naturally to the concept of a "limiting" or "critical" temperature beyond which the concept of hadron would cease to make sense. In 1973 the development of QCD as the theory describing strong interactions introduced the concept of asymptotic freedom[3][4]: at small distances quarks and gluons would be loosely interacting, or quasi-free.

In 1975, N. Cabibbo and G. Parisi[5]proposed for the first time the existence of a ``different phase of the vacuum in which quarks are not confined", and drew the first schematic diagram of hadronic matter with temperature and baryonic density as axes. Two regions were defined: at low-T and low baryonic density, ordinary hadrons, at high-T, high-μ_B a deconfined state: a new field of research was born! In the same year, J.C. Collins and M.J. Perry[6]developed the idea that matter under extreme density conditions, as in neutron star cores and in the early universe, is made of quarks instead of hadrons, which overlap and lose their identity. Their argument was based on the fact that quarks interact weakly when they are close together, and therefore at high densities could be described as a gas of free massless quarks. Finally, in 1980 E. Shuryak introduced [7]the term quark gluon plasma (QGP) to describe the state of nuclear matter in which quarks are deconfined, stressing the analogy between a classical plasma made by ionized atoms and the state made of colored strongly interacting objects.

In a nucleus-nucleus collision the energy of the incoming projectiles is dissipated in the relatively large volume defined by the overlap region of the two colliding nuclei, thus creating the conditions for a phase transition to deconfined quark matter by heating.

So, by colliding nuclei at high energies we compress and heat a system with a large number of quarks, to form a “bubble” of deconfined strongly-interacting matter, which in a very short time, of the order of 10^{-23}s, expands and cools until ordinary hadrons reconstitute, just as they did in the evolution of primordial Universe, some 10 millionth of a second after the Big Bang.

Over the last three decades, a vigorous experimental program, carried both in Europe and in the United States, has dramatically increased our understanding of nuclear matter under extreme conditions of temperature and density, allowing a real exploration of the phase diagram of strongly interacting matter. Our current understanding is graphically summarized in Fig. 3, with the trajectories allowed by the most important accelerators in operation or under construction: the two colliders LHC (in Europe) and RHIC (in the USA) and the fixed-target high-luminosity facility FAIR under construction in Germany.

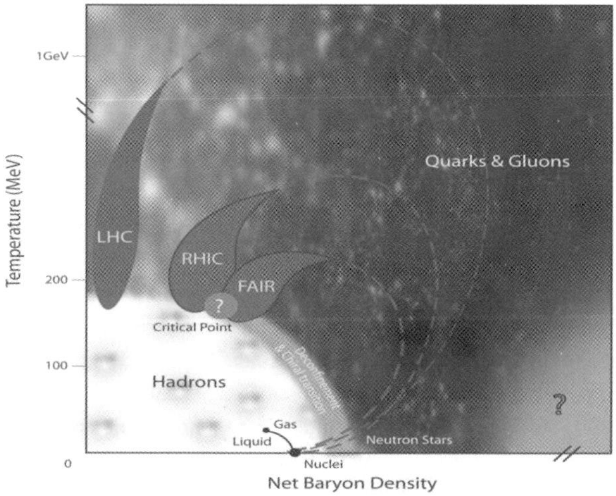

Fig.3 The phase diagram of nuclear matter, from Ref[1]

Together, these programs aim at providing answers to questions such as:

What are the properties of matter at the highest temperatures and densities?

What is the QCD equation of state? How can we test it?

What are the dominant microscopic mechanisms ofQCD non-equilibrium dynamics and thermalization?

How does hadronization proceed dynamically? How is it changed in dense QCD matter?

How can the QCD phase diagram be efficiently explored?

Clearly, a comparison of the experimental results with theory is not a trivial task. Since the transition to the deconfined phase is outside the limits of validity of the perturbative QCD, a number of different approaches has been developed to derive the equation of state and the transition temperature. Only at the LHC some aspects of the interactions will be accessible with perturbative methods. Effective lagrangians have been extensively used, and also potential models, and even calculations based on the AdS/CFT correspondence have been providing a fresh look and new insight. Still, the most important and widely used are QCD calculations on a space--time lattice. They generally predict a phase transition to occur, and are able to determine its temperature. Current results indicate a critical temperature aroundT_c~170 MeV, or about 10^{12} K, 100000 times higher than at the center of the Sun, corresponding to a critical energy density of ε_c~ 700 MeV/fm^3

In the following, I will give a brief overview of the main experimental results obtained so far at the LHC, just little over one year after the very first collision of Lead Nuclei was recorded in the experiments. Three experiments joined in this pioneering work: ALICE[9], a dedicated experiment specifically designed to study nuclear collisions, and both ATLAS [10]and CMS[11], the two general-purpose experiments of the LHC, primarily designed for the study of high-p_T processes in pp collisions. I will focus mostly on the ALICE results.

While I will not describe the experimental apparata, I would like to remind here that ALICE has a very different optimization compared to the other LHC experiments. For the understanding of nuclear collisions it is essential that as many of the produced particle as possible are measured and identified, for the widest possible range of transverse momenta (typically from 0.1 to 100 GeV/*c*) and for all particle types: electrons, muons, photons and hadrons. So, the two guiding principles in the ALICE design have been robust tracking, able to handle the very large number of tracks which characterize Heavy Ion Collisions, and Particle Identification capability. The first is provided by very high granularity tracking detectors providing 3-Dimensional space points (several hundred millions of points) operating in a moderate magnetic field and with very low material. To identify leptons, photons and hadrons, including short-lived particles, such as hyperons, D and B mesons, ALICE features essentially all known techniques: dE/dx, Cherenkov & Transition radiation detectors, Time Of Flight, calorimeters, muon filter and topology (secondary vertices, kinks).

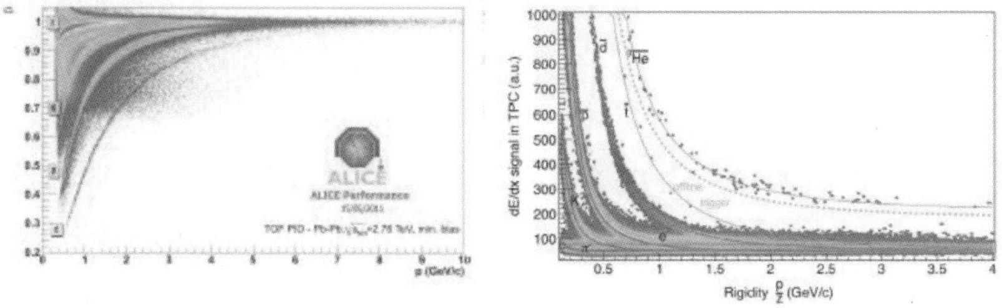

Fig. 4: PID in ALICE with TOF (left panel) and TPC (right panel) from ref.[12]

As an example of the ALICE PID capability, Fig. 4 shows the of the performance of the Time Of Flight (TOF) and of the Time Projection Chamber (TPC)'s energy loss measurement. Both are performing better than originally specified, a remarkable result in detector systems of such complexity observing events with thousands of tracks! The TOF has provided a time resolution below 100 picoseconds, and the TPC an energy loss measurement with a precision better than 6%.

In figure 5 (left panel) the momentum coverage of the various ALICE PID detectors is shown, demonstrating the extent of the coverage achieved. In addition, short-lived particles are identified through their decay topologies: in the right panel of figure 5 the impact parameter resolution expected in ALICE is shown, which is adequate for the measurement of D and B meson decays. K and Λ decays can be efficiently reconstructed for transverse momenta up to more than 10 GeV/*c*. The very fine granularity of the tracking detectors, providing a large number of space points per track, provides high efficiency tracking for charged particles even at the highest particle multiplicities foreseen at the LHC with Pb beams, with a lower limit in momentum lower than 200 MeV/*c*. The very long track length measured allows ALICE to achieve an excellent momentum resolution ($\delta p/p < 5\%$ at 100 GeV/*c*). Several smaller detectors (ZDC, PMD, FMD, T0 andV0) used for global event characterization and triggering are located at small angles. The combination of the central and forward detectors provides coverage for the measurement of charged multiplicity which extends from -3.4 to +5 in pseudorapidity. An Electromagnetic Calorimeter, EMCal, covers about a third of the azimuth and provides improved energy resolution and triggering for jets. A high-resolution crystal photon spectrometer, PHOS, is placed at a very large radial distance from the interaction point (4.6 m). The muon spectrometer covers the forward rapidities, between-2.5 and -4, and features a complex arrangement of absorbers, starting very close to the interaction point (90cm) to minimize muon background from weak hadronic decays, and fourteen planes of tracking and triggering chambers, using 2-D pad readout to handle the high multiplicities.

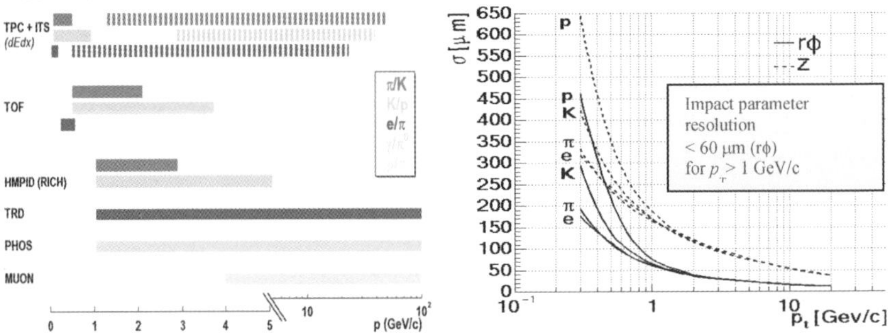

Figure 5: Left: particle identification in ALICE: momentum range and detectors used. Topological identification, not included in this plot, can reach over 10 GeV/*c* for K and Λ. Right: ALICE impact parameter resolution as a function of transverse momentum.

The hardware trigger in ALICE combines the input from detectors with fast trigger capability (T0, V0, ZDC, SPD, TOF, PHOS, EMCal, Muons, ACORDE). It operates at several levels (pretrigger, L0, L1 and L2) to satisfy the individual timing requirements of the different detectors, as the ALICE electronics is in general not pipelined. In addition, a software based High-Level Trigger (HLT) using a farm of up to 1000 multiprocessor PCs performs a preprocessing of the data allowing for a compression of the data volume by a factor of 3.5. It also allows real time data quality monitoring.

2. ALICE results

The first LHC heavy-ion results were already published during the 2010 lead–lead run, lasting for a month. In fact, there were three papers submitted almost simultaneously: the multiplicity measurement [13] and the measurement of v_2 coefficient of the azimuthal anisotropy [14] by ALICE, and by the ATLAS collaboration a paper on jet-energy imbalance [15]; these appeared together in one issue of Phys. Rev. Lett.. One of the most spectacular findings at RHIC was that the matter generated in heavy-ion collisions flows like a liquid with very low viscosity, almost at the limit of what is allowed for any material in nature. This tells us that the constituents of this QGP are quite different from freely interacting quarks and gluons, and led to the definition of the matter under study as sQGP, for strongly interacting Quark Gluon Plasma, a liquid very different from the expected gas of quasi-free quarks and gluons. The first LHC azimuthal anisotropy measurement [14] confirms the RHIC results: elliptic flow of particles with the same transverse momenta is almost identical at the two energies. The nearly-perfect fluid has been found to be opaque to even the most energetic partons (quarks and gluons), which appear as jets of particles from the collisions, an effect known as jet quenching. This is an interpretation of the reported strong jet-energy imbalance [15]. However, the physical mechanisms underlying these phenomena are not well understood. Another manifestation of jet quenching manifests is a reduced yield of high-transverse-momentum-particle in central collisions compared to that expected from the measurements in proton–proton reactions. To express such reduction a properly normalized ratio of yields in heavy-ion and in proton–proton interactions – the nuclear modification factor R_{AA} – is used. Since proton-proton collisions are rescaled according to the number of binary collisions, R_{AA} should be well below one at low transverse momentum, where particle production is rather proportional to the number of participant nucleons, and grow toward one for increasing transverse momenta. The suppression pattern observed by ALICE [16] gives factor about 7 (= $1/R_{AA}$) lower charged-particle production in lead–lead collisions at transverse momentum around 6GeV/c, a striking evidence of the energy loss in the extreme density matter.

The size of the pion-emitting source in central lead-ion collisions is deduced from the shape of the Bose-Einstein peak in the two-pion correlation functions [17]. The collective flow makes the size of the system appear smaller with increasing momentum of the pair. This behaviour is clearly visible for the radii measured in the ALICE experiment. The results for measurements of the radius of the pion source in three dimensions indicate a short duration for the emission, hence an "explosive" emission. Time when the emission reaches its maximum is 10–11fm/c, significantly longer than it is at RHIC, see fig. 6 (left side). Moreover, the product of the three components at low pair-momentum, an estimate of the homogeneity volume of the system at decoupling, is twice as large as at RHIC, see fig. 6 (right side).

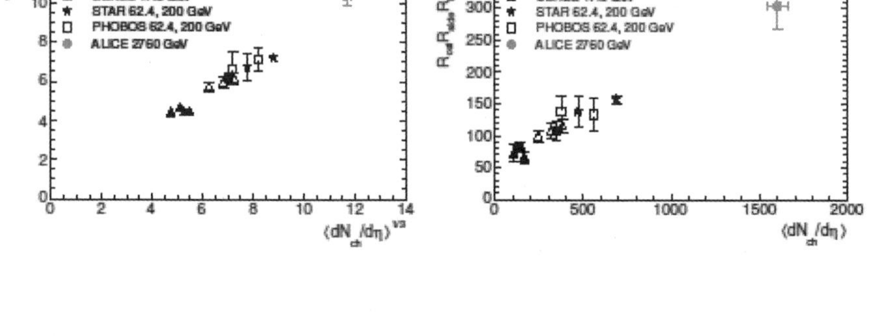

Figure 6. Bose–Einstein pion-interferometry results from different experiments [17]. Estimate of lifetime of emitting source as a function of cube root of particle density (left side). Volume of homogeneity region as a function of particle density (right side).

One of the crucial measurements for the characterization of the fireball produced in heavy-ion collisions are the spectra of identified hadrons, which encode the collective expansion velocity developed in the QGP and in hadronic stages. Moreover, the overall abundances of identified hadrons are believed to be fixed at hadronization, thus they indicate the temperature when chemical composition was established. The measured spectra [18] show an increase of about 10% in the radial-flow velocity when compared to RHIC results. At present, however, the yield ratios observed by ALICE seem to challenge both previous experiments and theory. While the K/π, Ξ/π and Ω/π ratios are compatible with the expectations from the thermal model with a temperature for the "chemical freeze-out" of about 165MeV, as in previous observations, the p/π ratio points to a significantly lower temperature. On the experimental side, there are indications of a similar effect at lower energies, which call for further investigations. On the theoretical side, a number of different possibilities are being investigated, none of them conclusive at the moment.

As mentioned above, the first results on elliptic flow were published [14] during the initial Pb–Pb run. Flow is a fundamental observable since it carries information on the equation of state and the transport properties of matter created in a heavy-ion collision. This observable relates final state anisotropies with features of the initial one, thus allowing study of the medium response and characteristics. The azimuthal anisotropy in particle production is a clear experimental signature of collective flow. It is caused by multiple interactions between the constituents of the created matter and the initial asymmetries in the spatial geometry of a non-central collision. The amount of elliptic flow measured by the v_2 coefficient, integrated over transverse momentum, increases by 30% compared to RHIC energies. However, this increase is entirely a consequence of the growth in transverse momenta. The overall azimuthal distribution of particles emitted in a heavy-ion collision can be described with a Fourier expansion, whose coefficients relate to different aspects of the initial state. ALICE measured also the higher harmonic coefficients of azimuthal distribution [19]; they exhibit much shallower centrality dependence than v_2, and the symmetry planes of v_2 and v_3 coefficients are uncorrelated (see fig. 7, left side). These observations point to fluctuations in the initial geometry as the origin of higher-order azimuthal asymmetries. The flow coefficients can also be studied with a different approach, using two-particle correlations in azimuthal angle [20]. A Fourier decomposition of the correlation function gives squares of different v coefficients, and describes well the structures seen in azimuthal correlations, see fig. 7 (right side), such as the "long-range ridge" (a small angle correlation between particles distanced in longitudinal direction) and the "Mach cone" (a correlation on two sides of back-to-back direction). The collective response to initial spatial anisotropy that causes elliptic flow economically explains these puzzling features, once event-by-event initial-state density fluctuations are considered. The ALICE experiment has analyzed azimuthal asymmetry also separately for different particle species, which constrains models incorporating a realistic initial state and hydrodynamic evolution.

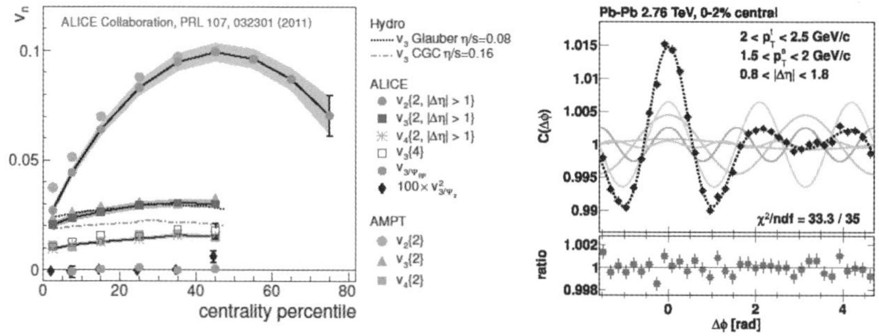

Figure 7. Measurements of azimuthal anisotropy in lead–lead collisions at LHC. Dependence of different v_n coefficients (see legend) on collision centrality [19] (left). Fourier decomposition of two-particle azimuthal correlation function in 2% most central collisions [20] (right).

The two-particle azimuthal correlations at higher transverse momenta measure particle-yield modifications in jet-like structures [21]. When compared to the expectation from proton–proton interactions, the observed yield around high-transverse-momentum particle is slightly higher, while the yield on the opposite side is reduced. This measurement assists the understanding of jet-quenching phenomena. The ALICE collaboration continues to contribute to this topic by studying reconstructed jets, underlying event fluctuations, modifications of jet-fragmentation function, jet-particle composition and providing detailed studies of the dependence of the energy loss on the parton traversing the QGP.

Heavy-flavour particles are recognized to be effective probes of a very dense and hot medium formed in nucleus–nucleus collisions; they are expected to be sensitive to its energy density, through the mechanism of in-medium energy loss. At LHC energies, heavy-flavour particles are copiously produced and thus provide ideal tools for QGP studies. The nuclear modification factor R_{AA} is well established as a sensitive observable for the study of the interaction of hard partons with the medium. Parton energy-loss is caused by the strong interaction, hence the amount of energy loss depends on colour charge of parton. Therefore, quarks are predicted to lose less energy than gluons. In addition, heavy quarks (up to some high momentum), which are slower than light, cannot emit gluons within the so-called "dead-cone" around their trajectory; this effect is expected to reduce the energy loss of heavy quarks with respect to light ones. Thus, a pattern of gradually decreasing suppression (i.e. increasing R_{AA}) should emerge when going from the light-flavour hadrons (e.g. pions), which mainly come from gluons, to the heavier D and B mesons: $R_{AA}(\pi) < R_{AA}(D) < R_{AA}(B)$. The measurement and comparison of these different probes provides, therefore, a test of the colour-charge and mass dependence of parton energy-loss. The ALICE collaboration has measured the production of the charmed mesons D^0 and D^+, detecting their hadronic decays in lead–lead collisions [22]. In central collisions a large suppression with respect to expectations at large transverse momentum was found, indicating that charm quarks undergo a strong energy loss in the hot and dense state of strongly-interacting matter formed at the LHC. This is the first time that D meson suppression has been measured directly in central nucleus–nucleus collisions. The results show a suppression by a factor 4–5, almost as large as for charged pions, above 5 GeV/c (see fig. 8). At lower momenta, there is an indication of smaller suppression for D than for π mesons.

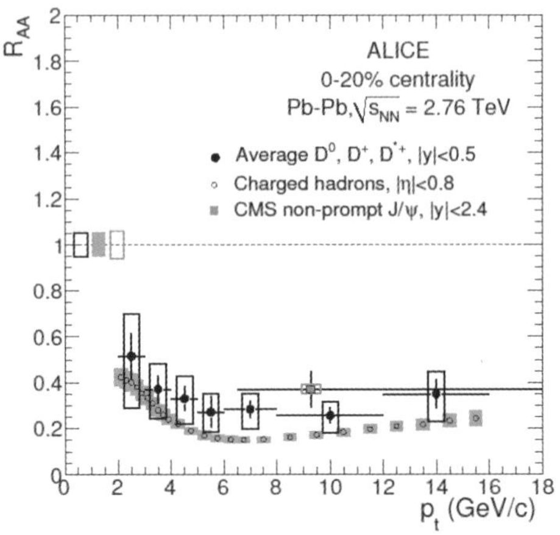

Figure 8. Charmed-meson nuclear modification factor as a function of transverse momentum compared to that of charged hadrons and of non-prompt (mostly from B-decays) J/ψ mesons (as measured by CMS).

Suppression of charmonium production was for a long time considered as one of the main probes for a deconfined medium. At large enough temperatures bound charm–anticharm states are supposed to be dissolved due to Debye screening. However, at LHC energies, new mechanisms of charmonium production in the QGP could occur because of a large number of charm quarks. Around one hundred charm–anticharm pairs are expected to be produced in a central lead–lead collision. Several dynamical transport models predict that charm and anticharm quarks could combine at later stages of the interaction, leading to an enhancement of charmonium production in the most central collisions. ALICE detects charmonium down to very low transverse momentum in two different regions: the central barrel in the dielectron channel and the forward muon arm in the dimuon channel. The detection at low transverse momentum is crucial because the recombination of the charm and anticharm quarks is expected to be the main production mechanism for charmonium below 3GeV/*c*. The different regions allow for the study of QGP with different charm densities. ALICE has studied the nuclear modification factor R_{AA} for J/ψ mesons as a function of collision centrality [23]. The results indicate that the J/ψ nuclear modification factor shows little dependence on centrality (see fig. 9), a trend that is different from that observed at lower energies. For central and mid-central collisions the J/ψ R_{AA} is larger at the LHC than that measured at RHIC. In a complementary study, the CMS collaborations at the LHC have measured a smaller value for the J/ψ nuclear modification factor at transverse momenta above 6.5 GeV/*c*. These observations hint at the recombination of charm and anticharm quarks in the QGP as the main mechanism for J/ψ production in central lead–lead collisions at LHC energies.

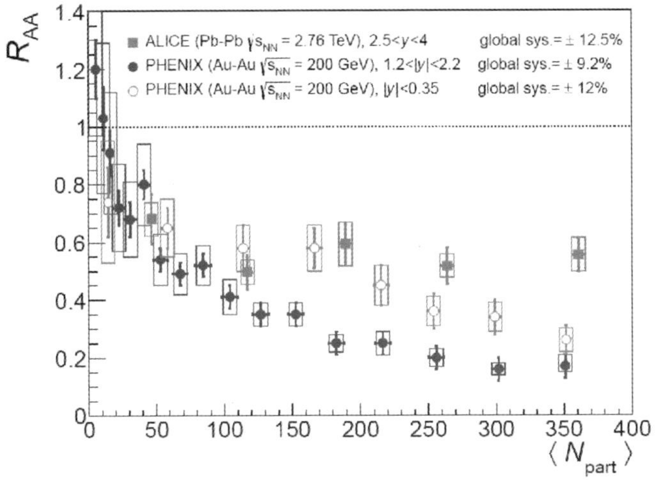

Figure 9. Nuclear modification factor for J/ψ as a function of collision centrality, expressed as number of participants. ALICE data are compared to those from the PHENIX experiment at RHIC.

It is clear from these first results that a coherent picture of the characteristics of the extreme density matter of which our Universe was made of in the first microseconds of its existence. Even more, the tool chosen to study it, heavy-ion collisions at the world's most powerful accelerator, has proven to be the right one, allowing for a continuous improvement of our understanding. The results shown here are but a small subset of the wealth of results obtained so far, and a was a mere hint to the ones coming. It is indeed a very exciting time for the physics of the Quark-Gluon Plasma.

3. The future of ALICE

In 2010 the ALICE detector collected with a minimum-bias trigger about 30 million of lead–lead collisions at centre-of-mass energy of 2.76TeV per nucleon pair. In 2011 there was a short period of proton–proton running at the same energy at the start of the LHC operation, to gather comparison data; these were used for normalization in practically all measurements described above. The rest of the proton running proceeded at (a standard) energy of 7 TeV. At the end of the year, the LHC switched again to the heavy-ion mode. This time the instant luminosity grew to above $10^{26}cm^{-2}s^{-1}$, and was above the design value for energy 2.76TeV per nucleon pair. ALICE used triggers for different types of lead–lead collisions: central, semi-central, dimuons, photons, jets, ultraperipheral, and others, and collected about 100 million events, inspecting over $0.1nb^{-1}$ of integrated luminosity.

The coming Heavy-Ion period, due for the beginning of 2013, will be dedicated to proton–lead collisions, to improve the baseline comparison, taking into account modifications of structure functions in nuclei. We expect to collect at least $0.03nb^{-1}$ of luminosity in the four weeks of pA running. Then the LHC operation will be paused for an upgrade necessary to increase the collision energy. After the restart the ALICE collaboration aims to complete its approved programme, collecting $1nb^{-1}$ of heavy-ion collisions at the higher collision energy (5.5TeV per nucleon pair in the centre of mass, being the design value). The intention is to achieve a significant part of this agenda before the second long shutdown for the preparation of the LHC luminosity increase, planned for 2018. Under discussion is an ALICE detector upgrade allowing for high-luminosity heavy-ion running after this period. The extended physics program justifying the LHC operation in heavy-ion mode beyond 2020, which would imply collecting over $10nb^{-1}$ of data, is being prepared.

After having confirmed the main discoveries obtained at RHIC, ALICE has entered an exciting phase of new measurements, allowing a much broader and deeper study of the QGP. At the same time, ALICE is already preparing, on the basis of what has been learnt so far, a next step in more detailed characterization of the extreme state of matter produced at LHC; new, unexpected discoveries may come!

References

[1] NUPECC Long Range Plan 2010: Perspectives of Nuclear Physics in Europe, published by the European Science Foundation
[2] R. Hagedorn, NuovoCim. Suppl. 3 (1965) 147-186
[3] D.J. Gross and F. Wilczek Phys. Rev. D 8(1973) 3633—3652.
[4] H. D. Politzer Phys. Rev. Lett. 30 (1973) 1346-1349.
[5] N. Cabibbo and G. Parisi Phys. Lett. B 59 (1975) 67-69
[6] J.C. Collins and M.C. Perry~M.C. Phys. Rev. Lett.34(1975) 1352—1356.
[7] E.V. Shuryak Phys. Rept. 61 (1980) 71—158
[8] B. Muller and J. Nagle, Ann. Rev. Nucl. and Part. Phys. **56** (2006) 93

[9] ALICE Collaboration, Aamodt K, et al., JINST 3:S08002 (2008).

[10] ATLAS Collaboration, Aad G, et al., JINST 3:S08003 (2008).

[11] CMS Collaboration, Adolphi R, et al., JINST 3:S08004 (2008).

[12] A. Kalweit for the ALICE Collaboration, J.Phys.G G38 (2011) 124073.

[13] K. Aamodt et al. (ALICE collaboration), Phys. Rev. Lett. **105** 252301 (2010).

[14] K. Aamodt et al. (ALICE collaboration), Phys. Rev. Lett. **105** 252302 (2010).

[15] G. Aad et al. (ATLAS collaboration), Phys. Rev. Lett. **105** 252303 (2010).

[16] K. Aamodt et al. (ALICE collaboration), Phys. Lett. **B696** 30 (2011).

[17] K. Aamodt et al. (ALICE collaboration), Phys. Lett. **B696** 328 (2011).

[18] M. Florisfor the ALICE collaboration, J. Phys. **G38** 124025 (2011).

[19] K. Aamodt et al. (ALICE collaboration), Phys. Rev. Lett. **107** 032301 (2011).

[20] K. Aamodt et al. (ALICE collaboration), Phys. Lett. **B708** 249 (2012).

[21] K. Aamodt et al. (ALICE collaboration), Phys. Rev. Lett. **108** 092301 (2011).

[22] B.Abelevet al. (ALICE collaboration), JHEP **1209** 112**(2012)**

[23] B.Abelev et al. (ALICE collaboration), Phys.Rev.Lett. **109**072301 **(2012)**

Highlights from BNL-RHIC

M. J. Tannenbaum *
Physics Department, 510c,
Brookhaven National Laboratory,
Upton, NY 11973-5000, USA
mjt@bnl.gov

November 13, 2012

1 Introduction

Since the AGS and SpS fixed target heavy-ion programs began operations in 1986, BNL and CERN have been the principal laboratories for measurements of high energy nucleus-nucleus collisions which create nuclear matter in conditions of extreme temperature and density [1, 2]. The kinetic energy of the incident projectiles is dissipated in the large volume of nuclear matter involved in the reaction. At large energy or baryon densities, a phase transition is expected from a state of nucleons containing confined quarks and gluons to a state of "deconfined" (from their individual nucleons) quarks and gluons, in chemical and thermal equilibrium, covering a volume that is many units of the confining length scale. This state of nuclear matter was originally given the name Quark Gluon Plasma (QGP) [3], a plasma being an ionized gas.

The startup of the Relativistic Heavy Ion Collider (RHIC) at BNL in the year 2000 provided a major advance in the field, leading to the discovery of the QGP at RHIC, which was announced on April 19, 2005. The results at RHIC [1] indicated that instead of behaving like a gas of free quarks and gluons, the matter created in heavy ion collisions at nucleon-nucleon c.m. energy $\sqrt{s_{NN}} = 200$ GeV appears to be more like a *liquid*. This matter interacts much more strongly than originally expected, as elaborated in peer reviewed articles by the 4 RHIC experiments [4, 5, 6, 7], which inspired the

*Research supported by U.S. Department of Energy, DE-AC02-98CH10886.

theorists [8] to give it the new name "sQGP" (strongly interacting QGP). These properties were quite different from the properties of the CERN SpS fixed-target heavy ion results which led to the claim of "new state of matter", in a press-conference [9] on February 10, 2000, that was neither peer-reviewed nor published. However, results in the past two years from Pb+Pb measurements at the CERN-LHC at $\sqrt{s_{NN}} = 2760$ GeV (2.76 TeV) confirm the RHIC discoveries [1, 2] and add some new information—notably with fully reconstructed jets [10, 11].

J/Ψ suppression, proposed by Matsui and Satz in 1986 [12] as the 'gold-plated' signature for deconfinement, suffered from the significant J/Ψ suppression observed in the Cold Nuclear Matter (CNM) of p+A collisions [13] ($A^{0.92}$) relative to the point-like scaling ($A^{1.0}$) from p-p collisions expected for the large mass scale. Although, "anomalous" J/Ψ suppression, i.e. more than the estimated CNM effect, was discovered at the CERN SPS heavy ion program and is its main claim to fame, the later development of J/Ψ measurements has not been concerned with J/Ψ suppression as a signature of deconfinement, but rather with the strong c.m. energy dependence of the CNM effect and the possibility of regeneration of J/Ψ from recombination of the large number of c and $\bar{c}$ quarks produced in the QGP (e.g. see ref. [1]). Nevertheless, the search for J/Ψ suppression (as well as thermal photon/dilepton radiation from the QGP) drove the design of the RHIC experiments [14] and the ALICE experiment at the LHC [15].

2 The RHIC machine in 2012

With the shutdown of the Tevatron at FERMILAB, on September 30, 2011 after 28 years of operation, RHIC at Brookhaven National Laboratory (Fig. 1) is the only hadron collider in the U.S. and one of only two hadron-colliders in the world, the other being the CERN-LHC. Also RHIC is the world's first and only polarized proton collider. RHIC is composed of two independent rings, of circumference 3.8 km, containing a total of 1,740 superconducting magnets. RHIC can collide any species with any other species and since beginning operation in the year 2000 has provided collisions at 15 different values of nucleon-nucleon c.m. energy, $\sqrt{s_{NN}}$, and six different species combinations. The performance history of RHIC with A+A and polarized p-p collisions is shown in Fig. 2.

This past year, the performance was even more outstanding than usual thanks to the new EBIS source and 3-dimensional stochastic cooling [16]. For the first time in a collider, Cu+Au and U+U collisions at $\sqrt{s_{NN}} = 200$

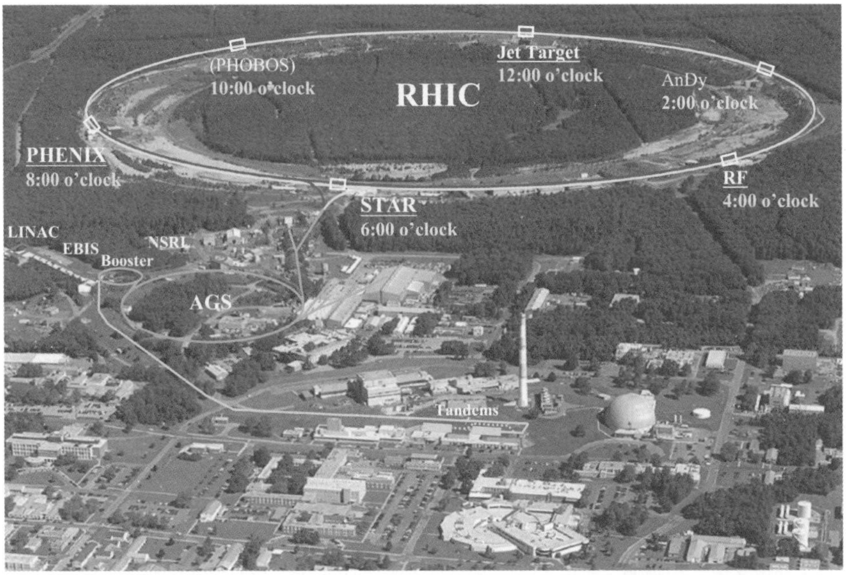

Figure 1: Aerial view of the RHIC facility [14]. The six crossing points are labelled as on a clock. The two principal experiments still running are PHENIX and STAR. Two smaller experiments PHOBOS [5] and BRAHMS [4] have been completed; a test run, AnDy, occupies the former location of BRAHMS. The LINAC is the injector for polarized protons into the Booster/AGS/RHIC chain; with the Jet Target used for precision beam polarization measurements. The TANDEM injector for Ions has been replaced by the Electron Beam Ion Source (EBIS) starting with the 2012 run.

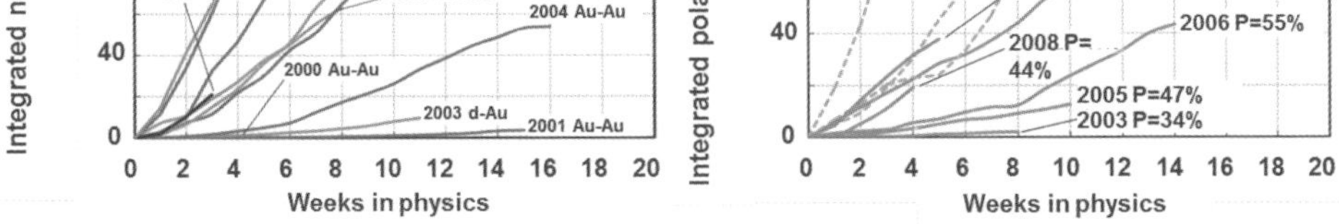

Figure 2: a)(left) Au+Au performance, where the nucleon-pair luminosity is defined as $L_{\mathrm{NN}} = A \times B \times L$, where L is the luminosity and A, B are the number of nucleons in the colliding species. b) (right) Polarized p-p performance. Courtesy Wolfram Fischer.

GeV were studied, with the purpose of: i) J/Ψ suppression with no-corona in central collisions (Cu+Au); ii) a large eccentricity of the overlap zone in central collisions by using a very large deformed nucleus with a prolate shape, like a rugby ball (U+U). Also, the polarized p-p runs in 2012 included both $\sqrt{s} = 200$ and 510 GeV with improved polarization and luminosity for the purposes of: iii) comparison data for the new silicon vertex detectors (200 GeV); iv) flavor-identified parton spin distribution functions using the parity violating single spin asymmetry in $W^{\pm}$ production (500 GeV) [17].

3 Issues in A+A versus p-p collisions

The principal difference in dealing with collisions of relativistic heavy ions, e.g. Au+Au, compared to p-p or e-p (or e-A) collisions at the same nucleon-nucleon c.m. energy, $\sqrt{s_{NN}}$, is that the particle multiplicity is $\sim$ A times larger in A+A central collisions than in p-p collisions as shown in actual events from the STAR and PHENIX detectors at RHIC (Fig. 3). This year,

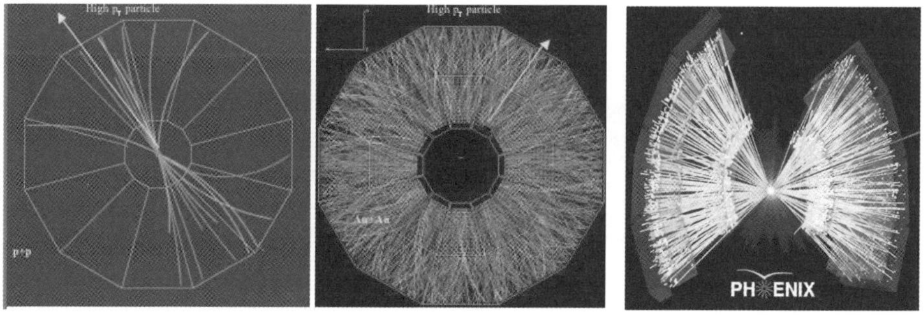

Figure 3: a) (left) A p-p collision in the STAR detector viewed along the collision axis; b) (center) Au+Au central collision at $\sqrt{s_{NN}} = 200$ GeV in STAR; c) (right) Au+Au central collision at $\sqrt{s_{NN}} = 200$ GeV in PHENIX.

both experiments have benefitted from incremental upgrades, with STAR at present making use of an EM calorimeter with a shower-maximum detector; and PHENIX with central and "forward" silicon vertex detectors for event-by-event identification of c and b quarks.

A schematic drawing of a collision of two relativistic Au nuclei is shown in Fig. 4a. In the center of mass system of the nucleus-nucleus collision, the two Lorentz-contracted nuclei of radius R approach each other with impact parameter b. In the region of overlap, the "participating" nucleons interact with each other, while in the non-overlap region, the "spectator" nucleons simply continue on their original trajectories and can be measured,

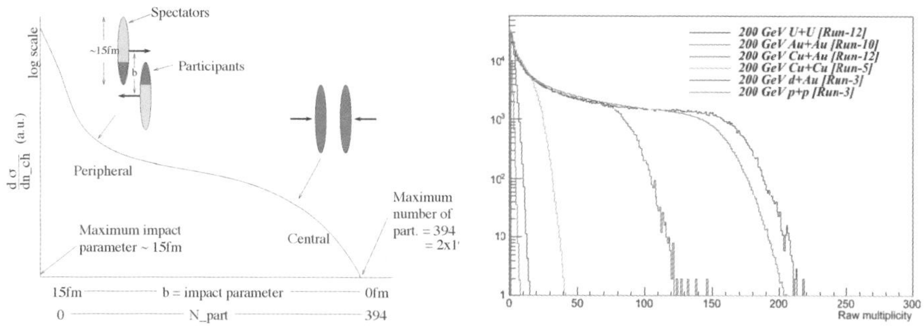

Figure 4: a) (left) Schematic of collision in the N-N c.m. system of two Lorentz contracted nuclei with radius R and impact parameter b. The curve with the ordinate labeled $d\sigma/dn_{\rm ch}$ represents the relative probability of charged particle multiplicity $n_{\rm ch}$ which is directly proportional to the number of participating nucleons, $N_{\rm part}$. b)(right) raw $n_{\rm ch}$ distributions in p-p to U-U collisions at $\sqrt{s_{NN}} = 200$ GeV from PHENIX [18].

in principle, by Zero Degree Calorimeters (ZDC), so that the number of participants, $N_{\rm part}$, can be directly determined. The degree of overlap is called the centrality of the collision, with $b \sim 0$, being the most central and $b \sim 2R$, the most peripheral. The impact parameter b can not be measured directly, so the centrality of a collision is defined in terms of the upper percentile of $n_{\rm ch}$ distributions, e.g. top 10%-ile, upper $10 - 20$%-ile, from which b and $N_{\rm part}$ can be deduced.

The energy of the inelastic collision is predominantly dissipated by multiple production of soft particles ($\langle p_T \rangle \approx 0.36$ GeV/c), where $n_{\rm ch}$, the number of charged particles produced, is directly proportional to the number of participating nucleons ($N_{\rm part}$) as sketched on Fig. 4a. Figure 4b shows the measured distributions [18] of the charged particle multiplicity, $n_{\rm ch}$, at mid-rapidity at $\sqrt{s_{NN}} = 200$ GeV for all the combinations of A+B collisions measured at RHIC. The increase of $n_{\rm ch}$ with both A and B is evident.

3.1 Collective Flow

Another unique feature of A+A collisions compared to either p-p or p+A collisions is the "flow" observed, which is a collective effect that can not be obtained from a superposition of independent N-N collisions. Immediately after an A+A collision, the overlap region defined by the nuclear geometry of participants is roughly almond shaped (see Fig 5a); with the shortest axis roughly along the impact parameter vector, which together with the

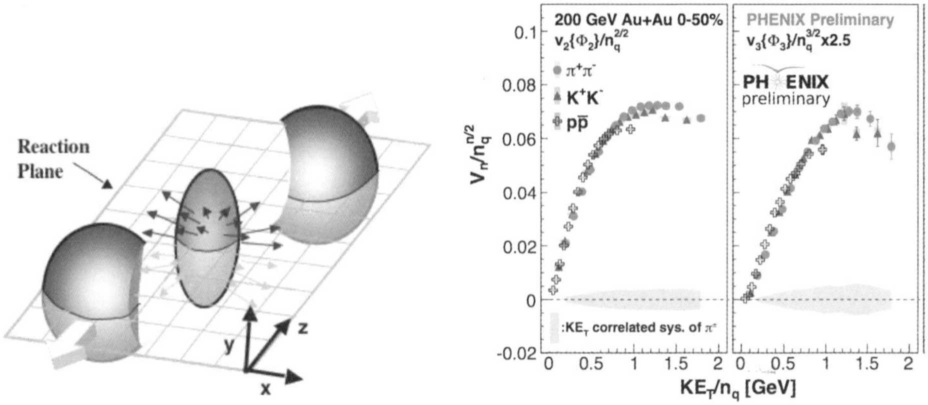

Figure 5: a) (left) Almond shaped overlap zone generated just after an A+A collision where the incident nuclei are moving along the $\pm z$ axis and the impact parameter vector is along the x axis. b)(right) Measurements of elliptical-flow (v_2) and triangular flow (v_3) for identified hadrons plotted as v_k, divided by the number of constituent quarks n_q in the hadron raised to the power $k/2$, as a function of KE_T/n_q [19].

beam (z) axis defines the reaction plane. Fluctuations in the distribution of participants from event-to-event do not respect the average symmetry of the almond, $\phi \rightarrow \phi + \pi$, so that the event plane of the participants, with angle Ψ_k, is relevant in the fourier decomposition of the azimuthal distribution:

$$Ed^3N/dp^3 = (d^3N/2\pi\, p_T dp_T dy)[1 + \sum_{k\geq 1} 2v_k \cos k(\phi - \Psi_k)] \qquad (1)$$

which is no longer restricted to even harmonics v_k by symmetry so that odd harmonics such as v_3 can be produced [1], which was observed (Figs 5b). The flow can also be seen in two-particle correlations. [1]

The fact that the flow persists for $p_T > 1$ GeV/c [20] and that the higher harmonics are not strongly damped [21, 22] implies that the viscosity is small, perhaps as small as a quantum viscosity bound from string theory [23], $\eta/s = 1/(4\pi)$ where η is the shear viscosity and s the entropy density per unit volume. This has led to the description of the "sQGP" produced at RHIC as "the perfect fluid" [8]. It was observed that the flow is not dominated by final state hadrons but is proportional to the number of constituent quarks n_q in the mesons and baryons, in which case v_2/n_q, as well as $v_3/n_q^{3/2}$ [19], as

[1] If two particles A and B are correlated to the event plane but not otherwise correlated, the analog of Eq. 1 is $dN^{AB}/d\phi^A d\phi^B \propto [1 + \sum_{k\geq 1} 2v_k^A v_k^B \cos k(\phi^A - \phi^B)]$.

a function of the transverse kinetic energy per constituent quark, KE_T/n_q, would be universal, as shown in Fig. 5b, where the power $k/2$ and the transverse kinetic energy, $KE_T \equiv m_T - m_0$, are suggested by the relativistic hydrodynamics of "ideal" fluids [24, 25].

4 Old and New RHIC results

One of the most important innovations at RHIC was the use of hard scattering as an in-situ probe of the medium produced in A+A collisions by the effect of the medium on outgoing hard-scattered partons produced by the

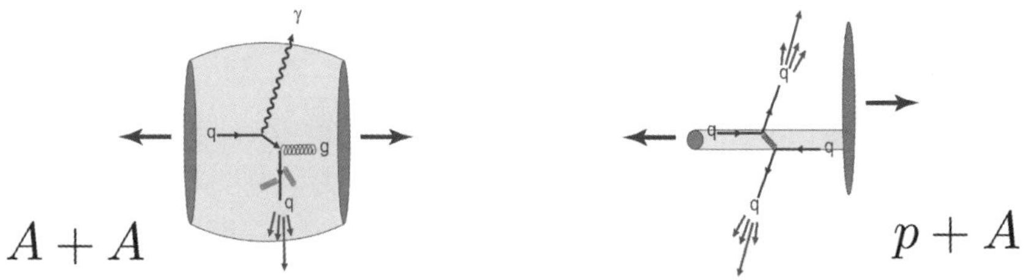

initial A+A collision. Measurements in p+A (or d+A) collisions, where no (or negligible) medium is produced, allow correction for any modification of the nuclear structure function from an incoherent superposition of proton and neutron structure functions.

Hard-scattering of partons in p-p collisions was discovered at the CERN-ISR in 1972 as shown at the ICHEP 1972 conference [26] by observation that the e^{-6p_T} dependence for pion production at low p_T breaks to a power-law for $p_T > 2$ GeV/c with a characteristic $\sqrt{s}$ dependence (Fig. 6a) which is more evident from the log-log plot of subsequent π^0 data (Fig 6b). This plot, as a function of $x_T = 2p_T/\sqrt{s}$ (Fig 6b), shows that the cross section for hard-processes (Fig 6c) obeys the scaling law, called"x_T-scaling":

$$E\frac{d^3\sigma}{d^3p} = \frac{1}{p_T^{n_{\rm eff}}} F(\frac{p_T}{\sqrt{s}}) = \frac{1}{\sqrt{s}^{n_{\rm eff}}} G(x_T) \tag{2}$$

for $\sqrt{s} \geq 53.1$ GeV, $x_T \geq 0.3$ ($p_T \geq 7$ GeV/c), where $n_{\rm eff}(x_T, \sqrt{s}) \sim 4-6$ gives the form of the force-law between constituents as predicted by Quantum Chromodynamics (QCD) with non-scaling structure and fragmentation functions and running coupling constant [1].

Figure 6a provided the first evidence that the electrically charged partons of e-p deeply inelastic scattering (DIS) interacted with each other much more strongly than electromagnetically [2]. Note that the parton degrees of freedom in p-p collisions become evident only for parton-parton scattering

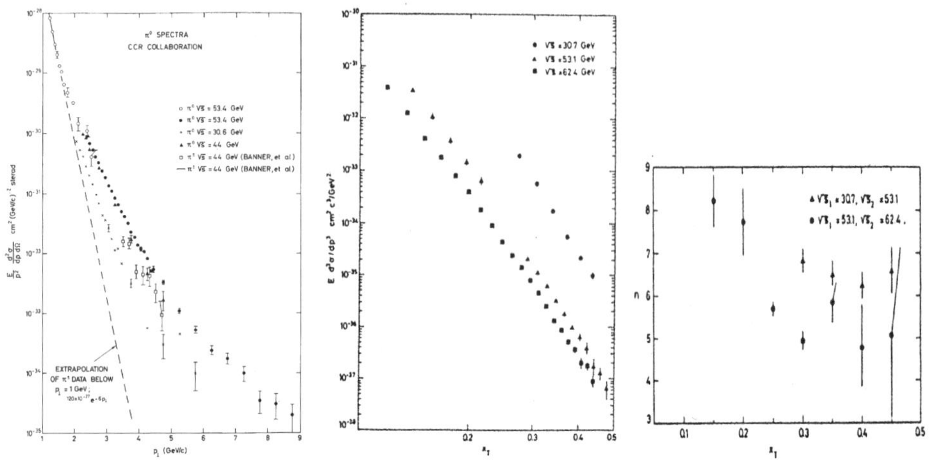

Figure 6: a) (left) Plot of invariant single particle inclusive cross sections vs. p_T at the CERN-ISR [26]. b) (center) Log-log plot of CCOR [27] invariant cross sections vs $x_T = 2p_T/\sqrt{s}$. c) (right) CCOR [27] $n_{\rm eff}(x_T, \sqrt{s})$ derived from the combinations indicated. The systematic normalization error at $\sqrt{s} = 30.6$ GeV has been added in quadrature. There is an additional common systematic error of ± 0.33 in $n_{\rm eff}$.

at large $Q^2 \approx 2p_T^2 \gtrsim 5 - 8$ (GeV/c)^2, while in DIS, only $Q^2 > 1$ (GeV/c)^2 is required.

The use of hard-scattering to probe the thermal or "soft" medium produced in RHI collisions was stimulated by pQCD studies [28] of the energy loss of partons produced by hard scattering, "with their color charge fully exposed", in traversing a medium "with a large density of similarly exposed color charges". The conclusion was that "Numerical estimates of the loss suggest that it may be significantly greater in hot matter than in cold. *This makes the magnitude of the radiative energy loss a remarkable signal for QGP formation*" [28]. In addition to being a probe of the QGP the fully exposed color charges allow the study of parton-scattering with $Q^2 \ll 1 - 5$ (GeV/c)^2 in the medium where new collective QCD effects may possibly be observed.

The discovery, at RHIC [29], that π^0's produced at large transverse momenta are suppressed in central Au+Au collisions by a factor of ~ 5 compared to point-like scaling from p-p collisions is arguably *the* major discovery in Relativistic Heavy Ion Physics. For π^0 (Fig. 7a) [30] the hard-scattering in p-p collisions is indicated by the power law behavior p_T^{-n} for the invariant cross section, $Ed^3\sigma/dp^3$, with $n = 8.1 \pm 0.05$ for $p_T \geq 3$ GeV/c. The Au+Au

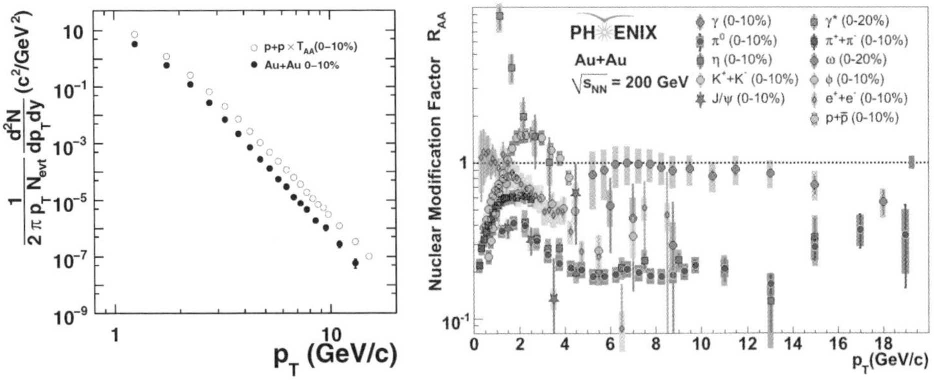

Figure 7: a) (left) Log-log plot of invariant yield of π^0 at $\sqrt{s_{NN}} = 200$ GeV as a function of transverse momentum p_T in p-p collisions multiplied by $\langle T_{AA} \rangle$ for Au+Au central (0–10%) collisions compared to the Au+Au measurement [30]. b) (right) $R_{AA}(p_T)$ for all identified particles so far measured by PHENIX in Au+Au central collisions at $\sqrt{s_{NN}} = 200$ GeV.

data at a given p_T can be characterized either as shifted lower in p_T by δp_T from the point-like scaled p-p data at $p'_T = p_T + \delta p_T$, or shifted down in magnitude, i.e. suppressed. In Fig. 7b, the suppression of the many identified particles measured by PHENIX at RHIC is presented as the Nuclear Modification Factor, $R_{AA}(p_T)$, the ratio of the yield of e.g. π per central Au+Au collision (upper 10%-ile of observed multiplicity) to the point-like-scaled p-p cross section, where $\langle T_{AA} \rangle$ is the average overlap integral of the nuclear thickness functions:

$$R_{AA}(p_T) = \frac{d^2N^{\pi}_{AA}/dp_T dy N_{AA}}{\langle T_{AA} \rangle \, d^2\sigma^{\pi}_{pp}/dp_T dy} \quad . \tag{3}$$

The striking differences of $R_{AA}(p_T)$ in central Au+Au collisions for the many particles measured by PHENIX (Fig. 7b) illustrates the importance of particle identification for understanding the physics of the medium produced at RHIC. Most notable are: the equal suppression of π^0 and η mesons by a constant factor of 5 ($R_{AA} = 0.2$) for $4 \leq p_T \leq 15$ GeV/c, with suggestion of an increase in R_{AA} for $p_T > 15$ GeV/c; the equality of suppression of direct-single $e^{\pm}$ (from heavy quark (c, b) decay) and π^0 at $p_T \gtrsim 5$ GeV/c; the non-suppression of direct-γ for $p_T \geq 4$ GeV/c; the exponential rise of R_{AA} of direct-γ for $p_T < 2$ GeV/c [31], which is totally and dramatically different from all other particles and attributed to thermal photon production by many authors (e.g. see citations in reference [31]). For $p_T \gtrsim 4$ GeV/c, the hard-scattering region, the fact that all hadrons are suppressed, but direct-γ are not suppressed, indicates that suppression is a medium effect on outgoing color-charged partons likely due to energy loss by coherent Landau-Pomeranchuk-Migdal radiation of gluons, predicted in pQCD [28], which is sensitive to properties of the medium. Measurements of two-particle correlations [1] confirm the loss of energy of the away-jet relative to the trigger jet in Au+Au central collisions compared to p-p collisions. However, there are still many details which remain to be understood, such as the apparent suppression of direct-γ for $p_T \gtrsim 18$ GeV/c, approaching that of the π^0. Interesting new results this year extend and clarify these observations.

An improved measurement of π^0 production in Au+Au and p-p collisions by PHENIX [32] now clearly shows a significant increase of R_{AA} (decrease in suppression) with increasing p_T over the range $7 < p_T < 20$ GeV/c for 0-5% central Au+Au collisions at $\sqrt{s_{NN}} = 200$ GeV (Fig. 8a). Interestingly, despite more than a factor of 20 higher $\sqrt{s_{NN}}$, the ALICE R_{AA} data from LHC [33] are nearly identical to the RHIC measurement for $5 < p_T < 20$ GeV/c. However, since the exponent of the power-law at LHC ($n \sim 6$) is flatter than that at RHIC ($n \approx 8$, Fig. 7a), a $\sim 25\%$ larger shift $\delta p_T/p_T$ in the spectrum from p-p to A+A is required at LHC to get the same R_{AA} (Fig. 8b), likely indicating $\sim 25\%$ larger fractional energy loss at LHC in this p_T range.

Measurements by STAR (Fig. 9a) [34] of the evolution of charged hadron suppression with $\sqrt{s_{NN}}$ in Au+Au collisions, using the variable $R_{CP} = (R_{AA}^{0-5\%}/R_{AA}^{60-80\%})$ which does not require the p-p cross section (see Eq. 3)

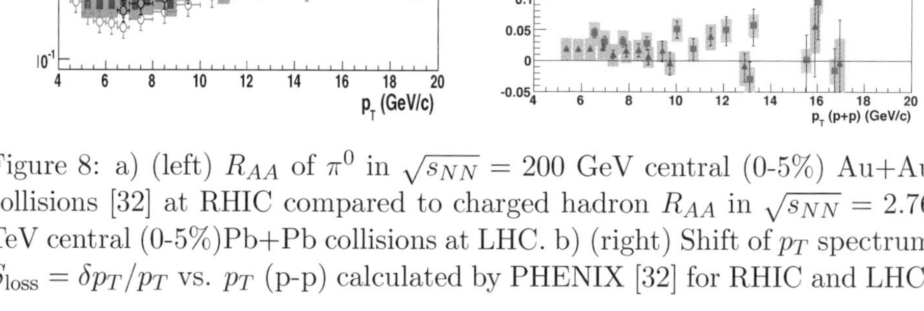

Figure 8: a) (left) R_{AA} of π^0 in $\sqrt{s_{NN}} = 200$ GeV central (0-5%) Au+Au collisions [32] at RHIC compared to charged hadron R_{AA} in $\sqrt{s_{NN}} = 2.76$ TeV central (0-5%)Pb+Pb collisions at LHC. b) (right) Shift of p_T spectrum $S_{\rm loss} = \delta p_T/p_T$ vs. p_T (p-p) calculated by PHENIX [32] for RHIC and LHC.

but is usually smaller than R_{AA} [4], show the transition from suppression ($R_{\rm CP} < 1$) to enhancement ($R_{\rm CP} > 1$) for $\sqrt{s_{NN}} > 27$ GeV.

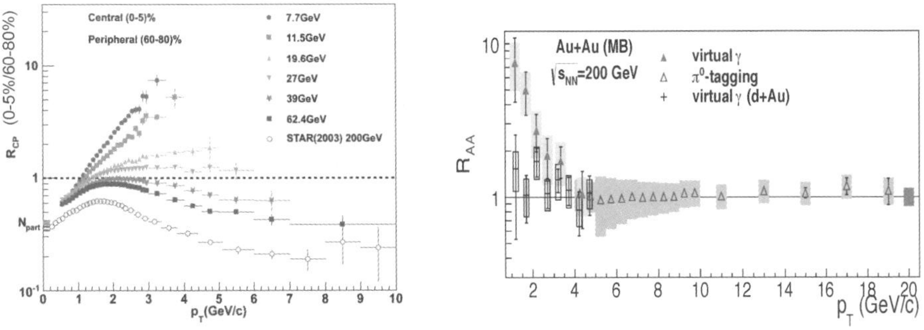

Figure 9: a) (left) STAR $R_{\rm CP}(p_T)$ for $h^{\pm}$ as a function of $\sqrt{s_{NN}}$ in Au+Au collisions [34]. b)(right) PHENIX $R_{AA}(p_T)$ of direct-γ in d+Au and Au+Au minimum bias collisions at $\sqrt{s_{NN}} = 200$ GeV [35].

Improved measurements of direct-γ production in p-p, dAu and Au+Au collisions by PHENIX [35] show several interesting results. In Fig. 9b, new measurements of $R_{AA} \approx 1$ for d+Au using internal conversions in the thermal region, $p_T < 4$ GeV, reinforce the uniqueness of the exponential rise of the Au+Au minimum bias photon spectrum, thus confirming that the exponential for $p_T < 4$ GeV/c in Au+Au is a hot matter effect, i.e. thermal photon production. Also in Fig. 9b, improved measurements of real direct-γ in Au+Au collisions, by eliminating background from γ-rays associated with a second γ in the π^0 mass range (π^0 tagging), no longer show an "apparent suppression" but are consistent with $R_{AA} = 1$ out to 20 GeV/c.

My favorite direct-γ result this year is the improved PHENIX measurement in p-p collisions at $\sqrt{s} = 200$ GeV out to $p_T = 25$ GeV/c [36], in excellent agreement with pQCD. A more direct way to show this without a detailed theory calculation is to use x_T scaling. Figure 10a shows x_T scaling for all presently existing direct-γ data[2], with $n_{\rm eff} = 4.5$, very close to the pure-scaling parton-parton Rutherford scattering result of $n_{\rm eff} = 4.0$ (Fig. 10b). The deviation of the data points in Fig. 10b from the universal curve for $\sqrt{s} > 38.7$ GeV is an illustration of the non-scaling of the coupling constant, structure and fragmentation functions in QCD—what I like to call "QCD in action". For $\sqrt{s} \leq 38.7$ GeV, the deviation of the data from the

[2]This includes the PHENIX p-p direct-virtual-γ measurement down to $p_T \approx 1$ GeV/c, further confirming the absence of a soft production mechanism for direct-γ in p-p collisions [1].

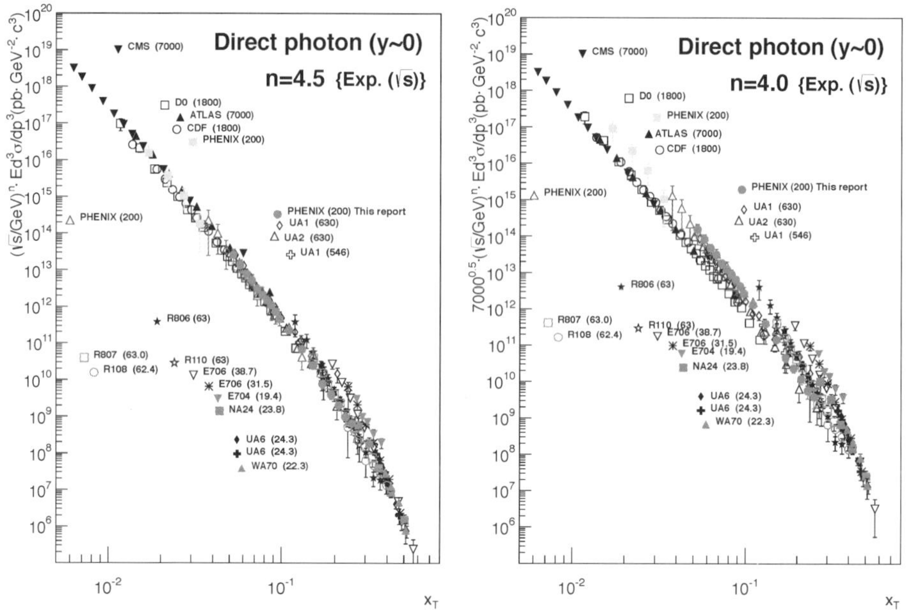

Figure 10: a)(left) $\sqrt{s}^{n_{\rm eff}} \times Ed^3\sigma/dp^3$, as a function of $x_T = 2p_T/\sqrt{s}$, with $n_{\rm eff} = 4.5$, for direct-γ measurements in p-p and $\bar{\rm p}$-p experiments at the ($\sqrt{s}$ GeV) indicated. [36]. b)(right) same as (a) with $n_{\rm eff} = 4.0$

universal curve in Fig. 10a (and from pQCD) is claimed to be due to the k_T effect (transverse momentum of the quarks in a nucleon).

Since the p_T^γ of a direct-γ can be measured very precisely, the fragmentation function of the jet from the away quark in the reaction $g+q \to \gamma+q$ can be measured by the direct-$\gamma-h$ correlations (where h represents charged hadrons opposite in azimuth to the direct-γ) because the p_T of the away-quark at production is equal and opposite to p_T^γ, thus known to high precision (modulo a small k_T-smearing effect). This year, improved measurements by PHENIX [37] in both p-p and Au+Au collisions (Fig. 11) now indicate a significant modification of the fragmentation function in Au+Au (0-40%) central collisions compared to p-p (Fig. 11a), with an enhancement at large $\xi = -\ln z_T$ (low $z_T = p_T^h/p_T^\gamma$) and a suppression at small ξ (large z_T) which is more clearly seen as $I_{AA}(\xi)$, the ratio of the fragmentation functions in Au+Au/pp (Fig. 11b). As shown in Fig. 11c,d, restricting the away-side azimuthal range reduces the large $\xi > 0.9$ ($p_T^h \lesssim 3$ GeV/c) enhancement but leaves the suppression at small $\xi < 0.9$ relatively unchanged, which shows that the large ξ enhancement is predominantly at large angles, similar to the effect observed by CMS with actual jets. [38].

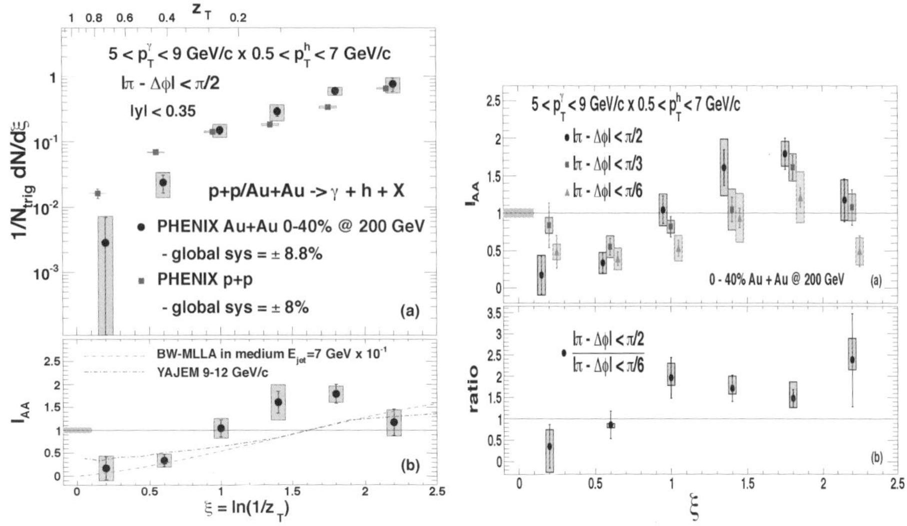

Figure 11: a)$\xi = -\ln(p_T^h/p_T^\gamma) = -\ln z_T$ distributions of hadrons opposite in azimuth ($|\pi - \Delta\phi| < \pi/2$) to a direct-$\gamma$ trigger with $5 < p_T^\gamma < 9$ GeV/c in p-p and $0-40\%$ central Au+Au collisions at $\sqrt{s_{NN}} = 200$ GeV from PHENIX [37]. b) Ratio of the two distributions, $I_{AA}(\xi)$. c) (right)-(top) $I_{AA}(\xi)$ as in (b) when the away-side azimuthal range is restricted as indicated. d) (right)-(bottom) Ratio of I_{AA} for $|\pi-\Delta\phi| < \pi/2$ to $|\pi-\Delta\phi| < \pi/6$.

One of the most exciting discoveries at RHIC, now confirmed at the LHC, is the suppression of heavy quarks by the same amount as light quarks for $p_T \gtrsim 5$ GeV/c as indicated at RHIC (Fig. 7b) by direct single-$e^{\pm}$ from heavy quark (c, b) decay; and at the LHC by D mesons from c-quarks [39], and non-prompt J/Ψ from b-quarks [40] (Fig. 12a). The discovery at RHIC was a total surprise and a problem since it appears to disfavor the radiative energy loss explanation [28] of suppression (also called jet-quenching) because heavy quarks should radiate much less than light quarks or gluons.

Many explanations have been offered including some from string theory; but the explanation I prefer was by Nino Zichichi [41] who proposed that since the standard model Higgs Boson, which gives mass to the Electro-Weak vector Bosons, does not necessarily give mass to Fermions, "it cannot be excluded that in a QCD coloured world (a QGP), the six quarks are all nearly massless". If this were true it would certainly explain why light and heavy quarks appear to exhibit the same radiative energy loss in the medium. This idea can, in fact, be tested because the energy loss of one hard-scattered parton relative to its partner, e.g. $g + g \to b + \bar{b}$, can be measured by experiments at RHIC and LHC using two particle correlations in which both

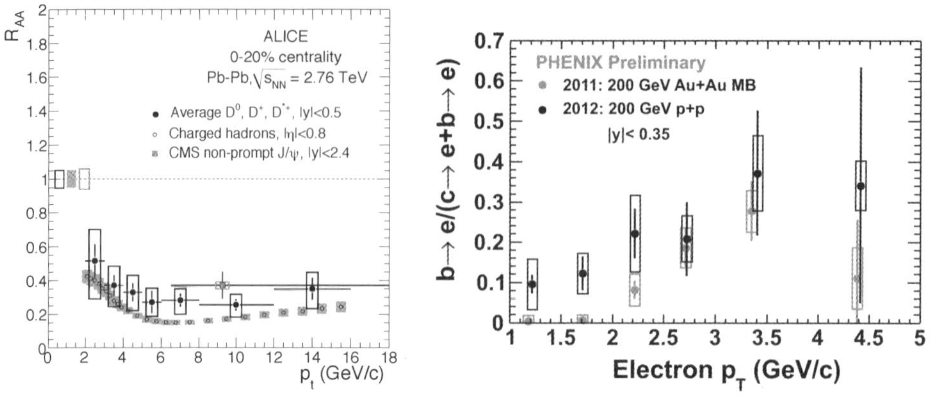

Figure 12: a) (left) R_{AA} of ALICE [39] D-mesons, charged hadrons, and CMS [40] non-prompt J/Ψ, in central (0-20%) Pb+Pb collisions at $\sqrt{s_{NN}} = 2.76$ TeV b) (right) b-quark fraction $F_b = b \rightarrow e/(c \rightarrow e + b \rightarrow e)$ of direct single-$e^{\pm}$ in p-p and Au+Au from PHENIX measurement of the Distance of Closest Approach (DCA) of the displaced vertex.

the outgoing b and $\bar{b}$ are identified by measurement of the Distance of Closest Approach (DCA) of their displaced decay vertices in silicon vertex detectors. When such results are available, they can be compared to π^0-charged hadron correlations from light quark and gluon jets, for which measurement of the relative energy loss has been demonstrated at RHIC [1].

Of course, measurement of the Yukawa couplings to Fermions of the candidate 125 GeV Higgs Boson at the LHC may be available by the end of 2012; but, already this year, the first direct measurement of b-quarks in p-p and Au+Au collisions at RHIC by their displaced vertices was made in the new PHENIX Silicon VTX detector [42]. Figure 12b shows the measurement of the b-quark fraction $F_b = b \rightarrow e/(c \rightarrow e + b \rightarrow e)$ of direct single $e^{\pm}$ in p-p and Au+Au collisions at $\sqrt{s_{NN}} = 200$ GeV, using the PYTHIA c and b quark p_T distributions in p-p collisions to calculate the DCA distributions of the $e^{\pm}$ in both p-p and Au+Au. The fact that the Au+Au measurements for all p_T^e are below the p-p measurements indicates clearly that the b-quark p_T spectrum is modified in Au+Au compared to p-p. However, the correct conditional DCA distribution requires the actual (modified) b-quark p_T spectrum in Au+Au, which must be obtained by iteration. Once the iteration has converged, the $R^b_{AA}(p^e_T)$ can be calculated from the measured $R_{AA}(p^e_T)$ of the direct-single-$e^{\pm}$ by the relation $R^b_{AA}(p^e_T) = R_{AA}(p^e_T) \times F^{AA}_b/F^{pp}_b$. For example, if the final $F^{AA}_b = F^{pp}_b$ then $R^b_{AA}(p^e_T) = R_{AA}(p^e_T)$.

References

[1] Discoveries and methods at RHIC have also been discussed by the present author in ISSP proceedings for the years 2008 (arXiv:0906.0745), 2009 (arXiv:1006.5701), 2011(arXiv:1201.5900), as well as in a textbook to appear soon [2].

[2] Jan Rak and Michael J. Tannenbaum, *High-p_T Physics in the Heavy Ion Era* (Cambridge University Press, Cambridge UK, 2013).

[3] E. V. Shuryak, Phys. Repts. **61** (1980) 71–158.

[4] I. Arsene *et al.* (BRAHMS Collab.), Nucl. Phys. A**757** (2005) 1–27.

[5] B. B. Back *et al.* (PHOBOS Collab.), Nucl. Phys. A**757** (2005) 28–101.

[6] J. Adams *et al.* (STAR Collab.), Nucl. Phys. A**757** (2005) 102–183.

[7] K. Adcox *et al.* (PHENIX Collab.), Nucl. Phys. A**757** (2005) 184–283.

[8] D. Rischke and G. Levin, eds., Nucl. Phys. A**750** (2005) 1–171.

[9] Press Conference, *A New State of Matter Created at CERN.*

[10] A. Aad, *et al.* (ATLAS Collab.), Phys. Rev. Lett. **105** (2010) 252303.

[11] S. Chatrchyan, *et al.* (CMS Collab.), Phys. Rev. C**84** (2011) 024906.

[12] T. Matsui and H. Satz, Phys. Lett. B**178** (1986) 416.

[13] D. M. Alde, *et al.* E772 Collab., Phys. Rev. Lett. **66** (1991) 2285–2288. M. J. Leitch, Eur. Phys. J. C**43** (2005) 157–160 and references therein.

[14] *The Relativistic Heavy Ion Collider Project: RHIC and its Detectors*, Nucl. Instrum. Methods A**499** (2003) 235–880.

[15] *The Large Hadron Collider: Accelerator and Experiments*, JINST **3** (2008) S08001–S08007.

[16] M. Blaskiewicz and W. Fischer, CERN Courier **52** (Oct. 2012) 17.

[17] G. Bunce, *et al.*, Ann. Rev. Nucl. Part. Sci. **50** (2000) 525.

[18] J. T. Mitchell *et al.* (PHENIX), Proc. QM2012, Nucl. Phys. A .

[19] Yi Gu *et al.* (PHENIX), Proc. QM2012, Nucl. Phys. A .

[20] D. Teaney, Phys. Rev. C**68** (2003) 034913.

[21] B. H. Alver, *et al.*, Phys. Rev. C**82** (2010) 034913.

[22] A. P. Mishra, *et al.*, Phys. Rev. C**77** (2008) 064902.

[23] P. K. Kovtun, *et al.*, Phys. Rev. Lett. **94** (2005) 111601.

[24] N. Borghini and J.-Y. Ollitrault, Phys. Lett. B**642** (2006) 227.

[25] e.g. see R. Lacey, *et al.* (PHENIX Collab.), J. Phys. G**38** (2011) 124048.

[26] R. L. Cool, *et al.*, in *Proc. XVI Int. Conf. High Energy Physics*, Eds. J. D. Jackson and A. Roberts (NAL, Batavia, IL, 1973), Vol 3, p. 317.

[27] A. L. S. Angelis, et al. (CCOR), Phys. Lett. B**79** (1978) 505.

[28] See R. Baier, D. Schiff and B. G. Zakharov, Ann. Rev. Nucl. Part. Sci. **50** (2000) 37–69, and references therein.

[29] K. Adcox, *et al.* (PHENIX Collab.), Phys. Rev. Lett. **88** (2002) 022301.

[30] S. S. Adler, *et al.* (PHENIX Collab.), Phys. Rev. C**76** (2007) 034904.

[31] A. Adare, *et al.* (PHENIX Collab.), Phys. Rev. Lett. **104** (2010) 132301.

[32] A. Adare, *et al.* (PHENIX Collab.), arXiv:1208.2254v1.

[33] K. Aamodt, *et al.* (ALICE Collab.), Phys. Lett. B**696** (2011) 30.

[34] Xin Dong *et al.* (STAR), Proc. QM2012, Nucl. Phys. A .

[35] A. Adare, *et al.* (PHENIX Collab.), arXiv:1208.1234v1.

[36] A. Adare, *et al.* (PHENIX Collab.), Phys. Rev. D**86** (2012) 072008.

[37] J. Frantz, *et al.* (PHENIX), Proc. QM2012, Nucl. Phys. A .

[38] G. Roland *et al.* (CMS), Proc. QM2012, Nucl. Phys. A .

[39] B. Abelev, *et al.* (ALICE Collab.), JHEP **09** (2012) 112.

[40] S. Chatrchyan, *et al.* (CMS Collab.), JHEP **05** (2012) 063.

[41] A. Zichichi, Nucl. Phys. A**805** (2008) 36c.

[42] R. Nouicer, *et al.* (PHENIX), Proc. QM2012, Nucl. Phys. A .

CHAIRMAN: M. J. TANNENBAUM

Scientific Secretaries: M. Csanád, Á. Lukács

DISCUSSION

- Csanád:

Thank you for showing a lot of nice results on how the research of QGP is going on. I am wondering what your opinion would be if someone would try to find the QGP in proton-proton collisions?

- Tannenbaum:

Well, thank you, good question. But you are lucky because I actually wrote an article on that with Richard Weiner, it is the only article we wrote together, so it is straightforward to find it. It is only on the arxiv, arXiv:1010.0964. You should look at it. When the CERN heavy ion run began, ATLAS immediately could see the jets being completely different than in p+p. Right before that, Gunther Roland, from CMS, who also found v3 at RHIC, said that he found some longitudinal correlations in p+p collisions at LHC that nobody had ever seen before. I thought everybody sees longitudinal correlations in p+p. The first string model, the Lund Model and color strings have been there since the introduction of color. Richard Weiner believed that there is a QGP in p+p but I did not believe it. It was interesting that we collaborated on an article, especially as we did this by telephone and by email and had different opinions on the subject. So here are my most important condition for finding the QGP in p+p; I don't remember them all. The one I really remember is that you have to see the suppression of pi0's. How do you see that? You have to measure direct photons and pi0's in fully inclusive events, the standard QCD processes, you plot pi0/direct-gamma. Then you measure these at central (meaning high multiplicity) p-p collisions where CMS found the effect. In high multiplicity events, the spectrum of course gets steeper because of energy conservation, but the same kinematic effect will be there in both pi0 and the direct photon spectrum. You would have to look for the difference (e.g. a suppression) in the pi0/direct-gamma ratios for central and fully inclusive collisions. You have to measure things like that to convince me that there is QGP in p+p, and do all the checks that we have done in AA. So, in summary, is it possible that there is QGP in p+p? Yes! Will I accept some lame measurement as proof? No, you really have to prove it scientifically.

- Davis:

You mentioned the Uranium nucleus has this football shape and that there are different orientations in which you can collide it. You said that this does not actually affect your results. I am wondering, why could it affect and why does it not? And was that surprising to you?

- Tannenbaum:

That is a very good question. Why doesn't it surprise me? There are two reasons. The main reason is on my poster for QM2012. The answer is the following. There is this famous equation that people use that I showed in my talk. The multiplicity that we measure per number of participant-pairs is not equal to a constant, but it increases with the number of participants. Kharzeev and collaborators wrote a formula that the average multiplicity in nucleus-nucleus collisions would be the average multiplicity in p-p collisions times a fraction x times the number of participants plus a fraction 1-x times the number of collisions. It is on the PHENIX plot from PRL86, 3500 (2001). The number of participants represents soft physics, the number of collisions is for hard physics. It is known from measurements that were made at the ISR, and especially at UA1 and UA2, that the multiplicity coming from jets is very small compared to soft particles. If you just look at inclusive single particle pT distributions in p-p collisions, the effect of jets is not visible until roughly 6 orders of magnitude down in cross section. We know that at c.m. energy 540-630 GeV the jet effect is negligible on total multiplicity, and we measure only at 200 GeV; but somehow some distinguished theorist puts it in a formula and everybody starts using it Let me tell you another interesting story. When we in PHENIX published our first paper which was on multiplicity in 130 GeV collisions, and which was the first measurement to plot the multiplicity/participant-pairs as a function of the number of participants; I was the senior physicist in the experiment and I was wildly against putting the mentioned formula in the paper (PRL86, 3500), as it is manifestly wrong: ``How can you put this formula in a paper?" ``Well, a distinguished theorist said it"; and I got overruled by all the young people, and I did not want to get into a fight, so I said I know I'm right, but fine, put it in the paper. The answer is that when we measure down to cm energy 7.7 GeV, the dN/deta/(0.5Npart) vs Npart distribution lies on the shape same curve as at 200 GeV and 2.76 TeV, where jet production is larger by orders of magnitude. Thus the shape of dN/deta/(0.5 npart) has nothing to do with hard scattering, it represents the number of quark participants per nucleon participant. This is easily calculated in a simple Glauber model, based on nucleons made up of three quarks independently distributed radially so as to give the correct proton radius and with quark-quark scattering cross section adjusted to give the correct p-p cross section. Also it's a lesson in statistics. That's what one can see from the plot of dN/deta/(0.5Npart) for all the reactions from p-p to CuAu to AuAu to UU. If I know what the pp distribution is and I convolute it Npart times I get a distribution roughly like the all the AA distributions. If I also convolute it Ncoll times which is roughly 4 time the number of participants, I get another distribution of the same shape which extends 4 times larger in multiplicity than is shown on the plot. The distribution that you would see would be a linear sum of the two distributions i.e. a distribution with a sharp drop at Npart and a long nose that extends out to Ncoll times the mean multiplicity in p-p. You don't see any distribution that looks like that; do you? One of the best young physicists on the experiment (a tenured faculty member who believed Kharzeev's formula) came to me and asked me why it looked so simple. The Npart-Ncoll formula was a very firmly held belief in the community. Anyhow, so that is why I'm not surprised that U+U is very similar to all other A+A collisions: the only difference between different orientations in U+U collisions is that in one of them (tip-tip) you get more collisions; but since the results depend on the number of quark-participants, that is irrelevant. But the beautiful thing I would like to do, is the following. As you know, the PHENIX detector is very limited in solid angle: two wedges, each 90 degrees in azimuth, located nearly back-back in azimuth. I thought, if we trigger on just transverse energy distributions, we select configurations in which the particles come out in the plane of our detector. This means that due to the azimuthal asymmetry of emission, a configuration in which the short axis of the uranium is pointed to our detector acceptance and the long axis is perpendicular in azimuth to our detector should be selected. So we should see that effect in anisotropic flow measurements in our forward v2 detector. But that is to be determined yet. Its like when quarks first were proposed. If you think that there is something inside the nucleon or the nucleus, I like to make it do tricks. So that's one trick that I can do with the Uranium nucleus---trigger on energy coming out in the horizontal plane and find out that this causes the uranium nuclei to be oriented with their long axis perpendicular to the horizontal plane and their short axis in the plane. That would be great. I hope that answers your question.

- *Gnesi:*

In your talk you showed a beautiful distribution of photon energies and you told that they are probably due to blackbody radiation. It is a very important point, and since it is not possible to see the peak of the distribution due to detector thresholds, you need to fit them with a blackbody radiation function to determine the temperature. Did anyone do that, and if yes, what temperature came out? Is it related to the critical temperature Hagedorn or not?

- *Tannenbaum:*

The fact that the pion cross-section as a function of pT is an exponential, exp(-6pT), is called the Cocconi formula, which means that the mean pT is 2/6=0.333 GeV/c and the inverse slope (an estimator of the temperature) is 1/6=0.167 GeV. This is known since the 1940's. Cocconi was another famous Italian physicist who came to the US. Anyway, all distributions for hadron production that you see in p+p are exponential. The only one I can show you on my slides is the pi0, but the other hadrons are similar as well. For the pi0 pT spectrum on my slide, which we measured down to 500 MeV/c, it is power-law at high momenta, but at low momenta pT< 1.5 GeV/c, it is exponential. We call this the soft physics. However, these exponentials are, I think, due to forward jets. When the protons interact, there are a lot of particles going forward. In fact Zichichi uses that because you don't count the energy that goes forward so the rest is the effective energy. But the forward particles also make forward-jets and these contribute to the mid-rapidity soft transverse momentum distributions. But there is no soft process that makes photons. The way I know this is from Drell-Yan, which is a similar process to direct-photon production: the transverse momentum distribution is not exponential at low pT, it does not go like the pi0 that I showed, but it turns over. We indeed found this to happen to the photon pT spectrum: it is not exponential in p-p.We measured low pT direct photon production with internal conversions [see PRL 104(2010) 132301] in both p-p and Au+Au collisions. This eliminates the pi0 background. The best fit to the p-p spectrum is the function used in Drell-Yan [Ito, et al, PRD 23, 601 (1984)] which is not exponential but becomes flat as pT goes to zero. It is different from the hadron pT spectra in p-p collisions which are exponential. What is interesting is that the Au+Au data can be fit with an exponential plus the Drell-Yan type of formula. In Au+Au the direct-photon spectrum at low pT is exponential, very different in shape from p-p collisions. This is why I believe that the low pT photons in Au+Au are thermal: there is no other reaction where you see that. If you read our paper [PRL 104(2010) 132301] you see that a lot of people have done fits to our measurement. If you fit our data to an exponential, you get 220 MeV in temperature. But remember, these are photons from the medium. The medium is expanding, cooling, interacting and radiating photons. The transverse momentum distribution is an average is over the whole history of the QGP. Thus the actual temperature at the collision is much higher than the value from our exponential fit. So in order to derive the temperature in the medium from this, you need a theory calculation. There were several, but I don't discuss theory, I don't even talk about the Higgs-boson until I see one (which is getting closer... but anyway). But we at RHIC are now recognized in the Guinness Book of Records as having achieved the highest man-made temperature.

- *Gnesi:*

But that is not much higher than the critical temperature. MJT: You tell me what the critical temperature is. IG: 150-160 MeV. MJT: How do you know that? IG: Well, it's the Hagedorn temperature.

- *Tannenbaum:*

The actual temperature of the QGP is more like 300-400 MeV. What we extracted is just the inverse slope that is averaged over the whole history. But as I said, there is a whole bunch of papers that give you the temperature from 200--600 MeV. Do I believe them? No, I don't, but until we can find a better way to measure it, that is fine. Anyway this is the hottest that's been seen. However, in fact, if we would measure the inverse pT slope of heavier particles such as protons, we would get an even larger value; but that is well known. For the temperature estimator, you have to use the inverse slope as a function of transverse mass not that of transverse momentum.

- *Russo:*

Q1. Besides the flow effect, which you have told us about, is there any other effect that you expect to see in the U-U collisions, that is not present in collisions of spherical shaped nuclei?

Q2. Is the determination of centrality different for these collisions?

- *Tannenbaum:*

Q2. So, we go back to the plot of multiplicity distributions for Au-Au and U-U. We don't know how to do the centrality yet. I mean, that there is always something you have to learn. So, for instance, normally, as I showed you, if you take this distribution, if everything is monotonic in the impact parameter, which may not be the case in UU, you can just take the upper percentile, and you get the centrality. Now, for UU collisions, the multiplicity may or may not be monotonic with impact parameter due to the different possible orientations, so we have to figure out how to obtain the centrality. We have ways to do it: we have forward calorimeters, and we are looking at central transverse energy, but we haven't figured out how to do it yet. Fortunately, our management was kind. Back in the good old days, if you said, "I would like to measure this", they would say: Has anybody ever done it before? Is it interesting?" Yes! Well, this is one of those cases. Also, the accelerator wanted to do U+U. They did not make us write a hundred page paper on how we would determine the centrality. So we were given a very short run, and we are going to figure out how to do it.

As to Q1, the only advantage that we saw in the UU is that you have the two different geometries; so maybe you can separate the Ncoll effect from the Npart effect. The other thing that you do not get in PbPb or AuAu collisions is the fact that, in the most central collisions, the overlap region is not a circle with no obvious direction or axis, but has an axis. So, in central collisions, you would get v2 that does not come from fluctuations. Those are the two reasons why we did UU.

- *Leontsinis:*

I would like to know if there are any Bose-Einstein correlation studies that could give you a good proof of the Quark Gluon Plasma? Having such a large multiplicity, you could do also third or fourth order as well.

- *Tannenbaum:*

Yes. The answer is yes, these were all done. They won. There was a big problem. You have to look in 2001 papers to see that. Some guys still do it now, in fact, maybe STAR. We are now doing the energy scan, maybe STAR has something new. The answer is, you get what you expected. Nobody has done it for years, and now people are doing it again, but I don't have any new exciting results, just the very old ones. You just think about it, 2001, that's eleven years ago that we had those results.

- *Carlucci:*

I would like to deepen a bit the impact of RHIC on the ALICE experiment, because the last week I listened to an analyst talk in which they told us that the indication they got from RHIC did not permit to explore some ranges of the transverse momentum. I would like to understand better, since I am not an experimentalist.

- *Tannenbaum:*

You know, as they say, it's nothing personal, it's only energy conservation. They have 10 times the c.m. energy that we do. So for instance, I would like to show you two things. The one thing that is totally unique at the LHC, so far, is the clear observation of jets. But, first, I'll answer a question that Tapia was going to ask me. In the comparison of RAA from PHENIX pi0 with the ALICE charged hadron measurement, the ALICE measurement took one month, and our data on the same plot is what we have been measuring for, say, a year. This is not the elapsed time since the measurement, but just the data taking time. So they just get a lot more events, because the pT spectrum does not fall like the 8th power, it is the 6th power, so you just get a lot more high-pT events. So they get better statistics, and, obviously, they will go out to pT=50 GeV/c and they see from 5-20 GeV/c that RAA rises, there is less suppression. It is very interesting to find out if the suppression keeps decreasing with increasing pT and to what value; and whether the effect is the same at RHIC as at LHC. If we ever got up to 30 GeV/c that would be fantastic, but it drops like the 8th power, so to go from 20 to 30, you need (1.5)^8 more integrated luminosity combined with detector improvement. That's what is difficult. The other thing they do at LHC is jets, and I'll sneak in something new before my time is up. We proposed to get into jet measurements in our next detector. But, let me note, that the original PHENIX detector that we made was a very good detector, and we accomplished what we wanted, or probably more: namely single electron, J/Psi at mid-rapidity, pi0s; and now we want to do is build a basically conventional collider detector, a hadron calorimeter, an electromagnetic calorimeter, and take out one muon arm, and take out the existing calorimeters, and we already have a vertex detector as the basis of a tracker. To show you the scale of it, lets go back to my comparison of PHENIX to a wedge of CMS. Remember, that PHENIX is 5m to the front face of the calorimeter and another meter in depth for 6m in radius. Now, what we want to do is something like a compact `4pi' detector, so that we will be able to measure jets. Another good thing is that our measurements of triggered correlations will also be improved. At present we don't have good acceptance because we can only see particles in two 90 degree back-to-back wedges in azimuth with limited rapidity acceptance $|y|<0.35$. The new detector will have full azimuthal coverage with $|y|<1$, a gain of a factor of 5-10, all within a 2.5m radius. So, that's what we want to do, more the younger guys than I, so we hope that we will be able to do it.

- Inguglia:

I am involved in the ARACHNID collaboration, and we are testing some I.N.MAPS technology for vertex detectors. This is also under study at SUPER-B and ALICE, for the upgrades. I am wondering what are the technologies that you are evaluating for the super-PHENIX, and if you have any flexibility in that.

- Tannenbaum:

Yeah. Do we want collaborators? You bet! Do you want to come? Come right away. We really need a collaboration, and we have not submitted the proposal yet. If you want to sign it, we're ready. We also need money, because we are limited to 20 million US dollars at the moment, so we can't build the whole thing we want from that, so we need outside sources. For example, PHENIX was approved as a 30 million dollar experiment and we actually spent $100 million, because we had some additional U.S. money plus huge outside sources, from Japan, from Russia, from other countries. So, we can do that. I can't really address the technology well, except that we already have a new Si vertex detector, and we want to add additional layers of silicon at larger radii to complete the tracking inside the solenoid. Now I know there are a lot of new technologies, so I don't know which one we are going to use. The guys from RIKEN in Japan built this detector, and so, yes, we are open to that.

- Inguglia:

Yes, because it is very cheap with its monolithic active piece.

- Tannenbaum:

You must send me all this information.

- Inguglia:

OK, I will do that in a minute.

Origin and Status of the Gran Sasso INFN Laboratory

Lucia Votano
Gran Sasso National Laboratory of the INFN, Italy
E-mail: lucia.votano@lngs.infn.it

ABSTRACT

The Gran Sasso National Laboratory of INFN (LNGS) is the largest underground laboratory for astroparticle physics in the world. Located in Italy between the cities of L'Aquila and Teramo, 120 Km far from Rome, is a research infrastructure mainly dedicated to astroparticle and neutrino physics. It offers the most advanced underground facility in terms of dimensions, complexity and completeness of its infrastructures. LNGS is one of the four national laboratories run by the INFN (Istituto Nazionale di Fisica Nucleare). The scientific program at LNGS is mainly focused on astroparticle, particle and nuclear physics. The laboratory presently hosts many experiments as well as R&D activities, including world-leading research in the fields of solar neutrinos, accelerator neutrinos (CNGS neutrino beam from CERN to Gran Sasso), dark matter, neutrinoless double beta decay and nuclear cross section of astrophysical interest. Associate sciences like earth physics, biology and fundamental physics complement the activities. The laboratory is operated as an international science facility and hosts experiments whose scientific merit is assessed by an international advisory Scientific Committee. A review of the main experiments carried out at LNGS will be given, together with the most recent and relevant scientific results achieved.

1. A brief history of the Gran Sasso Laboratory

The Gran Sasso National Laboratory of the INFN (Istituto Nazionale di Fisica Nucleare) was completed in 1987 and after 25 years, thanks to the farsightedness of the planning system, of the scientific conduct and of the continuous support of the INFN, the Laboratory is still the biggest and most advanced infrastructure in the world dedicated to astroparticle physics. The then President of INFN Antonino Zichichi conceived in the late 70s the idea of building a large underground Laboratory under the Gran Sasso massif, equipped with technological plants and facilities comparable with the ones used in high-energy accelerator laboratories. At that time the tunnel under the Gran Sasso mountain of the Rome-Teramo highway was under construction and this was a unique opportunity for the excavation of large halls of an underground laboratory at a reasonable price. The 70s were also the dawning of the astroparticle physics that was taking the first steps towards the extraordinary developments that we have seen in the following decades. In 1979 the proposal to build a large, high technology underground laboratory under the Gran Sasso massif was presented to the Senate Commission for Public Works. The Italian Parliament approved the 'Gran Sasso Project' and its first funding in 1982, followed by a second financing in 1984. By 1987 the civil engineering works were completed and the first experiments had begun to be commissioned. In the same year Enrico Bellotti took office as the first Director of the Laboratory that at the same time had been officially set up as one of INFN National Laboratories. In 1989 the first experimental apparatus, MACRO, started its data taking. The second Director was Piero Monacelli, followed by Alessandro Bettini and Eugenio Coccia. I have been the fifth Director and the sixth, Stefano Ragazzi, is in office since October 15, 2012. Looking back on those past decades, the advances we have made in understanding the fundamental laws of nature and the evolution of the universe as well as the extraordinary growth of astroparticle physics can be clearly observed. An ever increasing number of physicists joined this sector, which represents one of the most fascinating and challenging deal of the physics research. Technologies developed in accelerator apparatuses were first adopted; later on the search for very rare events, the need of increasing sensitivity and efficiency and the complexity of the analysis have called for the development of ever more cutting edge technologies. Astroparticle physics would not have been able to make such a massive and rapid progress without the great infrastructures necessary for this kind of study that only the facilities of underground laboratories can offer. Having planned and built such a large and well equipped laboratory as early as the late 1970s, has brought Italy to have a leading role in this field, since then.

Figure 1: Luciano Maiani, Antonino Zichichi, Piero Monacelli, Marco Napolitano during the visit of the Italian Republic President Eugenio Scalfaro.

Since the very beginning LNGS hosted first class experiments in the astroparticle field: LVD, a permanent observatory for the low energy neutrinos coming from gravitational stellar collapses in our galaxy still in operation at LNGS; GALLEX/GNO, a radiochemical experiment that, measuring the flux of neutrinos coming from the pp reaction of the Sun, gave a decisive contribution to solving the famous puzzle of solar neutrinos; MACRO, devoted to the hunt for magnetic monopoles that measured, together with the Japanese experiment Kamiokande, the effect of atmospheric neutrinos oscillation; GENIUS-TF, HDMS, DAMA, first generation search for dark matter particles; DBA, D-BGS, MIBETA, HMbb, CUORICINO, searching for neutrinoless double beta decays.

2. Gran Sasso National Laboratory

LNGS is one of the four National Laboratories of INFN - the Italian public research

Institute for subnuclear and nuclear physics - and a research infrastructure with no equal all over the world, as well as the most advanced underground Laboratory in length, complexity and completeness of the infrastructures. Located between L'Aquila and Teramo - about 120 km far from Rome - the underground structures are on one side of the highway tunnel (10 km long) which crosses the Gran Sasso massif (A24 Teramo-Rome Highway), and consist of three huge experimental halls (each one approximately 100 m long, 20 m large and 18 m high) linked by service tunnels, for a total volume of 180000 m^3 and a surface of 18000 m^2. The easy access to the Laboratory from the highway allows the transport of huge and heavy apparatus directly into the halls as well as the constant provision of supplies needed for the functioning both of the experiments and the Laboratory itself, and finally ensures an easy turnover of the personnel working underground. The Laboratory is equipped with all technical and safety plants required in order to run large and complex experiments and to ensure proper working conditions. For instance ventilation plants allow, in normal conditions, a change of air in less than 3 hours, and air radon content is kept in the 20÷80 Bq/m ^{3} range. The 1400 meter-rock thickness above provides a cosmic ray flux reduction by one million times; moreover, the flux of neutrons in the underground halls is 2.92×10^{-6} n cm^{-2} s^{-1} below 1 keV, and 0.86×10^{-6} n cm^{-2} s^{-1} over 1 keV, thousand times less than on the surface, due to the very small amount of uranium and thorium of the dolomite calcareous rock of the mountain. The gamma background is dominated by the U and Th content in the wall concrete (on average 1.05 ppm U, 0.656 ppm Th and 0.54 % of K). Outside, next to the highway tollgate of Assergi, an area of more than 23 acres belonging to a National Park of exceptional environmental and naturalistic value on the slopes of Gran Sasso hosts the external laboratories and facilities, the Computing Centre, the Directorate and various Offices. The on site personnel of LNGS including staff, temporary contracts and post docs is of the order of 110 people; besides them, more than 950 scientists from 29 different Countries take part in the LNGS experimental activities and constitute its large international scientific community.

3. LNGS Scientific Activities

Many of the experiments currently underway or soon to start at LNGS are looking at fundamental and fascinating questions which are still open despite the relevant progress in the astroparticle physics field. What is the Universe made of and in particular Dark Matter? In this area are we any nearer opening a door to a physics that goes beyond the Standard Model of particle physics? What is the intimate nature and what are the latest characteristics of the neutrino and what role have they played in the evolution of the Universe? What do we know of the interior of the Sun and the stars, of the mechanisms that produce energy and how they evolve and die? What do we know about the interior of the Earth, can we find out more through our study of neutrinos? The present activities of LNGS can then be grouped into the following main lines of research: Dark matter searches, Neutrino physics, Nuclear Astrophysics and Associate Sciences including Environmental Radioactivity for Earth Sciences, Geophysics, Fundamental Physics, Biology. The underground laboratory presently hosts many experiments and is quite crowded; Fig. 2 shows the actual occupancy of the underground laboratory and its evolution in the last few years: GERDA and ICARUS are the last experiments put in operation and precisely in 2010, XENON 1T, DS 50, LUNAMV and ERMES-WORLD are the new experiments or facilities recently approved by INFN, already in construction phase or expected to start soon the construction.

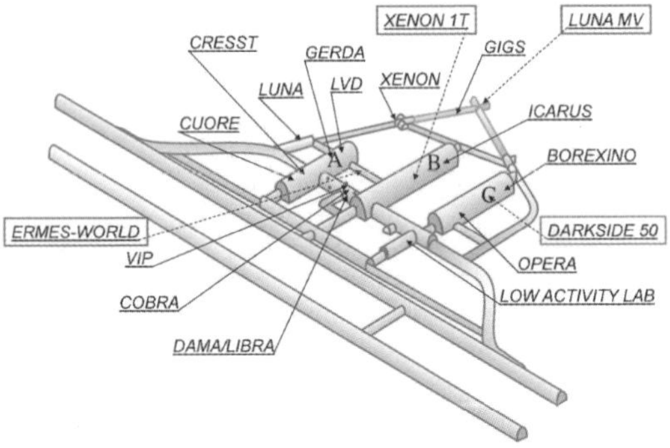

Figure 2: Present occupancy of the underground laboratory.

4. Neutrino Physics

The study of the properties of neutrino is of prime interest in particle physics and one of the main research topics of the present scientific program of the Laboratory. Neutrino physics has provided by now the only experimental evidence of phenomenon not included in the Standard Model of elementary particles by measuring neutrino oscillations and showing that neutrinos do have a mass, although very small. Besides, the study of a phenomenon as the neutrino-less double-beta decay could allow us to find out if the neutrino overlaps with its antiparticle, thus providing a very significant answer towards the comprehension of the evolution of the Universe. Massive Neutrinos may have played a relevant role to explain the slight asymmetry between matter and antimatter, which has generated the Universe of matter, the way we know it. Neutrino physics is therefore a window open to a new theory of elementary particles and to the comprehension of the evolution of the Universe. Finally, neutrinos from the cosmos are very important messengers which transport information fundamental to understand the functioning of the stars as energy sources, their evolution and what happens when they "turn off". LNGS activities range among various aspects of the neutrino physics study and various neutrino sources, both natural (the Sun, stars and the Earth) and artificial (particle accelerators), are used.

5. Neutrinoless Double-Beta Decay

Neutrino oscillation experiments have clearly shown that neutrinos do oscillate and that they are massive, though this is not enough to determine the nature of such mass (Dirac or

Majorana) and to determine the absolute mass scale. Neutrino oscillation experiments indeed can only measure the absolute value of the difference of the square of the neutrino masses while two different hierarchical mass arrangements of neutrino masses (Direct and Inverted), besides the obvious quasi-degenerate option, are possible. Neutrinoless double-beta decay ($2\beta0\nu$) is a process by which two neutrons in a nucleus can undergo beta decay by exchanging a virtual Majorana neutrino, emitting an electron each. This would violate lepton number conservation ($\Delta L = 2$) and would necessarily require neutrinos to be Majorana particles. From an experimental point of view, the only available information is carried by the daughter nucleus and the two possible emitted electrons and then the possible measurable quantities are: sum of the electron energies, single electron energy and angular distributions, identification and/or counting of the daughter nuclei. In most cases the experimental signature rely on the electron sum energies and on the discrimination of the different distribution expected for: a continuous bell distribution for $2\beta2\nu$, and a sharp line at the transition energy for $2\beta0\nu$. Energy resolution and level of measured background are the critical issues. In 2001 evidence for a $2\beta0\nu$ signal has been claimed by a small subset (KHdK) of the HdM ("Heidelberg-Moscow") collaboration at Gran Sasso National Laboratory that has raised some criticisms. At present, a number of experiments in the search of neutrino-less double-beta decay are in different phases of realization at LNGS and the related program exploits complementary approaches concerning isotopes and technique: GERDA, CUORE, COBRA, LUCIFER.

6. GERDA

The GERDA detector [1] is designed to search for $2\beta0\nu$ decay of ^{76}Ge using an array of high-purity germanium detectors, enriched (85%) in ^{76}Ge. Bare Ge diodes are operated in a 65 m^3 cryostat filled with liquid argon (LAr) which acts as a coolant for the detectors and as a shield against the radioactivity of the cryostat and of the underground environment. A 10 m diameter tank built all around the cryostat and full of ultrapure water represents a further shield. The Ge diodes are both source and detector for $2\beta0\nu$ decay. The expected signal is a peak at the Q-value of the decay ($Q_{\beta\beta}$= 2039 keV) in the spectrum of the sum of the two-electron energy. The water is instrumented with PMTs to detect the Cherenkov light of muons passing through the water tank. The GERDA experiment is foreseen to proceed in two phases. For Phase I, reprocessed enriched HPGe detectors from the past HdM and IGEX experiments and from the Genius Test-Facility have been deployed. In Phase II new diodes, using the BEGe technology and so providing more powerful discrimination between single and multi-site events, will be added to increase the active mass up to 37.5 kg. A simultaneous background reduction down to 10^{-3} cts/(kg keV y) will allow to increase the sensitivity of the half-life of the process by one order of magnitude for a total exposure of 100 kg y. In this case it would be also possible to probe the effective neutrino masses (m_{ee}) from 90 to 150 meV, depending on the matrix element. The construction of the experiment was completed in summer 2010 and the commissioning in autumn 2011. Since November 2011 Phase I data taking has started and GERDA is presently operating eight enriched coaxial detectors (approximately 15 kg of ^{76}Ge). Moreover seven new enriched BEGe detectors have been produced, 5 of which have been tested and deployed in GERDA at the beginning of July 2012. The background reached is approximately 17×10^{-3} cts/(keV kg y), which is about a factor of 10 lower than for previous experiment HdM and close to the design goal of 10×10^{-3} cts/(keV kg y). This is sufficient to confirm or refute the positive claim made in the past from the KHDK group. Depending on the overall performance the sensitivity will reach $2\text{x}10^{25}$ years at the beginning of 2013, which is practically the goal for Phase I.

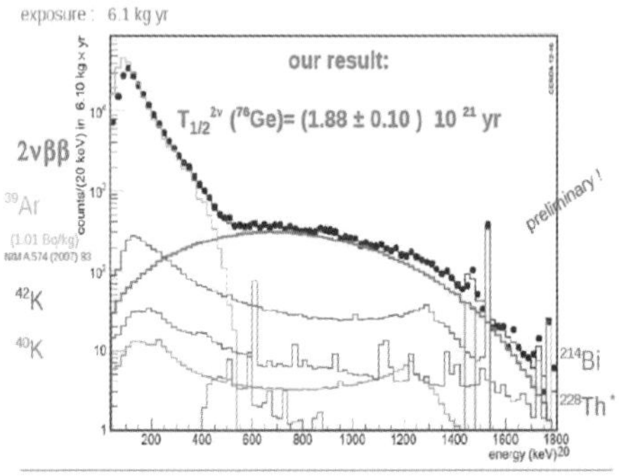

Figure 3: GERDA $2\beta2\nu$ preliminary results

7. CUORE

The experiment Cuore [2] is designed to search for $2\beta0\nu$ in the isotope ^{130}Te and represents the most ambitious development of the 'TeO$_2$ bolometers' technique, in which INFN has more than 20 years experience. In the bolometer detectors, the energy from particle interactions is converted into heat and measured via the resulting rise in temperature. Operated at a temperature of about 10 mK, these detectors have an energy resolution of a few keV over their energy range, extending from a few keV up to several MeV. The expected signature of $2\beta0\nu$ in ^{130}Te is a peak in the measured energy spectrum at 2527 keV, the Q-value for the transition $^{130}\text{Te} \rightarrow {}^{130}\text{Xe}+2\beta^-$. A background level of the order of 0.01 cts/(kg keV y) is expected by extrapolating the CUORICINO [3] background results and the dedicated CUORE R&D measurements. Considering the high mass of the experiment, the CUORE predicted limit for the half-life, after 5 years of operation, is 9.7×10^{25}y (90% C.L.) corresponding to m (51-120) meV. CUORE will therefore cover all the degenerate region of the neutrino mass pattern and will allow a close look at the IH region of neutrino masses. CUORE will consist of a rather compact cylindrical structure of 988 cubic TeO$_2$ crystals of 5 cm side (750 g), arranged into 19 separated towers (13 planes of 4 crystals each) and operated at a temperature of 10 mK. CUORE is in the construction phase at

LNGS and is expected to start operation in 2015. The first tower CUORE-0 will start data taking by autumn 2012.

8. Solar neutrinos and Borexino

Borexino is a large unsegmented liquid scintillator detector built for real time measurement of low-energy solar neutrinos. The inner part of the detector is a Stainless Steel Sphere (SSS) 13.7 m in diameter that represents both the container of the scintillator and the mechanical support of the 2200 PMTs for the light collection. Within the sphere, two nylon vessels separate the scintillator volume in three shells of increasing radiopurity. The SSS is enclosed within a large tank filled with ultrapure water acting as a shielding against external background and as a Cherenkov muon tracker. The experiment can then rely on an optimal energy and spatial resolution together with a very low energy threshold (40 keV). The main goal of Borexino is the direct detection of the monochromatic ^{7}Be solar neutrinos; recently, the improved understanding of the detector characteristics due to the calibration campaigns made it possible to significantly refine the ^{7}Be analysis and to publish the measurement of the ^{7}Be rate with an accuracy better than 5% rejecting the hypothesis of no oscillation for ^{7}Be solar neutrinos at 4.9 σ C.L. [4]. The stability of the detector allowed also to study the day-night effect of the ^{7}Be signal and to show the complete absence of any asymmetry in the data. This result, combined with the other solar neutrino data, allows a rejection of the MSW-LOW solution of the neutrino oscillation based on solar data alone and without relying on the assumption of CPT symmetry in the neutrino sector [5]. Thanks also to the development of a very effective procedure to identify and subtract the ^{1}1C cosmogenic background based on threefold coincidences [6] and the positronium formation study [7], Borexino is now able to explore the 1-2 MeV region with unprecedented sensitivity. This led to the first observation of solar neutrinos from the basic pep reaction [8]. In addition, the upper limit, the lowest ever published, for the CNO production in a star has been established. Borexino is also a very sensitive detector for geo-neutrinos too. Geo-neutrinos are electron antineutrinos produced in β decays of ^{40}K and of several nuclides in the chains of long-lived ^{238}U and ^{232}Th present in the Earth crust and mantle [9].

9. LBL neutrinos and the CNGS project

The CNGS (Cern Neutrino Gran Sasso) project consists of an artificial neutrino beam produced by the protons accelerator SPS of CERN and directed towards Gran Sasso. The ν_μ beam from CERN is optimized for the observation of ν_τ-CC interactions, the average energy being 17 GeV. The two experiments of LNGS devoted to CNGS neutrino detection are OPERA [10] in operation since 2006 and ICARUS [11] since 2010.

10. OPERA

Neutrino oscillations at the 'atmospheric scale' have been studied up to now through atmospheric and accelerator neutrinos in disappearance mode, while the direct observation of flavour transition through the detection of the corresponding lepton has never been observed before OPERA first evidence. The goal of the experiment is the detection of neutrino oscillations in direct appearance mode through the study of $\nu_\mu \leftrightarrow \nu_\tau$ channel, by seeking tau neutrinos in the beam CNGS originally constituted by muon neutrinos only.

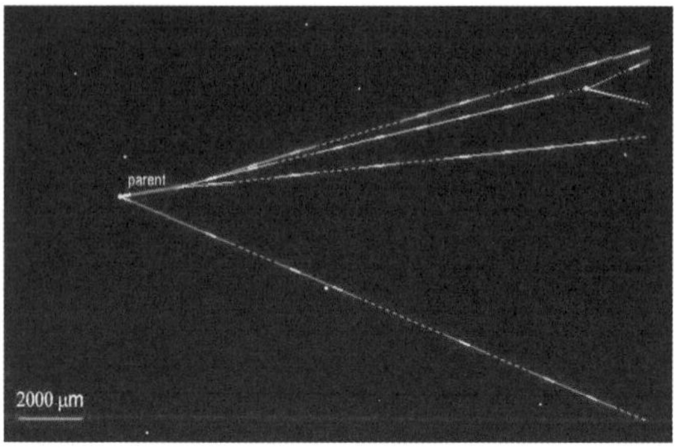

Figure 4: Second tau neutrino in the OPERA experiment - June 2012

The identification of the τ lepton via the topological observation of its decay requires a micrometric resolution and a large target mass of more than a kton to maximize the neutrino interaction probability. To accomplish these requirements the detector concept is based on the Emulsion Cloud Chamber (ECC) technique combined with real-time detection techniques ("electronic detectors"). The ECC basic unit in OPERA is a "brick" made of 56 lead plates, 1 mm thick, providing the necessary mass interleaved with 57 nuclear emulsion films. In total, 150000 bricks provide the overall mass of 1.25 kton. The electronic detectors are used to trigger the neutrino interactions, to locate the brick in which the interaction occurred, to identify muons and measure their charge and momentum. The detector is composed of a veto plane followed by two identical supermodules (SM), each one consisting of a target section of about 625 t made of 75.000 bricks, and of a scintillator Target Tracker Detector (TT), followed by a muon spectrometer. With a total CNGS beam intensity of 22.5×10^{19} protons on target, about 24300 neutrino events would be collected. OPERA should observe Ĩ0 ν_τ-CC events for the present δm^2_{23} allowed region with a background of less than one event. The observation of the first ν_τ candidate event in the experiment has been reported in June 2010 [12] while at the neutrino conference in KYOTO a second ν_τ event has been reported.

11. ICARUS

The detector is an innovative apparatus consisting of a big mass (760 t) of ultra pure liquid Argon, at a temperature of -186 ^{o}C. In particular conditions and by means of proper devices this liquefied gas is able to act as an extraordinary particle detector, allowing a 3D event visualization of any interaction of charged particles inside its volume, combined with a good calorimetric reconstruction and a complete electronic event processing. The detector consists of a large cryostat split into two identical, adjacent half- modules with internal dimensions $3.6\times3.9\times19.6\ m^3$. Each half-module houses two TPCs separated by a common cathode, with a drift length of 1.5 m. Ionization electrons, produced by charged particles along their path, are drifted under uniform electric field (ED = 500V/cm) towards the TPC anode made of three wire planes, oriented at different angles with respect to the horizontal direction, facing the drift volume. There are 53248 channels in total, with a 3 mm pitch. Combining the wire coordinate on each plane at a given drift time, a three-dimensional image of the ionizing event can be reconstructed with a resolution of about 1 mm^3. The absolute time of the ionizing event is provided by the prompt ultra-violet scintillation light emitted in LAr and detected through an array of 74 Photo Multiplier Tubes (PMTs), installed in LAr behind the wire planes. The detector layout is completed by a cryogenic plant made of a liquid Nitrogen cooling circuit and a system of LAr purifiers. The commissioning of ICARUS was successfully completed in 2010 [13]. Since then the experiment has been taking data smoothly on the CNGS neutrino beam with very high detector live-time, thanks to further improvements on both cryogenics plant and DAQ systems. In particular new electronics for PMT trigger system has been installed allowing to increase the cosmic trigger efficiency, all the Collection view wires have been equipped with new chips called SuperDaedalus and improvements to the DAQ system allowed to significantly reduce the event building time. Such a massive liquid argon experiment running in an underground laboratory is, so far, the most important milestone for the LAr-TPC technology towards the design of a much more massive multikiloton LAr detector.

12. Study of neutrinos from Supernovae

The LVD experiment is an observatory for the low energy neutrinos from gravitational stellar collapses in our galaxy [14]. Indeed, a sudden neutrino burst accompanies the last instants of life of some massive stars before they turn off. The apparatus consists of about 1000 t of liquid scintillator in which neutrinos may interact, distributed in 840 tanks. LVD has been observing the galaxy since 1992 and has given an upper limit to the rate of gravitational stellar collapses at a distance smaller than 20 kpc of 0.15 events per year at 90% confidence level. LVD is part of the SNEWS, the early warning world network of all the experiments able to detect stellar collapses, and very important for astronomical observation of supernovae through different messengers.

13. Dark Matter search at Gran Sasso

There is compelling evidence from astrophysical and astronomical observations that about one-quarter of the energy density in the universe is composed of non-baryonic and non-relativistic (cold) massive component larger than the one observable through telescopes, due to a not-yet-identified particle. One of the most common candidate for CDM are Weakly Interactive Massive Particles (WIMPs) predicted by supersymmetric theories (SUSY), searched for at LHC looking for their appearance in collisions. Direct detection of DM aims to observe the scattering of DM particle off target nuclei. WIMPs-target nuclei interactions produce exponentially falling differential energy spectra with energy deposition up to several tens of keV. Very low rates are expected (few counts/(kg day) to a few counts/(kg yr)) and then the direct search for dark matter candidates can only be performed in low background underground laboratories and through experiments making use of effective means to distinguish signals from background. LNGS houses experiments hunting for dark matter candidates through their direct detection, each based on a different technology. These experiments put Gran Sasso laboratory at the world forefront in such a class of studies.

14. DAMA/LIBRA

The DAMA/LIBRA experiment is mainly based on the development and use of low background scintillators and the main aim is the direct detection of DM particles in the galactic halo by investigating the model independent annual modulation signature [15]. The experiment has been operational since 2003 with 250 kg of extremely radio-pure NaI(Tl) crystals. The apparatus has the unique feature (as well as the previous DAMA/NaI) that gamma calibrations are regularly performed in the same conditions as the production runs, as the presence of a peak at 3.2 keV from internal ^{40}K contamination in the crystals allows to study the response of the detector near the software energy threshold. The high light yield and other response features have allowed working in a safe and reliable way down to 2 keV; in incoming years this software energy threshold will be lowered thanks to the use of the new PMTs, installed at the end of 2010, and of other optimisations. This can allow the improvement of its performance and sensitivity. Since January 2011 the DAMA/LIBRA experiment is again in data taking in the new configuration. The most recent published results [16] confirmed the annual modulation of the very low energy signals induced in the detector, already observed in the previous lower mass experiment DAMA. Such modulation is identical to the one that the relative motion of the Earth through the huge amount of dark matter halo of our galaxy is supposed to cause. No systematic or side process able to simultaneously satisfy all the many peculiarities of the signature and to account for the whole measured modulation amplitude is available. The cumulative exposure with the former DAMA set-up is 1.17 t×y corresponding to 13 annual cycles in total. The recent positive hint with lower statistics of an annual modulation from CoGeNT experiment (SUL), if interpreted in terms of a low mass WIMP candidate and taking into account the many uncertainties existing in the models and in the comparisons, could be compatible with the DAMA result. The absence of positive signals by other experiments produces a lively debate inside the scientific community as well as the production of theoretical models trying to conciliate different results.

15. XENON100 and XENON 1t

After the successful results of the 10 kg scale detector XENON10, a second-generation experiment exploiting the two-phase time projection chamber (TPC) technique based on liquid xenon (LXe) is operating at LNGS. The detector XENON100 contains 170 kg of xenon, 65 kg of which constitute the active part while the remaining acts as a shield. A key feature of the detector is its ability to localize events with millimetre resolution in 3 dimensions which further allows to select only the innermost 48 kg as ultra-low background. After the result published in 2011 from 100.9 live days of data, acquired between January and June 2010, the present improved running conditions of XENON100 allowed to further push the sensitivity of XENON100 that smoothly continues data taking. Very recently the XENON collaboration has published the analysis of 225 more days of data accumulated in 2011 and 2012 [17]. They see no evidence for the existence of WIMPs: the two candidate events being observed are statistically consistent with one event being expected from background radiation. Compared to their previous 2011 result, the world-leading sensitivity has again been improved by a factor $\tilde{5}$.

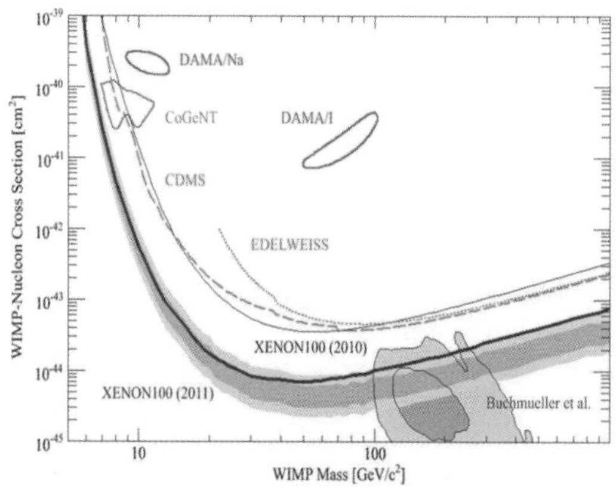

Figure 5: 2011 results: XENON100 Dark Matter Limit (90% CL)

16. CRESST

The CRESST experiment is based on the bolometer technique with CaWO4 crystals at a temperature of few mK as well as on the simultaneous detection of scintillation light and the heat resulting from the interaction of a particle with the crystals. One advantage of the cryogenic detectors developed for CRESST is that they can measure the deposited energy calorimetrically, independently on the type of interaction. Combining the calorimetric measurement of the deposited energy with a measurement of scintillation light, a potentially high discrimination of the nuclear recoils from radioactive background can be obtained and the type of recoiling nucleus can be determined in a multi atomic target. In 2011 CRESST has completed a long run and has submitted a paper with the analysis of data with a total exposure of 730 kg d [18]. An excess of events has been found in the acceptance region where a WIMP signal would be expected. A new run with several detector improvements aiming at a reduction of the overall background is expected to start soon.

17. DARKSIDE-50

Cryogenic TPC employing a two-phases argon medium as active volume are a quite promising technique for WIMP detection. The ratio between the scintillation light in the liquid argon (the primary signal) and the ionization in the gas and the analysis of the temporal shape of the primary signal enable to discriminate the signal produced by a dark matter particle form backgrounds. The Liquid Argon Technology for the search of dark matter particles has been pioneered at LNGS by WARP experiment, now it continues with DarkSide-50. The DarkSide Collaboration aims to ultimately reach sensitivity to WIMP-nucleon cross sections of 10^{-10}pb and below with a multi-ton detector. The program is based on the scalability and high performance of liquid argon (LAr) TPC detectors. DarkSide-50 is based on past experience and introduces innovative features aiming to operate in a 'background-free' mode for more than three years and thereby achieve a significant science result in spite of its relatively small size. DarkSide-50 will also serve as a prototype for future multi-ton detectors. To accomplish the scientific goals, several important technical innovations have been proposed mainly consisting in the use of underground argon very low in ^{39}Ar (UAr), surrounding the detector with a neutron veto within a thick water shield which is instrumented as a Cherenkov veto for cosmic-ray muons [19], novel TPC detector features, including efficient wavelength-shifting and light collection system with high quantum-efficiency (QE) photomultiplier tubes operating immersed in liquid argon, and very low internal radioactivity. DarkSide-50 will be deployed, enclosed in its liquid scintillator neutron veto, in the Borexino Counting Test Facility in the hall C at LNGS. The CTF is a 10m high, 11m diameter water tank that will serve both as a cosmic-ray muon veto and as a shielding against radiogenic and cosmogenic backgrounds. The detector is in construction phase at LNGS and the Collaboration aims to start commissioning in 2013.

18. Nuclear Astrophysics and LUNA experiment

Aim of the LUNA experiment is the measurement of thermonuclear fusion cross sections relevant for stellar nucleosynthesis. In order to start to measure this unexplored nuclear astrophysics region, the collaboration LUNA has developed since many years two accelerators, 50kV and 400kV, in the underground laboratory of Gran Sasso, in the interferometric area. The qualifying characteristics of both the accelerators are the very high beam current and the very small energy spread/dispersion/outflow. High pure solid and gaseous targets have been developed, while the silicon, ultrapure germanium and a segmented BGO detector have been used. The main relevant results obtained by LUNA are the only existing measures of cross sections inside the Gamow peak of the Sun: ^{3}He(^{3}He,2p)^{4}He e d(p,γ) ^{3}He. For instance, the first reaction is fundamental in the pp fusion cycle, with effects on the flux of solar neutrinos. The measurement of ^{14}N(p,γ)^{15}O [20] reaction permitted to obtain two significant results: the prediction of the flux of solar neutrinos produced in the cycle CNO has been reduced of a factor 2, while the estimate of the more ancient globular clusters age has been increased of about a billion years. At present only the 400 KV accelerator is being used and the last experimental activity has been focussed on the $^{2}H(\alpha,\gamma)^{6}Li$ and $^{17}O(p,\gamma)^{18}F$ reactions. Moreover, the project of a new accelerator LUNA 3.5MV, with an experimental programme mainly devoted to the He burning reactions, has been submitted and approved by LNGS. A detailed project of the preparation of the "B node" of the interferometric area, assigned to the project, was accomplished, together with the design of the shielding against neutrons with respect to both the rest of the laboratory and the internal rock "walls" where water uptake points of the Teramo aqueduct are present.

19. Conclusion

The Gran Sasso National Laboratory of INFN, the largest underground laboratory in the world for astroparticle physics, holds the leadership in massive experiments with record performance and low-level background. The present scientific program of LNGS includes a very broad spectrum of competitive experiments (astroparticle, particle and nuclear physics), including the world-leading ones in the fields of solar neutrinos, accelerator neutrinos, double beta decay, dark matter and nuclear astrophysics.

20. References

[1] Letter of Intent, 2004, hep-ex/04040390; Proposal to the Lngs, 2004
[2] C. Arnaboldi et al., Astropart. Phys. 20(2003) 91.
[3] C. Arnaboldi et al., Phys. Rev. Lett. 95 (2005) 14501.
[4] G. Alimonti et al., Astroparticle Physics 16 (2002) 205.
[5] G. Bellini et al., Phys. Lett. B 707 (2012) 22.
[6] G. Bellini et al., Journal of Instrumentation, doi:10.1088/1748-0221/6/05/P05005.
[7] D. Franco, G. Consolati and D. Trezzi, Phys. Rev. C 83 (2011) 015504.
[8] G. Bellini et al., Phys. Rev. Lett. 108 (2012) 051302.
[9] G. Bellini et al., Phys. Lett. B 687 (2010) 299.
[10] R. Acquafredda et al. JINST 4:P04018,2009.
[11] C. Rubbia, CERN-EP/77-08 (1977).
[12] N. Agafonova et al. Phys. Lett. B691, 138 (2010).
[13] M. Antonello et al., LNGS Annual Report 2010, June 2011, p. 107.
[14] LVD Collaboration, Il Nuovo Cimento A105 (1992) 1793
[15] K.A. Drukier et al., Phys. Rev. D 33 (1986) 3495.
[16] R. Bernabei et al., Nucl. Phys. B (Proc. Suppl.) 212-213 (2011) 307
[17] XENON100 Collaboration arXiv:1207.5988v1
[18] Angloher et al., The European Physical Journal C, Volume 72, article id. 1971
[19] A. Wright,et al., NIM A 664:18,26 (2011)
[20] M. Marta, A. Formicola, et al, Phys.Rev.C 83 (2011) 045804

CHAIRMAN: L. VOTANO

Scientific Secretaries: S. Hemmer, D. Thomas

DISCUSSION

- *Leontsinis:*

Are there Axion searches at the Gran Sasso laboratory?

- *Votano:*

All the dark matter experiments looking for annual modulation, like DAMA, are model-independent and thus also a search for axions.

- *Karakostos:*

The beam sent from CERN towards CNGS arrives with a spatial coverage of around 800 m. Does this affect other experiments?

- *Votano:*

Since different experiments study neutrinos in different energy ranges, the neutrino beam sent from CERN does not interfere with other experiments. Also since the time of arrival is very precise, beam events are clearly identified. This also allowed the cross check of the neutrino speed from four different experiments.

- *Inguglia:*

Can you comment on the fact that the spokesperson of the OPERA experiment had to resign after the correction of the results regarding superluminal neutrinos?

- *Zichichi and Votano:*

The resignation of Antonio Ereditato is an internal issue. (Zichichi refers to his paper on the subject for more information.)

- *Darosti:*

How well is the background in the region of interest understood in the GERDA experiment compared to the Heidelberg-Moscow experiment? And how is the analysis in GERDA different from the Heidelberg-Moscow experiment?

- *Votano:*

The GERDA experiment has a very good control on the background. The experiment is blinded in the signal region. In the 400 keV region around the Q-value of the decay of ^{76}Ge, the background is a factor 10 less than it was for the Heidelberg-Moscow experiment. More details can be obtained from recent papers of the GERDA experiment.

- *Darosti:*

In the OPERA experiment, the study of the neutrino oscillation from $\nu_\mu \rightarrow \nu_\tau$ events requires a good identification of events, e.g CC interactions of ν_e. How well are all the background sources understood?

- *Votano:*

There are many background sources for the measurement of neutrino oscillation. All these sources are taken into account by the experiment including measurement of charm events both in

MC and in data. More details can be obtained from the talk of Dario Autiero on the status of the Opera experiment.

- *Klopot:*

Could you give some details on the dark matter program at the Gran Sasso laboratory? What kind of signature are the experiments looking for?

- *Votano:*

They are looking for scattering on nuclei. Most experiments use a double signature, e.g. the ratio between ionisation and scintillation light in liquid Xenon or liquid Argon. There are defined regions for this ratio where you expect normal interaction and dark matter events, respectively. This allows to distinguish signal from background events. Since in case of liquid Argon also the shape of the signal can be used for the analysis, this technology should have some advantage with respect to the liquid Xenon technology.

- *Klopot:*

So you suppose that a certain dark matter model is correct and then compare the measured cross sections with the ones expected from the model?

- *Votano:*

Yes. The only model-independent analysis is the annual modulation since this is due to the change of the relative velocity between the detector and the dark matter cloud. Another analysis tool could be to consider the direction of the interacting particles, but this is not done in any of the Gran Sasso experiments.

- *Noferini:*

From the slides shown, it can be seen that the measurement of neutrino speed gives a time difference greater than zero. Is the measurement consistent with systematic uncertainties?

- *Votano:*

The time deviation of neutrinos with respect to light is greater than zero but it is less than the systematic uncertainty. Within one standard deviation, there is no effect and the speed of neutrinos is not confirmed to be greater than the speed of light.

- *Aschieri:*

What is the status of the experiment on the violation of the Pauli exclusion principle?

- *Votano:*

The VIP experiment is looking for atomic decay and the violation of the Pauli exclusion principle. The collaboration has finished the first measurement and published the limit. At the moment, the detector is being upgraded.

Seminars on Specialized Topics

D. AUTIERO

"Status of Opera"

The Slides of the Lecture can be found at:

http://www.ccsem.infn.it/issp2012/

CHAIRMAN: D. AUTIERO

Scientific Secretaries: E. Schlieckau, M. Görner

DISCUSSION

- *Csanàd, Loránd:*

How do you measure the position where the muon and the neutrino are created in the decay tunnel and how does it affect the measurement of the neutrino velocity?

- *Autiero:*

The muon and the neutrino are created by the decay of a relativistic pion. This happens at an unknown point in the decay tunnel. If you compute the difference, assuming the particle has the speed of light, from the graphite target to OPERA and a real simulation (taking into account the speed of the relativistic pion) this gives an effect much smaller than 1ns. It's only a second order effect among relativistic particles. The same kind of effect you have on the other side where neutrinos are interacting in the rock and are sending muons in the detector. Also here you have a second order effect where you have a muon of which you don't know where it was produced.

- *Zhang:*

Could you tell us more about the future of OPERA?

- *Autiero:*

Data taking should be completed by the end of this year. In the 2012 run we will probably collect a bit more than the annual nominal statistics and will have 84% of the proposal statistics collected by the end of this year. It is not yet decided what will happen after the stop for the LHC upgrade. It could be that data taking will be stopped or it will be asked for an extension.

- *Karakostas:*

Since the target tracker provides only the information which brick(s) participate in an specific event, how do you find the correct signature afterwards as the readout is not done continuously?

- *Autiero:*

The electronic detector identifies a list of bricks (ordered by probability) in/with which the neutrino might have interacted. Starting with the brick with the highest probability, you first analyze the box which contains two emulsion films at the surface of the brick. If you find a pattern of particles confirming the observation of the electronic detector, the brick is developed and the neutrino interaction vertex is reconstructed. Otherwise, you go to the next brick in the list.

- *Karakostas:*

Is it possible that one brick contains two events?

- *Autiero:*

That is extremely rare because you have to take into account that there are 150,000 bricks in the target and there are only a few 1,000 events per year.

- *Inguglia:*

How do you achieve the precision of the order of microns in the bricks and how does the read out system work?

- *Autiero:*

A brick contains 56 layers which have to be aligned to each other. There are several techniques to do that. Before dismounting a candidate brick, the surface of the emulsion brick is marked with x-rays, so that you get a relative alignment of the emulsion plates. Then, it is exposed to cosmic rays under controlled conditions to have muon tracks that go perpendicular through all emulsion layers. These muon tracks will be used for a final alignment to know the relative position of the layers in the brick to micron level.

- *Inguglia:*

Are there data points missing in the plot of the time difference between LVD and OPERA as a function of the operating year? (slide 34 of the presentation[1])

- *Autiero:*

If there is space between the data points, it is an artifact of the plot. Data taking was continuous and the error bars of the grouped data points should touch each other. When you compute a distribution like that you have to take into account the live time of the two detectors. There were some periods where data were not available.

- *Zichichi:*

The origin of this plot is a continuous measurement (see also[2]). The important point is that the high energy horizontal muons allow the synchronization between LVD and OPERA in a completely independent way wrt. the distance of CERN to Gran Sasso.

- *Leontsinis:*

What is the source of the high uncertainty in the momentum measurement (e.g. for the daughter up to 50%, see slide 43 of the presentation[1])?

- *Autiero:*

This measurement is performed by looking at the multiple scattering and measuring the relative position of the points between the different emulsion layers. This technique saturates at high momenta because an accuracy much better than a micron would be needed. As the pion was high energetic you have this large uncertainty. At lower momenta you have smaller uncertainties.

- *Mager:*

Given that your results (could) disfavor general relativity but GPS (you use for the synchronization of your clocks) strongly relies on that, wouldn't it be better to use different techniques (like optical fibers)?

- *Autiero:*

We do believe in general relativity at this level. GPS and general relativity are based on experimental confirmations and work very well and with high accuracy. We never said that general relativity is wrong, we just measured an effect we did not understand and, if confirmed, it could perhaps be explained in a more exotic way. An independent cross-check of GPS using optical fibers has been discussed. To use the existing academic network, you need to install boxes to bypass the normal data flow at each node in a deterministic way. Given the structure of the network, this can be done in France but since Italy is a long country the installation would have introduced some attenuation of the network. Therefore, the final conclusion was that a completely independent optical fiber had to be used (leading from the Alps to Gran Sasso). This is very expensive and practically too complicated to be set up.

- *Mager:*

What about synchronizing two atomic clocks at CERN and transport one to Gran Sasso?

- *Autiero:*

This can be done. You would have to take into account effects related to the transportation of the clock. This has already been done in other experiments with other techniques, like the time transfer with a two way path through other satellites (independent of GPS). These techniques have all been applied in parallel and are found to be consistent by other experiments.

- *Knobloch:*

How do you want to combine the results of the four Gran Sasso experiments (see slide 36 of the presentation[1]) to one measurement of the neutrino velocity, taking into account correlations of the ~12 systematic uncertainties? As at least five systematic uncertainty sources are in common (and probably dominate the total uncertainty), why are the quoted total uncertainties so different?

- *Autiero:*

Large parts of the systematic uncertainties are in common (like the proton tagging at CERN and the time transfer between CERN and Gran Sasso). Just the last part (the time measurement in the experiment) is different. These numbers presented here are preliminary numbers submitted at the Kyoto Conference. For the other experiments we know that some common uncertainties are missing. OPERA was more conservative with these numbers than the other experiments. For an overall result one should combine things properly and the common systematic uncertainties should be taken into account by all experiments. This will be done.

- *Knobloch:*

Do you use experimental data for the determination of your charm background in the tau measurement?

- *Autiero:*

This was done and the most precise results are from the CHORUS experiment. The charm background prediction in comparison to the proposal changed up to a maximum factor of two, depending on the tau decay channel. This was mainly driven by the measurement of the production cross section and the fragmentation of the different charm hadrons. Some hadrons affect the background more than others. For example, the effect of the Lambda_c baryon is smaller than the effect of the D mesons. The latest CHORUS data[3] resulted in a reshuffling in the fragmentation between the channels and we take this into account.

- *Knobloch:*

Will there be a problem with the lifetime of the emulsion layers in view of a potential extension of the OPERA running time?

- *Autiero:*

This will not be a problem for an extension after the stop although there will be a little more "fog" (isolated points that you see in the emulsions) which will make the track reconstruction more complicated. This is continuously monitored and will not be the limiting factor to take one or two years more data.

- *Dorosti Hasankiade:*

How was the pulse length of ten microseconds chosen and was there an optimization procedure involved?

- *Autiero:*

This is a geometrical factor. The SPS has a circumference equivalent to 22 microseconds (length in meter divided by the speed of light). There is a little gap in the middle technically related to how the protons are injected into the SPS. Furthermore, the machine is emptied in two shots (10.5 microseconds each), corresponding each to half of the filled length. So there is no optimization, it just corresponds to the complete filling of the machine.

- *Dorosti Hasankiadeh:*

Would you gain precision in the measurement of the time delay by using longer pulses (like e.g. 20 microseconds) because of the higher number of neutrinos?

- *Autiero:*

This can be done by extracting the whole SPS content in one shot but leads to problems with heating up the target. Therefore, the extraction is done in two pulses (separated by 50 milliseconds). For the measurement of the neutrino velocity, short pulses are preferable as they allow the measurement of the time of flight for each recorded event individually. But then you have a beam that is practically empty (with few short and isolated pulses) and you loose a huge factor of statistics (up to a factor of 30 wrt. normal data taking). For the search of neutrino oscillations you do not want to pay that price.

References

[1] http://www.ccsem.infn.it/issp2012/docs/autiero_erice.pptx, D. Autiero, 29th of June 2012, Status of OPERA, invited talk at the 50th course of the International School of Subnuclear Physics, Erice, Italy

[2] A. Zichichi,"Time-Shift in the OPERA set-up: proof against superluminal neutrinos without the need of knowing the CERN-LNGS distance and Reminiscences on the origin of the Gran Sasso Lab, of the 3rd neutrino and of the 'Teramo Anomaly'", arXiv:1206.2840 [hep-ex]

[3] A. Kayis-Topaksu et al, "Measurement of charm production in neutrino charged-current interactions", New J.Phys. 13 (2011) 093002, CERN-PH-EP-2011-109

Alessandro Bettini
University of Padua. Department of Physics; INFN. Padua division (Italy)
Laboratorio Subterráneo de Canfranc (Spain)
bettini@pd.infn.it
bettini@lsc-canfranc.es

The Origin and Status of the Third Neutrino

1. The three lepton families

The known fermions come in three families, each made of a doublet of quarks, of charge +2/3 q and $-1/3$ q (q is the proton charge), and a doublet of leptons, one of charge $-q$ and one neutral. The neutral leptons are collectively called neutrinos, but are three different particles, distinguished by an additive quantum numbers called lepton (electron, muon or tau) flavours.
The charged leptons are distinguished by their masses and lifetimes, hence cannot mix, neutrinos only by their lepton flavours. Neutrinos are produced in states of definite flavour, as ν_e, ν_μ or ν_τ, in pairs with an antiparticle of the same and opposite flavour. Elementary interactions conserve flavours. So that, by definition, the electron-neutrino (ν_e) is the neutral lepton produced with a positron (e^+), the ν_μ is the one produced with a μ^+ and the ν_τ is the one produced with a τ^-.
Neutrinos cannot be detected directly, rather we detect them through their neutral or charged current interactions. In the latter case, the identification of the charged lepton gives the flavour of the neutrino: if it is an electron (e^-) it was a ν_e etc.
It has been established in experiments at accelerators, in which the ratio L/E between distance L from production to interaction points and neutrino energy E is not very large, that neutrinos born with a flavour produce charged leptons of the same flavour. This property is assumed to be absolutely valid in the Standard Model. However, experiments in underground laboratories at larger values of L/E have shown that this is not the case. Neutrinos oscillate between one flavour and another. The evidence has gradually grown in the last four decades, by studying the ν_es produced by the fusion reactions in the core of the Sun and the ν_μs indirectly produced by the cosmic rays collisions in the atmosphere. Confirmations came from experiments with artificial neutrino sources: proton accelerators (producing mainly ν_μ) and nuclear power reactors ($\bar{\nu}_e$). There two types of experiments. In a disappearance experiment the flux of neutrinos of a certain flavour is known at production; if the flux is measured at a (large) distance and found to be less than expected, the oscillation to another flavour is inferred. In an appearance experiment a flavour not present at production is searched.
Oscillations may happen when neutrinos of definite flavour are not stationary states (mass eigenstates). The latter, ν_1, ν_2 and ν_3, have definite masses, m_1, m_2 and m_3. The two bases are linked by an orthogonal transformation that can be expressed in terms of three rotations, through angles that we shall call θ_{12}, θ_{23} and θ_{13}, and of phase factors. If neutrinos are Dirac particles, as assumed in the Standard Model, all but one of the phase factors can be absorbed, as in the case of quarks, in the wave functions of the states. However, neutrinos might be completely neutral, namely 'Majorana particles'. In this case two more phases, which we shall call Majorana phases α and β, are physically observable.
In conclusion, writing $c_{ij}=\cos\theta_{ij}$, and $s_{ij}=\sin\theta_{ij}$, the unitary transformation $\nu_l=\sum_{1=1}^{3}U_{li}\nu_i$ is

$$\begin{pmatrix}\nu_e\\ \nu_\mu\\ \nu_\tau\end{pmatrix}=\begin{pmatrix}1&0&0\\0&c_{23}&s_{23}\\0&-s_{23}&c_{23}\end{pmatrix}\begin{pmatrix}c_{13}&0&s_{13}e^{i\delta}\\0&1&0\\-s_{13}e^{-i\delta}&0&c_{13}\end{pmatrix}\begin{pmatrix}c_{12}&-s_{12}&0\\s_{12}&c_{12}&0\\0&0&1\end{pmatrix}\begin{pmatrix}1&0&0\\0&e^{i\alpha/2}&0\\0&0&e^{i\beta/2}\end{pmatrix}\begin{pmatrix}\nu_1\\ \nu_2\\ \nu_3\end{pmatrix}$$

Notice that Majorana phases are irrelevant for the oscillation and matter effects. They are observable in an extremely rare, not yet observed phenomenon, the neutrino-less double beta decay (see §5).

Experiments have measured cross sections and energy spectra relevant for the oscillation phenomena. The frequency of the oscillation in vacuum is proportional to the absolute value of the squares of two neutrino masses. The flavour conversion in matter, in the Sun in particular, depends also on the sign of the difference.

The values of two of the larger mixing angles, θ_{12} and θ_{23}, have been known with increasing accuracy since many years. Information on θ_{12} is mainly due to solar neutrinos and reactor antineutrinos, that on θ_{23} to atmospheric neutrinos, reactor antineutrinos and accelerator neutrinos. This year two short baseline experiments on $\bar{\nu}_e$ from power reactors, Daya Bay[1] and RENO[2], have definitely established a non-zero value of the third angle, θ_{13}, at more than 5 standard deviations. Both experiments measured the $\bar{\nu}_e$ flux and spectrum in locations near to the reactors and about 2 km far from them (the distance of the maximum of the oscillation for the most probable energy of neutrinos), observing their disappearance.

A global fit to these measurements, e.g. Fogli et al.[3] gives

$$\sin^2\theta_{12} = 0.307^{+0.018}_{-0.016}; \quad \sin^2\theta_{23} = 0.386^{+0.026}_{-0.021}; \quad \sin^2\theta_{13} = 0.0241^{+0.0025}_{-0.0024}$$

and for the square mass differences

$$\left|\Delta m^2\right| \equiv \left| m_3^2 - \left(\frac{m_1 + m_2}{2}\right)^2 \right| = 2430^{+60}_{-100}\ \text{meV}; \quad \delta m^2 \equiv m_2^2 - m_1^2 = 75.4^{+2.6}_{-2.3}\ \text{meV}$$

Fig. 1 shows schematically the neutrino square-mass spectrum. It consists of a singlet (ν_3) and a doublet (ν_1, ν_2). The approximate flavour composition of the eigenstates is also shown. We do not know either the absolute scale or whether the mass of the singlet is larger or smaller than that of the doublet.

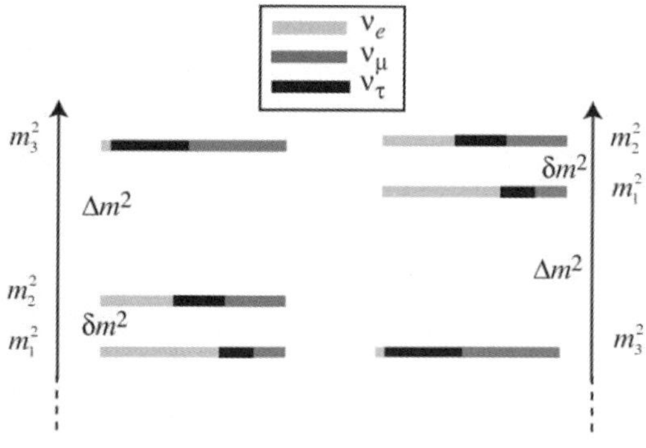

Fig. 1. Neutrino square-mass spectrum. The flavour contents of the eigenstates are also shown

2. Lepton identification

As mentioned above, the definition of the flavour of a neutrino requires experimentally the identification of the nature of the charged particle it produces and of the other particles that are also present in the final state of the interaction.

There are different types of detectors. The main categories are the trackers, which show, within a range of resolutions, the images of the tracks of the charged particles that go through, and the calorimeters that absorb and measure all the energy of the particle.

Electrons and charged pions produce very similar tracks, but can be distinguished by the different shape of the shower they produce. An electron penetrating in a dense medium will soon radiate a photon, which in turn will convert in electron-positron pair, which will radiate more photons, etc. A pion will behave similarly, but producing a shower containing also hadrons on top of electrons and photons. The difference in shape between the two showers can be enhanced and detected with proper techniques, as we shall see, particularly in their initial phases. A high discriminating power is needed to reliably identify the electrons that are produced very rarely compared to the pions.

Muons can be distinguished from hadrons due to their much higher penetrating power. Differently from pions, muons go through metres of material without interacting, leaving a long, straight track.

Tau leptons have a lifetime much shorter than the more common hadrons, about 0.3 ps; at a few GeV energy, which are typical in the experiments under discussion, the lengths of their tracks between production and decay points, are of the order of hundreds micrometres. Consequently, extremely high spatial resolution, of the order of a micrometre is needed, as provided by nuclear emulsions. Moreover, also large target masses are necessary to secure an appreciable rate of neutrino interactions. A powerful technique, known as emulsion cloud chamber (ECC), has been developed by the Nagoya group initially lead by K. Niu. Fig. 2 shows, as an example, the configuration in the DONUT experiment that we shall discuss later. One mm thick Fe sheets, providing the mass, are interleaved with emulsion sheet pairs, providing the images of the track segments. A short track, a kink between the mother and the daughter, is the signature of the tau lepton. Short lifetime hadrons, such as the charmed ones, give the same topology, but are much rarer than the pions.

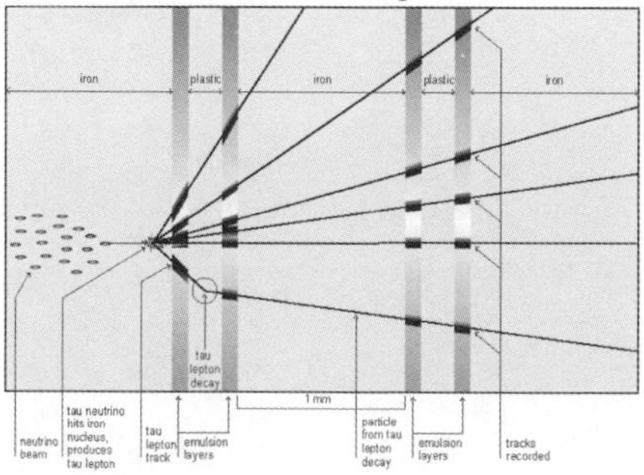

Fig. 2. The Emulsion Cloud Chamber (ECC) of the DONUT experiment at Fermi Lab

3. The history of the third lepton family

In 1897 J. J. Thomson[4] discovered the electron, the first elementary particle, by developing the technique of deflecting charged particles in a magnetic field in a vacuum and building the first mass spectrometer.

Forty years later in 1937, J. Street and E. Stevenson[5] and C. Anderson and S. Neddermeyer[6] discovered a penetrating component in the cosmic rays. Surprisingly, it was not the mediator of the nuclear forces, the pion, predicted by Yukawa, but was a lepton, as experimentally shown in 1947 by M. Conversi, E. Pancini and O. Piccioni[7].

Almost another decennium later, F. Reines[8] using the Savannah River power reactor as a source, finally discovered "the" neutrino that had been introduced as a "desperate hypothesis" by Pauli back in 1930, when only the electron was known, to explain the apparent non-conservation of energy and angular momentum in beta decay. It became later known that it was the "anti" of "one" of the three neutrinos (electron-anti-neutrino).

In 1962 M. Schwartz, L. Lederman, J. Steinberger et al.[9] discovered the muon-neutrino at BNL AGS proton accelerator.

In the same years, A. Zichichi[10] developed the PAPLEP (Proton-AntiProton into LEpton Pairs) experiment at the CERN Proton Synchrotron (PS). It was the beginning of the search for the 3rd sequential lepton family, imagined as a replica of the first two, made of a "Heavy Lepton" (HL) and its neutrino (ν_{HL}). For a complete history see Wu et al.[11]. The idea was to look for lepton pairs of opposite sign. These can be produced directly or result from the decays of a HL anti-HL pair, produced by the reaction $p+\bar{p}\rightarrow HL^{+}+HL^{-}$, shortly followed by the decays $HL^{+}\rightarrow e^{+}\nu_{e}\bar{\nu}_{HL}$ and $HL^{-}\rightarrow\mu^{-}\bar{\nu}_{\mu}\nu_{HL}$ or $HL^{+}\rightarrow\mu^{+}\nu_{\mu}\bar{\nu}_{HL}$ and $HL^{-}\rightarrow e^{-}\bar{\nu}_{e}\nu_{HL}$. The presence of the intermediate Heavy Leptons can be inferred when one of these decays is into an electron the other into a muon, while in the direct production the leptons have the same flavour. A further signature, due to the presence of undetected neutrinos, is the "acoplanarity", meaning that the plane defined by the momenta of the detected leptons does not contain the direction of the beam.

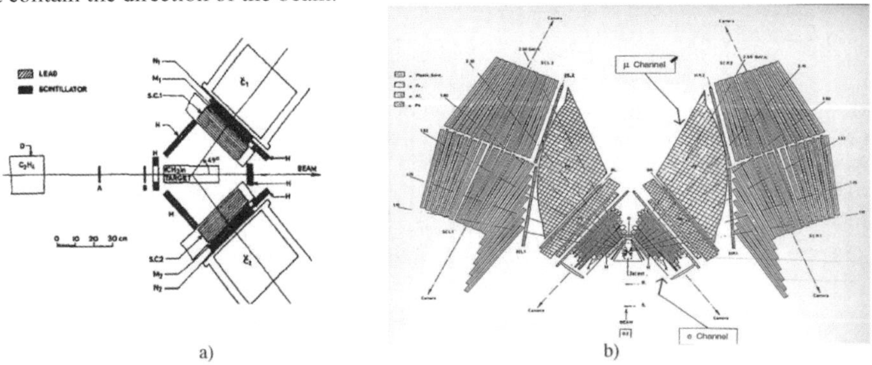

a) b)

Fig. 3. a) The principle of the double arm spectrometer. b) the full design of PAPLEP

Fig. 4. The PAPLEP apparatus at the CERN PS. Circa 1963

The experimental challenges were several.

- The detector should have had a large acceptance to collect a good fraction of the expectedly rare process: a large solid angle double arm spectrometer was built
- The very rare electrons had to be discriminated from the hadrons with high efficiency: the pre-shower was developed[12]
- The very rare muons had to be discriminated from the hadrons with high efficiency: the hadrons punch through phenomenon leading to the penetration, like the muons, of the Fe absorber was preliminarily investigated[13].

The PAPLEP principles are shown in Fig. 3a), while Fig. 3b) shows its design. Fig. 4 is a photo of PAPLEP, a very large device for the time.

The detector elements closest to the target in both arms are two equal copies of the early development of the shower, or pre-shower, apparatus. As shown in Fig. 5, it is a sandwich of five repeated elements, each composed of Pb layer controlling the early development of the shower, a plastic scintillator sampling the energy deposit and an Al-plates spark chamber visualizing the tracks. The novel method combines in this way visual and non-visual approaches[12], each providing 10^{-2} rejection to obtain an overall rejection of the pions of a few 10^{-4}.

PAPLEP did not find the HL, but produced important results based on the lepton pairs detection, in particular on the time-like nuclear form factor[14],[15].

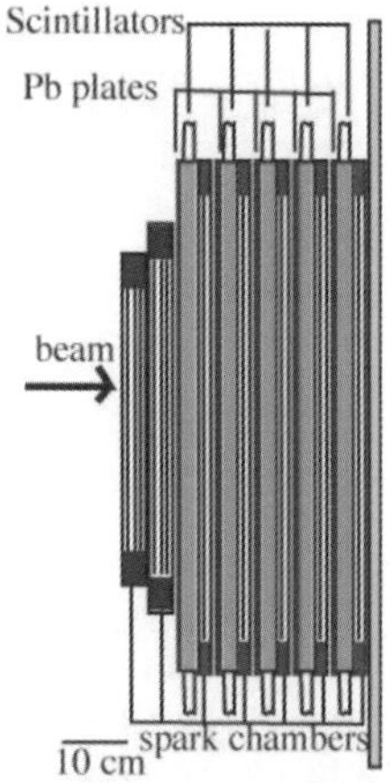

Fig. 5. The pre-shower scheme

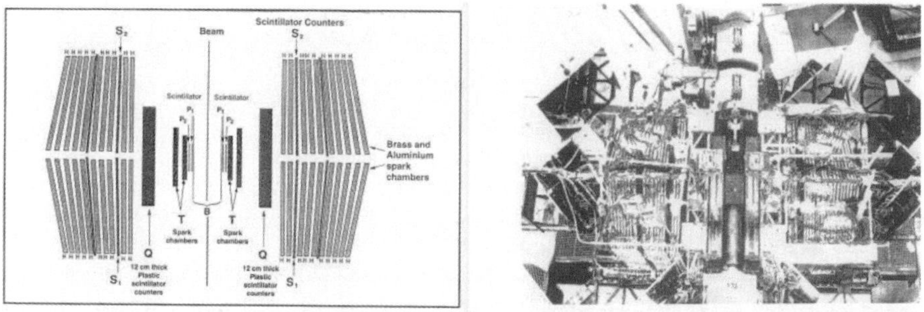

Fig. 6. The apparatus to search for Heavy Lepton at ADONE. Left: schematic. Right: a picture

A reason for the non-observation of the HL might have been the non point-like structure of the proton. Consequently, the next step of Zichichi was to move the search to the highest energy e^+e^- collider available, ADONE at Frascati[16].The apparatus, shown in Fig. 6 employs all the technologies for electron/pion and muon/pion discrimination developed in PAPLEP, adapted to the collider situation. The two equal arms are now on opposite sides of the collision point.

Once more, the HL was not found, and only a lower limit of 1 GeV could be established for its mass[17]. The reason is that the HL mass is 1.777 GeV, too large to be produced at ADONE, which had a maximum energy of 3 GeV. In 1974 a new e^+e^- collider, SPEAR at SLAC, had reached steady operation at 4.8 GeV, together with the detector shown in Fig. 7 left. The apparatus, which became later known as MARK I, had been built as a general purpose one with a broad physics programme, which were to discover a gold mine (the charm). Four layers of cylindrical wire chambers provided tracking in magnetic field produced by a solenoid. Outside the coils, lead-scintillator electromagnetic shower counters and tracking chambers for the muons were located.

As part of the facility programme, M. Perl and collaborators started searching for HL looking for acoplanar $e\mu$ pairs. However, the general-purpose detector had not been conceived with the necessary particle identification capability. A track was defined as an "electron" if it released more then four times the ionization minimum in a lead-scintillator sandwich. A "muon" was defined as a track penetrating 20 cm of Fe absorber. The resulting samples had estimated hadron contaminations of 18% and 20% respectively.

Under these conditions, the analysis had to rely on a statistical selection on acoplanarity. A statistically significant effect was found, published by Perl et al.[18] with the conclusion: "We have found 64 events of the form $e^+ + e^- \rightarrow e^\pm + \mu^{\mathrm{m}} + (\geq 2 \text{ undetected particles})$, of which we have no conventional explanation". Once more, the HL had escaped discovery, this time due to the insufficient particle identification.

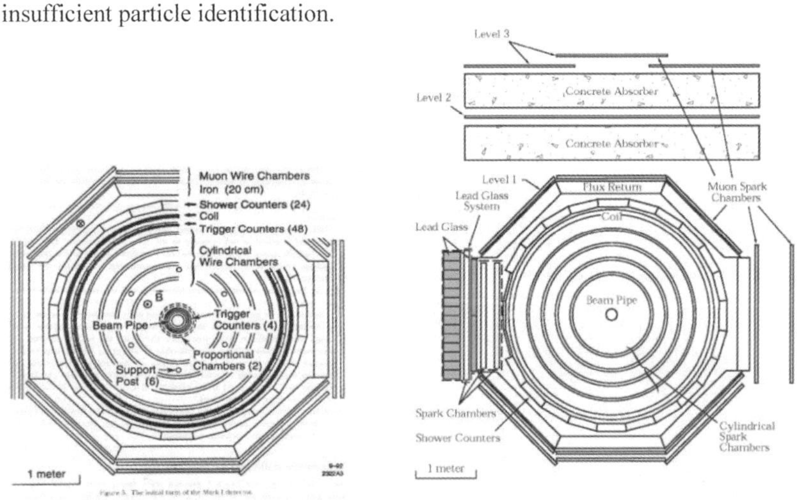

Fig. 7. The MARK I detector (sections perpendicular to the beams). Left: initial configuration in 1974. Right: with the additions of the hadron filter on top and the Pb glass wall on the left side

Already in the summer 1974 the muon identification was substantially improved, in a fraction of the solid angle, by adding thick concrete absorbers to filter the muons above the apparatus, and in 1976 a new "lead glass wall" photon detector was added by A. Barbaro Galtieri and her group to substantially improve the electron identification (Fig. 7 right). In their second article, Perl et al.[19] could reach the conclusion: "We present the properties of 105 events of the form $e^+ + e^- \rightarrow e^\pm + \mu^{\mathrm{m}} + \text{missing energy}$. …The simplest hypothesis compatible with all data is that these events come from the production of a pair of heavy leptons, the mass of the lepton being in the range 1.6 to 2.0 GeV". Finally, the discovery was consolidated in a third paper, one year later[20], titled "Properties of the proposed τ charged lepton". Notice that in this article, the name is changed from heavy lepton to tau, the initial letter of $\tau\rho\iota\tau o\nu$, "the third" in Greek, on the suggestion of P. Rapidis.

Soon, two other experiments, PLUTO and DASP at the e^+e^- DORIS collider at DESY confirmed the discovery.

The discovery of the other member of the 3rd lepton family, the tau-neutrino, had to wait another quarter of a century. As already discussed, the detector must have a large mass, because neutrino cross-sections are small and micrometre level spatial resolution, provided by emulsion techniques (Fig. 2).

In the DONUT experiment at Fermilab, the neutrino beam was produced using 800 GeV protons from the Tevatron interacting in a one-metre long tungsten "beam dump", which was 36 m upstream from the emulsion target. All the hadrons produced in the dump are absorbed, with the exception of those with picosecond lifetime. The latter are mainly charmed mesons. The primary source of ν_τ is the leptonic decay of a D_s meson into τ and ν_τ.

Fig. 8. Basic topologies of a) ν_μ interaction b) ν_τ interaction

The signature of the τ is a track with a kink, signifying a decay characterized by a large transverse momentum. Fig. 8 shows the difference between ν_τ and ν_μ interaction topologies. The already mentioned ECC technique was used. A charged particle spectrometer with electron and muon identification capabilities provided additional information.

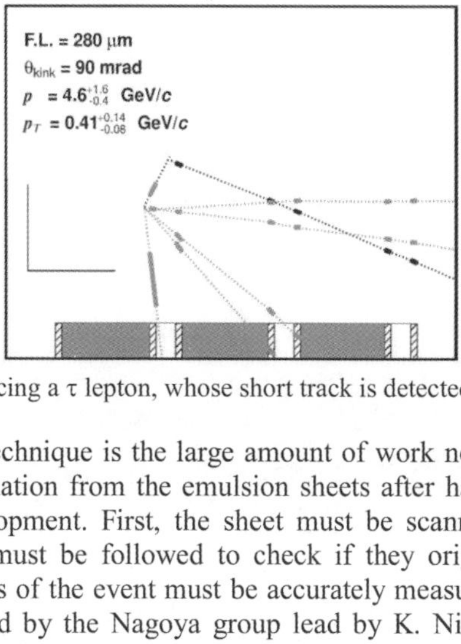

Fig. 9. A 3rd neutrino, producing a τ lepton, whose short track is detected in the DONUT experiment

A drawback of the ECC technique is the large amount of work needed at the microscope to extract the relevant information from the emulsion sheets after having exposed them to the beam and chemical development. First, the sheet must be scanned to find the interesting tracks, then these tracks must be followed to check if they originate from an interesting vertex, finally all the tracks of the event must be accurately measured. Automatic techniques were continually developed by the Nagoya group lead by K. Niwa to cope with the large amount of emulsion sheets of the DONUT experiment.
Kodama et al.[21] published in 2001 the discovery of the tau neutrino based on the observation of four ν_τ interactions with an estimated background of 0.34 events. Fig. 9 shows one of them. The short track of the τ is clearly resolved.

4. The third neutrino appearance

Experiments on atmospheric and accelerator ν_μs have proven that they disappear over long distances. While there is indirect evidence that they change flavour mainly into ν_τ, it is the running OPERA experiment at LNGS that is searching for direct evidence.
The history starts three decades ago. In 1979, A. Zichichi, then president of INFN, presented to the Italian Senate the Gran Sasso Project, to build a large, technologically advanced, laboratory under the Gran Sasso massive. The Project was soon approved with a first appropriation to ANAS (the Italian Government Road Department) in 1982. The laboratory halls were oriented, in particular, toward CERN, in order to be able in a future to host experiments on a neutrino beam from those accelerators. The draft presented by Zichichi to the Senate is shown in Fig. 10.

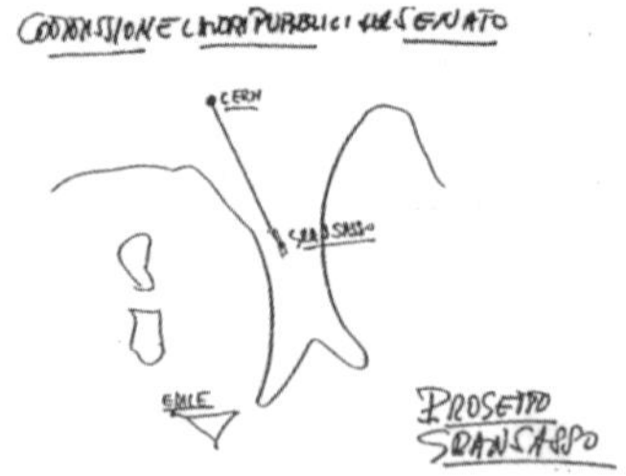

Fig. 10. Sketch of the neutrino beam from CERN to Gran Sasso by A. Zichichi (1979)

The CERN to Gran Sasso neutrino project started to become reality in 1997. After vivid discussions we took the decision for a nu-tau appearance programme. In 1997 Ereditato, Niwa and Strolin[22] proposed the OPERA experiment, based on the ECC technique. A joint CERN-INFN team designed a new neutrino beam, aiming at Gran Sasso (CERN Neutrinos to Gran Sasso, CNGS) optimised for ν_τ appearance. The Technical Design Report[23] was ready in 1998. In the August 2006 the large detectors at LNGS, LVD, OPERA and BOREXINO detected the first events produced by the neutrino beam.

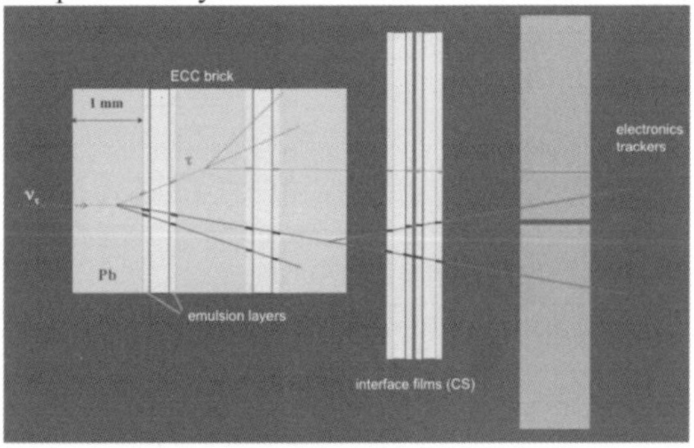

Fig. 11. The neutrino interaction detection principle of OPERA. Electronic trackers are used to identify the brick containing the interaction, which is then removed and processed.

The beam produced at CERN is mainly composed of ν_μ with no ν_τ. Consequently, the observation of any ν_τ at LNGS must be due to the appearance in the oscillation phenomenon. The (two) ECC tracking target are followed by (two) magnetic spectrometers. A very large target mass is necessary, because, even a distance of 732 km from CERN to Gran Sasso the appearance probability is small, being the 2.5 ms flight time only a small fraction of the oscillation period. Consequently, only 1-2% of neutrinos are expected to "oscillate". Considering in addition the very small neutrino cross section the conclusion is reached that the detector target mass needs to be considerably larger than 1000 t.

Fig. 11 shows schematically the components of each spectrometer. The OPERA ECC is composed of "bricks". Each brick is a sandwich of 1 mm thick Pb sheets and double, thin (50 μm) emulsion sheets on the two sides of a plastic sheet on each side of the lead. Overall, OPERA is made of 150 000 bricks, including about 110 000 m^2 emulsion films and 105 000 m^2 lead plates, for a total of about 1250 t. The automatic emulsion read-out techniques had to be further developed. OPERA is collecting and analysing data since 2008 and is planned to continue till the end of 2012. Already two candidate ν_τ were found (See lecture by Votano).

5. Are neutrinos and antineutrinos different particles?

The Standard Model assumes neutrino masses to be zero, but they are not, and that flavour lepton numbers are conserved, but the oscillation phenomena show this not to be true. These facts make it plausible that even the total lepton number might not be absolutely conserved. Being it the only quantum number that distinguishes neutrinos from antineutrinos, neutrino and antineutrino may be the same particle. In this case neutrinos are completely neutral particles, as, for example, photons and π° are. The properties of a spin ½ completely neutral particles are somewhat more un-usual. They obey, in particular, the Majorana 1937[24] rather than Dirac, equation. Let us recall that the Dirac equation is $(\gamma_\mu p^\mu - m)\psi = (E\gamma_0 - \mathbf{p}\cdot\boldsymbol{\gamma} - m)\psi = 0$, where the bi-spinor ψ has four components, namely four degree of freedom

$$\psi(x) = \begin{pmatrix} \psi_1 \\ \psi_2 \\ \psi_3 \\ \psi_4 \end{pmatrix} = \begin{pmatrix} \varphi \\ \chi \end{pmatrix}; \quad \varphi = \begin{pmatrix} \varphi_1 \\ \varphi_2 \end{pmatrix}; \quad \chi = \begin{pmatrix} \chi_1 \\ \chi_2 \end{pmatrix}$$

In the representation in which γ_5 is diagonal the Dirac equation, in terms of the two spinor components, the equation is

$$(E + \mathbf{p}\cdot\boldsymbol{\sigma})\chi - m\varphi = 0; \; (E - \mathbf{p}\cdot\boldsymbol{\sigma})\varphi - m\chi = 0$$

As it is well known, the two spinor components φ and χ correspond to the particle and the antiparticle.

The V-A structure of the charged current weak interactions implies that only the negative chirality projection of the four-component spinor field appears in the weak charged current. The bispinors of definite chirality are the (two) eigenstates of γ_5, called left and right, namely

$$\psi_L = \frac{1}{2}(1-\gamma_5)\psi, \qquad \gamma_5\psi_L = -\psi_L; \qquad \psi_R = \frac{1}{2}(1+\gamma_5)\psi, \qquad \gamma_5\psi_R = +\psi_R$$

The two projection can be expressed in terms of the two spinors, obtaining, for the left one

$$\psi_L = \frac{1}{2}\begin{pmatrix} \varphi - \chi \\ \chi - \varphi \end{pmatrix}$$

While chirality is a property of the bispinors, helicity, $h = \vec{\sigma}\cdot\vec{p}/|p|$, is a property of the (2-component) spinors. Consider for example the upper spinor, corresponding to the particle (as opposed to antiparticle) (for more details see for example [25])

$$\frac{1}{2}(\varphi - \chi) = \frac{1}{2}\left(1 - \frac{\mathbf{p}\cdot\boldsymbol{\sigma}}{E+m}\right)\varphi = \frac{1}{2}\left(1 - \frac{p_z}{E+m}\right)\varphi^{1/2} + \frac{1}{2}\left(1 + \frac{p_z}{E+m}\right)\varphi^{-1/2}$$

We see that the upper component of the left bi-spinor is not an eigenstate of helicity. For a massless particle, however, as the neutrinos of the Standard Model, $\frac{1}{2}(\varphi - \chi) = \varphi^{-1/2}$, namely massless neutrinos have definite helicity, $h=-1$. Similarly massless antineutrinos have helicity $h=+1$. This fact is at the origin of the widely used but confusing terminology to use "left" not only for negative chirality, which is correct, but also for negative helicity, which is not.

In the ultra-relativistic case, $m<<E$, the case of the real neutrinos produced by accelerator or reactors, the particle has a dominant negative helicity component and a very small positive helicity one.

In the Dirac theory the two φ and χ spinor components of the bispinor have different transformation properties. Question: is it possible to find a spinor φ_c constructed with the components of φ only (and hence without further degrees of freedom), which transforms like a χ instead than a φ and that can consequently take its place in the Dirac equation? And similarly for χ? The answer, found by Majorana, is

$$\varphi_c = i\sigma_2 \chi_M^*; \qquad \chi_c = i\sigma_2 \varphi_M^*$$

and the Majorana equation is

$$(E+\mathbf{p}\cdot\boldsymbol{\sigma})\chi - im_a\sigma_2\chi^* = 0\,; \; (E-\mathbf{p}\cdot\boldsymbol{\sigma})\varphi - im_b\varphi^* = 0$$

and the Majorana bi-spinor is

$$\psi_M = \begin{pmatrix} \varphi \\ i\sigma_2\varphi^* \end{pmatrix}$$

Notice that the Dirac and Majorana equations are identical if $m=0$.
To see that the Majorana field is self-conjugate, consider the charge conjugate of the Majorana bi-spinor

$$\psi_M^C = i\gamma_2\psi_M = \begin{pmatrix} 0 & -i\sigma_2 \\ i\sigma_2 & 0 \end{pmatrix}\begin{pmatrix} \varphi \\ i\sigma_2\varphi^* \end{pmatrix} = \begin{pmatrix} \varphi \\ i\sigma_2\varphi^* \end{pmatrix} = \psi_M$$

Majorana particles are completely neutral, namely they particle and antiparticles are the two helicity states of the same particles, similarly, say, to photon and "anti-photon".
Majorana neutrinos can in principle give origin to lepton number violation in CC weak interactions, but, in practice, there are no contradictions with the experiments, because the lepton number (quasi) conservation is a consequence of V-A character of the current.
When a massless neutrino hitting a nucleus produces a charged lepton l, the V-A nature of the current implies its helicity to be negative, which in turn implies it being a l^-. The lepton number appears to be conserved (see Fig. 12a) just as a consequence of that, namely no difference would be observed if it were not conserved. Lepton number conservation in weak interactions is not experimentally proven.
If neutrinos are massive, as they do, there are two possibilities, corresponding to the lepton number being conserved (neutrinos obey Dirac equation, Fig. 12b) or not (neutrinos obey the Majorana equation, Fig. 12c). In both cases the particles generated by the negative chirality field are no more in a helicity eigenstate, even if they are, in practice, "almost" so. The "wrong" helicity amplitude, proportional to m/E, is in the laboratory frame always extremely small, being the neutrino masses of the order of 100 meV and their energies of the order of the MeV or GeV.

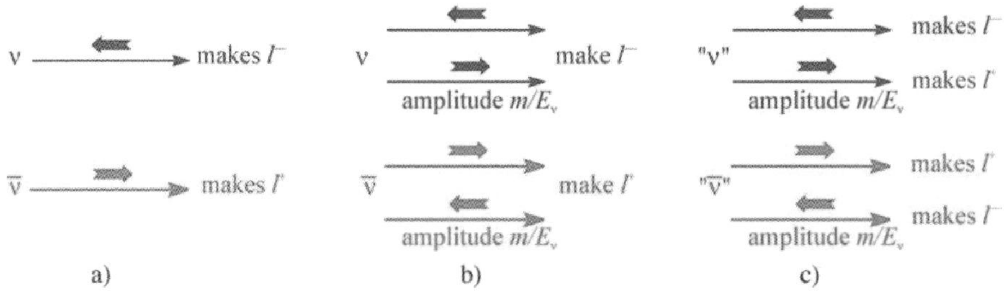

Fig. 12. Neutrino & antineutrino states in the target at rest frame. Long arrow = velocity direction, thick arrow=spin direction. a) massless neutrinos. b) Dirac massive neutrinos. c) Majorana neutrinos

It is however the effect of this tiny component that distinguishes the Dirac from the Majorana cases. If lepton number is conserved, this component produces leptons of the same sign as the dominant one (Fig. 12). On the contrary, a Majorana particle produces charged leptons of both signs, negative if its helicity is negative, positive if it is positive. Having those produced in the weak interactions a dominant helicity component, we call "neutrino" the one with dominant negative helicity, "antineutrino" the one with dominant positive helicity. The situation is pictured in Fig. 12c).

In principle, one might investigate whether neutrinos are Majorana particles searching for reactions like $\nu_\mu + N \rightarrow \mu^+ + N'$, as first suggested by G. Racah already in 1937[26], before the structure of weak interactions was known. However, the factor $(m/E)^2$ in the cross section, for energies of the order of GeV (to be above threshold) is as small as 10^{-20}, making the detection impossible. There is however a possibility, the neutrino-less double-beta decay, as first suggested by W. H. Furry in 1939.

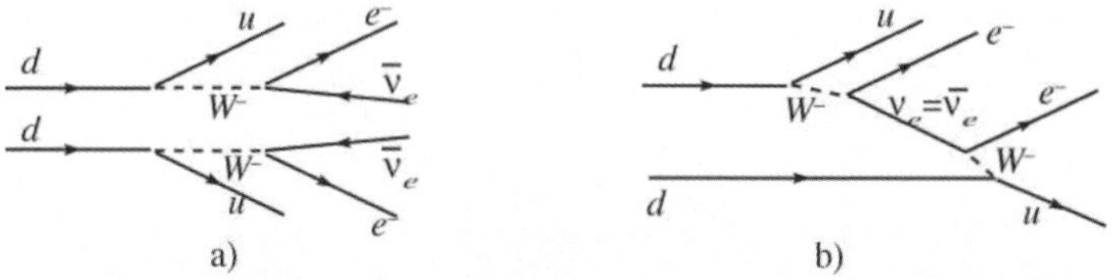

Fig. 13. Feynman diagrams for a) the $2\nu2\beta$ decay and b) $0\nu2\beta$ decay

As it is well known, beta decay is energetically forbidden for even-even nuclei. The open channel is a second order weak process, the two-neutrino double beta ($2\nu2\beta$) decay, in which two nucleons experience beta decay contemporarily, $(A,Z)\rightarrow(A,Z+2)+2\bar{\nu}_e+2e^-$. The underlying process at the quark level is shown in Fig. 13a).

The extremely long $2\nu2\beta$ half-lives of several nuclei have been measured. For example, it is for ^{76}Ge 150±10 10^{19} yr, for ^{100}Mo, 0.71±0.04 10^{19} yr and for ^{130}Te, 90±10 10^{19} yr.

If neutrinos are Majorana particles, the neutrino-less double beta ($0\nu2\beta$) decay, $(A,Z)\rightarrow(A,Z+2)+2e^-$, is possible. It is shown at the quark level in Fig. 15b). The process violates the lepton number by two units.

The matrix element in Fig. 13b) can be reliably calculated in terms of the relevant element of the mass matrix of the three neutrinos, called Majorana mass, M_{ee}

$$M_{ee} = |\sum_i U_{ei}^2 m_i| = |c_{13}^2 c_{12}^2 m_1 + c_{13}^2 s_{12}^2 m_2 e^{i\alpha} + s_{13}^2 m_3 e^{i\beta}|$$

It is the sum of three addenda, one for each neutrino mass eigenstate. Notice that cancellations may happen between them due to the (unknown) phase factors.

The measured quantity is the half-life, related to the Majorana mass by the equation

$$T_{1/2}^{-1} = G_Z(Q_{\beta\beta})|M^{0\nu}|^2 M_{ee}^2$$

where G_Z is the phase space volume, depending on the nucleus, which is calculated without uncertainties and $M^{0\nu}$ is the "nuclear matrix element". Much progress has recently been done in understanding the nuclear physics effects and the present uncertainties have been reduced to a factor 2-3.

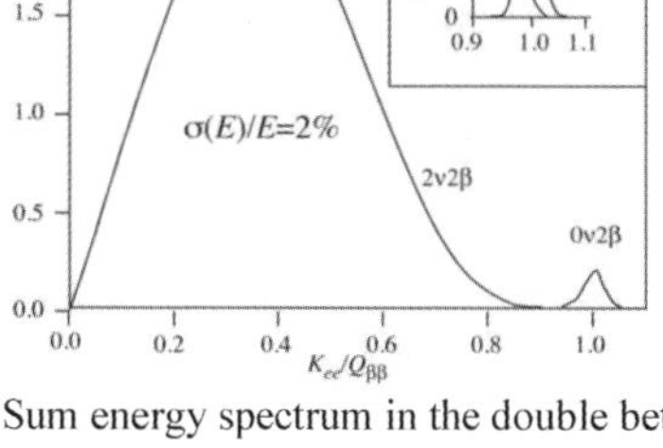

Fig. 14. Sum energy spectrum in the double beta decay

Notice that the differences between initial and final nuclear levels have been accurately measured. Consequently, the sum of the two electrons energies, called $Q_{\beta\beta}$, (which is typically a few MeV), is known, usually within a fraction of a keV.

Fig. 14 shows schematically the spectrum of the sum energy: the continuous part is the $2\nu2\beta$ decay, the small peak the $0\nu2\beta$, if it exists. Clearly, its size is not known, but its position is. In practice, other backgrounds are present. The "background index" b, in the region of interest, is the number of counts per unit exposure time, per unit detector mass and unit energy interval. The lowest background index in the last generation experiments has been $b=10^{-1}$ $yr^{-1}kg^{-1}keV^{-1}$. The present generation aims to improve to $b=10^{-2}$-10^{-3} $yr^{-1}kg^{-1}keV^{-1}$.

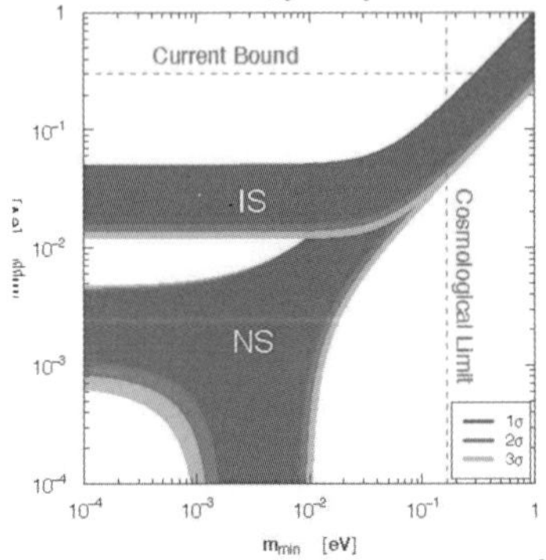

Fig. 15. M_{ee} vs. lightest neutrino mass[28]

The sensitivity of an experiment with background index b, sensitive mass M, live time T and energy resolution (FWHM) ΔE to M_{ee} scales as the figure of merit $F_M = \left(\frac{MT}{b\Delta E}\right)^{1/4}$. The fourth-root dependence implies that to gain, for example, an order of magnitude in the Majorana mass the sensitive mass must be increased by four orders of magnitude, if the background index is constant! If however the background is zero (or almost so), the figure of merit scales as $F_M \propto \sqrt[2]{MT}$. Another important parameter is the energy resolution. The region of interest of the spectrum has a width of the order of ΔE.

Moreover, the energy resolution is the only way to reduce the tail of the $2\nu2\beta$ decay under the peak, as shown in the insert of Fig. 14. It can be shown that, for a given ratio of the two half lives, the signal to (this) background ratio is $\frac{S}{B}; \frac{m_e Q_{\beta\beta}^5}{7(\Delta E)^6}\frac{T_{1/2}^{2\nu}}{T_{1/2}^{0\nu}}$, which depends on the 6th power of the energy resolution.

In conclusion, background free condition and energy resolution are the key features.

Fig 15 shows M_{ee} as a function of the smallest neutrino mass[28] for the two signs of Δm^2.

To fix the orders of magnitude consider the three nuclei employed by the experiments that just started running, EXO and KAMLand-Zen with (85%) enriched ^{136}Xe, GERDA-1 with (85%) enriched ^{76}Ge, or ready to do so, CUORE-0 with natural TeO_3 (the active ^{130}Te isotope abundance is 34%). At M_{ee}=50 meV, even with a very large exposure of 1 t yr, the numbers of expected signal events are small: 4.3, 4 and 2.5 respectively. These numbers should be compared with the expected background in the region of interest, namely the product $b\ \Delta E$. ΔE is 100 keV for EXO, 250 keV for KAMLand-Zen, 5 keV for CUORE and 4 keV for GERDA-1. The background indexes are not yet well known, given the initial phases of all the experiments; estimates range, in units 10^{-3}/(kg keV yr), between 1 and 300.

A review of the present status of the search for double beta decay by Gómez Cadenas et al.[29] has been published in 2012. Fig. 16 gives their estimate of the sensitivities of the next generation experiments, within a range of uncertainty due to both experimentally unknown parameters and nuclear matrix elements.

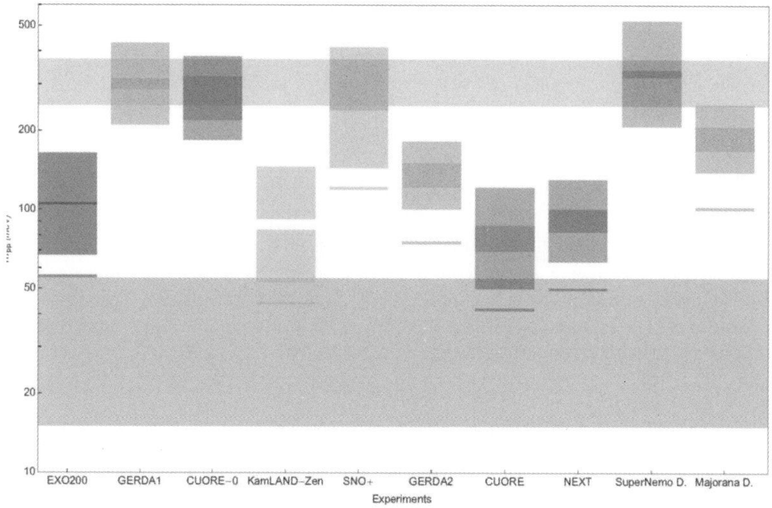

Fig. 16. From Gómez-Cadenas et al.[29]. Sensitivity of the different experiments to M_{ee} computed assuming a 5 years exposure, proper intervals for the nuclear matrix elements and for both "optimistic" and "pessimistic" experimental parameters (see [29]). A sensitivity line corresponding to a 10 years exposure, and to the most optimistic NME and experimental parameter set, is also shown

6. Conclusions

More than half a century is passed since the discovery of the first neutrino and more than 80 since the desperate hypothesis of Pauli. When the second neutrino was discovered, 50 years ago, the search for the third lepton family started. I hope to have given some flavour of the fascinating discoveries in neutrino physics.

Neutrinos are indeed the most elusive of the known particles and have always in store for us unpredicted properties. The first window on new physics, beyond the Standard Model, was opened in neutrino physics. It started in underground laboratories in the late 1960s with the solar neutrino puzzle, even before the Standard Model was established.

On the other hand, the difficulties of neutrino experiments have often induced premature enthusiasms, under evaluating the systematic uncertainties, overstressing the statistical significance, un-accounting for hidden experimental effects, etc. Examples are the Lubimov 30 eV and the 17 keV neutrino, the negative electron neutrino mass squared, the eV mass neutrinos, the recent two-sigma-or-so effects leading to poorly justified guesses of "sterile neutrinos", and more, till now.

Neutrino experiments are very difficult, but we have still a lot to learn from them.

References

1. F. P. An et al., "Observation of electron-antineutrino disappearance at Daya Bay," Phys. Rev. Lett. **108** (2012) 171803

2. J. K. Ahn et al. "Observation of Reactor Electron Antineutrino Disappearance in the RENO Experiment," Phys. Rev. Lett. **108** (2012) 191802

3. Fogli, G.L., et al. "Global analysis of neutrino masses, mixings and phases: entering the era of leptonic CP violation searches (2012). ArXiv: 1205.5254 v3

4. Thomson, J. J. Cathode rays. *Phil. Mag.* **44** (1897) 293

5. Street, J. C. & Stevenson, E. C.; New Evidence for the Existence of a Particle of Mass Intermediate Between the Proton and Electron. *Phys. Rev.* **52** (1937) 1003 (L)

6. Anderson, C. D. & Neddermeyer, S. H.; Note on the nature of cosmic particles. *Phys. Rev.* **51** (1937) 884; Cosmic-ray particles of intermediate mass. *Phys. Rev.* **54** (1938) 88
7. Conversi, M. et al.; On the disintegration of negative mesons; *Phys. Rev.* **71** (1947) 209 (L)
8. Cowan, C. L. et al.; Detection of the Free Neutrino: A Confirmation, *Science* **124** (1956) 103
9. Danby, G. et al.; Observation of High-Energy Neutrino Reactions and the Existence of Two Kinds of Neutrinos. *Phys. Rev. Lett.* **9** (1962) 36
10. Conversi, M. et al.; Search for time-like structure of the proton. *Phys. Lett.* 5 (1963) 195
11. C. S. Wu et al. "The origin of the third family" Academy of sciences – Bologna University – INFN – SIF (1997)
12. Massam, T. A new electron detector with high rejection power against pions. *Nuovo Cimento* **39** (1965) 464. See also CERN-63-26. Nuclear Physics Division, June 27, 1963
13. Buhler, A.; Range measurements for muons in the GeV region. *Nuovo Cimento* (1965) **35** 759
14. Conversi, M. et al. Proton antiproton annihilation into muon pairs. *Proceedings of the International Conference on High Energy Physics 1964.* Atommizdat, Moskow, 1966. Vol I, 857
15. Conversi, M. et al, The leptonic annihilation modes of the proton antiproton system at 6.8 $(GeV/c)^2$ time-like four-momentum transfer. *Nuovo Cimento* **40** (1965) 690
16. Bernardini, M. et al.; A proposal to search for leptonic quarks and heavy leptons produced by ADONE. INFN/AE-67/1, 20 March 1967
17. Alles Borelli, V. et al.. Limits on electromagnetic production of heavy leptons; *Nuovo Cimento Lett.* **4** (1970) 1156
18. Perl, M. L. et al.; Evidence for anomalous lepton production in e^+e^- annihilation. *Phys. Rev. Lett.* **35** (1975) 1489
19. Perl, M. L. et al.; Properties of anomalous eμ events produced in e^+e^- annihilation. *Phys. Lett.* **63B** (1976) 466
20. Perl, M. L. et al.; Properties of the proposed τ charged lepton. *Phys. Lett.* **70B** (1977) 487
21. Kodama, K. et al.; Observation of tau neutrino interactions. *Phys. Lett.* **B 504** (2001) 218
22. Ereditato, A, Niwa, K. and Strolin, P., INFN/AE-97/06, DPNU-97-07; H. Shibuya et al., CERN-SPSC-97-24, LNGS-LOI-8-97 (1997)
23. Elsner, K. et al.. The CERN neutrino beam to Gran Sasso (NGS): conceptual technical design. *CERN-98-02 ; INFN-AE-98-05* (1998)
24. Majorana, E. Teoria simmetrica dell'elettrone e del positrone *Nuovo Cimento*, **5** (1937) 171
25. Bettini, A. Introduction to elementary particle physics. Cambridge University Press. Cambridge UK (2008).
26. Racah, G.; *Nuov. Cim.* **14** (1937) 322
27. Furry, W. H.; *Phys. Rev.* **56** (1939) 1184
28. Bilenki, S. M. and Giunti, C.; Neutrinoless double-beta decay. A brief reviews. arXiv:1203.5250 (2012)
29. Gómez-Cadenas, J. J et al. The search for neutrinoless double beta decay. *Riv. Nuovo Cimento* **35** (2012) 29

CHAIRMAN: A. BETTINI

Scientific Secretaries: R. Knegjens, M. Ajaz

DISCUSSION

- *Tannenbaum:*

Suppose the ν_e turned out to be a Majorana particle? How would this affect the Standard Model and everything we know? What would have to be changed in all our models?

- *Bettini:*

Independently of being Majorana or Dirac particles, neutrino physics is beyond the SM, because they have non-zero mass and because flavour numbers are not conserved. The SM neutrinos have negative helicity. In practice they are always ultra-relativistic. The helicity components of the neutrinos (rather than anti-neutrinos) are consequently a dominant negative one and a very small, proportional to m/E, positive ("wrong") one. Vice-versa for the antineutrinos. The difference between Dirac and Majorana is in the "wrong" helicity component only, as I explained in my lecture. This component of your ν_e, when interacting with a target, produces e^- if Dirac, e^+ if Majorana. Take for example an experiment with 2 GeV beam energy and a reasonable m=200 meV. If ν_e were Majorana, the probability to observe e^+ would be $(m/E)^2 \approx 10^{-20}$. By far too small to be observable. Neutrino-less double beta decay ($0\nu 2\beta$) is the only process that might give an answer. In that case the modifications needed of the SM would be very important..

- *Zichichi :*

Interesting to point out that a particle that is massless, a la Weyl, and is its own anti-particle, a la Majorana, cannot be in 4-D, but must be in 10-D, as has been shown here in the past.

- *Bettini:*

The neutralino, the SUSY partner of the gauge bosons, which are completely neutral, must be completely neutral too. A completely neutral spin ½ particle is a Majorana particle.

- *Karakostas:*

In the case of Majorana neutrinos were observed, how should we treat the former experimental data?

- *Bettini:*

The differences between the two cases are too small, much smaller than the uncertainty in data. All data are fine as they are.

- *Hemmer:*

Would the existence of a sterile neutrino change the knowledge we have about mass differences and mixing angles?

- *Bettini:*

No. It would just add a new line and column to the mixing matrix, but not change the values of what has been measured so far.

- *Zhang:*

1. Is there any correlation between the PMNS matrix and the CKM matrix?

2. What would happen if we didn't observe 0ν2β decay? Is there any back up plan for experiment?

- Bettini:

1. Nobody knows. However they look very different. Quark mixing is small, lepton mixing is much larger and we don't know the reason. Theorists are looking into it, but no conclusion. Note that mixing angles run with energy scale, so it is not obvious that any symmetry possibly present at high energies would remain such at the energy we work at.

2. Neutrino of definite flavours have not definite masses. If they are Majorana there is a 3x3 mass matrix, of which, the element M_{ee} enters, at the second power, in the 0ν2β decay. Considering that $M_{ee} = \left|\sum_{i=1}^{3} U_{ei}^2 m_i\right|$ and that U_{ei} are complex numbers, cancellations can happen. They are such that M_{ee} may even be zero. So, it can be proved that the observation of 0ν2β proves that neutrino are Majorana, but the non-observation of that cannot prove they are not.

- Leontsinis:

Can experiments like IceCUBE contribute in neutrino studies? For example, to finding tau neutrinos?

- Bettini:

To neutrino physics they cannot contribute substantially. IceCUBE can see supernova neutrinos and possibly can give information on the evolution of the neutrino pulse. This might contribute to understand mixing and flavour change in the supernova.

- Slepukhin:

Some years ago the experiments KARMEN and Mini-Boone an anomolously small oscillation length. From this a fourth generation neutrino was hypothesized.

- Bettini:

The experiment you refer to is LSND. This experimental result was based on a background model extrapolated to the signal region. This method does not allow to reach an experimentally sound result.

- Carlucci:

In the case that neutrinos are found to be Majorana particles, will there be an immediate impact on any model beyond SM. For example SUSY or the See Saw mechanism? In other words is there any model that requires specifically Dirac neutrinos or Majorana neutrinos?

- Bettini:

In general theories, and in particular for the See Saw mechanism, Majorana mass terms are preferred. Sure you have to modify the SM if neutrinos have mass. This is easier if they are Majorana.

- Marzocca:

Are there cosmological limits on neutrino masses, and if so, what are they?

- Bettini:

Yes, there are upper linits on the sum of neutrino masses. These can be placed by studying how neutrinos influence the evolution of large scale structures. The current limit is Σm_i= 300-600 meV. The Planck mission may well improve the limit, or even find a positive evidence.

- *Minkowski:*

I would like to suggest that the original article by E. Majorana "Teoria simmetrica dell'eletrone e del positrone" be translated into English and that the formulas, presented in the real Majorana representation, be worked out in our modern notation comparable with the formulas presented in your lecture.

- *Zichichi :*

It has already been translated.

- *Melia:*

Following up Zhang's question on relating the quark and lepton mixing matrices - is there any experimental or theoretical reason why the generations are matched in the usual way i.e., electron and electron neutrino with up quark and down quark?

- *Bettini and Zichichi :*

There is no theoretical reason. Families are sorted by their masses.

- *Tannenbaum:*

Does Majorana neutrino violate lepton number? That is, given a beam of ν_e Majorana neutrinos do I expect it to produce an equal number of electrons and positrons?

- *Bettini:*

Yes to your first question, no to the second. Majorana neutrino is the component of negative helicity, antineutrino the one of positive helicitiy. The beam will produce (as it does in all the experiments) practically only electrons, and a fraction $(m/E)^2$ of positrons, which is too small to be observable (see above).

HIGH ENERGY PHYSICS AND GRAVITATIONAL WAVES

Eugenio COCCIA

GSSI (INFN) Gran Sasso Science Institute and University of Rome "Tor Vergata"

ABSTRACT

I discuss the possible relevance of gravitational-wave (GW) experiments for high energy physics. I examine three topics: the relic GW background, the GW bursts form instabilities in neutron stars, the GW signal from a supernova. The basic principles and the physics cases are reported.

1. Introduction

Gravitational waves (GWs) were predicted by Einstein as a consequence of his general theory of relativity [1]. They can be emitted by a variety of extreme astrophysical systems and propagate at the speed of light as ripples in the geometry of spacetime.

LIGO, GEO600 and Virgo [2] are a network of interferometric detectors aiming to make the first direct observations of GWs. They consist of two experimental sites in the US (Hanford, WA and Livingston, LA) and two in Europe (Hannover, Germany and Cascina, Italy). An additional experimental site to host an underground cryogenic detector named KAGRA (previously known as LCGT [3]) is under construction in Japan while a proposal to build a third LIGO site in India is currently being evaluated[4]. These detectors use kilometer-scale Michelson interferometry to measure the fractional differential change $\delta L/L$ in the distances of two orthogonally positioned pairs of masses[5]. Such changes are expected to occur upon the passage of a GW. The fractional distance change $\delta L/L$ between two such pairs of perpendicularly arranged masses defines $h(t)$, the GW strain as measured by the detectors. This is a direct measure of the strength of the local spacetime distortions (which is the GW itself) folded with coefficients of order unity that depend on the direction of the wave source and the orientation of the interferometer [6].

The network of GW detectors is currently (2012) undergoing upgrades which are expected to produce a factor of about 10 improvement in sensitivity [7,8]. These will be the advanced LIGO and Virgo detectors and are expected to come online in 2014-2015.

So it is a good time to look to the potentiality of these detectors for astrophysics, cosmology and particle physics. I focus here on the relevance of these detectors for high energy physics, discussing three topics: the relic GW background from the early Universe, the instabilities of compact objects, usually indicated as neutron stars but which could be stars made of quarks, and the Supernova scenario, in which much can be understood by a simultaneous detection of the GW and neutrino bursts.

2. Relic GW background

The energy range between the grand unification scale $M_{GUT} \sim 10^{16}$ GeV and the Planck scale $M_{Pl} \simeq 1.22 \times 10^{19}$ GeV is crucial for fundamental physical questions and for testing current ideas about grand unification, quantum gravity, string theory.

There are important experimental results in particle physics that can be translated into statements about this energy range, like the accurate measurement of Gauge coupling constant versus energy at accelerators and the measurement of proton lifetime. Here we take a cosmological point of view: informations on this energy range can only come from particles which decoupled from the primordial plasma at very early time [9,10]. Particles which stay in thermal equilibrium down to a decoupling temperature T_{dec} can only carry information on the state of the Universe at $E \sim T_{dec}$. All information on physics at higher energies has in fact been obliterated by the successive interactions.

The condition for thermal equilibrium is that the rate Γ of the processes that maintain equilibrium be larger than the rate of expansion of the Universe, as measured by the Hubble parameter H [11]. The rate is given by $\Gamma = n\sigma|v|$ where n is the number density of the particle in question, and for massless or light particles in equilibrium at a temperature T, $n \sim T^3$; $|v| \sim 1$ is the typical velocity and σ is the cross-section of the process. Consider for instance the weakly interacting neutrinos. In this case the equilibrium is mantained, e.g., by electron-neutrino scattering, and at energies below the W mass $\sigma \sim G^2_F \langle E^2 \rangle \sim G^2_F T^2$, where G_F is the Fermi constant and $\langle E^2 \rangle$ is the average energy squared. The Hubble parameter during the radiation dominated era is related to the temperature by $H \sim T^2/M_{Pl}$.

Therefore

$$\left(\frac{\Gamma}{H}\right)_{neutrino} = \frac{G_F^2 T^5}{T^2/M_{Pl}} \approx \left(\frac{T}{1\ MeV}\right)^3$$

Even the weakly interacting neutrinos, therefore, cannot carry information on the state of the Universe at temperatures larger than approximately 1 MeV. If we repeat the above computation for gravitons, the Fermi constant G_F is replaced by Newton constant $G = 1/M^2_{Pl}$ (using units $\hbar = c = k_B = 1$) and at energies below the Planck mass

$$\left(\frac{\Gamma}{H}\right)_{graviton} \sim \left(\frac{T}{M_{Pl}}\right)^3$$

The gravitons are therefore decoupled below the Planck scale. At the Planck scale the above estimate of the cross section is not valid and nothing can be said without a quantum theory of gravity. It follows that relic gravitational waves are a potential source of information on very high-energy physics. Gravitational waves produced in the very early Universe have not lost memory of the conditions in which they have been produced, as it happened to all other particles, but still retain in their spectrum, typical frequency and intensity, important information on the state of the very early Universe, and therefore on physics at correspondingly high energies, which cannot be accessed experimentally in any other way. It is also clear that the property of gravitational waves that makes them so interesting, i.e. their extremely small cross section, is also responsible for the difficulties of the experimental detection.

Incidentally, thinking in terms of cross-sections, one is lead to ask how comes that gravitons could be detectable, since the graviton-matter cross section is smaller than the neutrino-matter cross section, at energies below the W-mass, by a factor $G^2/G^2_F \sim 10^{-67}$ and neutrinos are already so difficult to detect. The answer is that gravitons are bosons, and therefore their occupation number per cell of phase space can be $n_k \gg 1$; in interesting cases, in the relic stochastic background we can have $n_k \sim 10^{40}$ or larger, and the squared amplitude for exciting a given mode of the detector grows as n^2_k. So, we should not speak of detecting gravitons, but rather classical gravitational waves. Neutrinos, in contrast, are fermions and for them $n_k \leq 1$.

There are various production mechanisms from which we might expect cosmological gravitational waves, including cosmic strings [12], amplification of vacuum fluctuations following inflation [13, 14], pre-Big-Bang models [15, 16], or the electroweak phase transition [17].

Whatever the production mechanism, the intensity of a stochastic background of gravitational waves (GWs) can be characterized by the dimensionless quantity

$$\Omega_{gw}(f) = \frac{f}{\rho_c}\frac{d\rho_{gw}}{df}$$

where $d\rho_{gw}$ is the energy density of gravitational radiation contained in the frequency range f to $f+df$ and ρ_c is the present value of the critical energy density for closing the Universe [11].
In terms of the present value of the Hubble constant H_o , the critical density is given by

$$\rho_c = \frac{3H_o^2}{8\pi G}$$

The value of H_o is usually written as $H_o = h_o \times 100$ km/(s Mpc), where h_o parametrizes the existing experimental uncertainty.

As a signal is expected to be much smaller than current detector noise, and because we assume both the detector noise and the signal to be Gaussian random variables, it is not feasible to distinguish the two in a single detector. We must therefore search for the stochastic GW background using two or more detectors. The optimal method is to cross-correlate the strain data from a pair, or several pairs of detectors [18].

There are also mechanisms, such as due to magnetars or rotating neutron stars which can give a component to the stochastic background in term of a superposition of unresolvable gravitational-wave signals of astrophysical origin.

Direct limits on the amplitude of the stochastic gravitational-wave background using the data from a two-year science run of the Laser Interferometer Gravitational-wave Observatory (LIGO) [19] have been reported in 2009. These results constrained the energy density of the stochastic gravitational-wave background normalized by the critical energy density of the universe, in the frequency band around 100 Hz, to be less than 6.9×10^{-6} at 95% confidence. These data ruled out models of early universe evolution with relatively large equation-of-state parameter [20], as well as cosmic (super)string models with relatively small string tension [21] that are favoured in some string theory models [22]. This search for the stochastic background improved upon the indirect limits from the Big Bang Nucleosynthesis and cosmic microwave background at 100 Hz.

New results were presented in 2012 with the first joint search for a stochastic background using data from the LIGO and Virgo interferometers [23]. In a frequency band of 600-1000 Hz, a 95% upper limit on the amplitude of $\Omega_{GW}(f) = 0.33\ (f/900\text{Hz})^3$. These new limits are a factor of seven better than the previous best in this frequency band.

The conclusion is that LIGO and Virgo are exploring the Universe at temperatures $T \sim 10^7$ GeV.

3. Compact stars

Neutron stars (NS) are the final state of stars which, after the exhaustion of their nuclear fuel and the subsequent explosion and ejection of the external layers, remain with a core more massive than the Chandrasekhar mass $M_{Ch} \simeq 1.4M_{\odot}$ (but still the core must be lighter than a critical value $M_{bh} < O(2-3)M_{\odot}$, beyond which a black hole instead forms). When $M_{Ch} < M_{core} < M_{bh}$, the core of the star collapses under its own weight, until it reaches a radius $R \simeq 10$ km, where the self-gravity of the star is now balanced by the neutron degeneracy pressure. As a consequence, the nuclear matter inside the star is compressed to extreme densities [24].

The internal structure of neutron stars depends strongly on the equation of state; however in the inner core, say $R < O(1)$ km, the density reaches values of order 1 GeV/fm^3. We are therefore in a regime governed by QCD at high density.

Neutron star perturbations can be described in terms of quasinormal modes, which can be excited by many possible mechanisms: accreting material, crust breaking, starquakes, onset of phase transitions, coalescence, etc. These perturbations can excite, in particular, the neutron star quasinormal modes described by spherical harmonics with l=2, which dominates the emission of GWs. Furthermore, if the NS is rotating, even the radial oscillation will induce a time varying quadrupole moment, and in this case energy will be liberated in GWs.

The frequency of the quasinormal mode depends, in a calculable way, on the equation of state in the NS interior. Therefore the value of the GW frequency carries important information on the internal NS structure.

It must be remarked that for GWs even an object such as the core of a NS is basically transparent, because of the smallness of gravitational cross-sections. Therefore, GWs generated inside the core, for instance as a consequence of a corequake, travel unaffected outside the NS. This is of course very different from electromagnetic waves, for which the NS interior is totally opaque, and is an excellent example of the fact that GW astronomy can potentially open up a completely new window on the Universe, unaccessible to electromagnetic observations.

A NS accreting matter at a steady rate from a companion can undergo periodically some structural changes which are accompanied by the emission of bursts of GWs. The magnetic field of the NS is not sufficiently strong to channel the accreting matter toward the poles, and therefore the accretion is spherically symmetric. Each time a layer of about 1 m of material is accreted (which happens in a time which, depending on the particular star, can be between a few hours and a few days) a thermonuclear flash takes place, and is observed as an x-ray burst. These sources therefore repeatedly emit x-ray bursts.

The mass radius relation of a NS depends on the equation of state, but it is always such that the larger is the mass, the smaller is the radius. Thus, when a NS accretes material, its new equilibrium radius decreases. If the NS were a fluid, a continuous accretion of matter would produce a continuous decrease of the radius. However, neutron stars have a solid crust, about 1 km thick and with a rigidity which, in the inner part, is huge by terrestrial standards. Therefore the radius will rather stays constant until sufficient material has been accreted so that the crust can be broken, and the evolution of the radius will rather be a sequence of jumps. In each jump a certain amount of energy is released. The important issue, for our purposes, is that the amount of energy released per jump, radiated away in GWs, depends on the composition of the NS and on the high energy physics holding in such compact objects.

The maximum energy that can be liberated in a single starquake is [25]:

$$\Delta E_{rad} \sim 10^{-9}\, M_{\odot} c^2 \left(\frac{R}{14km}\right)\left(\frac{M_{\odot}}{M}\right)\left(\frac{\theta_{max}}{10^{-3}}\right)$$

We have taken 14 km as a reference value for R, since this is the typical value of the radius for a fast rotating NS in a large range of masses. θ_{max} is the maximum strain angle that the crust can sustain without breaking.

Another possibility, giving possibly larger GW emissions, is that the starquake due to the crust breaking, rather than being the source of the GWs, might be the trigger for some more important structural changes inside the NS. In particular, the core of a NS can perform a phase transition from a hadronic to a deconfined quark-gluon phase. In the literature [27,28] it has been considered the possibility that an accreting NS acquires a sufficient mass to perform completely the transition from a hadronic core to a quark-gluon core. The energy gained in the transition, for a NS with a mass M= 1.5M, is of order 0.15 Mc^2 , and it has been observed that this large energy can excite quasinormal modes and be liberated in a GW bursts (with high efficiency if the NS rotates sufficiently fast). Particularly interesting appear the discontinuity g-modes [29], having typical frequency in the range 0.5–1.4 kHz and constituting an unique probe for density discontinuities, like the ones induced by phase transitions. One can imagine that the phase transition does not take place suddenly and completely in the whole core. Rather, each time a critical mass is accreted, a starquake takes place and transforms successive layers of the NS core from the hadronic to the deconfined phase. In this case the 0.15 solar masses will not be released in a single, very large bursts, but rather in a series of bursts. One can imagine variants of this scenario in which the phase transition involves strange quark matter, hybrid quark stars, or a phase of color crystallization. In all cases, unfortunately, a computation of the energy liberated by the phase transition in a NS layer as a consequence of the starquake appears a rather difficult task.

Another possible realization of a GW source is provided by the **r**-mode instability in strange stars (see Chapter 12 of Ref. [24] for an introduction to strange stars). As discussed in Ref. [30], in stars made of strange quark matter the r-mode instability has a dynamics quite distinct from the neutron star case. In particular, in an accreting strange star the evolution of the GW amplitude generated by the r-mode during its first year of evolution consists of a repeated series of bursts on a time scales from hours to months (see, in particular, Fig. 4 of Ref. [30]). The GW amplitude has been estimated to be [67]

$$h \sim 2 \times 10^{-20} \alpha \left(\frac{1\, kpc}{r}\right)\left(\frac{M}{1.4 M_{\odot}}\right)\left(\frac{1\, ms}{P_{rot}}\right)^{3}\left(\frac{R}{10\, km}\right)^{3}$$

where P_{rot} is the rotation period and α is the r-mode amplitude. During the first year of evolution of a young strange star the parameter α performs large oscillations from very low values, of the order of 10^{-15}, up to values of order one. These rapid variations therefore result in a series of GW bursts. The main feature of this source, is that within the model discussed in Ref. [30] this kind of activity can take place only for very young quark stars, i.e., in their first year of evolution.
The detection and study of these sources are a challenge for the next generation of interferometers.

4. The Supernova

A supernova is an explosion of a massive supergiant star. No events in nature surpasses its raw power: about 10^{53} erg s^{-1} (equivalent to 10^{46} J s^{-1}) is released as neutrinos from a 'core-collapse' supernova, which is as much instantaneous power as all the rest of the luminous, visible Universe combined. They give birth to the most exotic states of matter known: neutron stars and black holes. Supernovae have been at the forefront of astronomical research for the better part of a century, and yet no one is sure how they work [31].
We discuss here the scientific motivations for a joint analysis of GWs and low-energy neutrino data to probe the processes powering a supernova explosion [32].
Several mechanisms in a core-collapse supernova can give rise to GW bursts. Signals may last from milliseconds to seconds. While the astrophysical motivation for expecting GWs to accompany core-collapse supernovae is strong, the expected rate, GW strength and waveform morphology are uncertain. The expected energy going into GWs would be 10^{-10} – 10^{-4} of a solar mass, or 10^{44} - 10^{50} ergs. This is only a small fraction of the energy liberated in neutrinos: the core collapse of a massive star is expected to produces a huge flux of neutrinos. Actually nearly all (about 99%) of the binding energy $O(10^{53}$ ergs) of the resulting neutron star (or black hole) ends up in neutrinos escaping on a short timescale (a few tens of seconds). The neutrinos (and antineutrinos) produced are of all flavours, and have energies in the few to tens of MeV range. The burst of neutrinos from SN1987A

in the Large Magellanic Cloud, detected by various neutrino detectors, confirmed the baseline model of stellar collapse. On the contrary, the estimates of GW bursts associated with supernova rely on models. Most such models are not yet three-dimensional, do not incorporate the entire set of possibly relevant physics, and do not predict robust supernova explosions as observed in the electromagnetic universe. So, despite the availability of multiple potential explosion mechanisms and their associated multi-dimensional dynamics and GW signatures, the current picture is unlikely to be complete.

It is well worth to study the supernova signals with GWs and neutrino detectors with a multi-messenger approach.

The first important point is that GWs and neutrinos from core-collapse supernovae are emitted in the inner-most, high density region of the supernova core which cannot be probed electromagnetically. GWs and neutrinos thus are the only messengers that can carry ``live" dynamical information from deep inside a dying massive star and constrain the detailed, yet unknown, mechanism driving the core-collapse supernova explosion. Another point is that both GWs and neutrinos leave the core-collapse event on approximately the same time scale (seconds) in contrast to electromagnetic radiation, which can take hours or days to become visible to astronomers. Moreover, a visible supernova near the edge of sensitivity may be a rather poor tag of core collapse for both GW and neutrino detectors separately, in the presence of detector background, given what will likely be a large uncertainty on the occurrence itself of a collapse and on its timing [33]. Let me develop these considerations.

Several of the world's present neutrino detectors are sensitive to a neutrino burst from a galactic supernova. Super-K, a 50-kton water Cherenkov detector in Japan, would observe some ~8000 events for a core collapse at the center of the Milky Way, ~8.5 kpc away [34].

The LVD [35] and Borexino [36] scintillation detectors at Gran Sasso in Italy, and KamLAND [37] in Japan, would observe hundreds of interactions. The IceCube detector at the South Pole, although nominally a multi-GeV neutrino detector, would observe a coincident increase in count rate in its phototubes due to a diffuse burst of Cherenkov photons in the ice, and has sensitivity to a galactic supernova.

The duty factor of neutrino detectors in observing mode has typically been 90% or larger.

The Super-K, LVD, IceCube and Borexino detectors are also operating as part of the SNEWS (SuperNova Early Warning System) network [38,39], which has the aim of providing a prompt alert to astronomers in the case of a coincident supernova neutrino burst.

Uncertainties on the rate of supernovae in the nearby universe are significant. According to current rate estimates, there is about one supernova every 40 years in the galaxy, 3-4 supernovae per century in the local group (at a distance of ~1Mpc), about one supernova every other year between 3-5~Mpc, with an integral rate of one supernova per year out to the Virgo cluster. These rates are more likely to be lower than upper bounds.

A significant fraction of galactic or very close extragalactic supernovae (within 1~Mpc) could be optically silent, not only because of no or very weak explosions, but also because of dust extinction of the electromagnetic (optical) emission. We can use GWs and neutrinos to identify such core-collapse events that do not lead to a strong electromagnetic display. A neutrino and GW coincidence, in the absence of a strong prompt electromagnetic signal, would provide smoking-gun evidence for this weak or ``failing'' supernova scenario. It is worth noting that the rate of such events may be comparable to the rate of supernovae with optical signals [kochanek:08]. An example is SN 2008iz that exploded in M82 in late January 2008, but was discovered serendipitously in the electromagnetic spectrum (through radio observations) only more than a year later. Besides the two large spiral galaxies M31 (Andromeda) and the Milky Way, there are a number of smaller galaxies in the Local Group that may have a much higher core-collapse rate than presently expected. These are the Magellanic Clouds (SN1987A exploded in the LMC) and IC 10, IC 1613, and NGC 6822 (all within ~50-700 kpc). Combining GW data with neutrino observations would enhance the ability to find electromagnetically dark or obscured supernovae in our nearby universe.

There is an assured improvement offered to the core-collapse supernova reach by the joint analysis of the GW and neutrino detectors. This is due to the possibility of a tighter search window and lower background rate. Estimates have been done and are in progress in order to quantify this improvement. Let me report two examples from ref. [32].

1) In the all-sky GW burst search using the first-year data of LIGO's fifth science run, a false alarm rate of 1 in 100 years in the 64-200 Hz frequency band corresponds to a three-detector network signal-to-noise ratio (SNR) threshold of 8.5. For frequencies above 200 Hz, the corresponding SNR threshold is lower, but the interferometers' strain sensitivity is lower at this frequency band [40]. Requiring a coincidence of GW events within O(1s) of a neutrino signal tuned at a rate of 1 per day would allow GW detectors to operate at a false alarm rate of 3×10^{-5} Hz, which in turn corresponds to a SNR threshold of 3.5 [40], or an improvement of a factor of 2 in sensitivity. The distance reach of gravitational-wave detectors scales linearly with the inverse of SNR. Such potential improvement in gravitational-wave sensitivity, in a joint GW-neutrino search, will increase the science reach of the GW instruments relative to what they can achieve alone.

2) A gravitational-wave coincidence requirement also has the potential to improve the sensitivity of neutrino experiments by relaxing the criteria for detection. For example, Super-K's "distant" burst

search [34] requires two neutrino events (with energy threshold 17 MeV) within 20 seconds, which corresponds to approximately 8% probability of detecting a supernova in Andromeda. The accidental coincidence rate for this criterion is less than one per year; the single event rate at this threshold is about 1 per day. If one could achieve an acceptable accidental rate by requiring coincidence of a single neutrino event with a gravitational-wave signal, then the probability of a core-collapse event in Andromeda satisfying the search criterion would be about 35%, . Distant burst search parameters could be re-optimized with respect to current ones; the neutrino event energy threshold could potentially be reduced, further improving sensitivity.

References

[1] A. Einstein. Sitzungsberichte der Koniglich Preußischen Akademie der Wissenschaften (Berlin), page 688, 1916.

[2] LIGO, GEO, VIRGO: http://ligo.caltech.edu, http://geo600.aei.mpg.de/, http://www.virgo.infn.it

[3] 3. K. Kuroda and LCGT Collaboration. Class. Quantum Grav., 27, 084004, 2010.

[4] B. Iyer et al. LIGO-India, Proposal of the Consortium for Indian Initiative in Gravitational-wave Observations (IndIGO). Technical report, LIGO Scientific Collaboration, 2011. Available at https://dcc.ligo.org/cgi-bin/DocDB/ShowDocument?docid=75988.

[5] P. R. Saulson. Fundamentals of Interferometric Gravitational Wave Detectors. World Scientific, Singapore, 1994.

[6] K. S. Thorne. In S. W. Hawking and W. Israel, editors, 300 Years of Gravitation, Cambridge, UK, 1987. Cambridge University Press.

[7] B. P. Abbott et al. (LIGO Scientific Collaboration). Phys. Rev. D., 80(10), 102001, 2009.

[8] J. Abbott et al. (LIGO Scientific Collaboration and Virgo Collaboration). Phys. Rev. D. 81(10), 102001, 2010.

[9] M. Maggiore, Phys.Rept. 331 (2000) 283-367

[10] M. Maggiore, VIR-NOT-PIS-1390-113 IFUP-TH 58/97, March 1998

[11] E.W. Kolb and M.S. Turner, The Early Universe, Addison Wesley 1990.

[12] V. Mandic and A. Buonanno, Phys. Rev. D73, 063008 (2006), arXiv:astro-ph/0510341.

28

[13] R. Apreda et al., Nucl. Phys. B631, 342 (2002), arXiv:gr-qc/0107033.

[14] B. Allen and J. D. Romano, Phys. Rev. D59, 102001 (1999), arXiv:gr-qc/9710117.

[15] B. P. Abbott et al., Reports on Progress in Physics 72, 076901 (2009), arXiv:gr-qc/0711.3041.

[16] F. Acernese et al., Classical and Quantum Gravity 25, 114045 (2008).

[17] H. Grote for the LIGO Scientific Collaboration, Classical and Quantum Gravity 25, 114043 (2008).11.

[18] L. Grishchuk, JETP Letters 23, 293 (1976).

[19] B. P. Abbott et al., Nature 460, 990 (2009), arXiv:astro-ph/0910.5772.

[20] T. Regimbau and J. A. de Freitas Pacheco, Astron. Astrophys. 447, 1 (2006).

[21] V. Ferrari, S. Matarrese, and R. Schneider, MNRAS 303, 247 (1999), arXiv:astro-ph/9804259.

[22] X. Siemens, V. Mandic, and J. Creighton, Phys. Rev. Lett. 98, 111101 (2007).

[23] B. P. Abbott et al. Phys.Rev. D85 (2012) 122001

[24] N. K. Glendenning, Compact Stars (Springer-Verlag, Berlin, 2000).

[25] S. L. Shapiro and S. A. Teukolsky, Black Holes, White Dwarfs and Neutron Stars (Wiley, New York, 1983).

[26] E. Coccia, F. Dubath, M. Maggiore, Phys.Rev. D70 (2004) 084010

[27] K. S. Cheng and Z.G. Dai, Astrophys. J. 492, 281 (1998).

[28] G. F. Marranghello, C. A. Z. Vasconcellos, and J. A. de Freitas Pacheco, Phys. Rev. D 66, 064027 (2002).

[29] G. Miniutti et al., Mon. Not. R. Astron. Soc. 338, 389 (2003); Astrophys. J. 411, L33 (1993).

[30] N. Andersson, D. I. Jones, and K. D. Kokkotas, Mon. Not. R. Astron. Soc. 337, 1224 (2002).

[31] F. Woosley and H-T Janka arXiv:astro-ph/0601261

[32] Leonor, I. et al., Class. Quant. Grav. 27, 084019 (2010)

[33] G. Pagliaroli, et al. Phys. Rev. Lett. 103, 031102, 2009.

[34] M Ikeda et al. (Super-KAMIOKANDE Collaboration). ApJ., 669, 519, 2007.

[35] Agafonava N Y et al 2008 Astropart. Phys. 28 516

[36] Cadonati L, Calaprice F P and Chen M C 2002 Astropart. Phys. 16 361

[37] Piepke A et al 2001 Nucl. Phys. B 91 99

[38] Antonioli P et al 2004 New J. Phys. 6 114

[39] Scholberg K 2008 Astronomische Nachrichten 329 337

CHAIRMAN: E. COCCIA

Scientific Secretaries: T. Drozhzhova, Y. Klopot

DISCUSSION

- *Gnesi:*

If the relic gravitons carry an energy which is below the spacing of quantum gravitational level they would not be detected or maybe we could have information about the quantization of gravity?

- *Coccia:*

There are exotic theories which predict this background could extend to a frequency and be substantially large. The only way to say something about that is to correlate the output of two detectors that should be near enough to have the gravity force coherent on a two. The distance should be less then the wavelength you want to measure but not too close otherwise the seismic noise will be correlated too. The experiment is not easy however gravitons arrive with occupation number which is very high even they are weak. The effect on the detector can be understood not in terms of quantum energy but just in terms of classical waves arriving on the detector.

- *Drozhzhova:*

For a good measurement of gravitational waves it is important to keep the distance between three of LISA`s cosmic apparatus the same. Otherwise it wouldn`t be a clear interference figure. On the other hand we know about meteorites and asteroids falling on the Earth and may be it would be hazard of hitting one of LISA`s apparatus by it so that apparatus could be deviated from its Kepler`s orbit.Will be any protection mechanism provided for LISA?

- *Coccia:*

The protection of the free falling masses is the spacecraft itself which is able to do micro transfers, which keep the internal mass always in free-fall. You need to measure the distance of the internal mass with respect to the frame, i.e. the spacecraft. That means that if the spacecraft arrives too near an internal mass, the transfers act in a way to recover the original position. At the end what we have is that the free-falling masses around the Sun with the Spacecraft act as a total protection against any external disturbance.

- *Karakostas:*

Concerning the LISA project, isn`t the spacecraft going to affect the free falling of the masses?

- *Coccia:*

Inside the spherical shell gravity is zero. This effect is negligible. The spacecraft is designed not to have the Newtonian effect directly on the center of mass.

- *Karakostas:*

As you mentioned in your lecture, space-time perturbations propagating in vacuum with the speed of light are gravitational waves. Is the presence of matter going to affect them? What about Black Holes?

- *Coccia:*

There is a very high reciprocity between generation and absorption. It is very difficult to generate the gravitational waves. To generate it you need very high density and an equivalent motion with a large acceleration. To absorb gravitational waves you need very stiff material. A gravitational wave propagates through interstellar mediums, for instance. I leave it to dark matter – why not? This does not affect in any measurable way the direction or the shape of the signal. If the wave hits black holes or neutron stars the effect is very small on the dynamics of the neutron star. And the energy lost by the gravitational wave propagating from one neutron star to another is negligible at large distance.

- *Leontsinis:*

What did you do to reach the sensitivity that was calculated when designing the experiment?

- *Coccia:*

One possibility is to increase the power of the laser. At the beginning you need to have very low power of the laser, because of perturbations of the system (no so simple to explain); only when the system is stable you can apply the full power of the laser, so it's not only reduction of noise but also you can increase the power of the system by stabilizing all the other sources. Between one curve and the other there have been many effects. One was for instance the air conditioning in the building where the descriptor was, which was causing some trouble. You learn day-by-day how to reduce sources of noise.

- *Aushev:*

That is the smallest scale on which the gravitational constant is measured?

- *Coccia:*

The best result was from the group of Washington State. They measured the gravitational effect up to a distance of the order of half a mm. It requires experiments that should be very delicate. That is not easy.

Four Decades of Computing in Subnuclear Physics – from Bubble Chamber to LHC[a]

Jürgen Knobloch/CERN

Abstract

This manuscript addresses selected aspects of computing for the reconstruction and simulation of particle interactions in subnuclear physics. Based on personal experience with experiments at DESY and at CERN, I cover the evolution of computing hardware and software from the era of track chambers where interactions were recorded on photographic film up to the LHC experiments with their multi-million electronic channels.

Introduction

Since 1968, when I had the privilege to start contributing to experimental subnuclear physics, technology has evolved at a spectacular pace. Not only in experimental physics but also in most other areas of technology such as communication, entertainment and photography the world has changed from analogue systems to digital electronics. Until the invention of the multi-wire proportional chamber (MWPC) by Georges Charpak in 1968, tracks in particle detectors such as cloud chambers, bubble chambers, spark chambers, and streamer chambers were mostly recorded on photographic film. In many cases more than one camera was used to allow for stereoscopic reconstruction in space and the detectors were placed in a magnetic field to measure the charge and momentum of the particles. With the advent of digital computers, points of the tracks on film in the different views were measured, digitized and fed to software for geometrical reconstruction of the particle trajectories in space. This was followed by a kinematical analysis providing the hypothesis (or the probabilities of several possible hypotheses) of particle type assignment and thereby completely reconstructing the observed physics process.

From the 1970s on, most data were generated (or transformed into) digital form. In Table 1, typical experiments of each decade are listed. The number of scientists involved has increased from a few tens to several thousands. The increasing complexity of detectors required more and more sophisticated software to acquire, reconstruct and analyse the data and to perform the necessary simulations. It is interesting to note that the productivity of software developers stayed rather constant at 10 to 20 lines of code per developer and day.

In this article, I discuss computing aspects of these projects. The final chapters cover the specific topics event display simulation and data preservation.

[a] Lecture presented at the International School of Subnuclear Physics, 50th Course, "What we would like LHC to give us", Erice, Sicily, 23 June – 2 July 2012, slides of talk at: http://www.ccsem.infn.it/issp2012/

Institute	Collaboration	Years	Countries	Institutes	People	Developers	Lines of Code
DESY	Streamer Chamber	1969 - 1976	2	4	20	3	4 K
CERN	CDHS Neutrino	1976 - 1992	4	5	40	4	40 k
CERN	ALEPH @LEP	1985 - 2000	11	30	350	20	600 k
CERN	ATLAS @LHC	since 1991	38	176	3000	200	3-5 M
CERN	Grids for LHC	since 2000	56	340	13800		

Table 1: Typical experiments and projects of the last four decades

The Streamer Chamber at DESY

A streamer chamber had replaced the DESY 80 cm bubble chamber because it allowed to trigger on rare events and it allowed linking the obtained pictures of events to external electronic detectors such as a photon beam energy tagging system [1], lead glass shower detectors [2] or proportional chambers[3]. The interactions of high-energy photons and electrons took place in a small hydrogen or deuterium target inside the streamer chamber; therefore the interaction point (vertex) was not visible on the pictures. – see Figure 1.

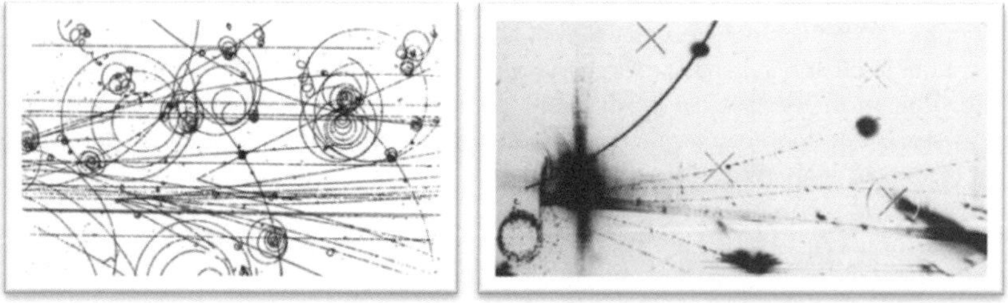

Figure 1: Track chamber pictures; left: CERN Bubble Chamber; right: DESY Streamer Chamber

In the early 1970s, the online computing infrastructure for the 20 experiments at the DESY electron synchrotron in Hamburg consisted of on-line computers Digital Equipment PDP-8 connected to the IBM-360 systems in the computer centre. A PDP-8 has an addressing space of 4k 12-bit words. The installations at DESY had four banks of core memory with 4k words each. An innovative operating system[4] developed at DESY provided functionality that became mainstream only years later with the advent of personal computers with windows software and networking. The PDP-8 minicomputers were connected through 24 parallel coaxial cables to a "Parallel Data Adapter (PDA)" in the central IBM 360/75 for data transfer at - for the time remarkable –rate of 100 kbytes/s. Raw data from the experiments were transferred to the mainframe and results of a first analysis in form of histograms sent back to the on-line computer. The PDP-8 was also used to edit programs and to submit them to the mainframe. Documents like doctoral theses could be edited on the PDP and printed on a connected IBM Selectric "golf ball" typewriter.

In the DESY computer centre, a Hierarchical Storage Manager (HSM) provided automatic migration and transparent access to data sets independent of whether they resided on disk or tape.

The beam line was designed with the help an analog computer. From the positions and field strengths of the quadrupole magnets, the horizontal and vertical beam envelopes were determined and could be visualized on a CRT and drawn on a plotter.

The film from the streamer chamber was measured in two different ways. The first one was to project the image on a measuring table. Manually, the operator placed a crosshair on some ten points for each track. These coordinates were digitized for each of the three stereo views and punched on cards to be read on the mainframe computers. The second way employed an automatic device, a flying spot digitizer, the Hough Powell Device (HPD)[5], that could perform many more measurements per track, leading to significantly improved track parameters and the process required much less human intervention.

The measured coordinates were handled by the CERN program THRESH for geometrical reconstruction. The program was modified at DESY to cope with the features of the streamer chamber such as invisible vertex and material distribution. The three stereoscopic views were corrected for optical distortions and demagnification, and corresponding tracks in the views were combined and reconstructed in space. The program GRIND[6] developed by R. Böck at CERN performed kinematical analysis evaluating probabilities for particle mass assignment to tracks.

In the seventies we started to use the most long-living program for physics: the numerical minimization program MINUIT[7] by Fred James. Re-implemented in object-oriented languages it is still used today for a large number of physics studies.

The Neutrino Experiment WA1 at CERN

Coming to CERN in mid-1976 meant that I had to give up certain habits that were standard at DESY:

- a hierarchical storage management system in the computer centre was still decades away at CERN,
- instead of central data recording, tapes recorded at the experiment were transported to the CERN computer centre – "bicycle on-line"!

In December 1976, the WA1 experiment started data taking in the CERN SPS neutrino beam. The detector consisted of 19 iron modules combining the functions of target, muon absorber, spectrometer magnet and hadron calorimeter interspersed with three-layer drift chambers. The 1250-ton detector was exposed to wide-band and narrow-band neutrino beams up to 400 GeV from 1976 to 1984 studying deep inelastic neutrino scattering on iron (and later also on hydrogen and deuterium). The experiment was a collaboration of CERN, Dortmund, Heidelberg, Saclay and (later) Warsaw, known as CDHS. Physics topics were amongst others: nucleon structure functions, electroweak couplings, QCD, and search for neutrino oscillations. One of the first publications could falsify an effect called "high-y anomaly" where a competing experiment at Fermilab had observed unexpected behaviour in antineutrino events. Two years after the discovery of charmed quarks, neutrino interactions offered a unique possibility to study the coupling of charm to other quark flavours by measuring elements of the Cabibbo–Kobayashi–Maskawa matrix. The most frequent events, charged current interactions, have a negative (leading) muon in the final state for neutrino interactions and a positive one for antineutrinos. Charm manifests itself in an extra muon in the final state mostly of opposite charge to the leading muon. As the magnetic field polarity was set to focus the leading muon, and the secondmuon generally has lower momentum and was bent away from the axis, the automatic reconstruction initially failed to reconstruct this short track reliably (see an example in Figure 2 – left side). For a first analysis, we reverted to a traditional bubble chamber procedure by measuring the track curvature on printed event displays using curve templates as in the early bubble chamber days. Later, of course, the reconstruction was improved to cope also with the difficult cases. Most computing was done on the CDC 7600 machines at CERN and at Saclay.

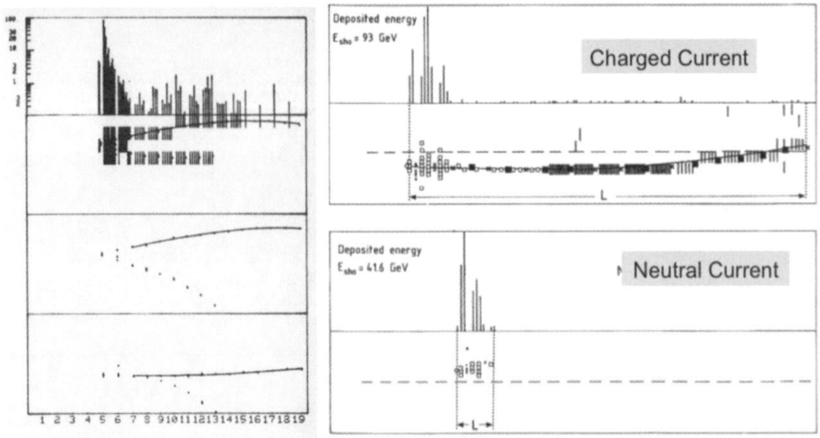

Figure 2: Typical neutrino events in the CDHS detector – left: an early di-muon event; top right: a charged current event; bottom right: a neutral current candidate. The events on the right side were recorded in the upgraded CDHSW detector. The histograms give the energy deposited in the scintillators of the calorimeter in units MIP (Minimum Ionizing Particle).

The ALEPH Experiment at LEP

At colliders such as LEP and LHC, several experiments began operation simultaneously and competed for first results. It was therefore vital that the computing systems (hardware and software) performed to requirements from the first data taking. This was achieved with remarkable success for all experiments at LEP and at LHC. Contrary to the previous experiments with very few sub-detector types, ALEPH[8] had about 10 sub-detectors constructed in the various institutes forming the collaboration. The software was developed in a distributed way between the participating institutes where the partner building the device generally provided the program components specific for that detector. This distributed development required new techniques in the management of software development. In ALEPH, we pioneered in using software engineering techniques[9,10]. External consultants trained the developers in the Structured Analysis, Structured Design (SASD) method. The data were structured according to the entity relationship model as proposed by P. Palazzi for the ADAMO system[11]. The design method allowed decomposing the software projects into components with clearly defined interfaces such that the parts could, in a first step, be developed independently. Though most software was still written in FORTRAN, our system automatically generated code from a formal Data Description Language (DDL) as well as documentation matching exactly the design. The documentation system also generated commented calling trees. Coinciding with the development of the documentation system was the advent of the World Wide Web (WWW) at CERN. This allowed automatic generation of a computing platform independent software documentation with cross-references (hyperlinks) between DDL, tree diagram and code. The system was automatically updated with each new release of the programs for simulation, reconstruction and physics analysis.

With the LEP era, the move from mainframe computers to farms of workstations (UNIX or VMS in the case of LEP, PC in the case of LHC) started to bear fruit. As an interlude, the UA1 experiment had successfully deployed systems of home-made processor boards – emulators. The four LEP experiments initially used different solutions. OPAL started out with a system of APOLLO workstations running UNIX. Their system was subsequently developed into the SHIFT (Scalable Heterogeneous Integrated FaciliTy)[12] system based on SGI and DEC machines in the computer centre. This system was eventually used by all experiments. In ALEPH, we decided to use two distinct operating systems: Digital Equipment VMS running mostly on farms of work stations and IBM/VM on mainframes. Both were used for program development, event processing and simulation. The long lifetime of the experiments at LEP (and even more so now at LHC) exceeds the useful lifetime of a computer platform – operating system – compiler combination. It has proven valuable to avoid the binding to a particular system to later allow moving to a rather different platform. In ALEPH, we could transport our software, documentation and production system rather painlessly to new systems like ULTIX or CRAY[13]. Though the vector performance of the CRAY supercomputer could not really be exploited to full extent for the reconstruction and simulation algorithms, the Cray still acted as key element in the ALEPH computing strategy as fast data server where the vector unit was efficiently used for data compression and conversion.

In ALEPH, we pioneered the reconstruction of events in almost real-time using off-the-shelf commercial processors[14]. A system of VAX workstations processed the events during data taking at the control station of the experiment. This allowed immediate evaluation of calibration and rapid verification of the correct performance of all detectors.

The ATLAS Experiment at LHC

The detectors at the Large Hadron Collider (LHC) at CERN pose new, extreme challenges for computing:

- the ATLAS detector[15] has 100 million electronic channels;
- bunch crossings every 25 ns delivers 40 MHz of data to the trigger system;
- though the trigger system reduces the rate to a few hundred Hz, the annual recorded data volume is several Peta bytes (PB);
- the ATLAS experiment is designed, constructed and operated by a world-wide collaboration of 3000 scientists in 174 institutes in 38 countries,
- 200 software developers have produced about 5 million lines of (mostly C++) code[16].

The computing requirements estimated in 1996 appeared enormous[17,18]; but it was expected that continued validity of Moore's law would ensure that the CPU and storage capacity required would be available about a decade later at LHC turn-on. In fact, the first full year of data taking was only in 2010 when the increase of capacity thanks to technology evolution had taken place such that it could even cope with the new more realistic requirements that exceeded the original plans by two to three orders of magnitude. The increased needs in CPU power and data storage volume were due to better understanding of reconstruction algorithms and to the need for substantial event simulation. Network bandwidth was a considerable concern in the early planning phase. Fortunately, during the LHC construction period, a worldwide privatization and deregulation of telecommunication services led to a massive progress in network performance together with a major cost reduction. Today, contrary to early fears, none of the LHC data needs to be transferred by transportable media such as tapes.

While at previous colliders like LEP the experiments employed disparate solutions for software and for the computing infrastructure, at LHC an attempt was made to find common solutions. Research and development projects were organized and monitored by the CERN Detector Research and Development Committee (DRDC) and later by the LHC Computing Board (LCB). Only projects approved by at least two LHC collaborations were supported centrally. Some of the projects aiming at "proof of concept" did provide valuable input for later developments. Examples are the RD41 (MOOSE) project aiming to evaluate object-oriented technology and RD45 studying the possibility to use a commercial Object-oriented Data Base Management System (ODBMS) for persistency management. Unfortunately the latter technology – though used for some time in the BaBar experiment at SLAC – could not convincingly be proposed for LHC because it depended on a single vendor who did not find sufficient support in the commercial domain. For decades, during the FORTRAN era, CERLIB provided common tools and packages used by all particle physics experiments and in numerous other sciences. Taking this success to the OO paradigm in the LHC era was the aim of the related LHC++ and CLHEP projects. These projects, though having a rigorous design philosophy were overshadowed by the ROOT[19] project, the object-oriented successor of PAW, which was highly successful in the LEP era. René Brun summarized HEP computing related to analysis tools such as PAW and ROOT in a recent article[20]. The ROOT project benefitted from excellent publicity and from the fact that it chose a subset of C++ as scripting language. Independent but related is the XRootD[21] package from SLAC providing high performance fault tolerant access to data repositories. XRootD is now gaining importance in all LHC collaborations. The most successful project under the LHC R&D developments was Geant4, the simulation framework for LHC discussed in a later chapter.

In experiments prior to LHC, the early data analysis was performed at the site of the accelerator. Only after some time (of about a year), data was also available at outside centres and competitive physics studies could be continued there. For LHC experiments with several thousand scientists it was inconceivable to have all collaboration members come to CERN for the initial running phase. Already the early estimates of the required computing capacity – modest compared to what was actually available at the start of physics running – exceed the possible resources at CERN by a large factor. Early ideas for LHC computing therefore aimed at a hierarchical model with a computing centre at CERN where all raw data were archived and the initial reconstruction was done complemented by several regional centres with significant computing and data storage capacity followed by computing installations at institutes for individual analysis. The groundwork for the topology of the LHC computing infrastructure was laid in the MONARC[22] project. The aim was to determine the requirements and characteristics of the regional centres and to understand the data access pattern during the analysis process. The project developed and implemented simulation tools to understand the characteristics of different implementation models and their principal parameters in terms of data handling, network and CPU capacity. The results of MONARC together with ideas of grid computing as formulated by Foster and Kesselman[23] formed the basis of the LHC computing grid.

The LHC Computing Grid and European Grids

Two projects leading to the current computing infrastructure for LHC were started in 2001:

- The first in a series of computing grid projects funded through the framework programmes of the European Commission, the European Data Grid (EDG).
- The LHC Computing Grid (LCG) project, based on a proposal by Delfino and Robertson[24], was approved by the CERN Council[25].

The LCG project is a collaboration between the four major LHC physics collaborations, twelve large centres (CERN as "Tier-0" and 11 "Tier-1" centres) and 38 federations of smaller "Tier-2" centres. The task of the Tier-0 at CERN is to provide the raw data recording, a first pass reconstruction and the data distribution to the Tier-1 centres where another copy of the data is permanently stored and where re-processing and physics analysis is done. The 140 Tier-2 sites concentrate on Monte Carlo production and end-user analysis. The basic principles of LCG (called now WLCG to emphasize its Worldwide character) are documented in a Technical Design Report[26].Apart from forming collaboration between computing sites, LCG ensures the network connections - initially at least 10 Gbit/s from Tier0 to each Tier-1. The grid middleware (the "operating system" of the grid) is developed and maintained jointly with the European and US grid projects. One major achievement is the interoperation of both middleware implementations.

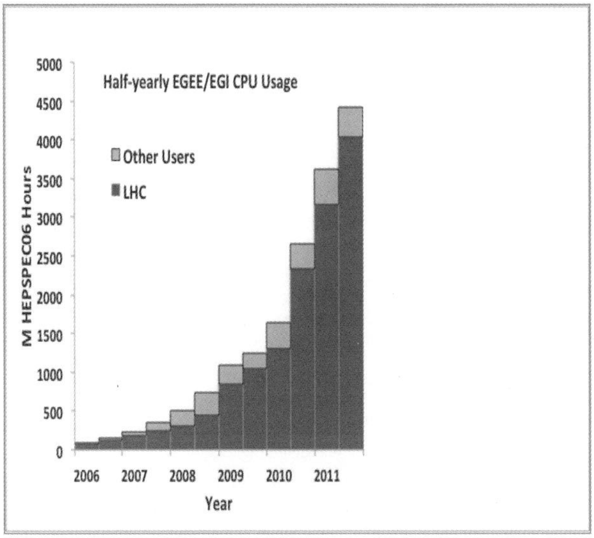

Figure 3: Evolution of CPU capacity of European Grid Projects EGEE and EGI

The European grid projects EDG, EGEE, and EGI had a strong CERN and HEP component but also constituted a fruitful partnership of many scientific disciplines including Archaeology, Astronomy & Astrophysics, Civil Protection, Computational Chemistry, Computational Fluid Dynamics, Computer Science, Condensed Matter Physics, Earth Sciences, Fusion, High Energy Physics, and Life Sciences. The rapid evolution of the grid resources provided through these projects is visualized in Figure 3 showing the CPU evolution from 2006 to 2011.

To set the scale, the last column in Figure 3 represents about 100,000 of the currently most powerful CPU cores. It is obvious that the resource usage (and its provision) is dominated by LHC, but e.g. in 2008 life sciences made significant use of the infrastructure for an application to evaluate drugs against the avian flu.

Event Display

In the era of photographic recording of interactions in track chambers, the film itself represented the visual image of the tracks. Human inspection was essential to understand the physics process behind each event. With the advent of electronic detectors, software reconstruction programs performed the reconstruction. Visualisation of the events was necessary to evaluate the performance (reconstruction efficiency) of the programs and to optimize their performance. A first generation of passive event displays evolved into interactive systems, where, through human interaction, coordinates could be assigned to tracks whose parameters were then determined using the integrated fitting procedure. For the CDHS neutrino experiment, we used terminals storing a vector image on the phosphor of the display screen (TEKTRONIX 4012 and 4014) connected to the interactive CERN FOCUS system on a CDC computer. The UA1 and UA2 experiments certified events constituting the discovery of the W and Z bosons on a 3D display system. Myers and Bettels had developed the Pions graphics system[27]based on an object-oriented database of 3-dimensional coordinates. The data could be visualized on Megatek graphics stations allowing dynamic scaling and rotation giving a 3D impression. H. Drevermann invented another techniquefor ALEPH [28,29]. A popular projection used in many publications of ALEPH events is the Fish-Eye view visualising in a single picture the sub-millimetre resolution of silicon vertex detectors and the muon chambers extending to 10 m from the interaction point. An example is given in Figure 4.

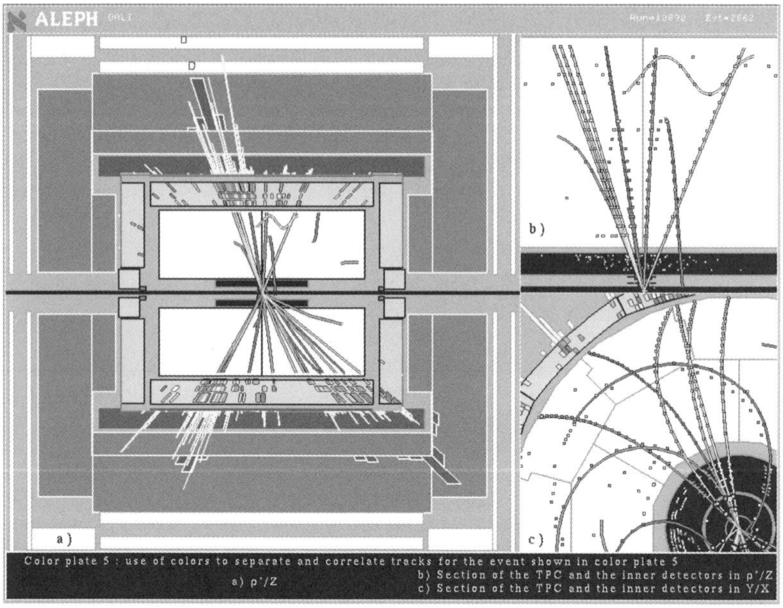

Figure 4: DALI event display of an event in ALEPH; the lower right view uses the non-linear radial representation – "fish-eye view"

Another innovative projection is the "v-plot" visualizing, without pattern recognition program, groups of coordinates belonging to individual tracks, indicating whether tracks form a common vertex and giving a measure of the track curvature. These techniques used in the ALEPH event display program DALI are now also implemented in the ATLANTIS event display program[30] for ATLAS shown in Figure 5.

Figure 5: ATLANTIS event display showing a Higgs candidate event in ATLAS (ATLAS Experiment © 2012 CERN)

Simulation

Simulation (Monte Carlo) methods played a major role in nuclear and subnuclear physics even before the advent of electronic computers. An early example was E. Fermi's invention, the FERMIAC[31], a mechanical device to reproduce the stochastic progression of neutrons through various materials. The idea of using computers to simulate particle interactions was first formulated by S. Ulam[32]. During the four decades discussed here numerous simulation programs have been developed. Many of them are based on the groundwork of H. Fesefeldt (GEISHA) for hadronic showers[33] and R. Nelson for Electron Gamma Showers (EGS)[34]. These and other physics codes are now part of simulation packages such as Geant4[35]. Physics generators simulate the primary interactions and feed then into the simulation program, which contains an extremely detailed description of the detectors, giving the distribution of materials and magnetic fields as well as the location and behaviour of the sensitive detector elements. In the simulation process, primary and secondary particles are propagated through the materials and fields of the detectors where interactions and decays are calculated in a stochastic way using pseudo random numbers. The output is formatted such that it can serve as input to the event reconstruction interchangeably with the real data. The simulation programs are used for the design of new detectors, the evaluation of detector acceptance and efficiency and, during analysis, for the estimation of the impact of known background processes on physics observations.

Geant4 was the first large international software project implemented in OO technology following a rigorous design process. It is now the package of choice for the LHC collaborations and of many other subnuclear and nuclear physics experiments. The implemented physics processes are very general and cover a wide energy range such that Geant4 is used for simulations beyond particle physics applications e.g. in the space and medical domains[36]. A recent example is using Geant4to evaluate the benefits of carbon ion therapy for cancer treatment[37].

The Monte Carlo programs are also used to calculate the induced radiation levels in particle accelerators, experiments and nuclear reactors. This field is dominated by the FLUKA[38] Monte Carlo package with its own physics processes.

Data Preservation

Experiments in subnuclear physics are global endeavours of the scientific community requiring enormous financial and human resources. As an example, over its lifetime the LHC project will have cost several 10^{10} Euros and more than 10,000 scientists will have contributed to construction, operation and exploitation during the about 30 years from project approval to end of data taking. The obtained data and derived results are unique. It is possible that years after the current generation of physicist have terminated the data analysis, new theoretical or experimental evidence needs to be verified or complemented with LHC data. To allow a new analysis of old data it is necessary but not sufficient to keep the data in a format and on media that are still readable. Equally important are the programs for analysis and the metadata describing the detectors and the running conditions. In general, also simulation programs with detailed detector description are required – with the possibility to implement new physics models. In a study I undertook for the 2003 ERPANET/CODATA Workshop on the Selection, Appraisal, and Retention of Digital Scientific Data in Lisbon[39], physicists in several collaborations at CERN and elsewhere agreed that data preservation is important but that any meaningful re-analysis would require the participation of scientists who were members of the original experimental team.

The importance of long-term data preservation is now recognized; it will, however, require significant human and financial resources. A study group DPHEP[40] has been created to evaluate technical and organisational aspects of long-term data usability beyond the lifetime of an active collaboration. The project is supported by physics labs and collaborations and has recently published a status report DPHEP[41] leading to a proposal of a global organisation for data preservation in HEP.

Summary and Conclusion

We have seen over the past four decades tremendous changes:

The size of physics collaborations has increased from order $\mathcal{O}(10)$ to several thousand. The computing requirements have vastly exceeded the possibilities of a single institution.40 years ago, one or very few physicists located within earshot did the software development; now, hundreds of developers working on a single package are spread around the globe.

The combined computing capacity of DESY and CERN in 1970 represented one per mill of the power of a single modern smart phone. Over four decades, resources in CPU and storage used for subnuclear physics have increased by eight orders of magnitude. Currently, there are now about 10^{18} bits of data storage for the LHC experiments.

To address the main question of this lecture series: "What we would like LHC to give us?" - computing has been essential to enable physics experiments answering the fundamental questions. This has become even more so at LHC than previously. Computing is as important as any other subsystem of the experiments. But also, LHC computing has required pushing technology to new heights enabling other sciences and domains to profit from the new technologies.

Acknowledgements

In the four decades covered in this article, I had the pleasure to work with wonderful colleagues in an inspiring atmosphere. Naming them all is impossible here – but without them, none of the projects described in this manuscript could have been done.

I greatly appreciate the hospitality of the Ettore Majorana Foundation and Centre for Scientific Culture (EMFCSC). My thanks go to Professor Antonino Zichichi for his kind invitation to give this lecture at the 50th Course of the International School of Subnuclear Physics. Professor Antonino Zichichi and Professor Gerard 't Hooft have led this lively meeting in the splendid city of Erice to a memorable event.

References

[1] W. Struczinski et al., Study of Photoproduction on Hydrogen in a Streamer Chamber with Tagged Photons for 1.6 GeV$<E_\gamma<$ 6.3 GeV, Nuclear Physics B108 (1976), pp. 45-74

[2]J.Knobloch, Elektroproduktionbeikleinem ε in einerStreamerkammer, Thesis, Hamburg, 1976

[3] P. Joos et al., Rho Production by Virtual Photons, Nuclear Physics B113 (1976) pp. 53-92

[4] F. Akolk et al., DESY On-Line System, Computer Physics Communication 4 (1972) pp. 275-278

[5] P. V. C. Hough and B. W. Powell, NuovoCimento 18, (1960)1184

[6] R. Bock, GRIND Manual, DD/EXP/62/1 0 (CERN, Geneva, 1962) (Unpublished)

[7]F. James, M. Roos, Minuit: A System forFunctionMinimizationand Analysis ofthe Parameter Errors andCorrelations, Comput.Phys.Commun. 10 (1975), pp. 343-367

[8] D. Decamp et al., ALEPH: A Detector for Electron-Positron Annihilations at LEP, Nuclear Instruments and Methods in Physics Research A294(1990) 121-178

[9] J. Knobloch, Methods of Software Development for the ALEPH Experiment, in New Computing Techniques in Physics Research, ed. D. Perret-Gallix, W Wojcik, ISBN 2-222-04514-2

[10] J. Knobloch, Reality of Software Engineering in HEP, in Computing in high energy physics. Proceedings, International Conference, CHEP'91, FSS-3, Tsukuba, Japan, March 11-15, 1991, ed. Y. Watase, F. Abe, ISBN 4-946443-09-6

[11] Z. Qian et al., Use of the ADAMO data management system within ALEPH, Computer Physics Communications,Volume 45, Issues 1–3, 1 August 1987, Pages 283–298

[12] J. P. Baud, et al.: SHIFT, The Scalable Heterogeneous Integrated Facility for HEP Computing. In: Proceedings of International Conference on Computing in High Energy Physics, Tsukuba, Japan, March 1991, Universal Academy Press, Tokyo

[13] M. Delfino, T. R. Edgecock, C. Georgiopoulos, J. Knobloch, The ALEPH Data Processing Chain at CERN: A Successful Combination of Three Heterogeneous Computer Architectures, in Computing in high energy physics. Proceedings, International Conference, CHEP'91, FSS-3, Tsukuba, Japan, March 11-15, 1991, ed. Y. Watase, F. Abe, ISBN 4-946443-09-6

[14] M.Delfino, A. Pacheco,J. Knobloch, Quasi-online data processing with the ALEPH event reconstruction facility : one year of total success, in Computing in high energy physics. Proceedings, International Conference, CHEP'91, FSS-3, Tsukuba, Japan, March 11-15, 1991, ed. Y. Watase, F. Abe, ISBN 4-946443-09-6

[15] ATLAS Collaboration, The ATLAS Experiment at the CERN Large Hadron Collider, JINST 3 S08003 (2008)

[16]ATLAS Fact Sheets, http://www.atlas.ch/fact_sheets.html

[17] ATLAS Collaboration, ATLAS Computing Technical Proposal, CERN/LHCC/96-43

[18] J. Knobloch, ATLAS Computing, in: Computer Physics Communications 110 (1998) 26-31

[19]ROOT website: http://root.cern.ch

[20] R. Brun, The Evolution of Software in High Energy Physics, in: Journal of Physics: Conference Series 396 (2012)

[21]XRootD website: http://xrootd.slac.stanford.edu

[22] MONARC Phase 2 report CERN/LCB 2000-001, March 2000

[23] I. Foster, C. Kesselman: The Grid: Blueprint for a new computing infrastructure. Morgan Kaufmann, San Francisco (1999). ISBN: 1-558660-475-8

[24] M. Delfino, L. Robertson (eds.): Solving the LHC Computing Challenge: A Leading Application of High Throughput Computing Fabrics combined with Computational Grids. Technical Proposal. CERN-IT-DLO-2001-03.

[25]Proposal for Building the LHC Computing Environment at CERN. CERN/2379/Rev. 5 September 2001.

[26]J. Knobloch (ed.), LHC Computing Grid; Technical Design Report, CERN-LHCC-2005-024

[27] J. Bettels,, D.R Myers, The Pions Graphics System, Computer Graphics and Applications, IEEE , vol.6, no.7, pp.30-38, July 1986

[28] H. Drevermann et al., Graphical Concepts for the Representation of Events in High Energy Physics, Int. J. Mod. Phys. C 1991.02, pp328-330.

[29] H. Drevermann, D.Kuhn, B.S.Nilsson, 'Is there a Future for Event Display?', Proc. of 1992 CERN School of Computing, L'Aquila, Italy, CERN Yellow Report 93-03.

[30] N. Konstantinidis et al., The Atlantis event visualisation program for the ATLAS experiment, CHEP04 Interlaken proceedings, pp. 361-364; http://www.hep.ucl.ac.uk/atlas/atlantis/

[31]http://en.wikipedia.org/wiki/FERMIAC

[32]R. Eckhardt, "Stan Ulam, John von Neumann, and the Monte Carlo method". Los Alamos Science, (1987), Special Issue (15): pp. 131–137.

[33]H. Fesefeldt, GEISHA, RWTH Aachen report PITHA 85/2, 1985

[34]W.R. Nelson, H. Hirayama D.W.O. Rogers, The EGS4 code system Report SLAC-265 (Stanford, CA: Stanford Linear Accelerator Center), 1985

[35]GEANT4 website: http://geant4.cern.ch/

[36] M.G. Pia and J. Knobloch, Particle physics software aids space and medicine, CERN Courier Vol. 42, Nr. 5, June 2002

[37] A. Lechner, V. Ivanchenko, J. Knobloch, Validation of recent Geant4 physics models for application in carbon ion therapy, Nuclear Instruments and Methods in Physics Research Section B, Volume 268, Issue 14 (2010) pp. 2343-2354

[38]FLUKA website: http://www.fluka.org/

[39]J. Knobloch in Final Report ofThe Selection, Appraisaland Retention of Digital Scientific Data, ERPANET / CODATA Workshop, BibliotecaNacional, LisbonDecember 15-17, (2003) www.erpanet.org/events/2003/lisbon/LisbonReportFinal.pdf

[40] DPHEP website: http://www.dphep.org/

[41] Z. Apokov et al., Status Report of the DPHEP Study Group: Towards a Global Effort for Sustainable Data Preservation in High Energy Physics, DPHEP-2012-001, http://arxiv.org/pdf/1205.4667.pdf

CHAIRMAN: J. KNOBLOCH

Scientific Secretaries: V. Begun, M. Mager

DISCUSSION

- *Sleptsov:*

What is the present situation of using neural networks in analysis of experimental data?

- *Knobloch:*

I am not currently in one of the existing experiments. We have used neural networks already at the time of ALEPH. I think you should rather ask experimentalists, Peter Jenni, or one CMS talk is coming, you may ask there.

- *Csanad:*

Do you think the codes of old experiments are directly used in new ones? If yes, does this affect the similarity of results at different experiments? Wouldn't be this be a problem as the competition would be ceased?

- *Knobloch:*

That have several aspects, one aspect, for instance, is data preservation. People want to go back to the data. I have seen a presentation at the computing conference, showing a plot of how long after the end of the experiment there is the analysis going on. For example, there are still publications from ALEPH, ten years after they have stopped in 2002. I have never seen so much the re-use of the code from one experiment in another one, except for the case, when someone has written a piece of code and moves to next generation experiment. He, of course, than re-use his knowledge of what he has done in previous experiment. But re-use of the same code is not very frequent I believe.

- *Knegjens:*

Do you have an opinion or insights into good computing, entering public domain, like the Internet has, and if it could be compared and competitive with cloud computing? For example, if some future start-up could submit jobs on the GRID?

- *Knobloch:*

Cloud computing is taking-off in the industry. Big internet companies like Amazon and Google had their own computer centers. They discovered that they can sell their computing power. You go to the Amazon, enter your credit card number and get as much of computer power as your credit card would allow you. We have considered that for our computing tasks. Just for CPU cycles this could be interesting to us, but the real problem is massive data storage, in particular the long term data storage that cloud computing does not provide us at the moment. I have shown you, there is research going on using cloud computing, but it is currently in the experimental-testing level.

- *Lopes de Sa:*

There has been a lot of pressure of making collider data publicly available, like after satellite experiments that do that. Has there been any decision or at least conversation, in the LHC experiments, and if there is, what type of data would be public and how long after it has been collected?

- *Knobloch:*

There is a law in United States that enforces the publication of publicly funded research, so the astronomers and other scientists do that. It comes together with data preservation, because to make the data publicly available to outside is not very useful. Just the data is not enough. You need some meta-data and the knowledge, or you have to go to rather low level, for example you can put all the four-vectors of an event, but what you would do with it then? Preparing a seminar on data preservation I have asked physicists and they answered that no way that anybody from outside of an experiment can ever make any sense of our data.

- *Gazis:*

I have enjoyed your nice presentation of historical evolution of the detectors and computing power. You mention in your talk that supercomputing power was not influencing very importantly the lab analysis. What about the famous CRAY computer? It was too much promising but the benefit was not so big, right?

- *Knobloch:*

The use of CPU cycles at CRAY was far too expensive and we did not manage to use effectively vector capabilities of CRAY, because our programs have lots of branches, if statements and address calculations that do not fit to the vector processes that metrology or other data are. We did make good use of CRAY for trivial tasks like data compression or other tasks that run just on the data.

- *Liu:*

Have you made any attempt to involve GPUs in the computing framework?

- *Knobloch:*

I personally have not, but I know that there are people in the current experiments making efforts to use them, in particular for certain branches of simulation. Indeed, nowadays much of the compute power is in graphics cards. To exploit this power, however, special compilers are needed and the current C++ compilers that we use do not benefit of GPUs.

- *Karakostas:*

Since we are on the Peta-scale concerning the data storage and also facing the data preservation issue, are experimental methods like nanoparticles or 3-D optical data storage promising options for storage media?

- *Knobloch:*

The price of storage media is not a problem. When talking about storage the complication is always the right decision between storage on tapes on the one hand and storage on disks on the other hand. The advantage of the former is that once written to tape, data does not consume energy. Power is only needed once it is read again. Disks keep spinning and continuously consume power. The real problem is as follows: In my life at CERN I have seen several techniques coming and going. But for the massive storage needs that we are facing today, you need something that is well established and working reliably. For example, I remember that we had been investigating optical discs of 1 GByte in size, which was an enormous capacity at that time. But they disappeared and have never made it to the market. Therefore, I would rather be hesitant to jump on something that is not really well established.

- *Leontsinis:*

Are quantum computers considered for the future upgrades in order to deal with the increasing need of computing power?

- *Knobloch:*

One has to see how industry develops. In the early days computers were considered experimental equipment. I remember well, that when I was on shift and the computer broke down, we would go with an oscilloscope to identify and replace the faulty transistor. Today, this is not the case any longer. Computers became mainstream products and thus we have to go the way industry does. If some day quantum computers become mainstream products, we surely are going to use them. While I neither see the need nor the possibility to use quantum computers in the lifetime of LHC, they may well become part of the computing model in the longer term.

- *Mager:*

The increase of computing power as well as computing needs over the last decades that you have shown is quite impressive. But is it really true the experiments have become equally more complex or is it rather true that programs have become more complex. For instance does the step of switching to object oriented programming cause a lot of overhead?

- *Knobloch:*

I do not believe that object oriented programming necessarily leads to larger or more complicated programs. It forces people to structure their programs, if done properly. For the complexity of experiments I have shown how the number of different detectors types has increased from two in the neutrino experiments up to twelve in ALEPH. Correspondingly the number of ways of analyzing the data increases and complexity sets in when one needs to correlate the different detectors. In the LHC area, additional complexity is caused by large number of collaborators that want to do analysis. The amount of code is actually not such a big issue: the development speed stayed rather constant at 15-20 lines per programmer and day. By the way another constant that I have discovered is the processing time per event. Using GRIND to reconstruct a bubble chamber event took about 20 seconds. For ATLAS the full event reconstruction on a single core takes about 30 seconds.

- *Mager:*

Even if the number of lines did not change much, it makes surely a difference whether they are written in assembly or in C++.

- *Knobloch:*

Yes, otherwise I would have not been able to work on my PDP-8 with 16 k-words of memory.

- *Mager:*

So, how much computing time do we waste by not writing proper programs anymore?

- *Knobloch:*

Depends what you call wasted. On my first Macintosh I could do everything I thought I would need. Now, with some 4 GByte of RAM and CPU frequencies of several GHz, my life is not any easier. There is overhead generated by new gadgets and new possibilities, but one has to weigh this overhead with the gained convenience: you would not like to write the ATLAS programs in assembly. I would rather rely on someone writing a clever compiler that optimizes the code and that e.g. makes use of its graphics capabilities.

- *Apresyan:*

Is there any effort from current LHC experiments to collaborate with industry on exploring data mining techniques for new physics searches?

- *Knobloch:*

I am not aware of any concrete effort. There have been more informal connections to exchange ideas, e.g. in the form of seminars. At CERN we have OpenLab, which is a tight collaboration with industry on various topics, like networking and compilers. Within this framework one may imagine a research collaboration, but till today no experiment has gone this way.

- *Apresyan:*

Is this because the experiments do not want to share their results or because industry is not interested?

- *Knobloch:*

I would rather say that it took the experiments much time to come to this point where their analyses are working well and are producing results. At this level there is a natural hesitation of moving to yet another system. But maybe I am too conservative here.

- *'t Hooft:*

Imagine that tomorrow a new computer system will suddenly be on the market with 100 times the processing speed and 100 times the memory, or imagine the contrary that in the next ten years there will be no improvement. What would the effect be on the LHC?

- *Knobloch:*

Fortunately we know what will happen on the computing market in the next few years, which excludes the second scenario. Computing power will steadily increase over the next years. If, however, there is some exceptional increase one has to reevaluate the situation, in particular the associated costs. Here not only the cost of the device, but also the cost of adaption has to be taken into consideration. Especially the latter shall not be underestimated.

- *'t Hooft:*

But suppose that you would have 100 times more or 100 times less computing time and tapes, would it change the way you run the detectors?

- *Knobloch:*

The situation with 100 times less computing power and store is difficult to answer: It is true that, when the experiments were conceived, the demands were indeed much lower than what is required today. However, today there is a wide consensus – even with the funding agencies – that the demands are well justified.

The LAA Project and the Consequences on LHC

Horst Wenninger
ERICE - ISSP 2012 International School of Subnuclear Physics

The LAA project was initiated by Prof. A. Zichichi and implemented at CERN in 1986 as "another CERN program of activities" [1]. The goal of the project was to prove the feasibility of new detector technologies that could be used in future multi-TeV hadron colliders.

The promotion of detector developments or new technologies for the next generation of accelerators by Prof. Zichichi has always been motivated by the access to new physics.

Two examples from the nineteen-sixties to illustrate this are:

1) the development of the time-of flight technique was needed for the search for anti-matter, to identify anti-deuteron in a very high backgrounds of negative particles (discovery in March 1965) [(2)].

2) the pioneering work by A. Zichichi and his group developing the preshower method was aiming at and allowed to detect leptons in a high background level of hadrons. marking the start of the search for the 3rd leptonic family [(3)].

The physics motivation behind the present project to studying the properties of the "new world" which should be the source of totally unexpected phenomena, produced in a collision between heavy lead nuclei at extreme energy, is described in a Conference paper by A. Zichichi in honour of Murray Gell-Manns 80th birthday celebration [(4)]. The technological development for this project will be to improve particle identification, time-of-flight detection, and synchronization. The "new world" is the Quark-Gluon-Coloured-World (QGCW). In the extremely high energy collisions between heavy ions, many QCD open-colour-states should be produced, far higher in number than the baryons and mesons known so-far, since these baryons and mesons have to obey the condition of being QCD-colourless.

To appreciate the importance of the LAA project for CERN we need to look back to the 1980th, when CERN envisaged to enter the era of lepton collider physics.

From 1977 to 1983 Prof. A. Zichichi was President of the Instituto Nazionale di Fisica Nucleare (INFN) and all the great projects of European Physics, LEP at CERN (Geneva-CH), GRAN SASSO at INFN (Italy), HERA at DESY (Germany) and LAA at CERN, are linked to the name of Prof. Antonino Zichichi

Indeed, in April 1980, a progress report, published by the ECFA – LEP Working Group [5] chaired by Professor A ZICHICHI, was handed over to the CERN Management. This LEP tunnel infrastructure paved the way for 50 years of world-class subnuclear physics experiments by allowing to install high performance colliders, first LEP then LHC.

CERN LIBRARIES, GENEVA

CM-P00100391

ECFA/79/39
15/4/1980

ECFA EUROPEAN COMMITTEE FOR FUTURE ACCELERATORS

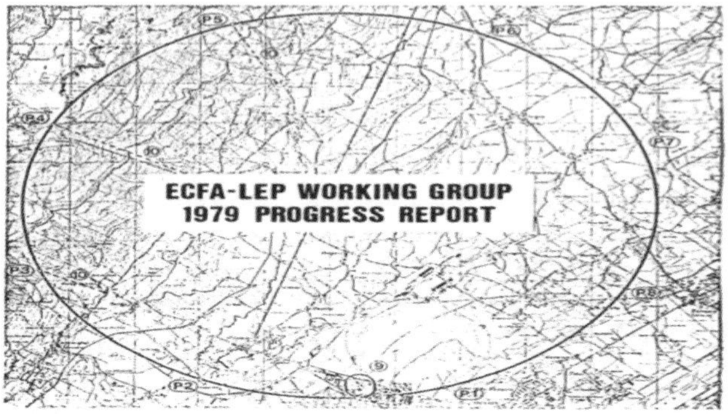

Edited by
A. Zichichi, Chairman
ECFA-LEP Working Group

Figure 1: Cover-page of the White Paper (ECFA-LEP Working Group) Report
In this Report it was recommended to construct a 27 km accelerator ring-tunnel adjacent to CERN between the French Jura and Geneva airport (see cover-page of the ECFA report above).

The Gran Sasso laboratory was proposed by Prof. Zichichi and built and

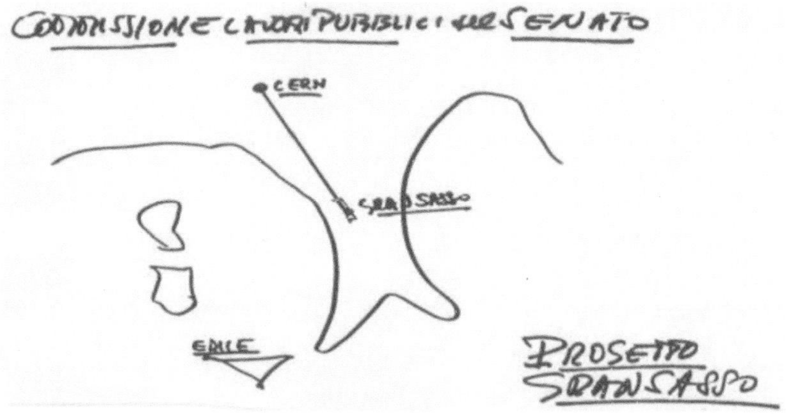

Fig. 11: Handwritten notes from the presentation by A. Zichichi to the Commission on Public Works of the Senate in a Session convened on short notice by the President of the Senate to discuss the proposal for Gran Sasso Project (1979).

at DESY the Italian participation allowed to start the HERA project

The idea behind the HERA project was that Prof. Zichichi wanted to interest Italian industry in the design and fabrication of superconducting magnets, still today one of the key industries for this technology.

When CERN then entered the LEP era within its constant funding levels from 1980 to 1989, CERN's resources were mainly dedicated to the SPS proton-antiproton program and to the LEP project. Not much was left for other activities: the ISR experiments, all bubble chambers and many fixed target experiments had to be closed down to free resources for LEP construction.

It was at that time when Prof. Zichichi proposed to CERN the LAA project permitting to invest in the future beyond LEP[6]. The project allowed the recruitment of 40 staff-members (technicians-engineers-physicists) dedicated to research, innovation and development. The LAA project was open to all physicists and engineers and in the end over 80 scientists worked for LAA. The project was presented in an open meeting at CERN in June 1987 and subsequently to CERN's Research Board.

All aspects of an LHC detector layout were considered in the project and, in view of the demands of the collider, special attention was paid to hermeticity, radiation hardness, rate capability, and momentum resolution of the detector assemblies. The LAA Project consisted of sub-projects (*High precision tracking, calorimetry, large area muon detection devices,* l*eading particle detection,* d*ata acquisition and analysis).*

Main achievements are described in CERN reports [7] and highlighted in a book with the title: "From the Preshower to the New Technologies for Supercolliders ", dedicated to Antonino Zichichi [8].

The LAA Project allowed CERN to invest in the future after LEP. The milestone dates of LEP and LHC are indicated below:

1982 : First studies for the LHC project
1989 : Start of LEP operation (Z/W boson-factory)

1994 : Approval of the LHC by the CERN Council
1996 : Final decision to start the LHC construction
2000 : Last year of LEP operation above 100 GeV
2002 : LEP equipment removed
2003 : Start of LHC installation
2005 : Start of LHC hardware commissioning
2008 : Start of (short) beam commissioning
Powering incident on 19th Sept.
2009 : Repair, re-commissioning and beam commissioning

Considering the history of LEP and LHC we understand that - although some detector solutions adopted for the LHC experiments were finally different from those developed during the R&D phase - the LAA work had a great influence on the design of the actual LHC detectors. The LAA R&D started 10 years before technical choices were approved for LHC detectors and over 20 years before LHC detector operation started for physics. Technical solution adopted by collaborations, clearly depended also on factors such as specific know-how and competences existing in those Institutes, which took the responsibility to build and finance the detector components.

Nevertheless, the LAA technologies had an impact on LHC: examples are the spaghetti electromagnetic calorimeter, multi-drift chambers, scintillation fiber trackers, micro-strip detectors, precision tracking and read-out electronics, IPSA tube (Imaging Silicon Pixel Array), silicon pixels detectors, CMOS chips and ASIC/VLSI chip detector read-out. In addition engineers, physicists, technicians, recruited for the LAA activities, helped LHC experiments and participate in the experiments still today.

I mention a specific example to demonstrate the importance of the LAA project for CERN in preparation for the Large Hadron Collider experiments: MICRO-ELECTRONICS, a technology strengthened at CERN thanks to the resources of the LAA project which allowed to built-up the know-how within the CERN Experimental Facility Division with the recruitment of trained electronic engineers. LAA allowed to finance experts staff, hardware & software tools, which were essential for the development and design of microelectronics, silicon strip and pixel detectors.

A recent article in the IEEE Solid-State Circuits Society News [(9)], published by Erik Heijne, physicist from CERN, outlines that the design of custom chips for silicon detector readout was started at SLAC in 1981 and in 1986 at CERN mainly thanks to the LAA project.

A first application of microelectronics tools at CERN was leading to the construction of a hermetic silicon detector for the UA2 collider experiment. The AMPLEX chip [(10)], developed by Pierre Jarron at CERN in collaboration with A. Gößling from the University of Dortmund, was used to read-out the silicon pad detectors signals.

Another more recent example is the so-called NINO chip [(11)] for the ALICE Time-of-Flight ASIC front end. This chip is also interesting for medical applications of PET detectors, equipment developed for subnuclear physics experiments and now used in most hospitals.

A complete and detailed description of the LAA achievements has been published by the World Scientific Series in 20th Century Physics entitled: "SUBNUCLEAR PHYSICS the first 50 years Highlights from ERICE to ELN" (A.Zichichi, et al.) [(12)].

From 1990 to 1996, the LAA project was complemented by a CERN Detector R&D program. LAA R&D groups participated in eight major proposals approved by the CERN DRDC [(13)]. Activities of the LAA R&D have been published in more than 350 papers and journals.

A most recent achievement of LAA teams concern the development of the MPRC (MULTIGAP RESISTIVE PLATE CHAMBER) for the ALICE Time of Flight systems [(14)].

Furthermore, an educational spin-off program using LAA technology, initiated by Prof. A. Zichichi, has been implemented in schools all over Italy. The EEE (Extreme Energy Events) project [(15)], proposed by collaborator and team leader Crispin Williams, who is responsible for this R&D works, allows Italian pupils to construct themselves large area physics detectors to monitor cosmic ray showers and combine local results via Grid computing with schools in other locations.

Figure 3: Location of schools all over Italy where large area physics detectors have been installed to monitor cosmic ray showers as part of the *Extreme Energy Events* Project– to bring pupils in touch with real physics experiments

Future technological developments concentrate on an ambitious new project, which was proposed by Prof Zichichi at the International School of Subnuclear Physics at ERICE, the ISSP 2006 [(16)].

The proposal is to probe a plasma of free quarks and gluons - the Quark–Gluon–Colored–World – produced in heavy ion collisions of extreme energies at LHC, by injecting beams of particles and photons to observe the different interaction results. The Quark–Gluon–Colored–World (QGCW) is expected to be totally different from our world made of QCD colorless baryons and mesons. The physics and details of this proposed project are outlined in a contribution by A. Zichichi to a conference in honour of MURRAY GELL-MANN'S 80th birthday [(17)].

The QGCW project aims at experiments to probe the quark gluon colored world. Figure 4 illustrates schematically the idea of the experiment.

We need timing and synchronization between: colliding heavy ions, forming the quark gluon colored world, and bombarding particle, simultaneously injected to probe the QGCW, and detectors, triggering on the emerging particles

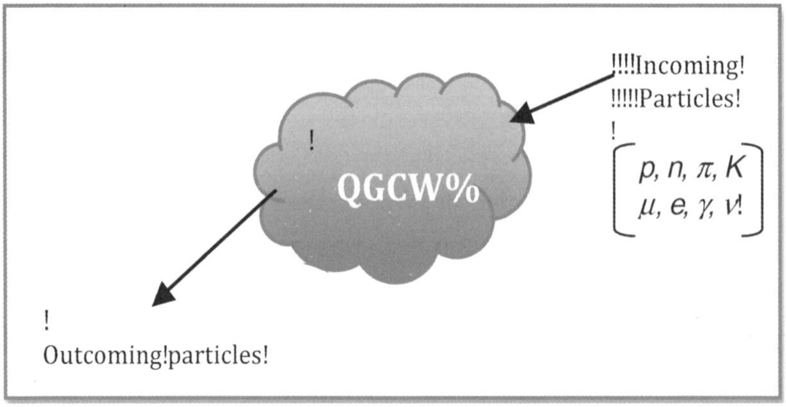

Figure 4: Schematic illustration of the way how to probe the QGCW

Advanced R&D has started all around Europe to prepare for the FAIR [18] project at GSI [19] and at CERN [20] for the LHC upgrade program, and this R&D allows to chose eventually the suitable technology for the preparation of future experiments to explore the entirely new world of states formed by free quarks in the QGCW.

Examples of R&D work proposed to prepare the future of the QGCW project:

The "Bunch-phase Timing System (BuTiS)" for FAIR concentrates on research of thermal stability properties of optical fibers [21], which are essential to carrying control and monitor signals between complex accelerator equipment.
Also in the LHC tunnel optical fibers are used for fast signal transmission through large distances to synchronize the accelerators, to take measurements of the beams, and to send controls to the LHC. Over the past 7 years CERN has developed special optical fibers that can resist the radiation levels of the LHC [22].
Intensive R&D on the detector to obtain an improvement of timing for the Time of Flight system using the MRPC combined with the NINO readout electronics[23].

To conclude:
Large science projects push innovation (LEP, LHC) and the LAA Project had an important impact on Innovation in Science at CERN.

Acknowledgements

It is a pleasure to thank Professor A. Zichichi and Professor G.'t Hooft, the organizers of the School for having permitted me to present the LAA project achievements.

References

1) CERN Council document (December 1986), *Resolution on the LAA programme of Activities*. CERN/1642.
2) Experimental Observation of Antideuteron Production,T. Massam, Th. Muller, B. Righini, M.

Schneegans, and A. Zichichi, Nuovo Cimento 39, 10 (1965).
3) *The Origin of the Third Family*, World Scientific Series in 20th Century Physics, Vol. 20, In honour of A. Zichichi on the XXX anniversary of the proposal to search for the third lepton at Adone
4) Murray Gell-Mann and the last Frontier of LHC Physics: The QGCW Project, A. Zichichi, 2010, Conference in Honour of Muray Gell-Mann's 80th birthday ceremony, Nanyang Executive Centre, Nanyang Technological University, Singapore
5) A. Zichichi Chairman, ECFA – LEP Working Group Progress Report (1979) *ECFA European Committee for Future Accelerators*, ECFA /79/39.
6) CERN Council document CERN/1642, December 1986, 82nd Session of Council *Resolution on the LAA programme of activities.*
7) CERN Internal Reports (1988-1989), *LAA Progress Report*, CERN/EF 89-14 & CERN/ LAA/SD 89-1. CERN Internal Report (1991), *The achievements of LAA* , CERN/ LAA/ 91-1
8) Wiik B., Wagner A., Wenninger H . (2002) In honour of Antonino Zichichi, *From the Preschower to the New Technologies for Supercolliders*" , World Scientific Series in 20th Century Physics, Vol 31.
9) Heijne, E.H.M. (2008), Gigasensors for an Attoscope: Catching Quanta in CMOS, *IEEE Solid-State Circuits society news*, Fall 2008, Vol13, N0.4 500
10) Pierre Jarron, *AMPLEX chip 1986 – 1988*, private communication, Erik Heijne
11) Anghinolfi F., Jarron P., Martemiyanov,A.N., Usenko E., Wenninger H., C.Williams, Zichichi A., M.C.S., (2004) *NINO: an ultra-fast and low-power front-end amplifier /discriminator ASIC designed for the multigap resistive plate chamber*, Nucl. Instrum. Methods Phys. Rev. A 533, 183-187
12) Zichichi A. Subnuclear Physics *The First 50 Years : Highlight from Erice to ELN,* World Scientific Series in 20th Century Physics – Vol 24
13) CERN Detector R&D Committee (DRDC) , Retrieved October 2011, web site CERN http://committees.web.cern.ch/Committees/DRDC/Projects.html
14) C. Williams, M.C.S., (2011) *Development of High Time Resolution Multi-gap RPCs for the TOF Detector of AILCE*, Proceedings of the 7th International Conference on Advanced Technology, page 563
15) Zichichi, A CENTRO FERMI, *The EEE Extreme Energy Events* , Retrieved October 2011, from http://www.centrofermi.it/eee.html
16) Zichichi A., (2006) Erice: International School of Subnuclear Physics (ISSP 6) Retrieved 2011, from http://www.ccsem.infn.it/issp2006/index.html
17) Zichichi, A., (2010) Murray Gell-Mann and the Last Frontier of LHC Physics: The QGCW Project *Procedings of the Conference in honour of Murray Gell-Mann's 80th Birthday,* Quantum Mechanics, Elementary Particles, Quantum Cosmology & Complexity, Nanyang Technological University, Singapore,24–26February 2010, World Scientific books
18) FAIR (Facility for Antiproton and Ion Research) at GSI Darmstadt, Germany, Retrieved from http://www.fair-center.de/index.php?id=1&L=1
19) GSI (Helmholtzzentrum für Schwerionenforschung) Darmsadt, Germany Retrieved from http://www.gsi.de/portrait/index.html
20) CERN (European Organization for Nuclear Research) , Geneva, Switzerland Retrieved from http://public.web.cern.ch/public/
21) *Campus-Wide Time Synchronization System-,, BuTiS*"(D.Beck,2011) Retrieved from *https://www-acc.gsi.de/wiki/Timing/TimingSystemButisInterface*
22)TIPP 2011 2nd International Conference on Technology and Instrumentation in Particle Physics Retrieved 2011 from http://conferences.fnal.gov/tipp11/

23) Report and talk by Crispin Williams at this conference " Improve MPRC "
Retrieved 2011 from *crispin.williams@cern.ch* - Indico - CERN

«ETTORE MAJORANA» FOUNDATION AND CENTRE FOR SCIENTIFIC CULTURE
TO PAY A PERMANENT TRIBUTE TO GALILEO GALILEI, FOUNDER OF MODERN SCIENCE
AND TO ENRICO FERMI, THE "ITALIAN NAVIGATOR", FATHER OF THE WEAK FORCES

INTERNATIONAL SCHOOL OF SUBNUCLEAR PHYSICS
50TH ANNIVERSARY CELEBRATIONS 2011-2012-2013
50th Course: ***WHAT WE WOULD LIKE LHC TO GIVE US***

ERICE-SICILY: 23 June – 2 July 2012

COMPLEXITY and the QGCW Project

Antonino Zichichi
INFN and University of Bologna, Italy
CERN, Geneva, Switzerland
Enrico Fermi Centre, Rome, Italy
World Federation of Scientists, Beijing, Geneva, Moscow, New York

Paul A.M. Dirac Lecture Hall – Patrick M.S. Blackett Institute
Erice, 30 June 2012

CONTENTS

COMPLEXITY AND THE QGCW PROJECT

1 SEVEN DEFINITIONS OF COMPLEXITY

People speak of 'Complexity' as a source of new insights in physics, biology, geology, cosmology, social sciences and in all intellectual activities which look at the world through the lens of a standard analysis in terms of either Simplicity or Complexity. **But 'Complexity'** is ill-defined, as shown by the existence of very many definitions of Complexity. Seven examples follow.

DEFINITION NUMBER 1

Complexity is a property of systems that are somewhere in between a completely random and a completely regular state, often described by a highly non linear set of equations but sometimes not describable by equations at all.

DEFINITION NUMBER 2 (Gerardus 't Hooft)

Bad ones:

1) Chaos.
2) The need for lengthy calculations.
3) The need for many distinct variables.

Better ones:

4) Unexpected difficulty when attempting to describe something in a precisely formulated theory.
5) What is left over after all systematic approaches failed.

But it could also be that: Complexity is an excuse for sloppy thinking.

DEFINITION NUMBER 3 (Richard Kenway)

The Complexity of a theory (problem) is the minimum amount of computer time and storage required to simulate (solve) it to a specified level of precision.

Definition Number 4 (Constantino Tsallis)

If we admit that biological or linguistic evolution, or financial dynamics are complex phenomena, then their typical dynamics is somehow between strong chaos (i.e. positive Lyapunov exponents) and simple orbits (i.e. negative Lyapunov exponents). In other words, Complexity (or at least some form of it) is deeply related to the edge of chaos (i.e. vanishing maximal Lyapunov exponent). Since the edge of chaos appears to be related paradigmatically to an entropy index 'q' different from unity, there must be some deep connection between Complexity and generalized entropies such as 'S_q'.

Definition Number 5 (Leonard Susskind)

From the mathematical point of view:

- A problem can be polynomial, which means that it is not to hard to predict surprises.
- A problem can be NP or NP-complete, which represent different degrees of difficulty in predicting surprises.
- •• Surprises means: UEEC event (see later).
- •• That degree of difficulty can be associated with the level of Complexity.

Definition Number 6

A system is 'complex' when it is no longer useful to describe it in terms of its fundamental constituents.

Definition Number 7

The simplest definition of Complexity:'*Complexity is the opposite of Simplicity*'. This is why we have studied the platonic Standard Model (Addendum 1) and its extension to the platonic Superworld (Addendum 2).

These seven definitions of Complexity must be compared with the whole of our knowledge (see later) in order to focus our attention on the key features needed to study our real world.

2 COMPLEXITY EXISTS AT ALL SCALES

The Logic of Nature allows the existence of a large variety of structures with their regularities and laws which appear to be independent from the basic constituents and fundamental laws of Nature which govern their interactions.

But, without these laws it would be impossible to have the real world which is in front of us and of which we are part of. A series of complex systems is shown in figure 1.

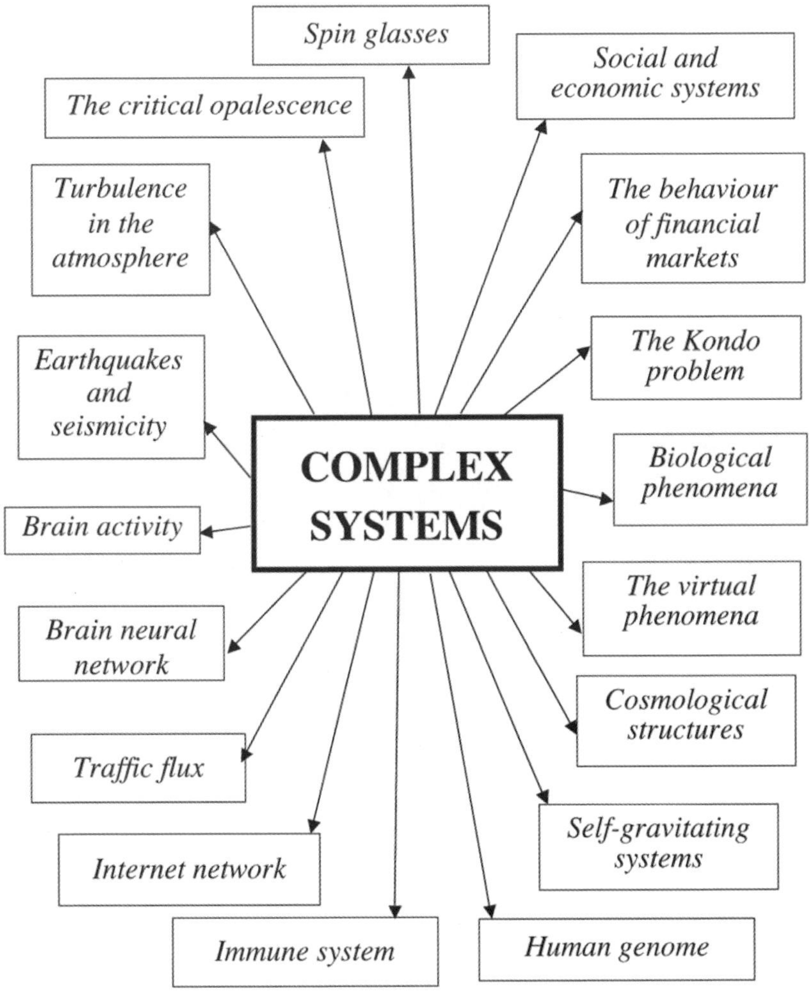

Figure 1

As you can see, we go from traffic flux, to the internet network, to earthquakes and seismicity, to social and economic systems, to the behaviour of financial markets, to the study of cosmological structures, and so on.

The experimental evidences for the **existence** of **Complexity are two**:

1) The Anderson-Feynman-Beethoven-type phenomena (**AFB**) i.e. phenomena whose laws and regularities ignore the existence of the Fundamental Laws of Nature from which they originate. This will be discussed in chapter 3.
2) The Sarajevo-type effects, i.e. Unexpected Events of quasi irrelevant magnitude which produce Enormous Consequences (**UEEC**), discussed in chapter 4.

The only certainty about Complexity is the existence of these experimentally observable effects. Nature tell us that AFB and UEEC exist at all scales, and therefore Complexity exists at all scales, as illustrated in figure 2.

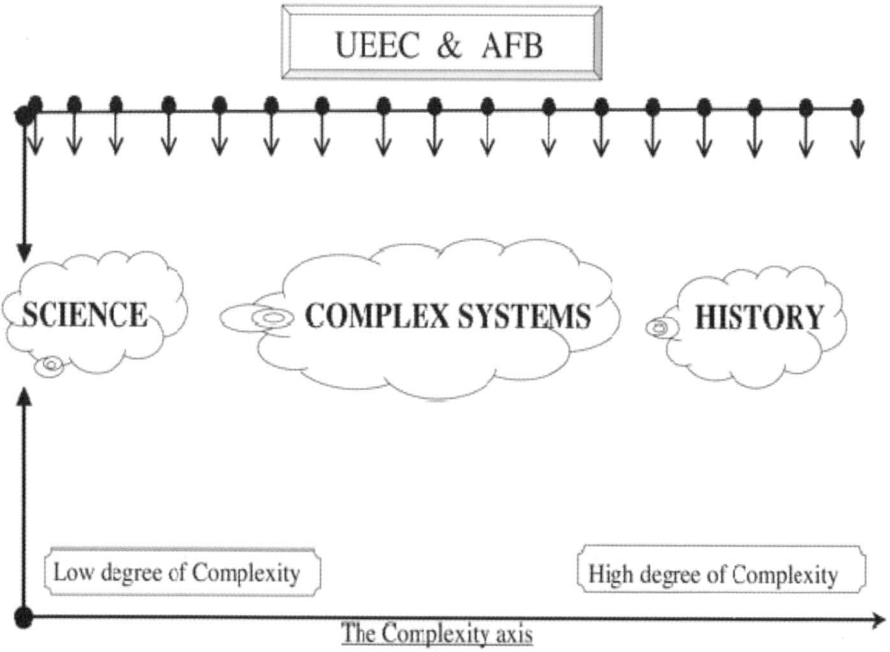

Figure 2

3 AFB PHENOMENA FROM BEETHOVEN TO THE SUPERWORLD

Beethoven and the laws of acoustics.

Beethoven could compose superb masterpieces of music without any knowledge of the laws governing acoustic phenomena.

But these masterpieces could not exist if the laws of acoustics were not there.

The living cell and QED.

To study the mechanisms governing a living cell, we do not need to know the laws of electromagnetic phenomena whose advanced formulation is QED (Quantum ElectroDynamics).

All mechanisms needed for life are, to a great extent, examples of electromagnetic processes. If QED was not there, Life could not exist.

Nuclear physics and QCD.

Proton and neutron interactions appear as if a fundamental force of nature is at work: the nuclear force, with its rules and its regularities.

These interactions ignore that protons and neutrons are made with quarks and gluons.

Nuclear physics does not appear to care about the existence of Quantum ChromoDynamics (QCD), the fundamental force acting between quarks and gluons at the heart of the subnuclear world.

Nuclear physics ignores QCD but all phenomena occurring in nuclear physics have their roots in the interactions of quarks and gluons.

In other words, protons and neutrons behave like Beethoven: they interact and build up nuclear physics without ‘knowing’ the laws governing QCD.

The most recent example of AFB type phenomenon comes from the frontier of our scientific knowledge: the Superworld. What we call the **World could *apparently* not care less about the existence of the Superworld**, whose foundation is not in the four dimensional space-time but in superspace with 43 dimensions.

4 UEEC EVENTS, FROM GALILEI UP TO SM&B

In figure 3 there is a sequence of **UEEC** events from Galilei to Fermi-Dirac and the 'strange particles'.

'UEEC' TOTALLY UNEXPECTED DISCOVERIES FROM GALILEI TO FERMI-DIRAC AND THE 'STRANGE' PARTICLES	
I	Galileo Galilei discovery of $F = mg$.
II	Newton discovery of $F = G\frac{m_1 \cdot m_2}{R_{12}^2}$
III	Maxwell discovers the unification of electricity, magnetism and optical phenomena, which allows him to conclude that light is a vibration of the EM field.
IV	Planck discovery of $h \neq 0$.
V	Lorentz discovers that space and time cannot be both real.
VI	Einstein discovers the existence of time-like and space-like worlds. Only in the time-like world, simultaneity does not change, with changing observer.
VII	Rutherford discovers the nucleus.
VIII	Hess discovers the cosmic rays.
IX	Dirac discovers his equation, which opens new horizons, including the existence of the antiworld.
X	Fermi discovers the weak forces.
XI	Fermi and Dirac discover the Fermi–Dirac statistics.
XII	The 'strange particles' are discovered in the Blackett Lab.

Figure 3

In figures 4, 5, 6 from Fermi-Dirac to the construction of the Standard Model and in figure 7 a synthesis of the UEEC events in what we now call the Standard Model and Beyond (**SM&B**).

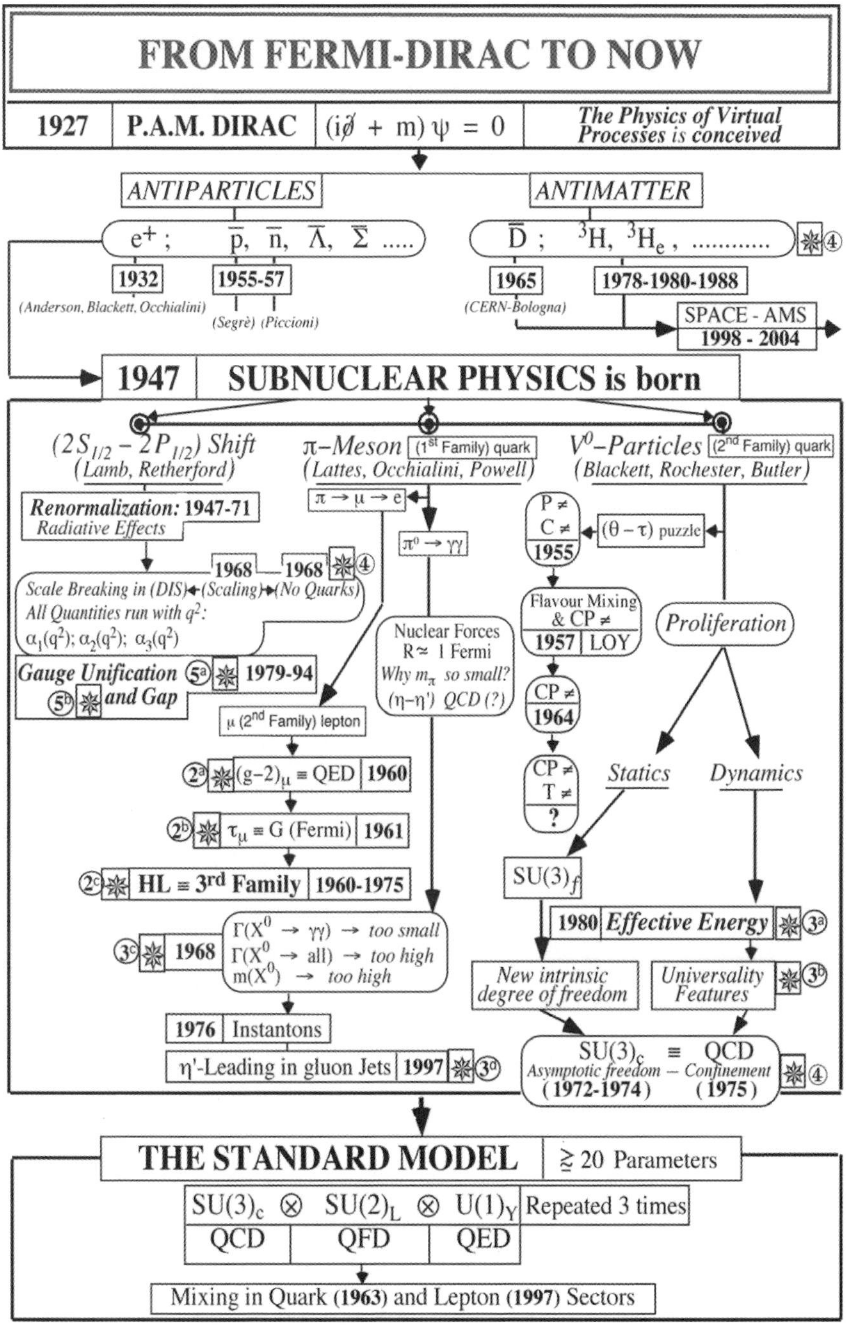
FROM FERMI-DIRAC TO NOW
1927
P.A.M. DIRAC
(i∂̸ + m) ψ = 0
The Physics of Virtual Processes is conceived
ANTIPARTICLES
ANTIMATTER
e+ ; p̄, n̄, Λ̄, Σ̄ .....
D̄ ; ³H̄, ³H̄e , ...
④
1932
(Anderson, Blackett, Occhialini)
1955-57
(Segrè) (Piccioni)
1965
(CERN-Bologna)
1978-1980-1988
SPACE - AMS
1998 - 2004
1947
SUBNUCLEAR PHYSICS is born
(2S1/2 – 2P1/2) Shift
(Lamb, Retherford)
π–Meson
(1st Family) quark
(Lattes, Occhialini, Powell)
V0–Particles
(2nd Family) quark
(Blackett, Rochester, Butler)
Renormalization: 1947-71
Radiative Effects
π → μ → e
π0 → γγ
P ≠
C ≠
1955
(θ – τ) puzzle
1968
1968
④
Scale Breaking in (DIS)
(Scaling)
(No Quarks)
All Quantities run with q2:
α1(q2); α2(q2); α3(q2)
Gauge Unification
5a
1979-94
5b
and Gap
Nuclear Forces
R ≃ 1 Fermi
Why mπ so small?
(η–η') QCD (?)
Flavour Mixing
& CP ≠
1957
LOY
Proliferation
CP ≠
1964
μ (2nd Family) lepton
2a
(g–2)μ ≡ QED
1960
2b
τμ ≡ G (Fermi)
1961
2c
HL ≡ 3rd Family
1960-1975
CP ≠
T ≠
?
Statics
Dynamics
SU(3)f
1980
Effective Energy
3a
3c
1968
Γ(X0 → γγ) → too small
Γ(X0 → all) → too high
m(X0) → too high
New intrinsic degree of freedom
Universality Features
3b
1976
Instantons
η'-Leading in gluon Jets
1997
3d
SU(3)c ≡ QCD
Asymptotic freedom – Confinement
(1972-1974) (1975)
④
THE STANDARD MODEL
≳ 20 Parameters
SU(3)c ⊗ SU(2)L ⊗ U(1)Y
Repeated 3 times
QCD
QFD
QED
Mixing in Quark (1963) and Lepton (1997) Sectors

Figure 4

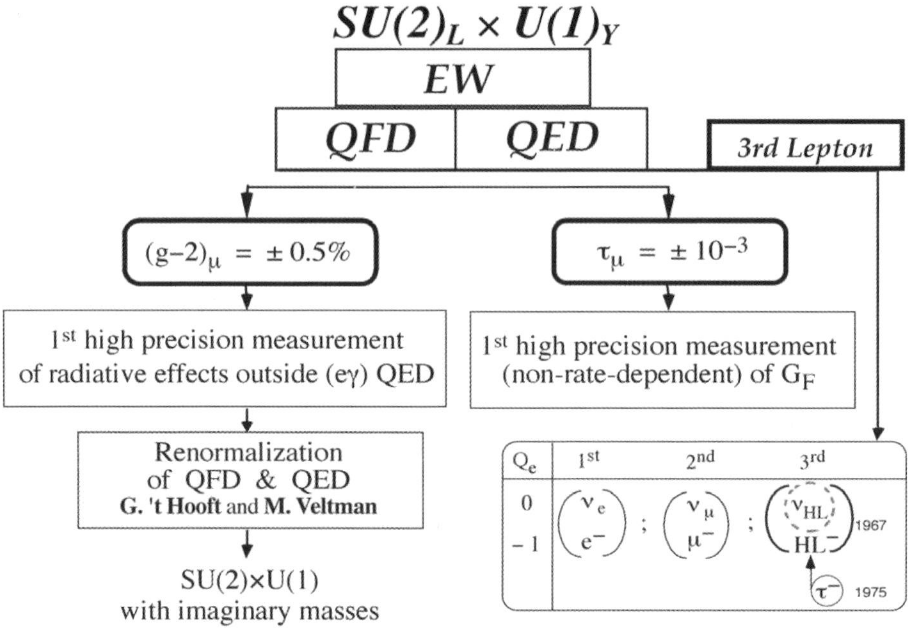

Figure 5: Details from figure 4, concerning $SU(2)_L$ and $U(1)_Y$.

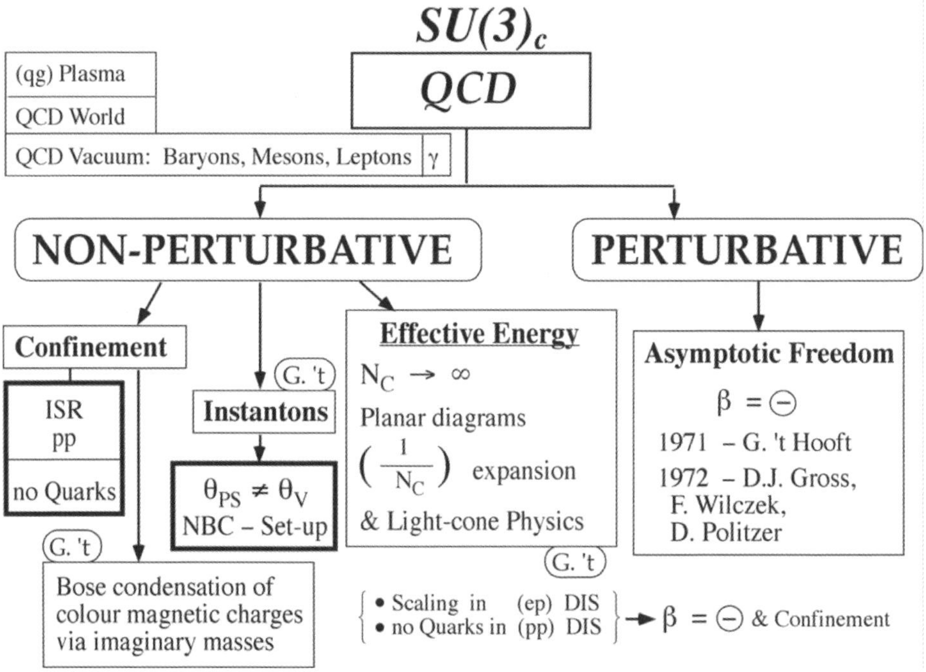

Figure 6: Details from figure 4, concerning $SU(3)_c$.

SM&B

THE STANDARD MODEL AND BEYOND

① RGEs (α_i ($i \equiv 1, 2, 3$); m_j ($j \equiv q, l, G, H$)) : $f(k^2)$.

- GUT ($\alpha_{GUT} \cong 1/24$) & GAP ($10^{16} - 10^{18}$) GeV.
- SUSY (to stabilize $m_F/m_P \cong 10^{-17}$).
- RQST (to quantize Gravity).

② Gauge Principle (hidden and expanded dimensions).

— How a Fundamental Force is generated: SU(3); SU(2); U(1) and Gravity.

③ The Physics of Imaginary Masses: SSB.

— The Imaginary Mass in SU(2)×U(1) produces masses ($m_{W^\pm}$; m_{Z^0}; m_q; m_l), including $m_\gamma = 0$.

— The Imaginary Mass in SU(5)⇒SU(3)×SU(2)×U(1) or in any higher (not containing U(1)) Symmetry Group ⇒ SU(3)×SU(2)×U(1) produces Monopoles.

— The Imaginary Mass in $SU(3)_c$ generates Confinement.

④ Flavour Mixings & CP ≠ , T ≠ .

— No need for it but it is there.

⑤ Anomalies & Instantons.

— Basic Features of all Non-Abelian Forces.

Note: q	≡	quark and squark;	m_F	≡	Fermi mass scale;
l	≡	lepton and slepton;	m_P	≡	Planck mass scale;
G	≡	Gauge boson and Gaugino;	k	≡	quadrimomentum;
H	≡	Higgs and Shiggs;	C	≡	Charge Conjugation;
RGEs	≡	Renormalization Group Equations;	P	≡	Parity;
GUT	≡	Grand Unified Theory;	T	≡	Time Reversal;
SUSY	≡	Supersymmetry;	≠	≡	Breakdown of Symmetry Operators.
RQST	≡	Relativistic Quantum String Theory;			
SSB	≡	Spontaneous Symmetry Breaking.			

The five basic steps in our understanding of nature. ① The renormalization group equations (RGEs) imply that the gauge couplings (α_i) and the masses (m_j) all run with k^2. It is this running which allows GUT, suggests SUSY and produces the need for a non point-like description (RQST) of physics processes, thus opening the way to quantize gravity. ② All forces originate in the same way: the gauge principle. ③ Imaginary masses play a central role in describing nature. ④ The mass-eigenstates are mixed when the Fermi forces come in. ⑤ The Abelian force QED has lost its role of being the guide for all fundamental forces. The non-Abelian gauge forces dominate and have features which are not present in QED.

Figure 7

These figures (4, 5, 6) cover the first fifty years of Subnuclear Physics, whose detailed description can be found in my book whose front page is reproduced below.

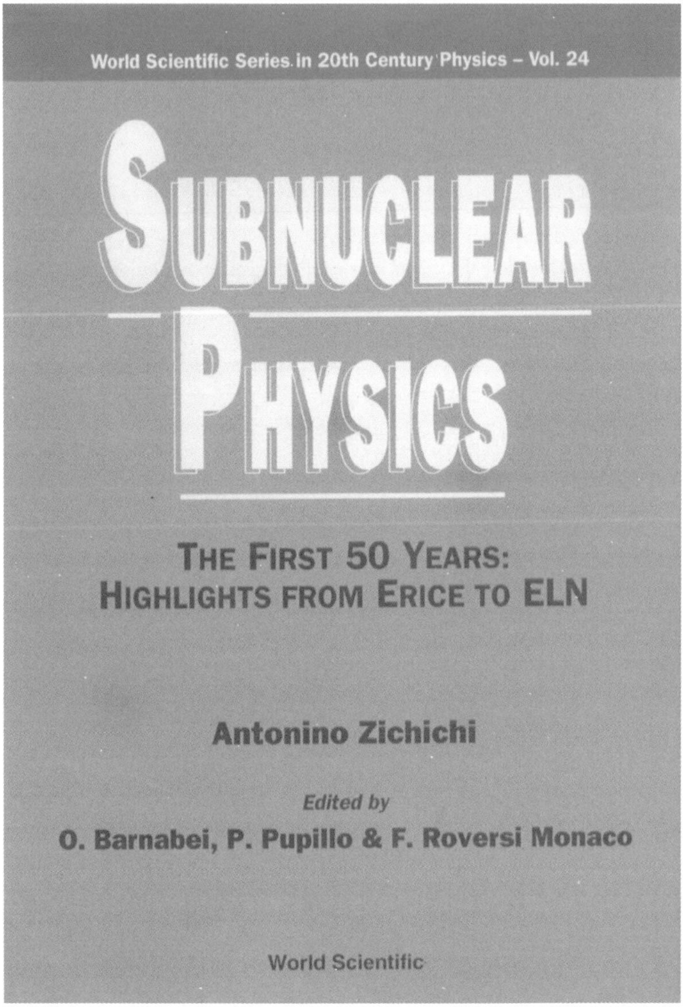

Let me dedicate some attention to discuss UEEC events in nuclear physics.

Nuclear Physics and the UEEC events.

It is considered standard wisdom the fact that nuclear physics is based on perfectly sound theoretical predictions. People forget the impressive series of UEEC events discovered in what I have decided to call the 'Yukawa gold mine'.

Let me quote just three of them:

1 The first experimental evidence for a cosmic ray particle believed to be the Yukawa meson was a lepton: the muon.

2 The decay-chain: $\pi \rightarrow \mu \rightarrow e$ was found to break the symmetry laws of Parity and Charge Conjugation.

3 The intrinsic structure of the Yukawa particle was found to be governed by a new fundamental force of Nature, Quantum ChromoDynamics: QCD.

As you know 2007 was the centenary of the birth of Hideki Yukawa, the father of theoretical nuclear physics. In 1935 the existence of a particle, with mass intermediate (this is the origin of 'mesotron' now 'meson') between the light electron, m_e, and the heavy nucleon (proton or neutron), m_N, was proposed by Yukawa [1].

This intermediate mass value was deduced by Yukawa from the range of the nuclear forces. Contrary to the general wisdom of the time, Yukawa was convinced that the particles known (electrons, protons, neutrons and photons), could not explain how protons and neutrons are bound into the extremely small dimensions of a nucleus.

In order to make this 'prediction', Yukawa needed the Heisenberg uncertainty principle: a totally unexpected theoretical discovery.

The origin of it was the totally unexpected discovery of the dual nature of the electron (wave and particle) and of the photon (wave and particle).

Heisenberg himself tried to explain the binding forces between the proton and the neutron, via the exchange of electrons, in order not to postulate the existence of a new particle. The very light electron, m_e, could not stay in the very small dimension of the nucleus.

The author of the uncertainty principle and father, with Dirac and Pauli, of Quantum Mechanics, did not realise this contradiction. The need for a new 'particle' was the reason. What no-one was able to predict is the 'gold-mine' hidden in the production, the decay and the intrinsic structure of this new 'particle'.

This 'gold-mine' is still being explored nowadays and its present frontier is the Quark-Gluon-Coloured-World (QGCW) [2].

I have recently described [3] the unexpected conceptual developments coming from the study of the production, the decay and the intrinsic structure of the Yukawa particle.

Let me just quote the most relevant UEEC events: chirality–invariance, spontaneous symmetry breaking, symmetry breaking of fundamental invariance laws (P, C, T), anomalies, and 'anomaly-free condition', existence of a third family of fundamental fermions, gauge principle for non-Abelian forces, instantons and existence of a pseudoscalar particle made of the quanta of a new fundamental force of Nature acting between the constituents of the Yukawa particle.

A few cases (seven) where I have been directly involved are summarised in figure 8. Each **UEEC** event is coupled with a **despite**, in order to emphasize the reason why the event is unexpected.

UEEC EVENTS
IN THE CONSTRUCTION OF THE
SM&B ≡ MY PERSONAL EXPERIENCE

① ***The 3rd lepton*, HL** (now called τ) with its own neutrino, ν_{HL} (now called ν_τ),
despite the abundance of neutrinos: ν_e and ν_μ.

② ***Antimatter***
despite S-matrix and C, P, CP, T breakings.

③ ***Nucleon Time-like** EM structure*
despite S-matrix.

④ ***No quarks** in violent (pp) collisions*
despite scaling.

⑤ ***Meson mixings***
$\theta_V \neq \theta_{PS}$: (51°) ≠ (10°) ≠ 0 **despite** $SU(3)_{uds}$.

⑥ ***Effective energy:*** the QCD-light
despite QCD.

⑦ ***The running** of α_1 α_2 α_3 versus **energy** at* **a point** E_{GU}.
(1979) (1991) **despite** straight line convergence.
EGM

Figure 8

The **SM&B** is the greatest synthesis of all times in the study of the fundamental phenomena governing the Universe in all its structures. The basic achievements of the **SM&B** have been obtained via UEEC events; moreover the **SM&B** could not care less about the existence of Platonic Simplicity.

An example is shown in figure 9 where the straight line (small dots) would be the Platonic simple solution towards the Unification of all Fundamental Forces.

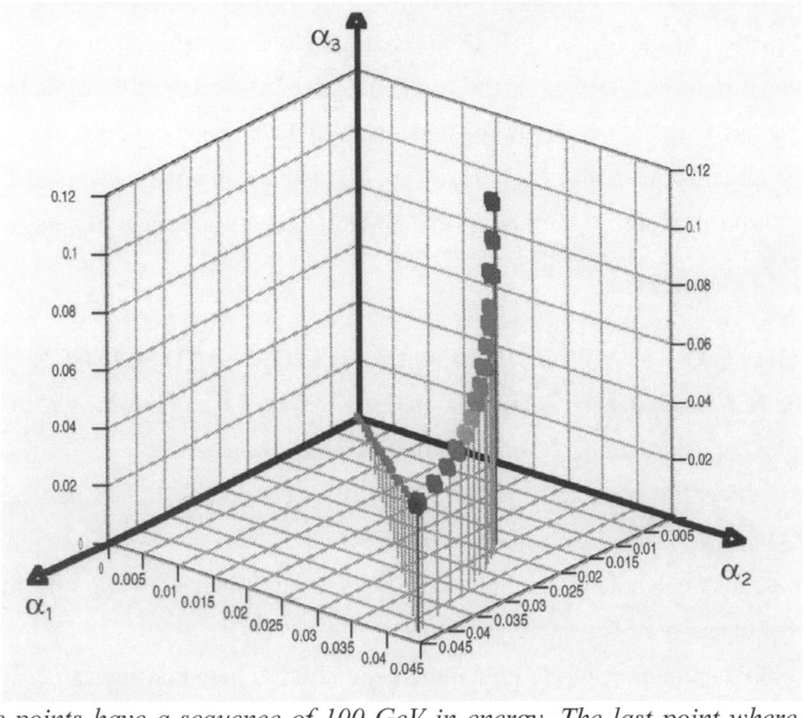

The points have a sequence of 100 GeV in energy. The last point where the 'ideal' platonic straight line intercepts the theoretical prediction is at the energy of the Grand Unification. This corresponds to $E_{GU} = 10^{16.2}$ *GeV. Other detailed information on the theoretical inputs: the number of fermionic families,* N_F *, is 3; the number of Higgs particles,* N_H *, is 2. The input values of the gauge couplings at the* Z^0*-mass is* $\alpha_3 (M_Z) = 0.118 \pm 0.008$*; the other input is the ratio of weak and electromagnetic couplings also measured at the* Z^0*-mass value:* $\sin^2\theta_W (M_Z) = 0.2334 \pm 0.0008$*.*

Figure 9

Nevertheless the effective unification is expected to be along the sequence of points (the big ones) computed using the Renormalization Group Equations (**RGEs**) [4].

Platonic Simplicity for the unification of all forces is again reported in Addendum 1 and Platonic Supersymmetry for the existence of the Superworld is discussed in Addendum 2. We will see how many times the platonic Simplicity is violated when we construct the Superworld.

In Addendum 3 there is a synthesis of UEECC events needed in the process of construction of the SM&B [5]. The numerous violations of Simplicity are the proof that Complexity exists at the fundamental level of scientific knowledge where we have proved that **AFB** phenomena and **UEEC** events are present.

The conclusion is that Complexity exists at the elementary level. In fact, starting from Platonic Simplicity, the **SM&B** needs a series of 'ad hoc' inputs [5].

5 THE TWO ASYMPTOTIC LIMITS: HISTORY AND SCIENCE

The real world seems to be characterized by two basic features, which are one on the opposite side of the other: *Simplicity* and *Complexity*.

It is generally accepted that *Simplicity* is the outcome of *Reductionism*, while *Complexity* is the result of *Holism*.

The most celebrated example of *Simplicity* is *Science* while the most celebrated example of *Complexity* is *History*.

Talking about asymptotic limits, the general trend is to consider ***History*** as the asymptotic limit of *Holism* and of *Complexity*; ***Science*** as the asymptotic limit of *Reductionism* and of *Simplicity*, as illustrated in figure 10 (page 467).

The Logic of Nature allows the existence of ***Science*** (the asymptotic limit of Simplicity) and of ***History*** (the asymptotic limit of Complexity), which **share a property, common to both of them**.

It is interesting to define ***Science*** and ***History*** in terms of this property, **probably the only one**, which they share; i.e. **Evolution**.

- Science is the Evolution of our Basic Understanding of the laws governing the world in its Structure ≡**EBUS**.

- History is the Evolution of the World in its Real Life ≡**EWRL**.

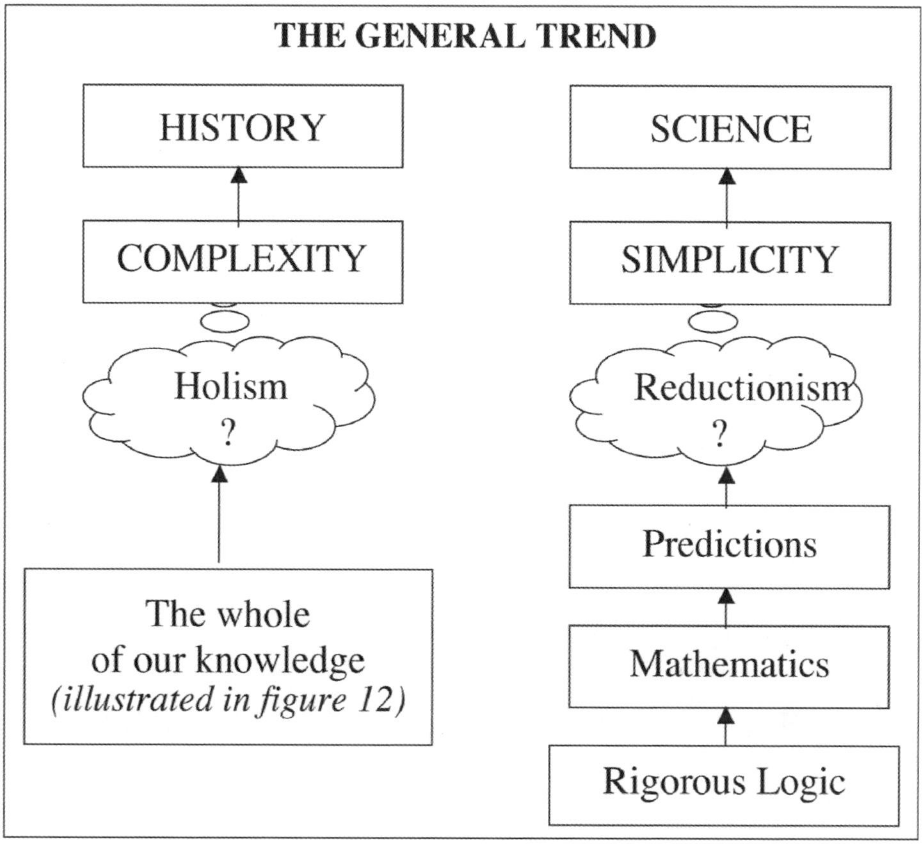

Figure 10

In Table 1 we compare these two supposedly asymptotic limits — History and Science — on the basis of 'What if?'; a condition elaborated by the specialists in what is now known as 'virtual history' [6].

On the basis of 'What if?' these specialists conclude [6] that the world would not be as it is, if one, or few, or any number of 'What if?' had not been as History tells us. This is not the case for Science. The world would have exactly the same laws and regularities, whether Galileo Galilei or somebody else had discovered $F = mg$ ($F \equiv$ force; $m \equiv$ mass;

$g \equiv$ acceleration due to gravity), and so on for all the other scientific discoveries.

It is in the consequences of 'What if?' that the two asymptotic limits of Simplicity and Complexity seem to diverge, despite the fact that the sequence of 'What if?' in Science belongs to the 'totally unexpected events' (UEEC) exactly like the others listed in the column of History.

'WHAT IF ?'

	In History ≡ EWRL		**In Science ≡ EBUS**
I	What if Julius Caesar had been assassinated many years before?	*I*	What if Galileo Galilei had not discovered that $F = mg$?
II	What if Napoleon had not been born?	*II*	What if Newton had not discovered that $F = G\frac{m_1 \cdot m_2}{R_{12}^2}$?
III	What if America had been discovered few centuries later?	*III*	What if Maxwell had not discovered the unification of electricity, magnetism and optical phenomena, which allowed him to conclude that light is a vibration of the EM field?
IV	What if Louis XVI had been able to win against the 'Storming of the Bastille'?	*IV*	What if Planck had not discovered that $h \neq 0$?
V	What if the 1908 Tunguska Comet had fallen somewhere in Europe instead of Tunguska in Siberia?	*V*	What if Lorentz had not discovered that space and time cannot be both real?
VI	What if the killer of the Austrian Archduke Francisco Ferdinand had been arrested the day before the Sarajevo event?	*VI*	What if Einstein had not discovered the existence of time-like and space-like real worlds? Only in the time-like world, simultaneity does not change, with changing observer.
VII	What if Lenin had been killed during his travelling through Germany?	*VII*	What if Rutherford had not discovered the nucleus?
VIII	What if Hitler had not been appointed Chancellor by the President of the Republic of Weimar Paul von Hindenburg?	*VIII*	What if Hess had not discovered the cosmic rays?
IX	What if the first nuclear weapon had been built either by Japan before Pearl Arbour (1941) or by Hitler in 1942 or by Stalin in 1943?	*IX*	What if Dirac had not discovered his equation, which opens new horizons, including the existence of the antiworld?
X	What if Nazi Germany had defeated the Soviet Union?	*X*	What if Fermi had not discovered the weak forces?
XI	What if Karol Wojtyla had not been elected Pope, thus becoming John Paul II?	*XI*	What if Fermi and Dirac had not discovered the Fermi–Dirac statistics?
XII	What if the USSR had not collapsed?	*XII*	What if the 'strange particles' had not been discovered in the Blackett Lab?

TABLE 1

6 THE BASIC POINTS ON THE CORRELATION BETWEEN COMPLEXITY AND PREDICTIONS

We have seen in chapter 1 that in the various attempts to identify the meaning of Complexity the problem of Predictions has been assumed to be granted by the mathematics needed to describe a problem in the complex system. It is therefore necessary to clearly establish the relation which exist between Complexity and Predictions.

In the previous chapters 3 and 4 we have discussed the experimental basis for the existence of Complexity, i.e., AFB and UEEC events.

We will now discuss the experimental evidence for the existence of predictions and the sequence which correlates UEEC and predictions.

Predictions.

The experimental evidences for the **existence** of **Predictions are** the very many results of scientific reproducible experiments.

Quantum Electro-Dynamics, QED, is the best example. The anomalous magnetic moments, in symbols (g–2), of the electron (e) and of the muon (μ):

$$(g-2)_{e,\,\mu}$$

are theoretically computed at an extraordinary level of precision (few parts in ten billion parts for the electron) and are experimentally verified to be correct. Could the

$$(g-2)_{e,\,\mu}$$

be theoretically predicted before the discovery of the Maxwell equations and the existence of Quantum Electro-Dynamics (QED)? The answer is obviously no.

The sequence which correlates UEEC events and Predictions is very clear.

Predictions at the **fundamental level of scientific knowledge** depend on **UEEC events**.

For example: it is the discovery of the laws governing electric, magnetic and optical phenomena (all totally unpredicted) which produced the mathematical structure called QED.

The mathematical structure was not discovered before the innumerable series of **UEEC** events was found in electricity, magnetism and optics. This series of UEEC events allowed Maxwell to express 200 years of experimental discoveries in a set of 4 equations.

The mathematical formalism comes **after** a totally unexpected discovery: an **UEEC event** which no one was able **to predict**.

In the whole of our knowledge rigorous predictions exist only in Science. These predictions are based on the mathematical description of a single UEEC event or a series of UEEC events. This description can either be the result of new mathematics (example the Dirac δ-function) or the use of existing mathematical formalism (example: the Einstein use of the Ricci tensor calculus). The UEEC event at the origin of the Dirac equation is the fact that the electron was not a 'scalar' particle but a spin ½ object.

The UEEC events at the origin of Einstein mathematical formulation of the gravitational forces are the discoveries of

$$\text{Galilei } (F = mg),$$

of

$$\text{Newton } \left(F = G \frac{m_1 \cdot m_2}{R_{12}^2} \right),$$

and of Lorentz that Space and Time could not be both real and that all electromagnetic phenomena obeyed a new invariance law, now called Loretz-invariance.

These are just two examples of the fact that the greatest steps in the **progress of Science** come from totally unpredicted discoveries. It is the mathematical formulation of these discoveries with allow Predictions to be made. Once made, these Predictions need experimental checks.

Even when we have a mathematical formalism coming from a series of UEEC events, if this formalism opens a new frontier, as it is the case for the Superworld, the experimental proof is needed to verify the validity of the new theoretical frontier.

Today we have a reasonable mathematical formalism to describe the **Superworld**, but in order to know if the Superworld exists we need the experimentally reproducible proof for its existence. And it could be that, while searching for the Superworld a totally unexpected discovery (UEEC) comes in. This is the reason why we need to perform experiments, as Galileo Galilei realized 400 years ago.

7 THE LESSON NEEDED FOR THE FUTURE

We have proved that AFB and UEEC – which are at the origin of Complexity, with its consequences permeating all our existence, from molecular biology to life in all its innumerable forms up to our own, including History – do exist at the fundamental level [7–10] and [5].

It turns out that Complexity in the real world exists, no matter the mass-energy and space-time scales considered.

Therefore the only possible prediction is that:

• **Totally Unexpected Effects** should **show up**.

• **Effects**, which are impossible to be predicted on the basis of **present knowledge**.

We should be prepared with **powerful experimental instruments, technologically at the frontier of our knowledge**, to discover Totally Unexpected Events in all laboratories, the world over (including CERN in Europe, Gran Sasso in Italy, and other facilities in Japan, USA, China and Russia). All the pieces of the Yukawa gold mine [3] could not have been discovered if the experimental technology was not at the frontier of our knowledge.

Example: the cloud-chambers (Anderson, Neddermeyer), the photographic emulsions (Lattes, Occhialini, Powell), the high power magnetic fields (Conversi, Pancini, Piccioni) and the powerful particle accelerators and associated detectors for the discovery – the world over – of the **SM&B** as synthetically reported in chapter 4. This means that we must be prepared with the most advanced technology for the discovery of totally unexpected events like the ones found in the Yukawa gold mine.

The mathematical descriptions, and therefore the predictions – for new phenomena to be discovered in the field opened by the given UEEC event – come after the UEEC event, never before.

Recall:

- The **discoveries in Electricity**, **Magnetism** and **Optics** (UEEC).
- **Radioactivity** (UEEC).
- The **Cosmic Rays** (UEEC).
- The **Weak Forces** (UEEC).
- The **Nuclear Physics** (UEEC).
- The **Strange Particles** (UEEC).
- The **3 Columns** (UEEC).
- The **origin of the Fundamental Forces** (UEEC).

The present status of Science is reported in figure 11.

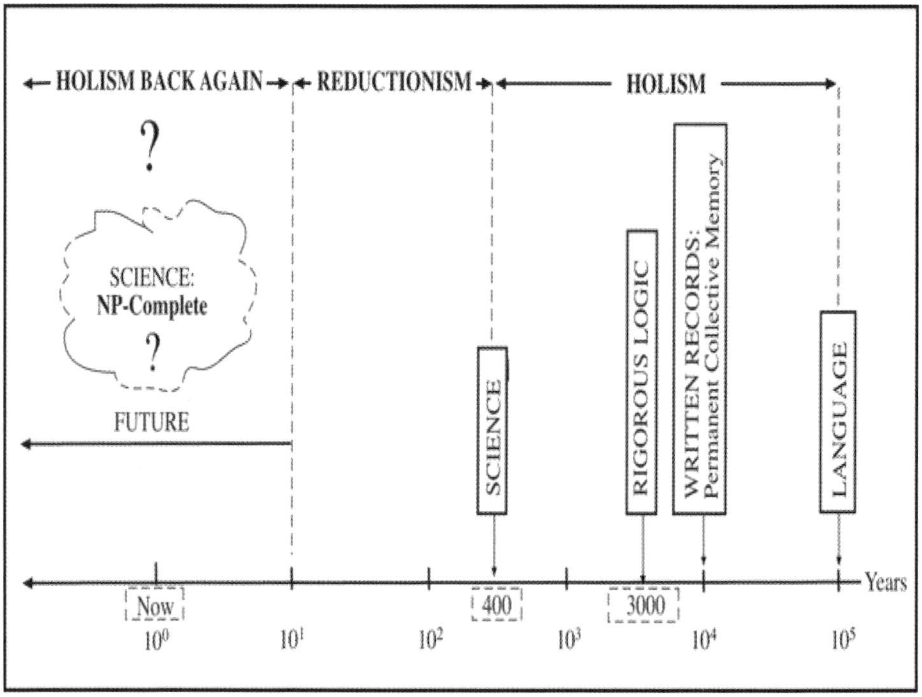

Figure 11

It could be that Science will be mathematically proved to be 'NP-complete'. This is the big question for the immediate future [11].

It is therefore instructive to see how Science fits in the whole of our knowledge as reported in figure 12.

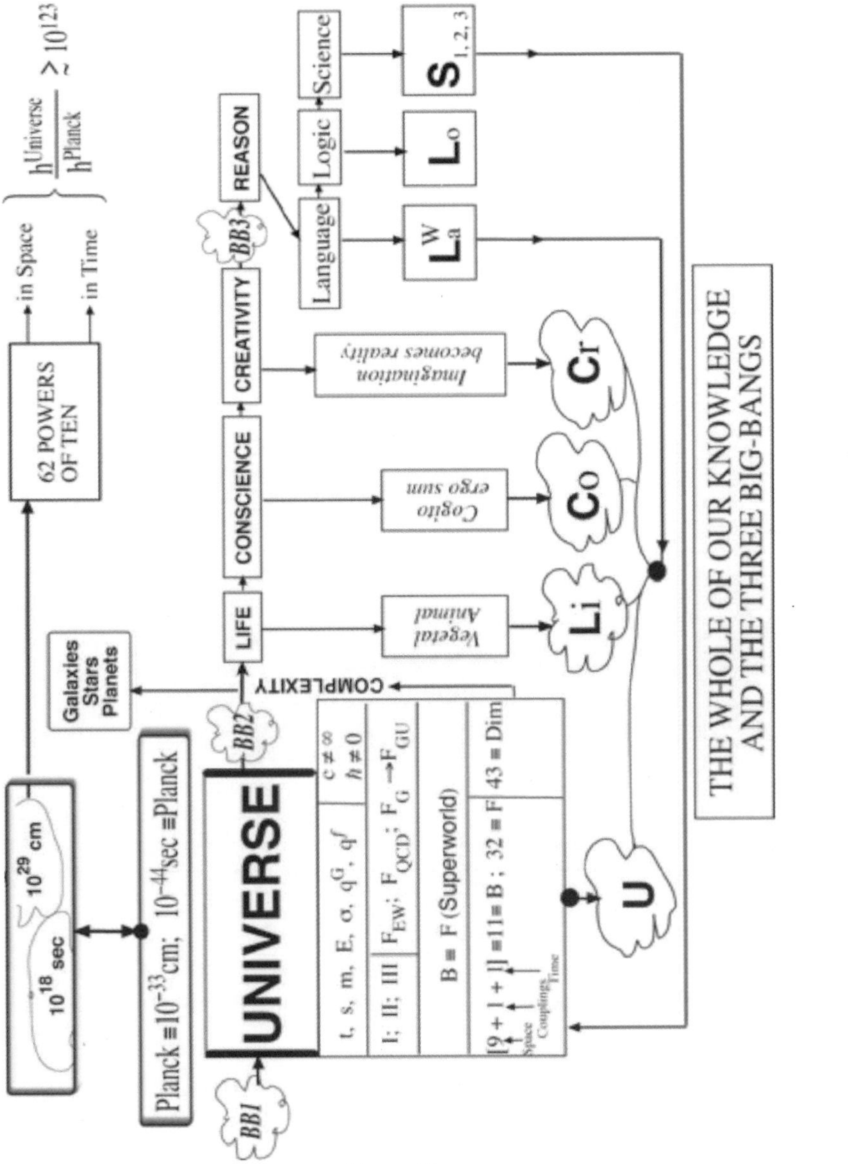

Figure 12

Let me point out that Science is the consequence of us being the only form of leaving matter endowed with Reason, from where the sequence of Language–Logic–Science has been originated [12]. The time-sequence of Language–Logic–Science is shown in figure 13.

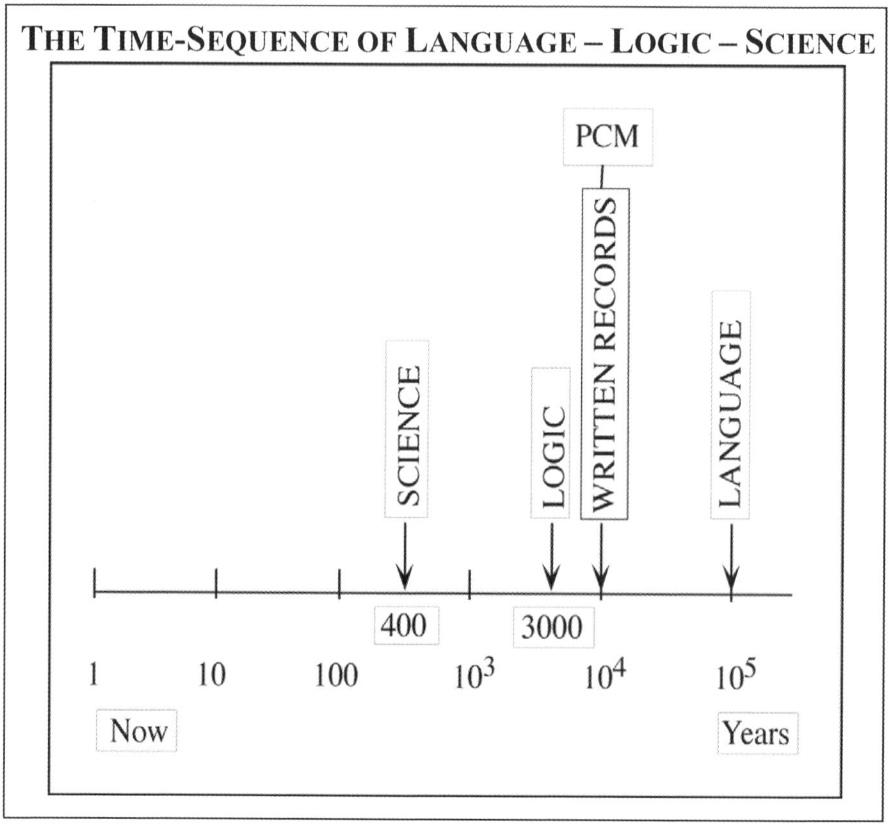

Figure 13

The experimental evidence is that UEEC events dominate our life as the evolution of the world in its real life (EWRL ≡ History) and the evolution of our basic understanding of the laws governing the world (EBUS ≡ Science) do show (see Table 1).

The present status of Subnuclear Physics is given in terms of Open Problems (Addendum 4) and of Ten Challenges (Addendum 5).

The next UEEC event must be outside these Ten Challenges.

We should be aware of the fact that it would be great if, for the first time in the 400 years of Galilean Science, the sequence of UEEC events could enjoy a formidable stop.

The final question is: why the greatest achievements of Science have always been originated by totally unexpected events?

8 FROM PLANCK TO COMPLEXITY

Four centuries of Galilean research work based on Reductionism, i.e. on the identification of the simplest elements in the study of Nature, has allowed us to get the greatest achievement of Science, i.e. the so called **Standard Model** and its extension (**SM&B**), illustrated before in figure 7.

This extension predicts GUT (the Grand Unification Theory), the existence of the Superworld and the resolution of the quantum-gravity problem via the powerful theoretical structure of RQST (Relativistic Quantum String Theory).

These developments started thirty years ago when a great scientific novelty came; all experimental discoveries obtained with our powerful accelerators were to be considered only matters of extremely low energy.

The scale of energy on which to direct the attention to understand the Logic that rules the world, from the tiniest structures to the galactic ones, had to be shifted at a much higher level: to the mass-energy named after Planck, E_{Planck}, something like seventeen powers of ten above the Fermi scale, E_{Fermi}, that already seemed to be an extremely high level of energy.

Figure 14

Now, after thirty years, it comes about the novelty of our time, illustrated in figure 14 (page 475): Complexity exists at the fundamental level [5]. In fact, **AFB** and **UEEC** events exist at all scales, as reported all along this lecture and summarized in figure 2 (page 456).

This result is corroborated from the mathematical structure (the only one) to be in a position of describing all that happens at the Planck scale: the Relativistic Quantum String Theory (**RQST**).

This mathematical structure produces innumerable minima of energy, named **Landscape**.

The theoretical discovery of the **Landscape** (Leonard Susskind) [13], has been followed by another formidable discovery in mathematical physics: the most rigorous model of RQST (Raphael Bousso and Joseph Polchinski) is **NP-complete** (Michael R. Douglas and Frederik Denef) [14].

This discovery corroborates all that we have put in evidence during the last five years [7–10]: **Complexity exists at the fundamental level** [5].

We do not know what will be the final outcome of String Theory.

What we know is that: '*The world appears to be complex at every scale.* ***Therefore*** *we must expect a continued series of surprises that we cannot easily predict'*.

But, with the advent of the LHC it will be possible to study the properties of the Quark-Gluon-Coloured-World (QGCW), which is a world totally different from all we have been dealing with since the origin of Science.

9 CONSEQUENCES FOR LHC: THE QGCW PROJECT

9.1 INTRODUCTION

With the advent of the LHC supercollider at CERN we propose to study the properties of the '**new world'** which is produced in a collision between heavy nuclei ($_{208}Pb^{82+}$)at the maximum energy so far available in our planet, i.e. 1150 TeV (1.15×10^{15} eV). The 'new world' is the **Quark-Gluon-Coloured-World** (QGCW).

We avoid in purpose to call it 'quark-gluon plasma' since, in the extremely high energy collision between heavy ions, many QCD open-colour-states should be produced.

The number of these QCD open-colour-states is by far higher that the number of baryons and mesons so far known, since these baryons and mesons have to obey the condition of being QCD-colourless.

In chapter 9.2 we discuss four basic new problems addressed by the QGCW project. In chapter 9.3 how to detect the properties of the new world (QGCW), in chapter 9.4 we discuss the two most difficult technical problems to solve in order to study the QGCW; in chapter 9.5 the synchronization issues, including four relevant details. The conclusions in chapter 10.

9.2 THE PHYSICS OF THE QUARK-GLUON-COLOURED WORLD (QGCW): THE NEW WORLD

The basic purpose of the project is to study the Quark-Gluon-Coloured World (QGCW) which is totally different from our world made of QCD vacuum with colourless baryons and mesons. We want to search for specific effects due to the fact that the colourless condition is avoided.

1st problem – In the QGCW there are all states allowed by the $SU(3)_c$ colour group. The number of possible states is by far more numerous than the number of colourless baryons and mesons which have so far been built in all Labs, since the colourless condition is not needed. **Question**: What are the consequences on the properties of the QGCW?

2nd problem – Light quarks versus heavy quarks. Are the coloured quark masses **the same** as the values we derive from the fact that baryons and mesons need to be in a colourless state? It could be that all six quark flavours are associated with nearly 'massless' states like those of the 1st family (u, d). In other words the reason why the 'top' quark appears to be so heavy ($\simeq 10^2$ GeV) could be due to the fact that it must satisfy some, so far unknown, condition related to the fact that the final state must be QCD-'colourless'.

We know that confinement produces masses of the order of a GeV. Therefore, according to our present understanding, the QCD 'colourless' condition could not explain the heavy quark mass, but since the origin of the quark masses is still not known, it cannot be excluded that in a QCD coloured world, the six quarks are all nearly massless.

If this was the case, the masses we measure are heavier than the effective coloured quark masses. In this case all possible states generated by 'heavy' quarks would be produced in the **QGCW** at much less temperature than the one needed in our world made with baryons and mesons, i.e. QCD colourless states.

Here again we should try to see if with masses totally different from those expected, on the basis of what we know about colourless baryons and mesons, new **effects** could be detected due to the existence of all six flavours at relatively low temperature in the QGCW world.

3rd problem – To search for effects on the thermodynamic properties of the QGCW. Are these properties going to be along the 'extensivity' and / or 'non-extensivity' conditions?

4th problem – **Derive the equivalent** Stefan-Boltzmann Radiation Law for the QGCW. The relation between **energy density** at emission **U**, and **Temperature** of the source **T**, is

$$U = cT^4$$

in classical Thermodynamics.

In the QGCW the correspondence should be

U $\equiv$ $p_\perp$ (transverse momentum)

T $\equiv$ average energy in the CM system.

In the QGCW the production of 'heavy' flavours should be studied versus $\langle p_\perp \rangle$ and versus $\langle E \rangle$. The expectation is

$$\langle p_\perp \rangle \equiv C \cdot \langle E \rangle^4$$

and any deviation would be extremely important.

The study of the properties of the QGCW should produce the correct mathematical structure able to correctly describe the QGCW; the same mathematical formalism should allow to go from QGCW to the Physics of Baryons and Mesons (PBM) and from here to a restricted component of PBM, namely Nuclear Physics, where all properties of the nuclei should find a correct description.

9.3 HOW TO STUDY THE NEW WORLD: QGCW

An example is illustrated in the figure 15 below where beams of known particles (p, n, π, K, μ, e, γ, ν) bombard the QGCW and a special set of detectors measures the properties of the outcoming particles.

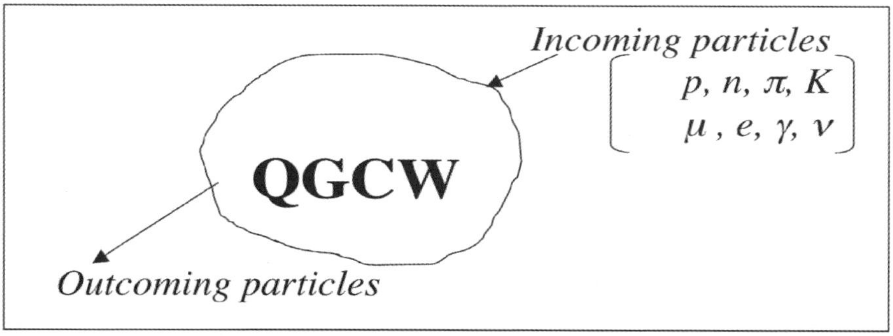

Figure 15

Totally unexpected effects should show up if Nature follows the Logic of Complexity at the Fundamental level.

9.4 THE NEW TECHNOLOGY NEEDED TO STUDY THE QGCW

The technology needed covers two fields: one is the accelerator technology, the other is the detectors technology.

The first step in the accelerator technology refers to the availability of a proton beam able to bombard the QGCW produced in the lead-lead collisions.

The LHC physics program foresees lead-lead collisions with a design luminosity of 10^{27} cm^{-2} s^{-1}. For this to be achieved an upgrade of the ion injector chain comprising Linac3, LEIR, PS and SPS machine is needed. Each LHC ring will be filled in 10 min by almost 600 bunches, each of $7{\times}10^7$ lead ions. Central to the scheme is the Low Energy Ion Ring (LEIR), which transforms long pulses from Linac3 into high brilliance bunches by multi-turn injection, electron cooling and accumulation.

The total collision energy between heavy ions, $_{208}Pb^{82+}$(fully stripped), is 1150 TeV. Table 2 shows the basic parameters of the lead-ion injectors.

NOMINAL PARAMETERS OF THE LEAD ION INJECTORS

	Linac3	LEIR	PS	SPS
Output energy	4.2 MeV/n	72.2 MeV/n	5.9 GeV/n	177 GeV/n
^{203}Pb charge state[1]	27+/54+	54+	54+/82+	82+
Output Bρ [Tm][1]	2.28/1.14	4.80	86.7/57.1	1500
# Batches to fill next machine	4–5	1	13,12,8	12
Bunches/ring		2 (1/8 PS)	4	52,48,32
Ions per pulse[2]	1.15×10^9	9×10^3	4.8×10^3	$<4.7\times10^9$
Ions/LHC bunch	1.15×10^9	2.25×10^2	1.2×10^2	9×10^7
Bunch spacing [ns]		352	100	100
ε^*_{rms} [μm] $= (\beta\gamma)_{rel}\, \sigma^2/\beta_{twitss}$	0.25	0.7	1.0	1.2
ε_l [eVs/u/bunch]		0.05	0.05	0.24
4 σ bunch length		200 ns	4 ns	2 ns
Repetition time [s]	0.2–0.4	3.6	3.6	~50

[1]Values before/after stripping.
[2]50 eμA × 200 μs Pb^{54+} Linac3 output after stripping.

TABLE 2

Once the lead-lead collision is available, the problem is to synchronize the 'proton' beam with the QGCW produced. This problem is at present under study.

The detector technology is also under intense R&D since the synchronization needed is at a very high level of precision; the status of the problems is reported in the following chapter 9.5.

9.5 SYNCHRONIZATION ISSUES RELATED TO THE QGCW PROJECT

The present limit of precision timing is given by the stability of master clocks and of optical fiber signal transmission lines.

Present timing signal stabilities is of order **nanoseconds.** Synchronization to the level down to **femto-seconds (10^{-15} sec)** between distributed equipment, accelerators RF systems, and between accelerator beam bunches will in future be required.

Synchronization at CERN LHC

The CERN network of accelerators requires the transmission, over long distances, of precise timing pulses. These timing pulses are derived from a master clock and therefore the ultimate time accuracy of the pulses is determined by the quality of the transmission of the clock frequency itself. A similar problem arises when two machines need to be synchronized RF-wise. In all cases the transmission of a CW wave in the frequency range of several hundred MHz over distances up to several kilometers is the key to a proper synchronization.

Optical fibers are used as transmission media and a 4 km optical fiber link was installed already in 1978 between the CERN SPS and its injector for the purpose of synchronizing the two machines. In preparation for the LHC, its timing distribution and its synchronization challenges, a special R&D project (RD12) had been set-up to study solutions for machine and experiments. A common solution to the timing, trigger and control (TTC) system requirement for machine and experiments leads to an important economies of scale and permits a rationalization of the requirements for development, operation, and support efforts.

The common systems allows to control the detector synchronization and delivers the fast signals and messages that are phased with the LHC clock, orbit or bunch structure. These include the bunch-crossing clock, level-1 trigger decisions, bunch and event numbers, as well as test signals and broadcast commands.

In the framework of the LHC common project RD12 the development and test of a multi-function optoelectronic TTC system was launched. The TTC system needed to meet the requirement of central signal broadcasting and local distribution at the different sub-detectors of the experiments.

A laser transmitter, modulator, encoder, VME-bus interface and machine interface have been developed as well as a subminiature radiation-hard optical fiber connector, active device mount and photo-detector / preamplifier.

A radiation-hard timing receiver ASIC is being designed which will generate the full range of decoded signals for electronics controllers from a single input and a PMC receiver module is being developed to facilitate initial applications.

The system incorporates programmable coarse and desktop facilities to compensate for different particle flight times and detector, electronics, propagation and test generator delays.

It can also transmit asynchronous slow controls and data such as individually addressed channel enables and calibration parameters to several thousand destinations. More details are published by the RD12 collaboration in their final report.

Synchronization challenges of the QCGW project

For the QCGW project it will be mandatory to improve the LHC timing system. One needs to provide a much shorter and predictable constant time delay distribution of RF signals than presently available at LHC. One should aim at carrying time information with resolution and stability to order of femtoseconds over distances of kilometers.

The R&D for the QCGW project needs to concentrate on research of **stability properties of optical fibers** for signal transport, on the development of an **improved microwave oscillator** (master clock), and on beam instrumentation.

LHC heavy ion particle bunches at high energies have rise-times of a few nanoseconds, the rise-time of proton bunches could be as low as 100 picoseconds, for photons and lasers one is in the femtosecond range. Therefore, referring to the rising or front edge of a bunch requires timing precision covering nanoseconds, picoseconds or even femtoseconds.

The key timing component in a synchrotron is the RF generator of the accelerating RF cavity. Charged particles moving through the cavity will see the longitudinal electrical field in the cavity gap. As a result, they are captured in the 'RF bucket' as bunches and thus are localized inside the synchrotron vacuum pipe.

By definition a 'bunch phase' is the momentary longitudinal position of the leading front or, in electrical terms, the rising edge of the moving particle bunch.

So, a bunch phase refers to the local position where it is momentary seen. The bunch phase can also be described in angular terms when referring it to a geometric reference point of the ring accelerator. To capture a bunch in a cavity implies that there is a relation between the bunch phase to the cavity RF phase. The presence of a charged bunch of particles is detected by a capacitive pick up in the vacuum pipe. These 'Beam Position Monitors' allow to measure the bunch phase in real-time and thus enables the cavity synchronization by controlling the phases of the RF-cavities. It is essential for proper acceleration to maintain phase differences of less than a few degrees between all cavities in the accelerator complex.

A precise reference signal common to all RF cavity stations is needed to operate with the same signal references during acceleration ramps. Otherwise, all RF-cavities will have arbitrary phases. At each group cavity system one has to select one reference signal for phase locking the cavity phases.

This is not only valid for a single accelerator synchronization. It is a rather direct extrapolation if one has to control a whole complex build out of synchrotrons and storage rings: Each of the named 'phase references' in a single accelerator system can be referenced to a wider spanned 'campus reference' in the same manner.

A system that transmits standard time- and frequency signals at the same time using signal-multiplexing techniques, could distribute more common time and frequency reference signals as well. These signals will all have the same predictable delay and link all local generated RF signals and additional time-

and trigger-signals (TTC) reference planes of the whole campus site together. Thus, a common coherent campus synchronous network is built. This has been the basic idea for the LHC synchronization scheme.

The communications industry has created an RF signal transport technique called 'optical RF link' [15]. These devices are intended for linking broadband RF signals between mobile telephone base stations.

These optical links utilize direct modulated laser diode sources, which feed optical single mode fibers of several kilometers length and fast photodiodes at the receiving end of the fiber. The bandwidth of the analog link transmitter/receiver pair is beyond 2GHz in base-band mode. The RF signals are transmitted in base-band, they cover the frequency band of 900-1900 MHz as 'block' without up/down converting in frequency. The capabilities of these links match with present accelerator signal transport tasks: (i) analog transmission of multiple RF signals, (ii) constant delay properties of the transport fibers, the thermal length deviation coefficient of 'from the stock fibers' is about 7 ppm/K.

For a 1000m run of optical fiber (5ns/mgroup delay) a deviation of 35ps/K can be derived. For the target value of 40 K temperature variation (over the whole length) the total delay change is 1.4 ns.

For the QCGW project this value will have to be improved.

Transporting femto-second accelerator timing signals over long distances requires a new approach. One has to overcome effects of time jitter, typically induced by microphony, electromagnetic noise and temperature dependent variation of fiber cable length.

A possible solution could be based on the idea of a reference signal feedback system to stabilize the signal transmission. We think of using an extremely stable laser as an optical oscillator.

The laser provides reference signals (master clock) and in combination with a piezo-electric fiber stretcher one stabilizes the signal frequency transmitted through a long signal transport fiber cable in feedback mode. This idea is under study for X-FEL pulses and will be worked out in more detail for the QCGW project [15].

ANNEX 1: DETAILS ON TIMING DISTRIBUTION AT THE LHC

Extract from **Timing Distribution at the LHC** by Bruce.Taylor@cern.ch

The timing signals for each ring of LHC will be encoded and transmitted over optical links from the RF system to the PCR, where beam-synchronous messages will be added. High power laser transmitters will then broadcast the signals over single-mode optical fibers to the four LHC experiments, to the test beam areas, and to the beam instrumentation located around the LHC ring and on the SPS transfer lines. At the experiment areas, trigger information and local synchronous commands and data will be added. The regenerated signals will then be broadcast over multimode passive optical networks to several thousand destinations.

At the LHC, 'fast' timing signals must be distributed to all experiments and to the beam instrumentation of the machine. These signals are derived from the LHC RF generators and will be synchronous with the circulating beams, so that their frequencies will vary a little during acceleration. At 7 TeV, the bunch clock frequency will be about 40.07897 MHz while the orbit frequency will be 11.2455 kHz.

They are distinct from the 'slow' LHC timing signals, having a granularity of 1 ms, which will signal machine events and distribute UTC time for data tagging and post mortem applications.

At the LHC, distributing correctly synchronized signals to several thousand electronics channels presents some interesting challenges. The R&D work has been done in the framework of the RD12 TTC Project [16], which comprised members from all the LHC experiments, the Microelectronics and Beam Instrumentation Groups and two industrial partners.

The unified approach to TTC distribution developed by RD12 provides for the broadcasting of the fast timing signals through all the transmission stages from the RF generators of the LHC machine to the outputs of the timing receiver ASICs at the experiment and beam instrumentation destinations. That general path will be followed in this review of the system.

Annex 2: Details on LHC Bunch Structure

Commencing with the timing of the LHC machine, it should first be noted that there has been an important change to the bunch structure described in the Yellow Book [17]. Initially it was proposed to accelerate trains of 84 bunches in the PS, 81 of which would be injected into the SPS, 3 being lost in the PS ejector. The difficulty with this configuration is that it would be dirty in the PS and SPS machines and there would be longitudinal stability problems in the PS with the 84-bunch trains.

Various solutions to these problems have been proposed and the one that has been retained as the current baseline foresees the acceleration of PS trains of 72 bunches which will be entirely injected into the SPS. In order to maintain an acceptable filling factor in the LHC with the 72-bunch PS trains, the SPS batches, which will be injected into the LHC, will comprise groups of 3, 3 and then 4 PS trains. So, whereas formerly we had quasi 12-fold symmetry in the LHC (the last PS train in the last SPS batch being suppressed to allow for the rise-time of the LHC extraction kicker), we now have quasi 4-fold symmetry with corresponding implications for the TTC synchronization algorithms [18].

As a result of this change the number of bunch crossings per orbit will be reduced from 2835 to 2808. Note that this applies only to ATLAS and CMS – since these experiments are diametrically opposite each other it is possible to phase the beams to make the LHC extractor gaps coincide at both of them. That is not possible at ALICE and LHCb, which will result in the loss of a further 188 bunch crossings per orbit at these experiments. It should also be noted that it is now expected that there may be quite a substantial initial running period with 75 ns instead of 25 ns bunch spacing. The expected rms collision length remains about 180 ps.

In the case of the LHC bunch structure for heavy ions, there may be several re-synchronizations during each orbit and there could be gaps, which are a non-integer number of bunch intervals in length.

But neither of these factors should be a cause for alarm, for in this case the bunch spacing concerned will be 100 ns or 125 ns. The TTC system will continue to distribute a 40.079 MHz clock during this mode of operation and the bunch crossings will remain in phase with this clock.

ANNEX 3: DETAILS ON CLOCK ARTEFACTS

On the other hand, at times there may be some artefacts in the distributed clocks. There could be a 1 ms hole in the SPS RF/5 clock occurring once before each transfer from the PS, because the LHC machine will be the master of the timing and the SPS has to be synchronized such that the SPS batches are injected into the correct part of the LHC orbit. That will be done by calculating back to the PS, so the PS and SPS have to be re-synchronized before each injection and during this time there may be an interruption in the clock.

The situation with colliding beams at the LHC will be more comfortable. In that case there could be a 1 ms hole in the bunch clock, which will occur once, and once only, before the very first injection from the SPS into the LHC. The reason for this is that a general RF system reset will be made prior to each LHC run in order to ensure that all the dividers have the correct phase and there may be an interruption to the clock while this is applied.

During these clock holes, the TTC system will continue to distribute a 40.079 MHz clock to the experiments. But developers should be aware that there may be a momentary phase perturbation when the system re-synchronizes with the real clocks when they are restored after the interruptions.

ANNEX 4: DETAILS ON DISTRIBUTION ARCHITECTURE

At present the RF timing generators are located in the BA3 Faraday Cage adjacent to the Prevessin Control Room (PCR).

Four clocks are available: the constant frequency 40.079 MHz LHC bunch clock, a pseudo LHC orbit signal obtained by dividing the clock by 3564, the real SPS orbit signal and the ramping SPS 40 MHz clock obtained by dividing the SPS RF by 5.

The PCR transmitters can each broadcast only one orbit and one clock signal simultaneously. The selected pair are encoded and used to modulate a high power laser, the output from which is split by a 1:32 optical tree coupler and broadcast via optical fibers to different destinations around the CERN sites.

At present these destinations include the test beam areas in the North and West halls and labs where beam instrumentation and TTC development work are being done. Finally they will include the LHC experiment areas. At the experiment areas the signals will be received by a TTC machine interface in which they are decoded.

10 CONCLUSIONS

We must be prepared with the most advanced technology for the discovery of totally unexpected events like the ones found in the Yukawa gold mine.

The occasion of the Yukawa Centenary (2007) has been of great value in order to draw attention to the impressive series of conceptual developments linked with his meson: chirality–invariance, spontaneous symmetry breaking, symmetry breaking of fundamental invariance laws, anomalies, and 'anomaly-free condition', existence of a third family of fundamental fermions, gauge principle for non-Abelian forces, instantons and existence of a pseudoscalar particle made of the quanta of a new fundamental force of Nature acting between the constituents of the Yukawa particle.

All the pieces of the Yukawa gold mine **could not have been discovered** if the experimental technology was not at the frontier of our knowledge, as already reported in chapter 7: the cloud-chambers (Anderson, Neddermeyer), the photographic emulsions (Lattes, Occhialini, Powell), the high power magnetic fields (Conversi, Pancini, Piccioni) and the powerful particle accelerators and associated detectors for the discovery – the world over – of the intrinsic structure of the Yukawa particle (quarks and gluons) which has brought us to QCD and now to the QGCW Project.

ADDENDUM 1
THE PLATONIC GRAND UNIFICATION

We report here again figure 9 from page 465 since this is the best examples of Platonic Grand Unification. The points have a sequence of 100 GeV in energy. The last point where the 'ideal' platonic straight line intercepts the theoretical prediction is at the energy of the Grand Unification. This corresponds to $E_{GU} = 10^{16.2}$GeV. Other detailed information on the theoretical inputs: the number of fermionic families, N_F, is 3; the number of Higgs particles, N_H, is 2. The input values of the gauge couplings at the Z^0-mass is α_3 $(M_Z) = 0.118 \pm 0.008$; the other input is the ratio of weak and electromagnetic couplings also measured at the Z^0-mass value: $\sin^2\theta_W(M_Z) = 0.2334 \pm 0.0008$.

The Platonic Grand Unification should be along the straight line (in blu) but Nature seems to follow the red points.

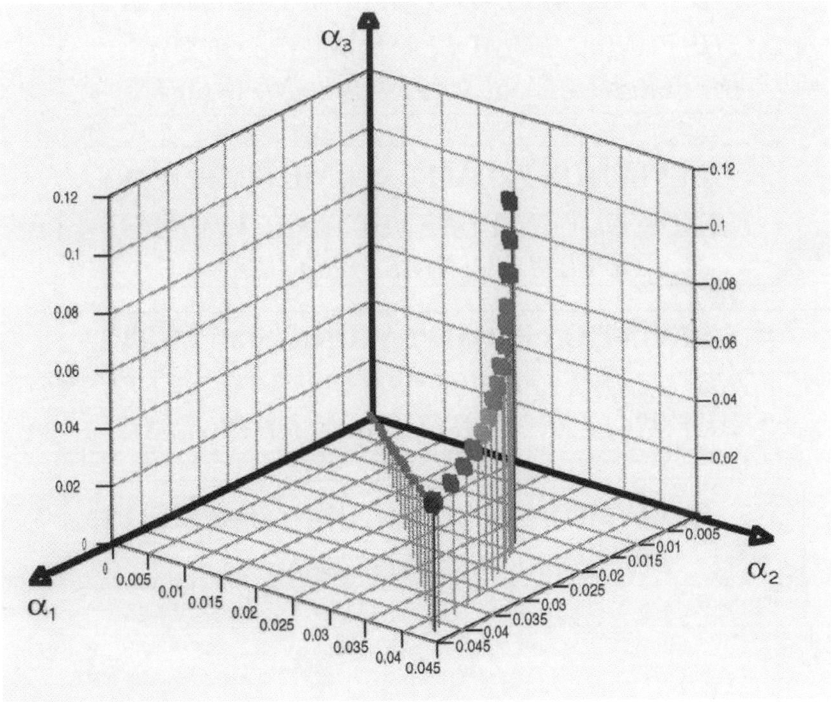

Figure 9 (from page 465)

ADDENDUM 2
THE PLATONIC SUPERSYMMETRY

THE PLATONIC CONCEPT OF SUPERSYMMETRY

The **Gauge Principle** should generate a
Gauge Force ⟶ **Gauge Bosons**
If *NATURE* was platonically *SUPERSYMMETRIC*
Supersymmetry Transformation should generate **Gauginos**

1st **DEVIATION FROM PLATONIC SIMPLICITY**
OUR FERMIONS ARE NOT THE GAUGINOS

2nd **DEVIATION FROM PLATONIC SIMPLICITY**
THE FUNDAMENTAL FERMIONS ARE OF TWO DIFFERENT CLASSES: LEPTONS AND QUARKS

3rd **DEVIATION FROM PLATONIC SIMPLICITY**
THERE IS NOT ONLY ONE BUT THREE FAMILIES OF FUNDAMENTAL FERMIONS

4th **DEVIATION FROM PLATONIC SIMPLICITY**
THE FUNDAMENTAL FERMIONS BECAME MIXED WHEN THE WEAK FORCES ARE SWITCHED ON: MIXINGS EXIST

5th **DEVIATION FROM PLATONIC SIMPLICITY**
THERE ARE DIFFERENT MIXINGS

Figure 16

ADDENDUM 3

EXAMPLES OF UEEC EVENTS IN THE CONSTRUCTION OF THE STANDARD MODEL AND BEYOND

There are very many UEEC events in the construction of the Standard Model and Beyond (SM&B). Let me select a group of problems where I have been directly involved: 1) the mass ≠ matter problem; 2) the mesonic mixings; 3) the Effective Energy and the Gribov QCD-light; 4) the Gap between the GUT energy and the string unification energy.

ADDENDUM 3–1

THE MASS≠MATTER PROBLEM

The fact that mass and matter had to be two different physical quantities, i.e. the mass ≠ matter problem, started with the Einstein discovery $E = mc^2$. The symbol 'm' originally was considered to represent 'matter' and thus the Einstein discovery become the problem of explaining the stability of matter. The meaning of 'm' had to be different from 'matter'. This is how the distinction between 'matter' and 'mass' come in the forefront of fundamental physics. Einstein proposed to solve the problem mass ≠ matter with the existence of just one 'charge', the electromagnetic one. But, seven decades and the sequence of UEEC events reported in figure 17 were needed to understand the stability of matter. The 'charge' needed for the mass ≠ matter problem is not the 'gauge' charge, like the electromagnetic one, but a set of 'flavour' charges; and these are 12, as reported in figure 18 where it is also specified that there are three classes of 'masses'.

THE INCREDIBLE SERIES OF UEEC EVENTS
NEEDED TO EXPLAIN THE STABILITY OF MATTER
SEVEN DECADES: FROM THE ANTIELECTRON TO ANTIMATTER
AND THE UNIFICATION OF ALL GAUGE FORCES

- **The validity of C invariance from 1927 to 1957.**

After the discovery by Thomson in 1897 of the first example of an elementary particle, the Electron, it took the genius of Dirac to theoretically discover the Antielectron thirty years after Thomson.

1927 → Dirac equation [19]; the existence of the antielectron is, soon after, theoretically predicted. Only a few years were needed, after Dirac's theoretical discovery, to experimentally confirm (Anderson, Blackett and Occhialini [20]) the existence of the Dirac antielectron.

1930-1957 → **Discovery of the C operator**[(charge conjugation) H. Weyl and P.A.M. Dirac [21]]; discovery of the P Symmetry Operator [E.P. Wigner, G.C. Wick and A.S. Wightman [22, 23]]; discovery of the T operator (time reversal) [E.P. Wigner, J. Schwinger and J.S. Bell [24, 25, 26, 27]]; discovery of the CPT Symmetry Operator from RQFT (1955-57) [28].

1927-1957 → Validity of C invariance: e^+ [20]; $\bar{p}$ [29]; $\bar{n}$ [30]; $K^0_2 \to 3\pi$[31] but see LOY [32].

- **The new era starts: C ≠ ; P ≠ ; CP ≠ (*)**.

1956 → Lee & Yang P ≠ ; C ≠ [33].

1957 → Before the experimental discovery of P ≠ & C ≠, Lee, Oehme, Yang (LOY) [32] point out that the existence of the second neutral K-meson, $K^0_2 \to 3\pi$, is proof neither of C invariance nor of CP invariance. Flavour antiflavour mixing does not imply CP invariance.

1957 → C.S. Wu et al. P ≠ ; C ≠ [34]; CP ok [35].

1964 → $K^0_2 \to 2\pi \equiv K_L$: CP ≠ [36].

1947-1967 → QED divergences & Landau poles.

1950-1970 → The crisis of RQFT & the triumph of S-matrix theory (i.e. the negation of RQFT).

1965 → Nuclear antimatter is (experimentally) discovered [37]. See also [38].

1968 → The discovery [39] at SLAC of Scaling (free quarks inside a nucleon at very high q^2) but in violent (pp) collisions no free quarks at the ISR are experimentally found [40]. Theorists consider Scaling as being evidence for RQFT not to be able to describe the Physics of Strong Interactions. The only exception is G. 't Hooft who discovers in 1971 that the β-function has negative sign for non-Abelian theories [41].

1971-1973 → β = − ; 't Hooft; Politzer; Gross & Wilczek. The discovery of **non-Abelian** gauge theories. Asymptotic freedom in the interaction between quarks and gluons [41].

1974 → All gauge couplings $\alpha_1\alpha_2\alpha_3$ run with q^2 but they do not converge towards a unique point.

1979 → A.P. & A.Z. point out that the new degree of freedom due to SUSY allows the three couplings $\alpha_1\alpha_2\alpha_3$, **to converge towards a unique point**[42].

1980 → QCD has a 'hidden' side: the multitude of final states for each pair of interacting particles: (e^+e^- ; $p\bar{p}$; πp; Kp; νp; pp; etc.)
The introduction of the Effective Energy allows to discover the Universality properties [43] in the multihadronic final states.

1992 → All gauge couplings converge towards a unique point at the gauge unification energy: $E_{GU} \cong 10^{16}$ GeV with $\alpha_{GU} \cong 1/24$ [44, 45] .

1994 → The Gap [4] between E_{GU}& the String Unification Energy: $E_{SU} \cong E_{Planck}$.

1995 → **CPT loses its foundations at the Planck scale (T.D. Lee)** [46].

1995-1999 → **No CPT theorem from M-theory (B. Greene)** [47].

1995-2000 → A.Z. points out the need for new experiments to establish if matter-antimatter symmetry or asymmetry are at work.

(*) The symbol ≠ stands for 'Symmetry Breakdown'.

Figure 17

Mass $\neq$ Matter

$$|m_i\rangle \equiv \text{Mass} \equiv \text{Antimass} \equiv |\overline{m}_i\rangle$$

i ≡ 1 (Intrinsic); **i ≡ 2** (Confinement); **i ≡ 3** (Binding)

$$C\,|m_i\rangle = |m_i\rangle \qquad \boxed{***}$$

i ≡ 1, 2, 3

$$|m_i Q_j\rangle \equiv \text{Matter} \neq \text{Antimatter} \equiv |m_i \overline{Q}_j\rangle$$

Q_j ≡ Flavour Charges

$$\mathbf{j} \equiv (\mathbf{u}\ \ \mathbf{d}\ \ \mathbf{c}\ \ \mathbf{s}\ \ \mathbf{t}\ \ \mathbf{b}) \equiv (1, 2, 3, 4, 5, 6)$$

$$(\nu_e\ e^-\ \nu_\mu\ \mu^-\ \nu_{HL}\ HL^-) \equiv (7, 8, 9, 10, 11, 12)$$

$$\nu_{HL} \downarrow \nu_\tau \qquad HL^- \downarrow \tau^-$$

$$C\,|m_i Q_j\rangle = |m_i \overline{Q}_j\rangle \qquad \boxed{***}$$

i ≡ 1, 2, 3 ; J ≡ 1, 2, 3, 4, 5, 6, 7, 8, 9, 10, 11, 12.

Figure 18

ADDENDUM 3–2

THE PSEUDOSCALAR AND VECTOR MESONIC MIXINGS

The problem started when experimental physics was dominated by bubble chambers and the 'mixing' was determined using mass–formulae: i.e. a tautology. I designed and built a non-bubble-chamber detector, NBC; it consisted of an original neutron missing mass spectrometer coupled with a powerful electromagnetic detector which allowed to clearly identify all final states of the decaying mesons into (e^+e^-) or ($\gamma\gamma$) pairs. The mass of the meson (be it pseudoscalar or vector) was measured by the neutron missing mass spectrometer. The two 'mixing angles', the pseudoscalar θ_{PS} and the vector θ_V, where directly measured (without using the masses) to be, not as expected by $SU(3)_{uds}$, i.e. $\theta_{PS} = \theta_V = 0$, but, $\theta_{PS} \neq 0$, $\theta_V \neq 0$ and totally different $\theta_{PS} \neq \theta_V$. Many years were needed and Gerard 't Hooft instantons to explain why $\theta_{PS} \simeq 10°$ and $\theta_V \simeq 51°$(figure 19).

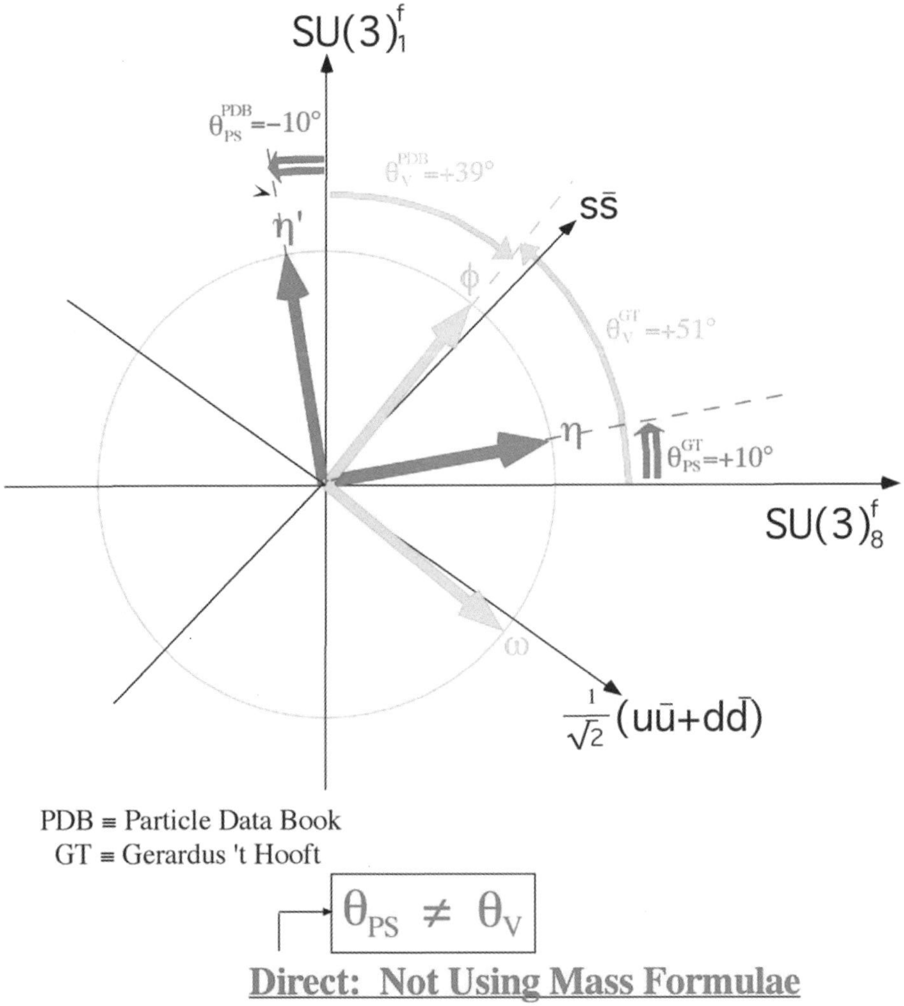

Figure 19

ADDENDUM 3–3

THE EFFECTIVE ENERGY AND THE GRIBOV QCD LIGHT

When the physics of strong interactions finally became the physics of quarks and gluons, QCD had a problem, defined by Gribov as being its 'hidden QCD side': i.e., the large number of different final states produced by different pairs of interacting particles, such as (πp, pp, $\bar{p}p$, Kp, e^+e^-, νp, μp, ep, etc.). I did not limit myself to suggesting that a totally different approach was needed to put all these final states on the same basis. I found what this basis could be and this is how the 'Effective Energy' became the correct quantity to be measured in each interaction. The 'Effective Energy' was not predicted by QCD. To perform this study, it was necessary to analyse tens of thousands of (pp) interactions at the ISR. This was done despite all the difficulties to be overcome. And this is how what Vladimir Gribov defined the 'QCD light' was discovered (figures 20 and 21). Gribov pointed out what follows. Newton discovered that QED light is the sum of different colours. In QCD we have quarks and gluons interacting and producing Jets made of many pions, as for example in the (pp) reaction

$$pp \rightarrow \pi + X$$

whose spectrum is shown in figure 20. The horizontal axis is for the fractional energy of the pion (also called Feynman x), while the vertical axis is for the number of pions having fractional energy x_F. The spectrum in figure 20 (page 496) is the sum (Σ) of all spectra shown if figure 21 (page 497) where each one corresponds to a single value of the 'effective energy' (defined in terms of $2E_{had}$).

$p\text{-}p \rightarrow \pi^+ + X$

Nominal Energy of the (pp) collision = $\sqrt{s}$= *24 GeV*

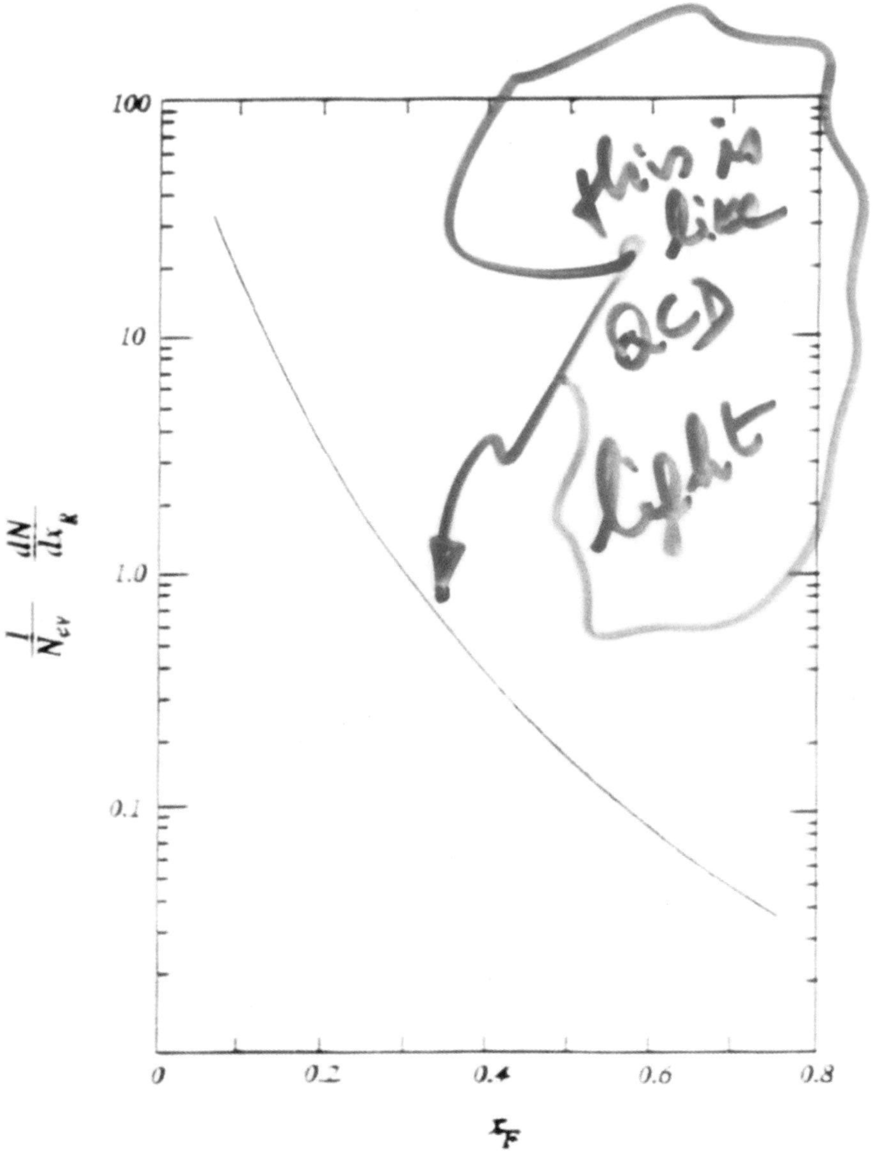

Figure 20

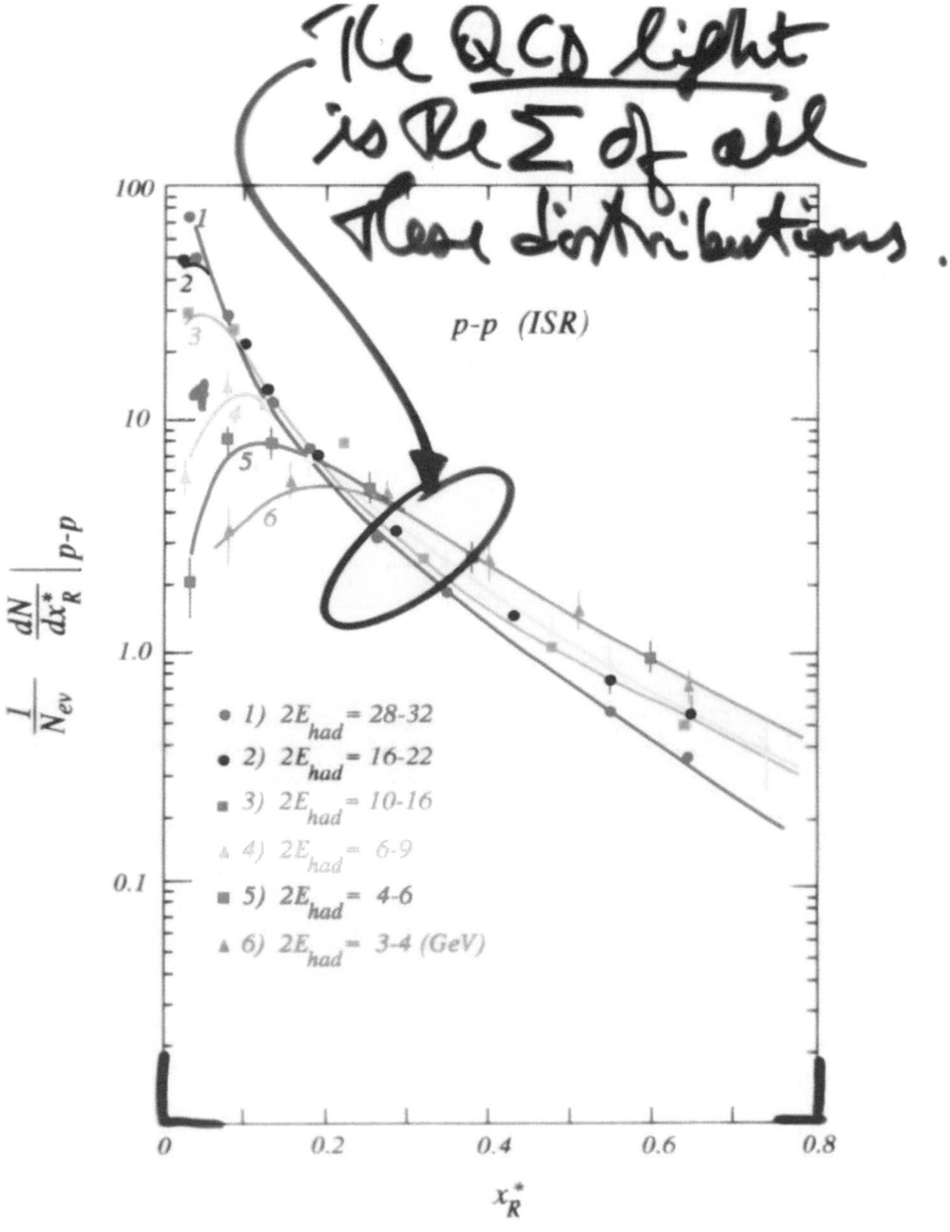

The QCD light is the Σ of all these distributions.
p-p (ISR)
100
10
1.0
0.1
$\frac{1}{N_{ev}} \frac{dN}{dx_R^*}\Big|_{p\text{-}p}$
1) $2E_{had}$ = 28-32
2) $2E_{had}$ = 16-22
3) $2E_{had}$ = 10-16
4) $2E_{had}$ = 6-9
5) $2E_{had}$ = 4-6
6) $2E_{had}$ = 3-4 (GeV)
0
0.2
0.4
0.6
0.8
x_R^*

Figure 21

ADDENDUM 3–4

THE GAP BETWEEN E_{GUT} AND E_{SU}

The exact use of the RGEs for the running of the three gauge couplings ($\alpha_1\alpha_2\alpha_3$) including the EGM (Evolution of Gaugino Mass) effects brings to the conclusion that the three gauge couplings ($\alpha_1\alpha_2\alpha_3$) converge at E_{GUT} which is two powers of ten below the String Unification Energy E_{SU}. This is shown in figure 22 (page 499) and the details in figure 23 (page 500).

The lines are the result of calculations executed with a supercomputer using a system of three weakly coupled differential non-linear equations:

$$\mu \frac{d\alpha_i}{d\mu} = \frac{b_i}{2\pi}\alpha_i^2 + \sum_j \frac{bij}{8\pi^2}\alpha_j\alpha_i^2 .$$

describing the evolution of all phenomena including the superworld, from the maximum level of energy (Planck scale) to our world at the minimum of energy.

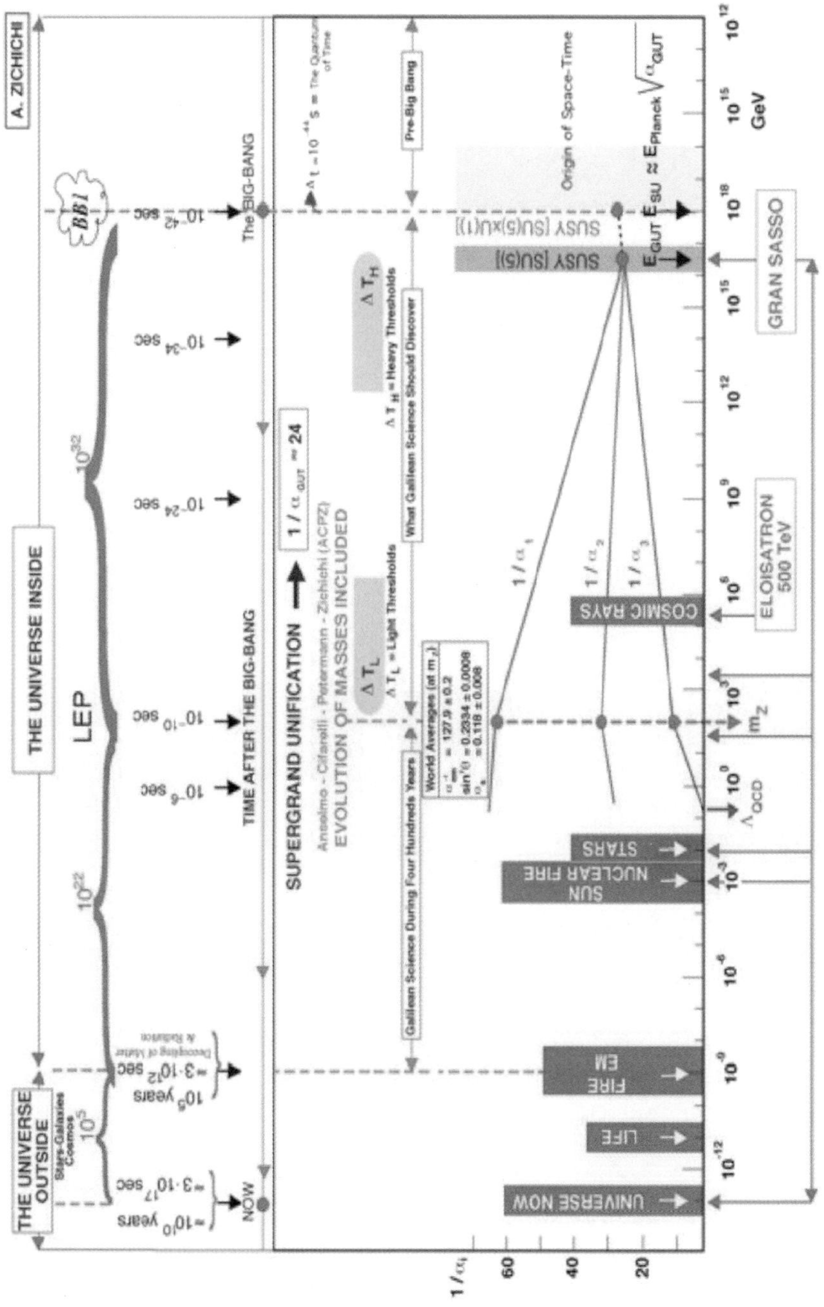
A. ZICHICHI
THE UNIVERSE INSIDE
THE UNIVERSE OUTSIDE
Stars-Galaxies Cosmos
LEP
TIME AFTER THE BIG-BANG
The BIG-BANG
NOW
SUPERGRAND UNIFICATION
EVOLUTION OF MASSES INCLUDED
Anselmo - Cifarelli - Petermann - Zichichi (ACPZ)
Galilean Science During Four Hundreds Years
What Galilean Science Should Discover
Pre-Big Bang
Origin of Space-Time
SUSY [SU(5)]
SUSY [SU(5)xU(1)]
UNIVERSE NOW
LIFE
FIRE EM
SUN NUCLEAR FIRE
STARS
COSMIC RAYS
ELOISATRON 500 TeV
GRAN SASSO
GeV

Figure 22

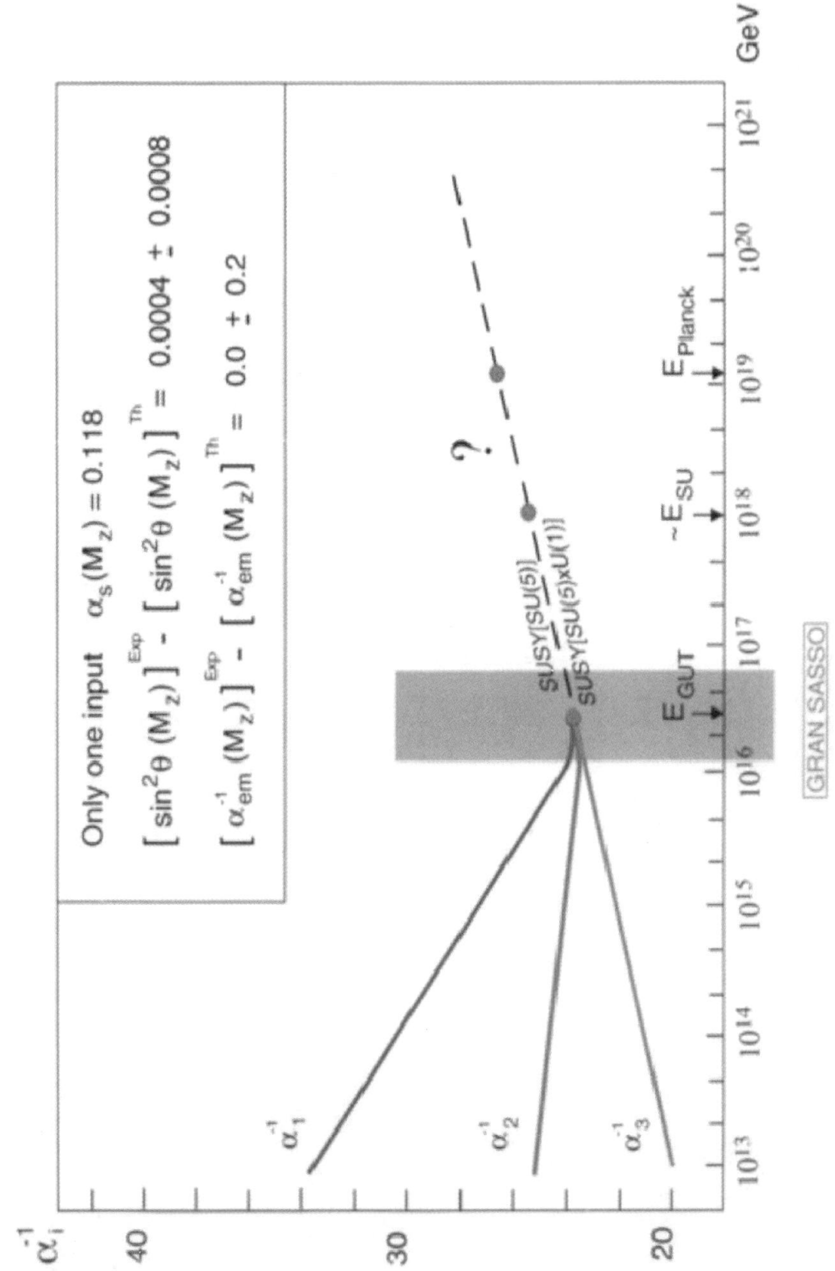

Only one input $\alpha_s(M_Z) = 0.118$
$[\sin^2\theta(M_Z)]^{Exp} - [\sin^2\theta(M_Z)]^{Th} = 0.0004 \pm 0.0008$
$[\alpha_{em}^{-1}(M_Z)]^{Exp} - [\alpha_{em}^{-1}(M_Z)]^{Th} = 0.0 \pm 0.2$
α_i^{-1}
40
30
20
α_1^{-1}
α_2^{-1}
α_3^{-1}
SUSY[SU(5)]
SUSY[SU(5)xU(1)]
?
E_{GUT}
$\sim E_{SU}$
E_{Planck}
10^{13}
10^{14}
10^{15}
10^{16}
10^{17}
10^{18}
10^{19}
10^{20}
10^{21}
GeV
GRAN SASSO

Figure 23

ADDENDUM 4

OPEN PROBLEMS IN SUBNUCLEAR PHYSICS

ADDENDUM 4–1

THE ORIGIN OF MIXINGS IN QUARK AND LEPTONIC STATES

- Is there an explanation for the flavour mixing mechanisms?
- Why do these mechanisms produce results that differ substantially in the quark and in the lepton sectors?
- Why does the mixing of states not exist in any other fundamental interaction?

ADDENDUM 4–2

THE ORIGIN AND STRUCTURE OF THE COMPOSITE STATES

- What is the role of instantons in the spectrum of hadrons in QCD? Where are the scalar hadronic states in QCD? Are there scalar states in the leptonic system?
- Why are there only fundamental particles with the minimum quark or lepton quantum numbers? Do elementary particles with higher, composite quantum numbers exist?
- Is there a fundamental reason why elementary fermions exist (quarks and leptons) but not elementary scalars in the same mass range?
- Do we really need sterile neutrinos? If yes, why? If not, why?

ADDENDUM 4–3

SYMMETRY LAWS AND THEIR BREAKING: SPONTANEUS AND DIRECT

- Why are the global discrete symmetries (C, P, CP, T) explicitly and not spontaneously broken, as it seems to be the case today?
- To what extent can we be confident that the Supersymmetry breaking threshold is not at the Planck Scale?
  (If this were the case it would be impossible to find at LHC any evidence for Supersymmetry).
- Are we really sure that the electro-weak symmetry breaking occurs at the Fermi scale due to a local maximum of the potential for some scalar field in the Lagrangian, also known as the "Higgs mechanism"?
  (If this were not the case, it would not be possible to find at LHC any evidence for Higgs particles).

ADDENDUM 5

THE TEN CHALLENGES OF SUBNUCLEAR PHYSICS

1 Non-perturbative QCD.

2 Anomalies and Instantons.

3 The Physics of NSSB (non-Spontaneous Symmetry Breaking: CP ≠, T ≠, CPT ≠)(*); Matter-Antimatter Symmetry.

4 The Physics of Imaginary Masses: SSB (part of this is the Higgs particle/particles).

5 The Physics of 43 dimensions (part of this is Supersymmetry).

6 Flavour mixing in the quark sector.

7 Flavour mixing in the leptonic sector.

8 The problem of the missing mass in the Universe.

9 The problem of the Hierarchy.

10 The Physics at the Planck scale and the number of expanded dimensions. Here the most interesting consequence would be that for a given expanded dimension it could be that the Planck scale goes down to the range of the Fermi scale, as illustrated in figure 24.

(*) The symbol ≠ means that a Symmetry law is non spontaneously broken as it happens with C, P, CP and T). [C (charge conjugation, i.e. interchange of charges with anti-charges); P (parity, i.e. interchange of left and right); T (inversion of the arrow of Time)]. The products CP and CPT mean the simultaneous Symmetry laws for all operations CP and CPT, respectively. The existence of Matter-Antimatter Asymmetry would be a proof of CPT ≠ .

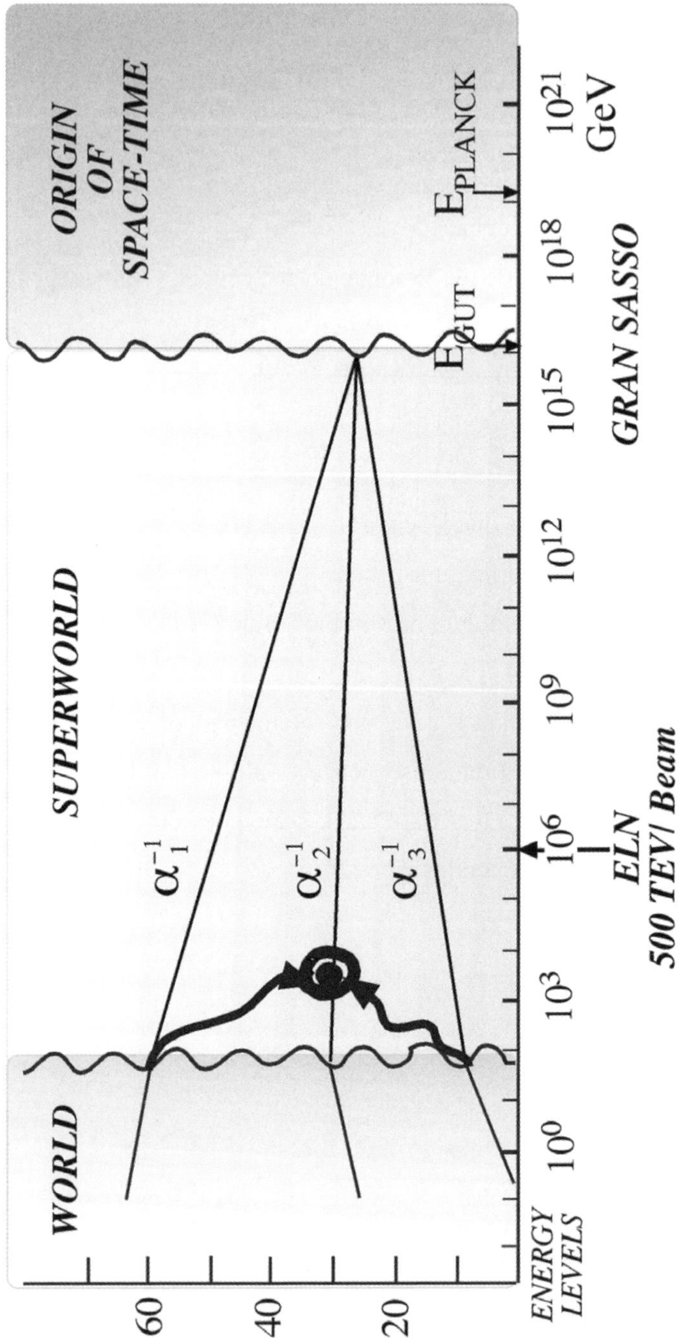
WORLD
SUPERWORLD
ORIGIN OF SPACE-TIME
α_1^{-1}
α_2^{-1}
α_3^{-1}
60
40
20
ENERGY LEVELS
10^0
10^3
10^6
10^9
10^{12}
10^{15}
10^{18}
10^{21}
GeV
E_{GUT}
E_{PLANCK}
GRAN SASSO
ELN
500 TEV/ Beam

Figure 24

REFERENCES

[1] *Interaction of Elementary Particles*
H. Yukawa, Part I, Proc. *Physico-Math. Soc. Japan* 17, 48 (1935);

Models and Methods in the Meson Theory
H. Yukawa, *Reviews of Modern Physics* 21, 474 (1949).

[2] *The QGCW Project*
A. Zichichi et al., CERN-LAA Preprint, October 2006; see also

Logical Reasoning in Experimental Physics: Past and Future
A. Zichichi, in *Gerardus 't Hooft Liber Amicorum to celebrate his 60th anniversary* (2006).

[3] *From the Yukawa Particle to the QGCW*
A. Zichichi, in Proceedings of the *'Symposium for the Centennial Celebration of Hideki Yukawa'*, International Nuclear Physics Conference, Tokyo, Japan, June 3-8 (2007).

[4] *A Study of the Various Approaches to M_{GUT} and α_{GUT}*
F. Anselmo, L. Cifarelli and A. Zichichi, *Nuovo Cimento* 105A, 1335 (1992).

[5] *Complexity exists at the Fundamental Level*
A. Zichichi in Proceedings of the 2004–Erice Subnuclear Physics School '*How and Where to Go Beyond the Standard Model*', World Scientific (2007) pages 251-333.

[6] *Virtual History: Alternatives and Counterfactuals*
Niall Ferguson ed., Basic Books, New York (1999).

The References [7–10] refer to the various occasions where I have presented papers on highly specialized topics and discussed the connection of these topics with Complexity. Each title on the upper part refers to the connection with Complexity while the specialized topic is reported in the detailed references.

[7] *Complexity at the Fundamental Level*
A. Zichichi
presented at:

- International Conference on 'Quantum [un]speakables' in Commemoration of John S. Bell, International Erwin Schrödinger Institut (ESI), Universität Wien (Austria), November 2000, '*John Bell and the Ten Challenges of Subnuclear Physics*'.

- 40th Course of the International School of Subnuclear Physics, Erice (Italy), September 2002, '*Language Logic and Science*'.
- 31st, 32nd and 33th Course of the International School of Solid State Physics, Erice (Italy), July 2004, '*Complexity at the Elementary Level*'.
- 42nd International School of Subnuclear Physics, Erice (Italy), August - September 2004, '*Complexity at the Elementary Level*'.
- Trinity College, Dublin (Ireland), February 2005, '*Complexity at the Elementary Level*'.
- Department of Physics, University of Padova (Italy), March 2005, '*Complexity at the Elementary Level*'.
- 43th Course of the International School of Subnuclear Physics, Erice (Italy), September 2005, '*Complexity at the Elementary Level*'.
- Italian Physics Society (SIF) XCI Annual National Congress, University of Catania (Italy), September 2005, '*Complexity at the Elementary Level*'.
- Desy, Hamburg, November 2005,'*Complexity at the Fundamental Level*'.
- 44th Course of the International School of Subnuclear Physics, Erice (Italy), September 2006, '*Complexity at the Fundamental Level*'.

[8] *The Logic of Nature and Complexity*
A. Zichichi
presented at:

- Pontificia Academia Scientiarum, The Vatican, Rome (Italy), November 2002, '*Scientific Culture and the Ten Statements of John Paul II*'; '*Elements of Rigour in the Theory of Evolution*'.
- The joint Session of:
  6th Course of the International School of Biological Magnetic Resonance; Erice (Italy), July 2003, '*Language Logic and Science*'.
- 2nd Workshop on Science and Religion of the Advanced School of History of Physics; Erice (Italy), July 2003, '*Language Logic and Science*'.
- 10th Workshop of the International School of Liquid Crystals; Erice (Italy), July 2003, '*Language Logic and Science*'.
- International School on Complexity, 1st Workshop on Minimal Life, Erice (Italy), December 2004, '*Evolution and Complexity at the Elementary Level*'.

[9] *Complexity and New Physics*
A. Zichichi
presented at:
- INFN-Alice Meeting, University of Catania (Italy), January 2005, '*Complexity at the Elementary Level*'.
- INFN Eloisatron Project 'The 1st Physics ALICE Week', Erice (Italy), December 2005, '*Complexity and New Physics with ALICE*'.
- 50th Anniversary of INFN Bologna - ALICE Week, Bologna (Italy), June 2006, '*Complexity at the Fundamental Level*'.

[10] *Complexity and Planetary Emergencies*
A. Zichichi
presented at:
- 27th Sessions of the International Seminars on Planetary Emergencies, Erice (Italy), August 2002, '*Language, Logic and Science*'.
- 28th Sessions of the International Seminars on Planetary Emergencies, Erice (Italy), August 2003, '*Language Logic and Science, Evolution and Planetary Emergencies*'.
- 36th Sessions of the International Seminars on Planetary Emergencies, Erice (Italy), August 2006, '*Complexity and Planetary Emergencies*'.

[11] *The Logic of Nature, Complexity and New Physics: From Quark-Gluon Plasma to Superstrings, Quantum Gravity and Beyond*
Proceedings of the 2006–Erice Subnuclear Physics School, World Scientific.

[12] *Language, Logic and Science*
A. Zichichi, proceedings of the 26th Session of the International Seminar on Nuclear War and Planetary Emergencies, Erice 18-26 August 2002 (World Scientific, 2003).

[13] *The Landscape and its Physics Foundations*
How String Theory Generates the Landscape
L. Susskind
Proceedings of the 2006–Erice Subnuclear Physics School, World Scientific.

[14] *Complexity and Landscape in String Theory*
M.R. Douglas and F. Denef
Proceedings of the 2006–Erice Subnuclear Physics School, World Scientific.

[15] Harald Klingbeil
GSI, private communication.

[16] RD12 collaboration: *Timing, Trigger and Control Systems (TTC) for LHC Detectors, spokesman*
Bruce G. Taylor 1992 – 2000 CERN/ Preprint 1999-12-01
Further references RD12
DRDC/90-62/P15, DRDC/91-22/P15 Add.1, DRDC/92-10/SR, DRDC/93-22/SR, DRDC/94-16/SR, LHCC/95-26/SR, LHCC/97-29/LEB/SR, LHCC/98-38/LEB/SR LHCC 2000-002/LEB/SR.

[17] *Yellow Book*
LHC Conceptual Design CERN/AC/95-05 (LHC), (October 1995).

[18] *Trigger, Timing and Control (TTC) Systems for the LHC*
LHC parameters and commissioning plan
Paul Collier/ CERN AB http://ttc.web.cern.ch/TTC/intro.html.

[19] P.A.M. Dirac, '*The Quantum Theory of the Electron*', *Proc. Roy. Soc.* (London) A117, 610 (1928); '*The Quantum Theory of the Electron, Part II*', *Proc. Roy. Soc.* (London) A118, 351 (1928).

[20] *The Positive Electron*
C.D. Anderson, *Phys. Rev.*43, 491 (1933);

Some Photographs of the Tracks of Penetrating Radiation
P.M.S. Blackett and G.P.S. Occhialini, *Proc. Roy. Soc.*A139, 699 (1933).

[21] H. Weyl, '*Gruppentheorie und Quantenmechanik*', 2nd ed., 234 (1931).

[22] E.P. Wigner, '*Unitary Representations of the Inhomogeneous Lorentz Group*', *Ann. Math.*, 40, 149 (1939).

[23] G.C. Wick, E.P. Wigner, and A.S. Wightman, '*Intrinsic Parity of Elementary Particles*', *Phys. Rev.*88, 101 (1952).

[24] E.P. Wigner, '*Über die Operation der Zeitumkehr in der Quantenmechanik*', *Gött. Nach.* 546-559 (1931). Here for the first time an anti-unitary symmetry appears.

[25] E.P. Wigner, *Ann. Math.*40, 149 (1939).

[26] J. Schwinger, *Phys. Rev.*82, 914 (1951).

[27] J.S. Bell, '*Time Reversal in Field Theory*', *Proc. Roy. Soc.* (London) A231, 479-495 (1955).

[28] To the best of my knowledge, the CPT theorem was first proved by W. Pauli in his article '*Exclusion Principle, Lorentz Group and Reflection of Space-Time and Charge*', in 'Niels Bohr and the Development of Physics' [Pergamon Press, London, page 30 (1955)], which in turn is an extension of the work of J. Schwinger [*Phys. Rev.* 82, 914 (1951); '*The Theory of Quantized Fields. II.*', *Phys. Rev.*91, 713 (1953); '*The Theory of Quantized Fields. III.*', *Phys. Rev.*91, 728 (1953); '*The Theory of Quantized Fields. VI.*', *Phys. Rev.* 94, 1362 (1954)] and G. Lüders, '*On the Equivalence of Invariance under Time Reversal and under Particle-Anti-particle Conjugation for Relativistic Field Theories*' [*Dansk. Mat. Fys. Medd.*28, 5 (1954)], which referred to an unpublished remark by B. Zumino. The final contribution to the CPT theorem was given by R. Jost, in '*Eine Bemerkung zum CPT Theorem*' [*Helv. Phys. Acta*30, 409 (1957)], who showed that a weaker condition, called 'weak local commutativity' was sufficient for the validity of the CPT theorem.

[29] *Observation of Antiprotons*
O. Chamberlain, E. Segrè, C. Wiegand, and T. Ypsilantis, *Physical Review*100, 947 (1955).

[30] *Anti-Neutrons Produced from Anti-Protons in Charge Exchange Collisions*
B. Cork, G.R. Lambertson, O. Piccioni, W.A. Wenzel, *Physical Review*104, 1193 (1957).

[31] *Observation of Long-Lived Neutral V Particles*
K. Lande, E.T. Booth, J. Impeduglia, L.M. Lederman, and W. Chinowski, *Physical Review*103, 1901 (1956).

[32] *Remarks on Possible Noninvariance under Time Reversal and Charge Conjugation*
T.D. Lee, R. Oehme, and C.N. Yang, *Physical Review*106, 340 (1957).

[33] *Question of Parity Conservation in Weak Interactions*
T.D. Lee and C.N. Yang, *Phys. Rev.*104, 254 (1956).

[34] *Experimental Test of Parity Conservation in Beta Decay*
C.S. Wu, E. Ambler, R.W. Hayward, D.D. Hoppes, *Phys. Rev.*105, 1413 (1957);

Observation of the Failure of Conservation of Parity and Charge Conjugation in Meson Decays: The Magnetic Moment of the Free Muon
R. Garwin, L. Lederman, and M. Weinrich, *Phys. Rev.*105, 1415 (1957);

Nuclear Emulsion Evidence for Parity Non-Conservation in the Decay Chain $\pi^+\mu^+e^+$
J.I. Friedman and V.L. Telegdi, *Phys. Rev.*105, 1681 (1957).

[35] *On the Conservation Laws for Weak Interactions*
L.D. Landau, *Zh. Éksp. Teor. Fiz.*32, 405 (1957).

[36] *Evidence for the 2π Decay of the K_2^0 Meson*
J. Christenson, J.W. Cronin, V.L. Fitch, and R. Turlay, *Physical Review Letters* 113, 138 (1964).

[37] *Experimental Observation of Antideuteron Production*
T. Massam, Th. Muller, B. Righini, M. Schneegans, and A. Zichichi, *Nuovo Cimento* 39, 10 (1965).

[38] *The Discovery of Nuclear Antimatter*
L. Maiani and R.A. Ricci (eds), Conference Proceedings 53, Italian Physical Society, Bologna, Italy (1995); see also A. Zichichi in '*Subnuclear Physics - The first fifty years*', O. Barnabei, P. Pupillo and F. Roversi Monaco (eds), a joint publication by University and Academy of Sciences of Bologna, Italy (1998); World Scientific Series in 20th Century Physics, Vol. 24 (2000).

[39] The first report on 'scaling' was presented by J.I. Friedman at the 14th International Conference on *High Energy Physics* in Vienna, 28 August-5 September 1968. The report was presented as paper n. 563 but not published in the Conference Proceedings. It was published as a SLAC preprint. The SLAC data on scaling were included in the Panofsky general report to the Conference where he says «... the apparent success of the parametrization of the cross-sections in the variable ν/q^2 in addition to the large cross-section itself is at least indicative that point-like interactions are becoming involved». '*Low q^2 Electrodynamics, Elastic and Inelastic Electron (and Muon) Scattering*', W.K.H. Panofsky in Proceedings of 14th International Conference on *High Energy Physics* in Vienna 1968, J. Prentki and J. Steinberger (eds), page 23, published by CERN (1968). The following physicists participated in the inelastic electron scattering experiments: W.B. Atwood, E. Bloom, A. Bodek, M. Breidenbach, G. Buschhorn, R. Cottrell, D. Coward, H. DeStaebler, R. Ditzler, J. Drees, J. Elias, G. Hartmann, C. Jordan, M. Mestayer, G. Miller, L. Mo, H. Piel, J. Poucher, C. Prescott, M. Riordan, L. Rochester, D. Sherden, M. Sogard, S. Stein, D. Trines, and R. Verdier. For additional acknowledgements see J.I. Friedman, H.W. Kendall and R.E. Taylor, '*Deep Inelastic Scattering: Acknowledgements*', *Les Prix Nobel 1990*, (Almqvist and Wiksell, Stockholm/Uppsala 1991), also *Rev. Mod. Phys.*63, 629 (1991). For a detailed reconstruction of the events see J.I. Friedman '*Deep Inelastic Scattering Evidence for the Reality of Quarks*' in '*History of Original Ideas and Basic Discoveries in Particle Physics*', H.B. Newman and T. Ypsilantis (eds), Plenum Press, New York and London, 725 (1994).

[40] *Quark Search at the ISR*
T. Massam and A. Zichichi, *CERN preprint*, June 1968;

Search for Fractionally Charged Particles Produced in Proton-Proton Collisions at the Highest ISR Energy
M. Basile, G. Cara Romeo, L. Cifarelli, P. Giusti, T. Massam, F. Palmonari, G. Valenti and A. Zichichi, *Nuovo Cimento* 40A, 41 (1977); and

A Search for quarks in the CERN SPS Neutrino Beam
M. Basile, G. Cara Romeo, L. Cifarelli, A. Contin, G. D'Alì, P. Giusti, T. Massam, F. Palmonari, G. Sartorelli, G. Valenti and A. Zichichi, *Nuovo Cimento* 45A, 281 (1978).

[41] A. Zichichi in '*Subnuclear Physics - The first fifty years*', O. Barnabei, P. Pupillo and F. Roversi Monaco (eds), a joint publication by University and Academy of Sciences of Bologna, Italy (1998); World Scientific Series in 20th Century Physics, Vol. 24 (2000).

[42] *New Developments in Elementary Particle Physics*
A. Zichichi, *Rivista del Nuovo Cimento* 2, n. 14, 1 (1979). The statement on page 2 of this paper, «*Unification of all forces needs first a Supersymmetry. This can be broken later, thus generating the sequence of the various forces of nature as we observe them*», was based on a work by A. Petermann and A. Zichichi in which the renormalization group running of the couplings using supersymmetry was studied with the result that the convergence of the three couplings improved. This work was not published, but perhaps known to a few. The statement quoted is the first instance in which it was pointed out that supersymmetry might play an important role in the convergence of the gauge couplings. In fact, the convergence of three straight lines ($\alpha_1^{-1}\alpha_2^{-1}\alpha_3^{-1}$) with a change in slope is guaranteed by Euclidean geometry, as long as the point where the slope changes is tuned appropriately. What is incorrect about the convergence of the couplings is that, with the initial conditions given by the LEP results, the change in slope needs to be at $M_{SUSY} \sim 1$ TeV as claimed by some authors not aware in 1991 of what was known in 1979 to A. Petermann and A. Zichichi.

[43] V.N. Gribov, G. 't Hooft, G. Veneziano and V.F. Weisskopf '*The Creation of Quantum ChromoDynamics and the Effective Energy*', L.N. Lipatov (ed), a joint publication by the University and the Academy of Sciences of Bologna, Italy (1998); World Scientific Series in 20th Century Physics, Vol. 25 (2000).

[44] *The Effective Experimental Constraints on M_{SUSY} and M_{GUT}*
F. Anselmo, L. Cifarelli, A. Petermann and A. Zichichi, *Nuovo Cimento* 104A, 1817 (1991).

[45] *The Simultaneous Evolution of Masses and Couplings: Consequences on Supersymmetry Spectra and Thresholds*
F. Anselmo, L. Cifarelli, A. Petermann and A. Zichichi, *Nuovo Cimento*105A, 1179 (1992).

[46] *Are Matter and Antimatter Symmetric?*
T.D. Lee, in Proceedings of the '*Symposium to celebrate the 30th anniversary of the Discovery of Nuclear Antimatter*', L. Maiani and R.A. Ricci (eds), Conference Proceedings 53, page 1, Italian Physical Society, Bologna, Italy (1995).

[47] *String Theory: the Basic Ideas*
B. Greene, Erice Lectures - Discussion 1999 in '*Basics and Highlights in Fundamental Physics*', A. Zichichi (ed), World Scientific (2001).

CHAIRMAN: A. ZICHICHI

Scientific Secretaries: D. Twigg, M. Zoller

DISCUSSION

- *Csànàd:*

As chaos is tiny events with enormous consequences, why do you say chaos is a bad definition of Complexity?

- *Zichichi:*

This is not my definition, but rather that of 't Hooft.

- *'t Hooft:*

Chaos is only one form of behavior that is sometimes complex. But that is somewhat different from what you might call Complexity such as the human brain or the extremely complex universe that we live in. There is really something much more delicate and complicated than what we see in chaos in nearly every example from classical mechanics, which shows instability that we characterize as chaos. I think in practice, what Prof. Zichichi means is something rather different. Even though we think the universe is governed by rather simple laws, its behavior is very complicated. This is something that we need to understand, so I think chaos is no substitute for Complexity. I also said that Complexity is rather loosely defined and often an excuse for sloppy thinking.

- *Carlucci:*

In your talk you showed a slide suggesting the possibility that Science may be NP-complete. How can we say this?

- *Zichichi:*

We had an interesting discussion here two years ago. It was proved by Michael R. Douglas and Frederik Denef that the best string-model constructed by Bousso and Polchinski was NP-complete. It was interesting for us to know that the most exact translation of string theory into a mathematical formalism was NP-complete. This does not mean that this would be the case for our final description of physics, nevertheless, if Science is NP-complete then we are lost. This is why Murray Gell-Mann has founded the Institute for Complexity in Santa Fe, to analyse possible solutions to this impossible problem. It is not proved that this is the case, but the problem is that people already say 'reductionism is over'. More precisely, they have been saying this for the last 10 years, but we keep silent. If we keep silent our physics will be demolished; it will no longer carry weight with the decision making fellows. To show that the reductionist approach is the only valid approach, we have to provide basic answers to fundamental questions. The most important thing is to prove that the experimental evidences for Complexity to exist are two and only two. Holism is not a new discovery. From the discovery of Language, through the rigorous logic of the construction of geometry, holism was going on, but no one understood anything until Galileo Galilei said: "*if you want to understand something, ask one question at a time*". This was the beginning of modern Science. Reductionism is only four hundred years old, while for ten thousand years holism has been going on. The question posed by my graph is *"should we go back to holism"*. I am arguing against this. We cannot remain silent, we must give strong answers to the argument.

- *Mager:*

You mentioned that Science might be NP-complete: a hard problem. Till today it is however unknown if P=NP. If this is proven to hold what would be the impact on physics?

- *Tannenbaum:*

What does NP-completeness mean?

- *Zichichi:*

In simple words, if a problem is NP-complete, this corresponds to the fact that the problem cannot be solved.

- *Karakostas:*

If we choose one definition of Complexity for a system under study, what would be the implications of a different outcome using another definition?

- *Zichichi:*

The outcome does not depend on the definition. There are about 70 definitions of Complexity of which 7 were chosen by me. The important point for me is the experimental evidence for the existence of Complexity. The answer is Anderson-Feynman-Beethoven phenomenology and Sarajevo-type events, UEEC (unexpected events with enormous consequences). These two experimental evidences appear at the most fundamental level of our knowledge. This can be seen in the comparison of Science and history. "The Science of Complexity" is a very serious matter. It is important to convince scientists, historians and the general public that reductionism is indeed to best choice we can make.

- *Marcozza:*

There is a definition of Complexity by Kolmogorov: "*Complexity is the minimal length of description needed to describe the macrostate of a system.*" What do you think about this definition?

- *Zichichi:*

This is a definition of Complexity in Language. My definition of Complexity in Science is that if there is a phenomenon where Anderson-Feynman-Beethoven phenomenology and UEEC events show up, then the phenomenon is complex.

- *'t Hooft:*

An experimental approach to Kolmogorov's definition is a competition for the shortest possible implementation of a phenomenon on a computer. If the shortest possible program that reproduces the effect on a computer is longer than the description of the effect itself then the phenomenon is complex, if the program is very short then the effect is simple.

- *Zichichi:*

This is not a reproducible experiment in the Galilean sense.

Special Sessions for New Talents

Patterns of flavour violation at the dawn of the LHC era

Maria Valentina Carlucci

Physik-Department, Technische Universität München,

James-Franck-Straße, D-85748 Garching, Germany

Abstract

The experimental success of the predictions of the Standard Model in the flavour sector suggests that New Physics should possess a highly non-generic flavour structure. Different approaches to this idea of Minimal Flavour Violation have been proposed and studied during the last years. At last, the LHC provides the possibility to test these patterns from different points of view, i.e. through the direct search of New Physics at the high energy frontier, and through the indirect constraints at the intensity frontier of precision measurements.

1 Introduction

Despite the Standard Model (SM) provides an experimentally successful description of particle interactions, there are well known reasons, from the divergence of the Higgs mass to the unexplained phenomena as the dark matter or the baryon-antibaryon asymmetry, to believe that it represents only the low-energy limit of a more general New Physics (NP) theory. On one hand, theoretical arguments based on a natural solution of the hierarchy problem suggest that the cutoff energy scale Λ should not exceed a few TeV. Another strategy to obtain clues about the value of Λ is to build effective non-renormalizable interactions that encode the presence of new degrees of freedom at high energy: they are suppressed by inverse powers of Λ, and they can be constrained by experiments. For example, $\Delta F = 1$ and $\Delta F = 2$ flavour-changing neutral current (FCNC) effective operators could in general induce large effects in processes that are not mediated by tree-level SM amplitudes, and the absence of evidence of sizable deviations in the very abundant and precise data in K and B physics leads to bounds above $10^2 - 10^3$ TeV for the cutoff scale. The large discrepancy between these two determinations of Λ is a manifestation of what in different specific NP frameworks (supersymmetry, technicolor, etc.) goes under the name of *flavour problem* [1]: it seems that NP presents the same flavour pattern as the SM, and this pattern has not been identified yet because the SM, which is the low-energy limit of the new theory, does not posses an exact flavour symmetry, but instead simply a flavour structure that could only be learnt from data. The formalization of this idea, i.e. the recognition or imposition of the SM flavour structure in a certain NP model, is know as *Minimal Flavour Violation* (MFV), and it has been developed and analyzed in different ways during the last years; here we present a review of the pragmatic approach called Constrained Minimal Flavour Violation (CMFV), and of a more formal approach that makes use of group methods and effective theories, and which at the same time allows more freedom.

The efforts performed during the last years in the theory and phenomenology of the flavour sector can be now finally supported from new data coming from the LHC, giving the possibility of obtaining more information about the flavour structure of the SM and beyond. Some of the most awaited measurements have represented another success for the SM: the CP-violating phase in the decay $B_s^0 \to J/\psi\ \phi$, previously suspected to show the presence of NP, has been found to be well consistent with the SM within current uncertainties [2], and the new constraints on the decays $B_{(s)}^0 \to \mu^+\mu^-$, of great interest because of their theoretical clearness and their large NP allowance, are now very close to the SM prediction as well [3]. Nevertheless, tensions of the order of 1-3σ that have emerged in the last years in flavour observables are still present, like the in the determination of β from $\epsilon_K - S_{\psi K_S}$ [4], in the determination of $|V_{ub}|$ [5], in the branching ratio of $B \to \tau\ \nu$ [4], in the like-sign dimuon charge asymmetry [6]. All these data draw a non-trivial picture of the situation of the SM and beyond, and the time of putting it in a corner through precision measurements and calculations seems to be very close.

2 Constrained Minimal Flavour Violation

CMFV can be seen as a brute-force method of extrapolating the flavour structure of the SM. It is defined by two conditions [7], [8]:

- all flavour changing transitions are governed by the CKM matrix with the CKM phase being the only source of CP violation;
- the only relevant operators in the effective Hamiltonian below the weak

scale are those that are also relevant in the SM.

2.1 Theory

The standard tool for the study of weak decays of hadrons is the Operator Product Expansion [9], which permits to derive an effective low energy theory for the weak interactions of quarks, characterized by a scale $\mathcal{O}(M_W)$, when they are bound by strong interactions of the much lower hadronic energy scale of $\mathcal{O}(1\text{ GeV})$. Using this technique the amplitude of a process can be calculated as

$$A(I \to F) = \langle F | \mathcal{H}_{\text{eff}} | I \rangle = \frac{G_F}{\sqrt{2}} \sum_i V^i_{\text{CKM}} C_i(\mu) \langle Q_i(\mu) \rangle \ , \qquad (1)$$

where Q_i are the local operators for the Dirac structures that contribute to the process and the CKM factors V^i_{CKM} and the Wilson coefficients $C_i(\mu)$ describe the strength with which they contribute. The scale μ separates the physics effects into short distance perturbative contributions, contained in $C_i(\mu)$, and the long distance contributions, contained in $\langle Q_i(\mu) \rangle$ and generally non-perturbative. This scale can be chosen arbitrarily, and it is standard to choose it of $\mathcal{O}(m_b)$ and $\mathcal{O}(1-2\text{ GeV})$ for B and K decays respectively. However, if the aim is to expose the short distance structure of flavour physics and in particular the NP contributions, it is much more useful to choose a scale $\mu_H \sim \mathcal{O}(M_W, m_t)$ as high as possible but still low enough so that below it the physics is fully described by the SM [10]. Hence the relevant Wilson coefficients are be obtained as

$$C_i(\mu) = \sum_j U_{ij}(\mu, \mu_H) C_j(\mu_H) \ , \qquad (2)$$

where $U_{ij}(\mu, \mu_H)$ are the elements of the renormalization group evolution matrix, and the coefficients $C_j(\mu_H)$ can be found in the process of matching the full and the effective theory, and they will be a linear combination of certain loop functions F_k:

$$C_j(\mu_H) = g_j + \sum_k h_{jk} F_k(m_t, \rho_{\text{NP}}); \qquad (3)$$

these functions will derive from the calculations of penguin and box diagrams containing the top quark and possible new heavy particles, and hence will depend on the parameters ρ_{NP} of the NP model; the other SM contributions are contained in the constant term. As a consequence, the process amplitude will take the form

$$A(I \to F) = P_0(I \to F) + \sum_k P_k(I \to F) F_k(m_t, \rho_{\rm NP}) \,, \tag{4}$$

where the coefficients P_i collect different contributions:

$$P_i(I \to F) \propto V^i_{\rm CKM} B_i \eta^i_{\rm QCD} \,, \tag{5}$$

where $V^i_{\rm CKM}$ denote the relevant combinations of elements of the flavour matrix, $\eta^i_{\rm QCD}$ stand symbolically for the renormalization group factors coming from $U_{ij}(\mu, \mu_H)$, B_i are non-perturbative parameters representing hadronic matrix elements $\langle Q_i(\mu) \rangle$.

The advantages of this approach, known as *penguin-box expansion* [11], become evident when one considers the properties of the contributions involved.

- In the SM, since the only source of flavour and CP violation is the mass matrix, that has been factored out, the master functions F_i are *universal* (i.e., process independent), and *real.*
- As there are no right-handed charged current interactions, in the SM only certain number of local operators is present, and hence only a particular set of parameters B_i is relevant. Similarly, if a careful treatment of the QCD corrections is performed, the factors $\eta^i_{\rm QCD}$ can be calculated within the SM independently from the choice of the operator basis in the effective weak Hamiltonian. In conclusion, the P_i coefficients are process dependent, but they depend only on the operator structure of the model.

As a consequence, the P_i are *model independent* within the whole class of CMFV, while the details of the single models are contained into the master functions F_i that preserve their universality and realness. Since the SM belongs itself to the class of CMFV, in these models the formulae of the observables will have the same form as in the SM with the only substitution $F_i(x_t) \to F_i(x_t, \rho_{\rm NP})$.

2.2 Phenomenology

2.2.1 Universal Unitarity Triangle

The analysis of the Unitarity Triangle (UT) provides a powerful test of the flavour pattern of CMFV in a model independent way; in fact, a triangle common to the whole class of CMFV models, known as *Universal Unitarity Triangle* (UUT) [7], can be built, and comparing it with the *Reference Unitarity Triangle* (RUT), can give information not only on the possible presence of NP, but also about its flavour structure (Fig. 1).

Once the parameters $\lambda \equiv |V_{us}|$ and $A = |V_{cb}|/\lambda^2$, unaffected by NP, have been determined, the determination of the apex $(\bar{\rho}, \bar{\eta})$ requires the knowledge of one side and one angle of the triangle, provided the CKM matrix is unitary. Two choices of sets with different characteristics are possible.

- **$\mathbf{R_b}$ and γ .** $R_b \propto |V_{ub}|$ is extracted by the inclusive and exclusive decays $B \to X_u \ell \bar{\nu}$, while $\gamma = \mathrm{Arg}(V_{ub})$ comes from $B^\pm \to D K^\pm$; these are all tree-level processes, very unlikely modified by new physics. The triangle obtained with this method is the RUT.

- **$\mathbf{R_t}$ and β .** With a very good approximation $R_t \propto |V_{td}/V_{ts}|$, while $\beta = \mathrm{Arg}(V_{td})$; the presence of the top quark implies that these can be only determined from loop processes. However, since in CMFV one simply has

$$\Delta M_{d,s} = \frac{G_F^2}{6\pi^2} \eta_B M_W^2 (\hat{B}_{B_{d,s}} F_{B_{d,s}}^2) m_{B_{d,s}} |V_{t(d,s)}|^2 F(x_t, \rho_{\mathrm{NP}}) \,, \qquad (6)$$

it is evident how the universality of the master function $F(x_t, \rho_{\mathrm{NP}})$ implies a model independent extraction of the ratio

$$\left| \frac{V_{td}}{V_{ts}} \right| = \xi \sqrt{\frac{m_{B_s}}{m_{B_d}}} \sqrt{\frac{\Delta M_d}{\Delta M_s}} \,, \qquad (7)$$

where ξ can be determined with non-perturbative methods and the other quantities are measurable. In an analogous way this ratio can also be extracted in for all the CMFV models from

$$\left| \frac{V_{td}}{V_{ts}} \right|^2 = \frac{\mathcal{B}(B_d \to \mu^+\mu^-)}{\mathcal{B}(B_s \to \mu^+\mu^-)} \,. \qquad (8)$$

On the other hand, the absence of new CP-violationg phases in the $B - \bar{B}$ mixing permits to write

$$\sin 2\beta = S_{\psi K_S} \qquad (9)$$

and hence allows to determine the angle β directly from the time-dependent asymmetry in the decay $B \to \psi K_S$. The triangle built in this way is common to all the models with CMFV and has been called UUT.

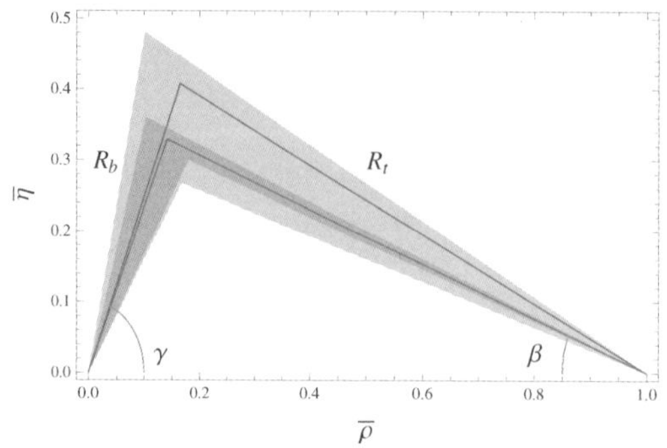

$\lvert V_{us}\rvert$	0.2252 ± 0.0009
$\lvert V_{cb}\rvert$	$(4.9 \pm 1.1) \times 10^{-3}$
$\lvert V_{ub}\rvert_{\mathrm{excl}}$	$(3.23 \pm 0.31) \times 10^{-3}$
$\lvert V_{ub}\rvert_{\mathrm{incl}}$	$(4.41 \pm 0.15^{+0.15}_{-0.19}) \times 10^{-3}$
γ	$(68^{+10}_{-11})^\circ$
ξ	1.237 ± 0.032
m_{B_d}	5279.58 ± 0.17 MeV
m_{B_s}	5366.77 ± 0.24 MeV
ΔM_d	$(3.337 \pm 0.033) \times 10^{-10}$ MeV
ΔM_s	$(116.4 \pm 0.5) \times 10^{-10}$ MeV
$S_{\psi K_S}$	0.668 ± 0.023

Figure 1: Comparison between the RUT (blue) and the UUT (red), obtained from the 2012 updated inputs [5], [12] in the table on the right. The shaded regions show the 1σ uncertainties.

2.2.2 Correlations and lower bounds

The flavour universality of the master functions F_i and the model independence of the parameters P_i suggest the idea of considering ratios between different observables in which at the same time the flavour pattern can be tested and the hadronic uncertainties are reduced [13]. Correlations like

$$\frac{\Delta M_d}{\Delta M_s} = \frac{M_{B_d}}{M_{B_s}} \frac{\hat{B}_d}{\hat{B}_s} \frac{F_{B_s}^2}{F_{B_s}^2} \left| \frac{V_{td}}{V_{ts}} \right|^2 r(\Delta M) \,, \tag{10a}$$

$$\frac{\mathcal{B}(B_s^0 \to \mu^+\mu^-)}{\mathcal{B}(B^0 \to \mu^+\mu^-)} = \frac{\tau(B_s)}{\tau(B_d)} \frac{M_{B_s}}{M_{Bs}} \frac{F_{B_d}^2}{F_{B_s}^2} \left| \frac{V_{td}}{V_{ts}} \right|^2 r(\mu^+\mu^-) \,, \tag{10b}$$

$$\frac{\mathcal{B}(B \to X_d \nu\bar{\nu})}{\mathcal{B}(B \to X_s \nu\bar{\nu})} = \left| \frac{V_{td}}{V_{ts}} \right|^2 r(\nu\bar{\nu}) \tag{10c}$$

have been indicated as *standard candles of flavour physics* [14], since in CMFV $r(\Delta M) = r(\mu^+\mu^-) = r(\nu\bar{\nu}) = 1$ and deviations from unity can be used to recognize and parametrize different patterns of flavour violation.

Less intuitive but still simple calculations permit also to find that the observables related to the meson mixing like ϵ_K and $\Delta M_{d,s}$ are not only correlated, but also can only be enhanced with respect to the SM [15].

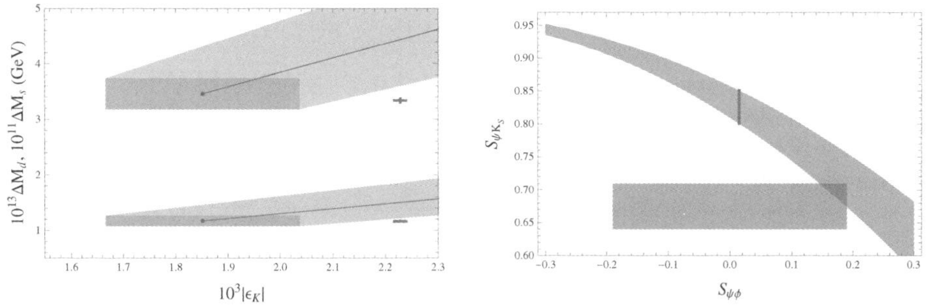

Figure 2: (Left) ΔM_d (red) and ΔM_s (red) as functions of ϵ_K in models with CMFV. The green crosses represent the data, while the darker regions the SM predictions. (Right) Correlation between $S_{\psi K_S}$ and $S_{\psi\phi}$ in the $2\text{HDM}_{\overline{\text{MFV}}}$, with a 2σ uncertainty in blue. SM is represented by the green line, while the red area is the 2σ experimental range.

2.3 Comparison with experiments

Since a large discrepancy is present between the inclusive and exclusive determination of the CKM matrix element V_{ub} [5], choosing one of the two values according to the impact on other observables can seems a criterion more meaningful than considering an average of the two. In the SM the exclusive measurement of V_{ub} implies the prediction for the asymmetry $S_{\psi K_S}$ to be in agreement with data but ϵ_K to be below data, while the inclusive brings ϵ_K in the right range at the price of an enhancement of $S_{\psi K_S}$.

Since $S_{\psi K_S}$ cannot receive new contributions in models with CMFV, in this framework the exclusive value of V_{ub} is preferred; on the other hand, as we have discussed, positive contributions to ϵ_K are allowed, and hence the $S_{\psi K_S} - \epsilon_K$ tension can be solved. Nevertheless, it has been shown that the enhancement of ϵ_K would determine a correlated enhancement of both ΔM_d and ΔM_s, which are already slightly above the experimental values [14] (Fig. 2, left).

The previous considerations, even if only qualitative, point out the difficulties that CMFV models have in accommodating the tensions in flavour data, due to the presence of few free parameters and strict correlations. More quantitative studies, as well as a complete analysis of more observables, could be already able to derive more conclusive statements about the viability of this flavour violation scheme.

3 Minimal Flavour Violation at Large

3.1 Effective field theory approach

The largest group of unitary quark field transformations that commutes with the SM gauge Lagrangian is

$$\mathcal{G}_q = (SU(3) \otimes U(1))^3 \ , \tag{11}$$

i. e. a $SU(3)$ symmetry and a phase symmetry for each electroweak multiplet:

$$SU(3)^3 = SU(3)_{Q_L} \otimes SU(3)_{U_R} \otimes SU(3)_{D_R} \ , \quad U(1)^3 = U(1)_B \otimes U(1)_Y \otimes U(1)_{\rm PQ} \ , \tag{12}$$

where the three $U(1)$ symmetries can be rearranged as the baryon number, the hypercharge, and the Peccei-Quinn symmetry. In the SM the Yukawa interactions break the flavour symmetry $\mathcal{G}_q$, but this can be formally recovered by promoting the Yukawa matrices to *spurions*, i. e. dimensionless auxiliary fields transforming as

$$Y_u \sim (3, \bar{3}, 1)_{SU(3)^3} \ , \qquad Y_d \sim (3, 1, \bar{3})_{SU(3)^3} \ . \tag{13}$$

One defines an effective theory as satisfying the criterion of MFV if all higher-dimensional operators, constructed from SM and spurion fields, are formally invariant under the flavour group $\mathcal{G}_q$ [16].

In practice, one can build effective couplings and higher-dimensional operators in which the only relevant non-diagonal structures are polynomials $\mathcal{P}(Y_u Y_u^\dagger, Y_d Y_d^\dagger)$ of the two basic spurions

$$Y_u Y_u^\dagger, \ Y_d Y_d^\dagger \sim (8, 1, 1)_{{\rm SU}(3)_q^3} \oplus (1, 1, 1)_{{\rm SU}(3)_q^3} \ . \tag{14}$$

As an example of this mechanism at work, we shortly discuss the application of this formulation of MFV to a generic model with two Higgs doublets, resulting in a NP scenario called $2{\rm HDM}_{\overline{\rm MFV}}$ [17].

3.2 The $2{\rm HDM}_{\overline{\rm MFV}}$

3.2.1 Imposing MFV

In a generic model with two-Higgs doublets, H_1 and H_2, with hypercharges $Y = 1/2$ and $Y = -1/2$ respectively, the most general renormalizable and gauge-invariant interaction of them with the SM quarks is

$$-\mathcal{L}_Y = \bar{Q}_L X_{d1} D_R H_1 + \bar{Q}_L X_{u1} U_R H_1^c + \bar{Q}_L X_{d2} D_R H_2^c + \bar{Q}_L X_{u2} U_R H_2 + {\rm h.c.} \ , \tag{15}$$

where $H^c_{1(2)} = -i\tau_2 H^*_{1(2)}$ and the X_i are 3×3 matrices with a generic flavour structure. By performing a global rotation of angle $\beta = \arctan(v_2/v_1)$ of the Higgs fields, the mass terms and the interaction terms are separated, but they cannot be diagonalized simultaneously for generic X_i and dangerous FCNC couplings to the neutral Higgses appear.

If the MFV hypothesis is imposed instead, the X_i are forced to assume a the particular structure

$$X_{d1} = Y_d \quad (16a)$$

$$X_{d2} = \mathcal{P}_{d2}(Y_u Y_u^\dagger, Y_d Y_d^\dagger) \times Y_d = \epsilon_0 Y_d + \epsilon_1 Y_d Y_d^\dagger Y_d + \epsilon_2 Y_u Y_u^\dagger Y_d + \dots \quad (16b)$$

$$X_{u1} = \mathcal{P}_{u1}(Y_u Y_u^\dagger, Y_d Y_d^\dagger) \times Y_u = \epsilon'_0 Y_u + \epsilon'_1 Y_u Y_u^\dagger Y_u + \epsilon'_2 Y_d Y_d^\dagger Y_u + \dots \quad (16c)$$

$$X_{u2} = Y_u \quad (16d)$$

that is renormalization group invariant [17]. At higher orders in $Y_i Y_i^\dagger$ FCNCs are generated, and in order to investigate them one can perform an expansion in powers of suppressed off-diagonal CKM elements, so that the effective down-type FCNC interaction can be written as

$$\mathcal{L}_{\text{MFV}}^{\text{FCNC}} \propto \bar{d}_L^i \left[\left(a_0 V^\dagger \lambda_u^2 V + a_1 V^\dagger \lambda_u^2 V \Delta + a_2 \Delta V^\dagger \lambda_u^2 V \right) \lambda_d \right]_{ij} d_R^j \frac{S_2 + i S_3}{\sqrt{2}} + \text{h.c.} \; , \quad (17)$$

where $\lambda_{u,d} \propto 1/v \ \text{diag}\,(m_{u,d}, m_{c,s}, m_{t,b})$, $\Delta = \text{diag}\,(0,0,1)$, and the a_i are parameters naturally of $\mathcal{O}(1)$; this structure shows a large suppression due to the presence of two off-diagonal CKM elements and the down-type Yukawas [17], demonstrating explicitly how MFV is effective and natural.

It is remarkable that the mechanisms of flavour and CP violation do not necessary need to be related: in MFV the Yukawa matrices are the only sources of flavour breaking, but other sources of CP violation could be present, provided that they are flavour-blind: this happens when the FCNC parameters a_i are allowed to be complex, as well as for the phases that can be present in the Higgs potential.

3.2.2 Comparison with experiments

With a more detailed analysis the following relevant properties have been found [17]:

- the impact in K, B and B_s mixing amplitudes scales with $m_s m_d$, $m_b m_d$ and $m_b m_s$ respectively;
- new flavour-blind phases can contribute to the B and B_s systems in the following way:

$$S_{\psi K_S} = \sin(2\beta + 2\phi_{B_d}) \; , \quad S_{\psi\phi} = \sin(2|\beta_s| - 2\phi_{B_s}) \; , \quad (18)$$

with $\phi_{B_s} = (m_s/m_d)\phi_{B_d}$; they are not present in the K system instead.

The previous observations imply that ϵ_K can receive only tiny new contributions while $S_{\psi K_S}$ could be in principle sizably modified; as a consequence, for the reasons that have been discussed in the previous section, the $2\mathrm{HDM}_{\overline{\mathrm{MFV}}}$ selects the inclusive $|V_{ub}|$. However, in this framework a suppression of $S_{\psi K_S}$ would determine a correlated enhancement of $S_{\psi\phi}$, an effect that was considered very welcome until last year, when LHCb put an end to the hopes of new physics in the B_s mixing phase, and that therefore puts this model in difficulty [14].

The flavour-blind phases of the Higgs potential imply instead $\phi_{B_s} = \phi_{B_d}$ [18], and could be used to remove the $S_{\psi K_S} - \epsilon_K$ anomaly, but the size of ϕ_{B_d} that is necessary would imply in turn $S_{\psi\phi} > 0.15$, which is 2σ away from the LHCb central value [14] (Fig. 2, right).

Acknowledgments

I would like to warmly thank Andrzej Buras for giving me the possibility to attend this school, Barbara Guerzoni for her availability and kindness, and Roberto Preghenella for the pleasant company during the days and the nights in Erice and Favignana. This work has been supported in part by the Graduiertenkolleg GRK 1054 of DFG.

References

[1] G. Isidori, Annales Henri Poincare **4** (2003) S97 [hep-ph/0301159].

[2] R. Aaij *et al.* [LHCb Collaboration], Phys. Rev. Lett. **108** (2012) 101803

[3] R. Aaij *et al.* [LHCb Collaboration], Phys. Rev. Lett. **108** (2012) 231801

[4] E. Lunghi and A. Soni, Phys. Lett. B **697** (2011) 323

[5] J. Beringer *et al.* [Particle Data Group Collaboration], Phys. Rev. D **86** (2012) 010001.

[6] V. M. Abazov *et al.* [D0 Collaboration], Phys. Rev. D **82** (2010) 032001

[7] A. J. Buras, P. Gambino, M. Gorbahn, S. Jager and L. Silvestrini, Phys. Lett. B **500** (2001) 161

[8] M. Blanke, A. J. Buras, D. Guadagnoli and C. Tarantino, JHEP **0610** (2006) 003

[9] K. G. Wilson, Phys. Rev. **179** (1969) 1499.

[10] A. J. Buras, Acta Phys. Polon. B **34** (2003) 5615 [hep-ph/0310208].

[11] G. Buchalla, A. J. Buras and M. K. Harlander, Nucl. Phys. B **349** (1991) 1.

[12] J. Laiho, E. Lunghi and R. S. Van de Water, Phys. Rev. D **81** (2010) 034503 [arXiv:0910.2928 [hep-ph]].

[13] A. J. Buras, Phys. Lett. B **566** (2003) 115 [hep-ph/0303060].

[14] A. J. Buras and J. Girrbach, Acta Physica Polonica B 43 (2012) 1427

[15] M. Blanke and A. J. Buras, JHEP **0705** (2007) 061

[16] G. D'Ambrosio, G. F. Giudice, G. Isidori and A. Strumia, Nucl. Phys. B **645** (2002) 155 [hep-ph/0207036].

[17] A. J. Buras, M. V. Carlucci, S. Gori and G. Isidori, JHEP **1010** (2010) 009 [arXiv:1005.5310 [hep-ph]].

[18] A. J. Buras, G. Isidori and P. Paradisi, Phys. Lett. B **694** (2011) 402 [arXiv:1007.5291 [hep-ph]].

Precise Measurement of the W Boson Mass with the DØ Detector

Rafael Lopes de Sá, for the DØ Collaboration

September 29, 2012

Abstract

We present a measurement of the W boson mass using DØ data collected from 2006 to 2009 corresponding to 4.3 fb^{-1} of integrated luminosity. This data set yields 1.68×10^6 $W \to e\nu$ candidates. We measure the mass using the transverse mass, electron transverse momentum, and missing transverse energy distributions. The transverse mass and electron transverse momentum measurements are the most precise and are combined to give

$$m_W = 80.367 \pm 0.012 \text{ (stat)} \pm 0.023 \text{ (syst) GeV} = 80.367 \pm 0.026 \text{ GeV}.$$

This is combined with an earlier DØ result determined using an independent Run II data sample which corresponds to 1 fb^{-1} to give

$$m_W = 80.375 \pm 0.023 \text{ GeV}.$$

1 Introduction

This note presents a measurement of the W boson mass using data collected from 2006 to 2009 with the DØ detector [1], corresponding to a total integrated luminosity of 4.3 fb^{-1}. This measurement is performed using the $W \to e\nu$ decay mode. Since the longitudinal momentum of the neutrino from the W decay is also unmeasurable, we use kinematic variables determined in the plane perpendicular to the proton beam direction.Three such variables are used in this measurement: the transverse mass, m_T, the electron transverse momentum, p_T^e, and the neutrino transverse momentum, p_T^ν. The transverse mass is defined as

$$m_T = \sqrt{2p_T^e p_T^\nu(1 - \cos(\phi_e - \phi_\nu))}, \tag{1}$$

where ϕ_e and ϕ_ν are the azimuthal angles of the electron and neutrino momentum, respectively. The $\vec{p}_T^{\,\nu}$ magnitude and direction are inferred from the event missing transverse momentum ($\vec{\not{E}}_T$).

The m_T and p_T^e measurements are complementary because the major cause of changes from the true to the measured spectrum for each quantity arises from different sources. For the m_T method, the major cause of the change is measurement resolution, while for the p_T^e method, the major cause is the W transverse momentum.

Because of complex detector acceptance and resolution effects, the distributions of the three measured variables cannot be calculated analytically. The measurement of m_W is obtained by a comparison of the spectra of the three different measurement variables with templates of the same variable distributions constructed from Monte Carlo (MC) simulation with varying input W masses. This requires templates with very high statistics ($\sim 10^9$ events) when evaluating systematic uncertainties to ensure that the statistical precision of the systematic uncertainty is negligible. To generate appropriate templates, a fast parametrized Monte Carlo simulation (PMCS) has been developed to produce large samples on a reasonable time scale and to provide a detailed description of the detector performance. We use $Z \to ee$ events to determine the parametrizations.Since the Z mass is known with a high precision from the LEP measurements [2], its value can be used to calibrate the electromagnetic calorimeter. Care must be taken to ensure that the calibrations at the Z pole are valid at the somewhat lower average energy of the electrons from W decay. Once this has been established, the W mass measurement is effectively a measurement of the ratio of W and Z masses.

2 Event Generation

The initial step in constructing templates for extracting the W mass is simulation of vector boson production and decay kinematics using the RESBOS [3] program coupled with the CTEQ6.6 parton distribution functions [4].

The W boson p_T spectrum has a significant impact on the simulated p_T^e and p_T^ν spectra so an accurate description of this is an important ingredient of the W mass measurement. RESBOS uses a next–to–leading order calculation for high W and Z boson p_T and a Collins–Soper–Sterman resummed cross section [5] at NNLL for low p_T. For large impact parameters, an extra non-perturbative form factor is introduced.

The non-perturbative form-factor in RESBOS uses the BLNY [6] form which defines a function of three parameters called g_1, g_2 and g_3. The values of these three parameters need to be determined from experimental data. The production of electroweak bosons at the Tevatron has limited sensitivity to g_1 and is quite insensitive to g_3, whereas the Z boson p_T distribution provides strong constraints on g_2. We use the global fit values listed in Ref. [6, 7] with $g_2 = 0.68 \pm 0.02$ GeV^2 and propagate the uncertainty on g_2 to the W mass measurement. The uncertainties from g_1 and g_3 are negligible.

3 Event Selection

The data used for this analysis were recorded in the period 2006-2009 and correspond to a total integrated luminosity of 4.3 fb^{-1} after data quality requirements.

The kinematic requirements for the W selection are:

- at least one electron reconstructed in the well-instrumented region of the central calorimeter(CC; $|\eta_{det}| < 1.05$, where η_{det} is the pseudorapidity measured with respect to the center of the detector), passing electron shower shape and energy isolation requirements,
- one track matching the electron with a match probability $P > 0.01$, having at least one silicon microstrip tracker hit and $p_T > 10$ GeV. The match probability is based on the difference in the η and ϕ values of the calorimeter cluster forming the electron and those of the track position at the calorimeter.
- electron $p_T > 25$ GeV,
- $\not{E}_T > 25$ GeV,
- $u_T < 15$ GeV and
- $50 < m_T < 200$ GeV.

Here u_T is the magnitude of the vector sum of the transverse component of the energies measured in calorimeter cells excluding those associated with the reconstructed electron.

The kinematic requirements for the Z selection are

- two electrons satisfying the calorimeter and track match requirements above. Both electrons must be reconstructed in the CC,
- both electrons have $p_T > 25$ GeV,
- $u_T < 15$ GeV and
- 70 GeV $\leq m_{ee} \leq$ 110 GeV.

Here m_{ee} is the invariant mass of the electron-positron pair. No events satisty both the W and Z selections.

These selections yield 1,677,394 candidate $W \to e\nu$ events and 54,512 candidate $Z \to ee$ events in which both electrons are in the CC. The W sample has a signal purity of 96%.

4 Parametrized Monte Carlo Simulation (PMCS)

The W mass is determined by comparison of data and simulation using distributions of m_T, p_T^e and $\not{E}_T$. The distributions, made using simulated events, require samples of 10^9 events, so the DØ standard full GEANT [11] simulation cannot be used. Instead PMCS was developed for this analysis. The electron trigger efficiency, reconstruction efficiency and energy response and resolution are simulated using parametric functions and binned look-up tables. The response and resolution of the hadronic recoil balancing the W boson transverse momentum is also simulated, again using a parametric model.

The initial input to the simulation uses parametrizations derived from detailed GEANT-based simulation of the DØ detector response to samples of single electrons, $Z \to ee$ events and $W \to e\nu$ events. This GEANT simulation does not describe the detector response with sufficient precision for the W mass measurement, so PMCS is fine-tuned using control data samples. The primary control sample is $Z \to ee$ events. In addition to the mean and width of the Z mass distribution, several other variables are used in the tuning. Among the additional variables are those first defined by UA2. The $\hat{\eta}$ unit vector (not to be confused with the pseudorapidity) is coincident with the bisector of the two electron directions in the transverse plane in $Z \to ee$ events (and $\hat{\xi}$ is perpendicular to $\hat{\eta}$). The $\vec{u}_T$ variable represents the hadronic recoil in W or Z production, and in W events, the component $u_{\|}$ is the projection onto the electron direction in the transverse plane. In Z events $u_{\|}$ can be defined using either electron.

4.1 Electron Response Simulation

The relationship between measured and true electron energy is given by

$$E = R_{EM}(E_0) \otimes \sigma_{EM}(E_0) + \Delta E(L, u_{\|}, \eta, \text{SET}) \tag{2}$$

where E_0 is the true electron energy, E is the reconstructed energy, $R_{EM}(E_0)$ is the response for a given E_0, σ_{EM} is the energy resolution for electromagnetic objects and ΔE is a correction for energy included in the reconstructed electron energy which is not related to the electron energy deposition. E_0 includes effects from photon radiation, including the loss of energy due to showers in uninstrumented material and the loss of photon energy from final state radiation at large angles to the electron. The ΔE correction is instantaneous luminosity (L) dependent and depends on the relative orientation of the electron and recoil. It is measured using both the energy observed in W candidates in a region separated from the electron by $|\Delta\eta| > 0.5$ and full simulation.

The increasing Tevatron instantaneous luminosity and improved precision of this result require several improvements to the electron energy determination. The main effect is corrected by a new model of the instantaneous luminosity dependence of the DØ calorimeter gains. The dependence of gain on instantaneous luminosity was predicted [12] before the start of Run II and has been studied in detail in the data used for this note. Though not lumionsity related, the increased W boson mass precision of this result also requires a refined η-dependent calibration of the calorimeter. These new electron energy corrections are

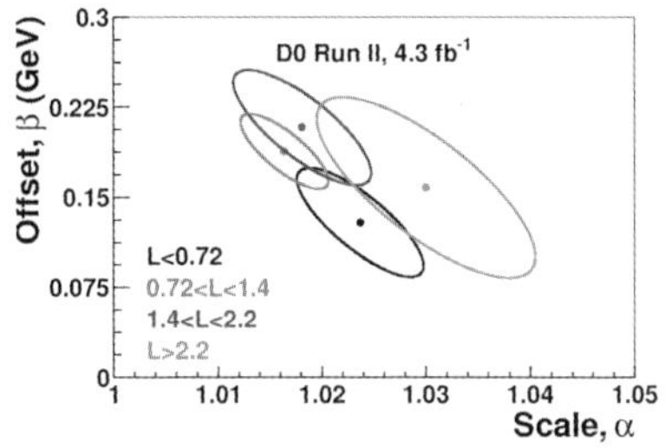

Figure 1: The results of the fits to the electron energy response determined in the four instantaneous luminosity ranges used in the analysis. The luminosities are given in units of $10^{32}\mathrm{cm}^{-2}\mathrm{s}^{-1}$.

applied to the data and affect the ΔE term in the response. The energy response of the EM calorimeter is modeled using

$$R_{EM}(E_0) = \alpha(E_0 - \overline{E}_0 \text{ GeV}) + \beta + \overline{E}_0 \text{ GeV} \tag{3}$$

where α is the response of the calorimeter to electrons and β is an offset. The constant $\overline{E}_0 \equiv 43$ GeV is the average electron p_T for Z decay, and it is used here to reduce the correlation between α and β and thus improves the numerical stability of the determination of these parameters. We determine α and β by fitting the Z mass distribution to that generated by PMCS for different α and β values. In obtaining these it is important that electrons from boosted Z's span a range of energies. In Z decay, when $\beta << (E_1 + E_2)$ with E_i the energies of the two electrons, the dielectron invariant mass after energy scale corrections can be expressed as [17]

$$m_{ee} \approx \alpha' m_{ee}^{(0)} + \beta' f_Z \tag{4}$$

with α' and β' the coefficients for the case $\overline{E}_0 = 0$ and $m_{ee}^{(0)}$ the measured dielectron mass before the energy scale corrections. The value of f_Z given by

$$f_Z = \frac{(E_1^{(0)} + E_2^{(0)})(1 - \cos\gamma)}{m_{ee}^{(0)}}. \tag{5}$$

with $E_i^{(0)}$ the energy before the scale correction and γ the angle between the two electrons. Thus β can be determined by a fit to the $m_{ee}^{(0)}$ versus f_Z two dimensional from $Z \to ee$ events. The fits use the exact expression for $m(\alpha, \beta, f_Z)$, not the approximation given above.

The data are divided into four instantaneous luminosity bins, and the values of α and β are determined for each bin. The fitted values are given in Fig. 1. The anticorrelation explains why the uncertainty on the W mass measurement from the electron energy calibration is much smaller than that of the β uncertainty alone.

The uncertainty on m_W arising from the α and β precision is determined by varying the parameters by the one sigma contour including the correlation and propagating the effect to the final W mass. The fit quality is good, so the α and β uncertainties are taken from the statistical uncertainties derived from the fit to $Z \to ee$ data. This is the dominant systematic uncertainty in the m_W measurement.

4.2 Recoil Response Simulation

The neutrino transverse momentum in W decay is inferred from the $\not{E}_T$. For data events, the $\not{E}_T$ is defined from the vector sum of the electron transverse momentum and the vector sum of the zero-suppressed calorimeter transverse energy deposition in each calorimeter cell excluding cells used to calculate the electron energy, cells in the coarse hadronic layer and cells in the inner cryostat detectors between the central and end calorimaters. It is modeled in the PMCS simulation as

$$-\vec{\not{E}}_T = \vec{p}_T^{\,e} + \vec{u}_T \tag{6}$$

in which $\vec{p}_T^{\,e}$ is the electron transverse momentum. The recoil, $\vec{u}_T$, is the vector sum of all transverse energy in the event not associated with the reconstructed electron. The recoil is modeled in PMCS using four components

$$\vec{u}_T = \vec{u}_T^{\,\mathrm{HARD}} + \vec{u}_T^{\,\mathrm{SOFT}} + \vec{u}_T^{\,\mathrm{ELEC}} + \vec{u}_T^{\,\mathrm{FSR}}. \tag{7}$$

These components are:

- $\vec{u}_T^{\,\mathrm{HARD}} = \vec{f}(\vec{q}_T; \vec{\zeta})$
  Here $\vec{u}_T^{\,\mathrm{HARD}}$ denotes the recoil component associated with the hard scattering of the $q\bar{q}$ pair that resulted in the $W(Z)$ boson. This hard component balances the transverse momentum of the vector boson. Here $\vec{q}_T$ is the generator level momentum of the boson. The $\vec{f}(\vec{q}_T)$ is an ansatz function used for smearing the $\vec{q}_T$.

- $\vec{u}_T^{\,\mathrm{SOFT}} = -\sqrt{\alpha_{MB}} \cdot \vec{\not{E}}_T^{\,\mathrm{MB}} - \vec{\not{E}}_T^{\,\mathrm{ZB}}$
  Here $\vec{u}_T^{\,\mathrm{SOFT}}$ represents the other interactions that contribute to the hadronic recoil. The first part is from the underlying event, and the second is the additional energy content in the event. The underlying event energy is defined to be that from the interactions of the remaining (spectator) partons of the same $p\bar{p}$ pair that produced the vector boson. It is modeled using data minimum bias (MB) events with only one reconstructed primary vertex. The additional energy content is associated with all the other interactions occurring in the $p\bar{p}$ pairs present in the same or previous beam crossings. Interactions contributing to detector noise are also included in this sub-component. It is modeled using ZB events.

- $\vec{u}_T^{\,\mathrm{ELEC}} = -\sum_e \Delta u_{\parallel} \cdot \hat{p}_T(e) + \vec{p}_T^{\,\mathrm{LEAK}}$
  The first term of this component gives a correction for the recoil energy that is parallel to the electron direction. The recoil energy present within the electron window is measured as part of the electron energy and is thus subtracted from the recoil energy.

This correction is determined in W data events using distributions of energy measured as described in Section 4.1. The second term accounts for energy from the electron that leaks outside the electron resconstruction window and thus appears in the recoil energy.

- $\vec{u}_T^{\,\mathrm{FSR}} = \sum_\gamma \vec{p}_T(\gamma)$
  This component contains the energy of final state radiation (FSR) photons (γ) that are far away from the electron and hence reconstructed as recoil energy.

These four components are derived using detailed GEANT simulation and data events. Because the simulation alone is not expected to reproduce the true hard recoil with sufficient precision, the recoil model is then further tuned using data $Z \to ee$ events. The sum of the recoil $\vec{u}_T$ and dielectron momentum $\vec{p}_T^{\,ee}$ vectors in Z events is projected on to the $\hat{\eta}$ and $\hat{\xi}$ axes defined earlier. The recoil model is then tuned using the imbalance η_{imb}. The η_{imb} mean is used to fine tune the recoil response, and the η_{imb} width is used to fine tune the recoil resolution in bins of p_T^{ee}. The tuning involves five parameters.

The systematic uncertainties arising from the recoil model are determined by propagating the uncertainties on the five tuning parameters derived from the fits. These uncertainties are dominated by the statistics of the Z control sample.

5 Mass Fits and Results

The W mass is determined by comparing each of the m_T, p_T^e and $\not{E}_T$ distributions from collider data to corresponding distributions generated using PMCS. The PMCS distributions (templates) are generated at a series of input W mass values with 10 MeV steps, with the backgrounds added to the simulated distributions. A binned likelihood between the data and each template is computed. The W mass is defined as the mass value corresponding to the minimum of the negative log likelihood. The measurements are performed separately for the m_T, p_T^e and $\not{E}_T$ distributions interpolated from the fixed mass fits.

A test of the analysis procedure was performed using events produced by the DØ GEANT Monte Carlo simulation treated as collider data. The methods used for the data analysis were applied to the simulated events in this Monte Carlo test, including performing the Z boson-based tuning using simulated events. For an input value $m_Z = 91.188$ GeV used in the tuning, the value returned from the post-tuning fit was 91.185 ± 0.004(stat). PMCS was separately tuned for this study to describe the full Monte Carlo. The differences between the fitted results and input W mass value of 80.450 GeV were $\Delta m_W = -0.002 \pm 0.005$(stat) GeV, $\Delta m_W = -0.002 \pm 0.005$(stat) GeV and $\Delta m_W = 0.005 \pm 0.006$(stat) GeV for the m_T, p_T^e and $\not{E}_T$ fits, respectively. The uncertainty arises from W statistics in the full Monte Carlo sample. No other sources of uncertainty were considered for this comparison.

During the PMCS tuning for the collider data analysis, the W mass returned from fits was blinded by the addition of an unknown constant offset until the analysis was finalized. This allowed the full tuning on the Z and W events and internal consistency checks to be performed without developing any knowledge of the final result. Once all distributions were of sufficient quality the results were unblinded. For an input value $m_Z = 91.188$ GeV used in the tuning, the value returned from the post-tuning fit was $91.193 \pm 0.017(\text{stat})$. After unblinding, the W mass fit results from data are given in Table 1.

Variable	Fit Range (GeV)	Result (GeV)	χ^2/d.o.f.
m_T	$65 < m_T < 90$	80.371 ± 0.013	37/49
p_T^e	$32 < p_T^e < 48$	80.343 ± 0.014	27/31
$\not{E}_T$	$32 < \not{E}_T < 48$	80.355 ± 0.015	29/31

Table 1: Results from the fits to data. The uncertainty is only that from the W sample statistics. The χ^2/d.o.f. values are computed over the fit range.

The distribution of the m_T variable showing the data and PMCS template with background for the best fit value is shown in Fig. 2. The figure also show the bin-by-bin χ values defined as the difference between the data and template divided by the data uncertainty.

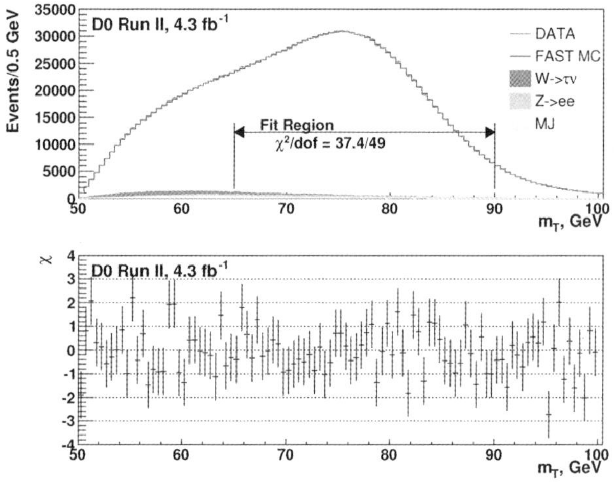

Figure 2: The m_T distribution for data and PMCS simulation with backgrounds added (top), and the χ value for each bin (bottom).

5.1 Systematic Uncertainties

The systematic uncertainties in the W mass measurement arise from a variety of sources, but can roughly be categorized as those arising from experimental sources and those arising from theory. The methods used to derive the systematic uncertainties have been described in the

Source	$\sigma(m_W)$ MeV m_T	$\sigma(m_W)$ MeV p_T^e	$\sigma(m_W)$ MeV $\not{E}_T$
Experimental			
Electron Energy Scale	16	17	16
Electron Energy Resolution	2	2	3
Electron Energy Nonlinearity	4	6	7
W and Z Electron energy loss differences	4	4	4
Recoil Model	5	6	14
Electron Efficiencies	1	3	5
Backgrounds	2	2	2
Experimental Total	18	20	24
W production and decay model			
PDF	11	11	14
QED	7	7	9
Boson p_T	2	5	2
W model Total	13	14	17
Total Systematic Uncertainty	22	24	29
W Statistics	13	14	15
Total Uncertainty	26	28	33

Table 2: Systematic uncertainties on the W mass results. The dominant systematic uncertainty comes from the electron energy scale, and this is determined by the statistical power of the Z event sample.

corresponding sections above. The systematic uncertainties are summarized in Table 2. The largest uncertainty, 16 MeV arises from the precision with which the electron energy scale is known. This is limited by the statistical power of the $Z \to ee$ sample, and it is expected to improve with more data.

5.2 Combination

The measurements from the three methods are correlated. Ensemble tests and standard uncertainty propagation methods are used to determine the correlation matrices for the individual uncertainties. The total correlation matrix including all uncertainties is

$$\rho = \begin{pmatrix} \rho_{m_T m_T} & \rho_{m_T p_T^e} & \rho_{m_T \not{E}_T} \\ \rho_{m_T p_T^e} & \rho_{p_T^e p_T^e} & \rho_{p_T^e \not{E}_T} \\ \rho_{m_T \not{E}_T} & \rho_{p_T^e \not{E}_T} & \rho_{\not{E}_T \not{E}_T} \end{pmatrix} = \begin{pmatrix} 1.0 & 0.89 & 0.86 \\ 0.89 & 1.0 & 0.75 \\ 0.86 & 0.75 & 1.0 \end{pmatrix} \quad (8)$$

The measurement based on met is of less weight than the other two measurements, and the correlations are large. We choose to combine just the mT and pTe measurements and obtain

$$m_W = 80.367 \pm 0.012 \text{ (stat)} \pm 0.022 \text{ (syst) GeV} = 80.367 \pm 0.026 \text{ GeV}.$$

The χ^2 probability of this combination is 2.8%. This result is combined with an earlier DØ measurement [16] to give the new DØ Run II result

$$m_W = 80.375 \pm 0.023 \text{ GeV}.$$

For the combination of this new measurement and the measurement in Ref. [16], the production model uncertainties are treated as 100% correlated between the two measurements, and all other uncertainties, driven by statistics, are assumed to be uncorrelated. The uncorrelated uncertainties are 0.022 GeV from this measurement and 0.041 GeV from Ref. [16]. The correlated uncertainties are 0.012 GeV (this measurement) and 0.013 GeV (Ref. [16]).

6 Conclusions

The W boson mass has been measured using the $W \to e\nu$ mode and 4.3 fb^{-1} of DØ data taken between 2006 and 2009. The mass was measured using three different kinematic variables m_T, p_T^e and $\not{E}_T$. The m_T and p_T^e measurements were combined to give the result

$$m_W = 80.367 \pm 0.012 \text{ (stat)} \pm 0.023 \text{ (syst) GeV} = 80.367 \pm 0.026 \text{ GeV}.$$

7 Acknowledgments

We thank the staffs at Fermilab and collaborating institutions, and acknowledge support from the DOE and NSF (USA); CEA, CNRS/IN2P3 and the CIMENT project, Grenoble (France); MON, Rosatom and RFBR (Russia); CNPq, FAPERJ, FAPESP and FUNDUNESP (Brazil); DAE and DST (India); Colciencias (Colombia); CONACyT (Mexico); NRF (Korea); NRF (South Africa); FOM (The Netherlands); STFC and the Royal Society (United Kingdom); MSMT and GACR (Czech Republic); BMBF and DFG (Germany); SFI (Ireland); The Swedish Research Council (Sweden); and CAS and CNSF (China).

References

[1] V.M. Abazov *et al.* (DØ Collaboration), Nucl. Instrum. Methods Phys. Res. A **565**, 463 (2006).

[2] C. Amsler et al., Physics Letters B **667**, 1 (2008) and references therein.

[3] C. Balazs and C.P. Yuan, Phys. Rev. D **56**, 5558 (1997).

[4] P. M. Nadolsky, *et al.*, Phys. Rev. D **78**, 013003 (2008).

[5] J.C. Collins, D.E. Soper, Nucl. Phys. **B**193, 381 (1981); **B** 213, 545E (1983); J.C. Collins, D.E. Soper, G. Sterman, ibid. **B**250, 199 (1985).

[6] F. Landry, R.Brock, P. M. Nadolsky, C.P. Yuan, Phys. Rev. D **67**, 073016 (2003);

[7] V. Abazov *et al.*, Phys. Rev. Lett. **100**, 102002 (2008).

[8] P. Golonka and Z. Was, Eur. Phys. J. **C45**, 97 (2006)

[9] H.L. Lai *et al.*, Phys. Rev. D **55**, 1280 (1997); J. Pumplin *et al.*, J. High Energy Phys. **10**, 046 (2003);

[10] T. Sjöstrand *et al.*, Comput. Phys. Commun. **135**, 238 (2001).

[11] R. Brun and F. Carminati, CERN Program Library Long Writeup W5013 (1993).

[12] H. Montgomery *et al.* (D0 Collaboration), FERMILAB-PUB-98-124-E, hep-ex/9804011 (1998).

[13] S. Jadach, J. H. Kuhn, and Z. Was, Comput. Phys. Commun. **64** (1990) 275; M. Jezabek, Z. Was, S. Jadach, and J. H. Kuhn, Comput. Phys. Commun. **70** (1992) 69; S. Jadach, Z. Was, R. Decker, and J. H. Kuhn, Comput. Phys. Commun. **76** (1993) 361-380.

[14] The Tevatron Electroweak Working Group, "Combination of CDF and DØ Results on the W Boson Mass and Width" FERMILAB-TM-2415 1st August 2008. More information is available at: http://tevewwg.fnal.gov.

[15] L. Lyons, D. Gibout, and P. Clifford, Nucl. Instrum. Methods in Phys. Res. A **270**, 110 (1988).

[16] V. M. Abazov *et al.* DØ Collaboration, Phys. Rev. Lett., **103**, 141801 (2009).

[17] V. M. Abazov *et al.* (D0 Collaboration), Phys. Rev. D **66**, 012001 (2002); B. Abbott *et al.* (D0 Collaboration), Phys. Rev. D **58**, 092003 (1998); B. Abbott *et al.* (D0 Collaboration), Phys. Rev. D **62**, 092006 (2000).

ITEP/TH-47/12

QFT and unification of knot theories

Alexey Sleptsov*

ITEP, Moscow, Russia

ABSTRACT

We discuss relation between knot theory and topological quantum field theory. Also it is considered a theory of superpolynomial invariants of knots which generalizes all other known theories of knot invariants. We discuss a possible generalization of topological quantum field theory with the help of superpolynomial invariants.

1 Introduction

At the present moment we know that topological QFT gives a physical description of knot theory and powerful tools to study knots and links. A great progress of 90's in this field is due to QFT consideration. At the same time knot theory does not give us something new about quantum field theory. Below we consider a new theory of knot invariants. We hope that exactly this theory can help us to find new structures in QFT.

2 Knots

We denote by $\mathcal{K}$ a knot, which is a mapping of S^1 into $\mathbb{R}^3$, $\mathcal{K} : S^1 \to \mathbb{R}^3$. On Figure 1 there are some examples of the simplest knots:

Figure 1: Examples: unknot, trefoil, eight-figure knot

Knots can be very complicated and tangled. The first question is to be asked how to distinguish them. In other words the problem is to construct invariants of knots, which are not changed under transformations preserving topology. If one considers a projection of a knot on a 2-d plane, these transformations can be reduced to compositions of the so-called Reidemeister moves. There are three types of them, which are depicted on Figure 2:

*email: sleptsov@itep.ru

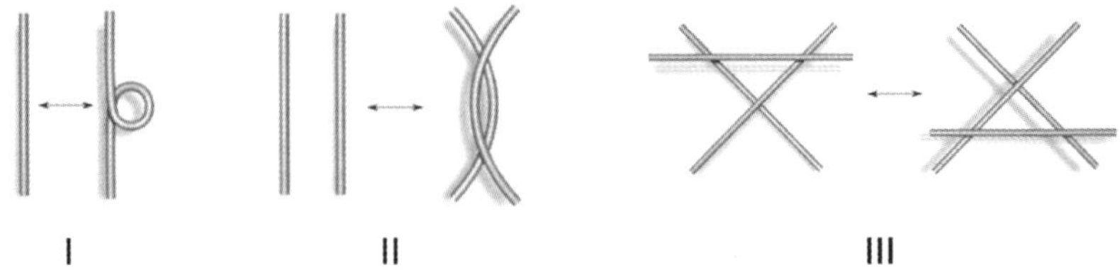

Figure 2: Three types of Reidemeister moves

Invariants also give us a classification of knots. Essentially, there are two types of invariants: numbers and polynomials. Number invariants was introduced by [1], they are known as Vassiliev invariants [2], [3]. In contrast to number invariants there are a lot of different types of polynomial invariants [4], we discuss it in the next section in details. In our below considerations we deal with polynomial invariants only. Note that there exists a relation between Vassiliev invariants and polynomial invariants [3, 2].

2.1 Torus knots

Torus knots are a special kind of knots embedded in the torus in $\mathbb{R}^3$. They are determined by two co-prime numbers $[m, n]$ (if m and n are not co-prime then it is a link. For simplicity we do not consider the case of links in this letter, because the generalization to the case of links is rather straightforward.). On Figure 3 it is given some particular examples of torus knots $T[m, n]$:

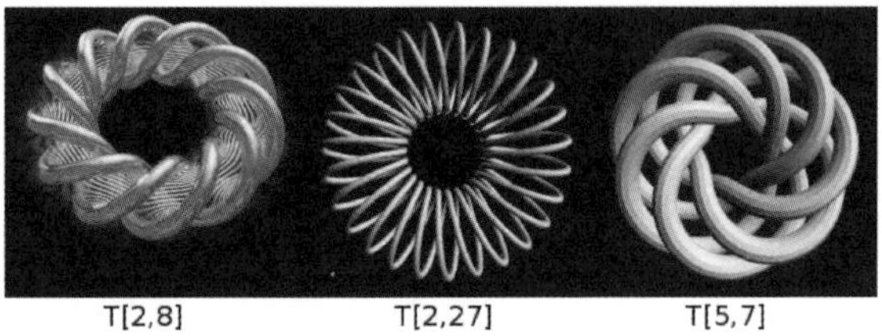

Figure 3: Examples of torus knots

For our needs it is convinient to divide torus knots and links into the following series $[m, mk + r]$, $k = 1, 2, 3, \ldots$:

$r = 0$	$[1, n]$	$[2, 2k]$	$[3, 3k]$	$[4, 4k]$	$[5, 5k]$	$[6, 6k]$	$\ldots$
$r = 1$		$[2, 2k + 1]$	$[3, 3k + 1]$	$[4, 4k + 1]$	$[5, 5k + 1]$	$[6, 6k + 1]$	$\ldots$
$r = 2$			$[3, 3k + 2]$	$[4, 4k + 2]$	$[5, 5k + 2]$	$[6, 6k + 2]$	$\ldots$
$r = 3$				$[4, 4k + 3]$	$[5, 5k + 3]$	$[6, 6k + 3]$	$\ldots$
$r = 4$					$[5, 5k + 4]$	$[6, 6k + 4]$	$\ldots$
$r = 5$						$[6, 6k + 5]$	$\ldots$
.	.	.	.	.	.	.	.

3 Various knot theories

The complete theory of knot invariants is not constructed yet. Moreover, there exist several various theories, which give different invariants of knots. Schematically, the net of various theories is given on Figure 4. The picture is not complete, we consider just some particular theories for illustrative purpose. These theories can be divided into two types: doubly graded theories, i.e. 2-variable polynomials, and 1-variable polynomials. Line stands by relation between two theories, however most of relations is implicit and has not been studied yet. Polynomial invariants in these theories are defined in mathematical way, for example, with the help of skein relations. From our point of view in this picture the central role is played by HOMFLY polynomials by the reason described in the next section.

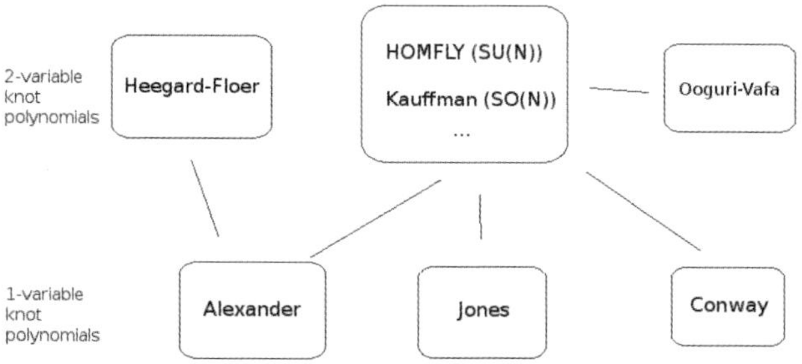

Figure 4: 1- and 2-graded theories

4 Topological quantum field theory (Chern-Simons theory)

In 1989 it was discovered by E.Witten [5] that quantum field theory can describe polynomial invariants of knots. For this reason one needs to consider topological QFT with the Chern-Simons action:

$$S_{CS} = \frac{g_{cs}}{4\pi} \int\limits_{\mathbb{R}^3} \mathrm{tr} \left(\mathcal{A} \wedge d\mathcal{A} + \frac{2}{3} \mathcal{A}^3 \right) \tag{1}$$

and the following vacuum expectation value

$$\langle W_R(\mathcal{K}) \rangle = \int \mathcal{D}\mathcal{A} \, \exp\{-2\pi \mathrm{i} \mathrm{S}_{\mathrm{CS}}\} \, \mathrm{W}_{\mathrm{R}}(\mathcal{K}), \tag{2}$$

where $\mathrm{W}_{\mathrm{R}}(\mathcal{K}) = \mathrm{tr}_{\mathrm{R}} \mathrm{Pexp}\{\oint_{\mathcal{K}} \mathcal{A}\}$. If we fix gauge in an appropriate way and take $SU(N)$ as the gauge group [6], then

$$\langle W_R(\mathcal{K}) \rangle (g_{cs}, N) = \mathrm{HOMFLY}(\mathrm{q}, \mathrm{A}), \; \mathrm{q} = \exp\left(\frac{2\pi \mathrm{i}}{\mathrm{g}_{\mathrm{cs}} + \mathrm{N}} \right), \; \mathrm{A} = \mathrm{q}^{\mathrm{N}}.$$

Thus HOMFLY polynomial invariants can be obtained as Wilson-loop averages in 3d Chern-Simons theory. Note that one can fix gauge by different ways. There are two distinguished gauges: holomorphic and temporal. Temporal gauge leads directly to polynomial invariants while holomorphic gauge leads to Kontsevich integral [7]. Explicit evaluation of HOMFLY polynomials from Kontsevich integral can be found in [8]. Anyway this approach gives us a physical description of knot invariants. And one of the problem is to extrapolate such description to other knot theories.

5 Unification

In 2005 it was suggested a hypothetical theory [9], which can unify all other known knot theories. It is a triply graded theory, which we call the theory of superpolynomial invariants (see a sketch on Figure 5). At the present moment there is no clear mathematical definitions of them. They are defined by a set of properties, which we consider in the next section. Since superpolynomials generalize HOMFLY polynomials, then it is natural to expect that there is a some new theory behind superpolynomials, which generalizes Chern-Simons theory. We hope that this theory can find out new aspects of quantum field theory and its generalizations. However at the present moment we are interested in exact formulas for superpolynomials, namely we have to construct superpolynomials and try to discover new structures behind them. Let us be more concrete.

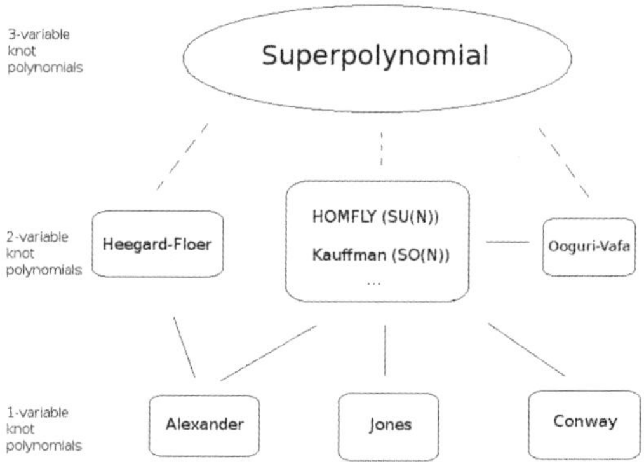

Figure 5: Unification of knot theories

6 Superpolynomials for torus knots

In this section we construct superpolynomials for a particular case of torus knots for fundamental representation. This section is based on papers [10] and [11]. First let us give a definition of superpolynomial invariant. Superpolynomial invariant $\mathcal{P}_{m,n}(A,q,t)$ of torus knot $T[m,n]$ is a polynomial in three variables A, q, t with the following properties:

1. all coefficients are positive;
2. $\mathcal{P}_{m,n}(A,q,t=q) = \text{HOMFLY(A, q)}$

These properties come from a lot of explicit calculations.

How to calculate superpolynomials? The point is to represent superpolynomial as a sum of MacDonald polynomials with appropriate coefficients over corresponding Young diagrams (about MacDonald polynomials see [12]). This ansatz comes from HOMFLY consideration [13], where one uses Schur polynomials $\chi_R\{p_k\}$ instead of MacDonald $M_R(\{p_j\},t,q)$. It is very natural because $M_R(\{p_j\},t,q)|_{t=q} = \chi_R\{p_j\}$ in agreement with property 2. Coefficients can be found from conditions 1 and 2. Also one should note that $\mathcal{P}_{m,n}(A,q,t) = \mathcal{P}_{m,n}(\{p_j\},t,q)|_{p_j=p_j^*}$, where $p_j^* = \frac{A^j-A^{-j}}{t^j-t^{-j}}$. Now let us give explicit formulas for superpolynomials taking into account that torus knots are given by $[m, mk+r]$:

$$\mathcal{P}_{m,r}\{p_j\} = \sum_{Q\vdash m} c_Q^{(m,r)} M_Q\{p_j\} = \sum_{\substack{Q\vdash m\\ l(Q)\le r}} h_Q^{(m,r)} L_Q\{p_j\} \tag{3}$$

$$h_Q = q^{\nu(Q)}(1-t)^{l(Q)-1}\hat{h}_Q$$

$$\boxed{\hat{h}_Q^{(m,1)} = \begin{cases} 1 & l(Q)=1 \\ 0 & \text{otherwise}\end{cases}} \quad \boxed{\hat{h}_Q^{(m,2)} = \begin{cases} 1 & l(Q)=1 \\ 1 & l(Q)=2 \\ 0 & \text{otherwise}\end{cases}} \tag{4}$$

$$\boxed{\hat{h}_Q^{(m,3)} = \begin{cases} 1 & l(Q)=1 \\ 1+t+(q-t)[\alpha]_q & l(Q)=2,3 \\ 0 & \text{otherwise}\end{cases}} \tag{5}$$

with $\alpha = \min(Q_1-Q_2, Q_2-Q_3)$, Q is a Young diagram, $\nu(Q)$ is a function on Q, $\nu(Q) = \sum_i Q_i(i-1)$. To restore k-dependence in formula (3) one should use evolution induced by cut-and-join operators [10]:

$$\boxed{\mathcal{P}_{m,mk+r}\{p_j\} = \sum_{Q\vdash m} c_Q^{(m,r)} q^{-k\nu(Q^T)} t^{k\nu(Q)} M_Q\{p_k\} = e^{k\hat{W}_{[2]}}\mathcal{P}_{m,r}\{p_j\}} \tag{6}$$

The last formula is in some sense a straightforward generalization of formula for HOMFLY polynomials for torus knots [13]. However we see that answers (3) are given in terms of Hall-Littlewood polynomials $L_R\{p_k\} = M_R(\{p_k\},t,q)|_{q=0}$. It is very surprising and it discovers new structures behind superpolynomials. Nevertheless it is not clear how to construct superpolynomials for higher representations and non-torus knots. We only begin to study unification of knot theories.

Acknowledgements

We are grateful to Petr Dunin-Barkowski, Andrei Mironov, Alexei Morozov, Shamil Shakirov and Andrey Smirnov for fruitful discussions and collaboration.

Our work is partly supported by Ministry of Education and Science of the Russian Federation under contract 8207, by RFBR grants 10-02-00509 and by joint grants 12-02-92108-Yaf-a, 11-01-92612-Royal Society.

References

[1] V.A.Vassiliev, Adv. Soviet Math., 1, 2369; Amer. Math. Soc., Providence, RI, 1990
M.Goussarov, Adv. Sov. Math., V.18, 1994
M.Kontsevich, Adv. Soviet Math. 16, Part 2, pp. 137-150, 1993

[2] P.Dunin-Barkowski, A.Sleptsov, A.Smirnov, arXiv:1112.5406, 2011

[3] S. Chmutov, S. Duzhin, J. Mostovoy, arXiv:1103.5628

[4] J.W.Alexander, Trans.Amer.Math.Soc. **30** (2) (1928) 275?306;
J.H.Conway, Algebraic Properties, In: John Leech (ed.), *Computational Problems in Abstract Algebra*, Proc. Conf. Oxford, 1967, Pergamon Press, Oxford-New York, 329-358, 1970;
V.F.R.Jones, Invent.Math. **72** (1983) 1 Bull.AMS **12** (1985) 103Ann.Math. **126** (1987) 335;
L.Kauffman,Topology **26** (1987) 395

[5] E.Witten, Comm.Math.Phys. **121** (1989) 351-399

[6] A.Morozov and A.Smirnov, the Nucl.Phys. **B835** (2010) 284-313, arXiv:1001.2003
A.Smirnov, hep-th/0910.5011, Proceedings of International School of Subnuclar Phys. in Erice, Italy, 2009

[7] J.M.F. Labastida, E. Perez, J.Math.Phys.39:5183-5198,1998, hep-th/9710176

[8] P.Dunin-Barkowski, A.Sleptsov, A.Smirnov, J. Phys. A: Math. Theor. 45 385204, 2012

[9] N.M.Dunfield, S.Gukov, and J.Rasmussen, math/0505662

[10] P.Dunin-Barkowski, A.Mironov, A.Morozov, A.Sleptsov, A.Smirnov, arxiv:1106.4305

[11] A.Mironov, A.Morozov, Sh.Shakirov and A.Sleptsov, JHEP, V.2012, N. 5 (2012), 70 arXiv:1201.3339

[12] I.G.Macdonald, Second Edition, Oxford University Press, 1995

[13] M. Rosso and V. F. R. Jones, J. Knot Theory Ramifications, **2** (1993) 97-112;
X.-S.Lin and H.Zheng, Trans. Amer. Math. Soc. **362** (2010) 1-18 math/0601267;
S.Stevan, Annales Henri Poincare, **11** (2010) 1201-1224, arXiv: 1003.2861;
A.Brini, B.Eynard and M.Mariño, arXiv:1105.2012

Hunting in Daya Bay Neutrino Experiment

ZHANG FeiHong
Institute of High Energy Physics, Beijing, P.R.China
E-mail: zhangfh@ihep.ac.cn

Abstract

The Daya Bay Reactor Neutrino Experiment has measured a nonzero value of θ_{13} with a significance of 7.7 standard deviation. Antineutrinos from six 2.9 GW_{th} reactors were detected in six well-calibrated antineutrino detectors deployed in two near (flux-weighted baseline 470 m and 576 m) and one far (1648 m) underground experimental halls. Using 139 days of data, 28909 (205308) electron antineutrino candidates were detected at the far hall (near halls). The ratio of the observed to expected numbers of antineutrinos at the far hall is $R = 0.944 \pm 0.007(\text{stat.}) \pm 0.003(\text{syst.})$. A rate-only analysis finds $sin^2 2\theta_{13} = 0.089 \pm 0.010(\text{stat.}) \pm 0.005(\text{syst.})$ in a three-neutrino framework.

1 Introduction

It is well established that the flavor of a neutrino oscillates with time. Neutrino oscillations can be described by the three mixing angles (θ_{12}, θ_{23}, and θ_{13}) and a phase of the PMNS matrix, and two mass-squared differences (Δm^2_{31} and Δm^2_{21}). Of these mixing angles, θ_{13} is the least known. Some experiments, such as CHOOZ, T2K, MINOS and Double CHOOZ, have indicated that θ_{13} may be nonzero.

For reactor-based experiments, an unambiguous determination of θ_{13} can be extracted via the survival probability of the electron antineutrino $\overline{\nu}_e$at short distances from the reactors,

$$P_{\text{sur}} \approx 1 - \sin^2 2\theta_{13} \sin^2(1.267\Delta m^2_{31} L/E)\,, \tag{1}$$

where $\Delta m^2_{31} = \Delta m^2_{32} \pm \Delta m^2_{21}$, and can be approximated by $\Delta m^2_{32} = (2.32^{+0.12}_{-0.08}) \times 10^{-3} eV^2$, E is the $\overline{\nu}_e$ energy in MeV and L is the distance in meters between the $\overline{\nu}_e$ source and the detector (baseline).

In March 2012, the Daya Bay collaboration first announced the discovery of a non-zero θ_{13} [1]. The near-far arrangement of antineutrino detectors (ADs), as illustrated in Fig. 1, allows for a relative measurement by comparing the observed $\overline{\nu}_e$ rates at various baselines. With functionally identical ADs, the relative rate is independent of correlated uncertainties and uncorrelated reactor uncertainties are minimized.

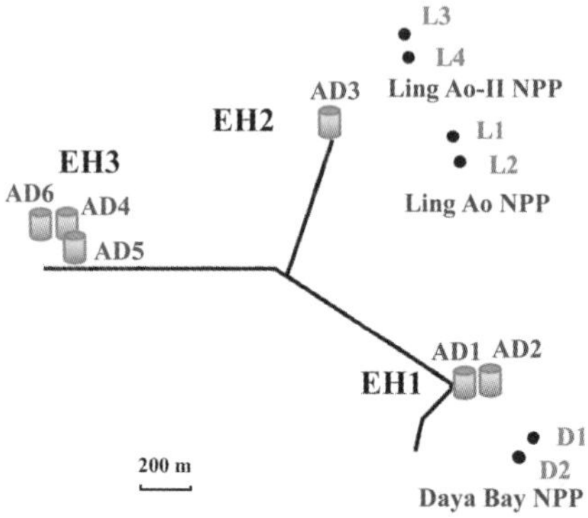

Figure 1: Layout of the Daya Bay experiment. The dots represent reactors, labeled as D1, D2, L1, L2, L3 and L4. Six ADs, AD1–AD6, are installed in three EHs.

2 Detectors

A detailed description of the Daya Bay experiment can be found in [2, 3]. The six pressurized water reactors are grouped into three pairs with each pair referred to as a nuclear power plant (NPP). The maximum thermal power of each reactor is 2.9 GW_{th}. Three underground experimental halls (EHs) are connected with horizontal tunnels. Two ADs are located in EH1 and one in EH2 (the near halls). Three ADs are positioned near the oscillation maximum in the far hall, EH3.

As shown in Fig. 2, the ADs in each EH are shielded with >2.5 m of high-purity water against ambient radiation in all directions. Each water pool is segmented into inner and outer water shields (IWS and OWS) and instrumented with photomultiplier tubes (PMTs) to function as Cherenkov-radiation detectors whose data were used by offline software to remove spallation neutrons and other cosmogenic backgrounds. The detection efficiency for long-track muons is >99.7% [2].

The $\overline{\nu}_e$ is detected via the inverse β-decay (IBD) reaction, $\overline{\nu}_e + p \rightarrow e^+ + n$, in a Gadolinium-doped liquid scintillator (Gd-LS). The coincidence of the prompt scintillation from the e^+ and the delayed neutron capture on Gd provides a distinctive $\overline{\nu}_e$ signature.

Each AD consists of a cylindrical, 5-m diameter stainless steel vessel (SSV) that houses two nested, UV-transparent acrylic cylindrical vessels. A 3.1-m diameter inner acrylic vessel (IAV) holds 20-t of Gd-LS (target). It is surrounded by a region with 20-t of liquid scintillator (LS) inside a 4-m diameter outer acrylic vessel (OAV). Between the SSV and OAV, 37-t of mineral oil (MO) shields the LS and Gd-LS from radioactivity. IBD interactions are detected by 192 Hamamatsu R5912 PMTs. A black radial shield and specular reflectors are installed on the vertical detector walls and above and below the LS volume, respectively. Gd-LS and LS are prepared and filled into ADs systematically to ensure all ADs are functionally identical [2]. Three automated calibration units (ACUs) mounted on the SSV lid allow for remote deployment of an LED, a ^{68}Ge source, and a combined source of ^{241}Am-^{13}C and ^{60}Co into the Gd-LS and LS liquid volumes along three vertical axes.

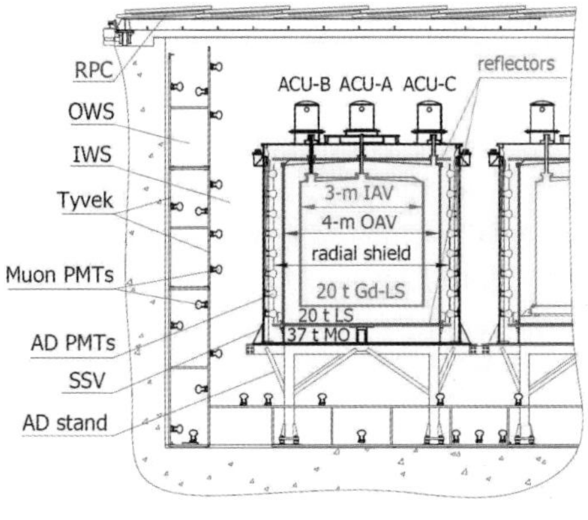

Figure 2: Schematic diagram of the Daya Bay detectors.

3 Calibration

In Daya Bay experiment, the energy information is collected from Analog-to-Digital Converters(ADC), and the time information is collected from Time-to-Digital Converters(TDC). Both ADCs and TDCs must be well calibrated.

3.1 ADC Calibration

ADC of each readout channel is calibrated with data collected in dedicated low-intensity LED runs. The calibration is verified using singles collected continuously during regular data running. In either case, samples dominated by single photo-electron (SPE) hits are selected and fit after baseline subtraction [2].

3.2 TDC Calibration

TDCs measure the interval time between trigger signals of the whole AD and hit signals on each PMT. Two components in TDC values need to be corrected. One is the time offset, which is due to the intrinsic property of each PMT and front-end electronics board(FEE). Data collected in center-located high-intensity LED runs are used for time offset calibration. We use median TDC value as a reference time. After time-of-flight subtraction, the difference between the TDC value of each channel and the reference time give us the time offset. To verify the time offset calibration, the TDC value of a reference channel, into which a mimic PMT signal generated simultaneously from the LED pulse generator was sent, is also used as the reference time.

Time walk effect is the other component in TDC values which need to be corrected. LED intensity-scan runs are used for this calibration. We fit the correlation between ADC values and TDC values to get a function for describing the time walk effect. LEDs are located at different positions inside the ADs to verify the time walk function. Consistent results are achieved.

4 Data Analysis

The analysis is based on the data from December 24, 2011 to May 11, 2012. The IBD candidates are obtained by searching for the coincidence of a positron-like signal and a neutron-like signal. Both correlated and uncorrelated backgrounds are carefully considered. The details are showing below.

4.1 IBD Selection

In order to suppress backgrounds associated with cosmic-rays, events closely following muons are vetoed before IBD selection. Inner water shield(IWS) and outer water shield(OWS) triggers with NHIT > 12 were classified as 'WS muon candidates' or μ_{WS}. Events in an AD within ± 2 μs of a μ_{WS} with energy $>$20 MeV and $>$2.5 GeV were classified as muons (μ_{AD}) and showering muons ($\mu_{\rm sh}$), respectively, for vetoing purposes. An instrumental background due to spontaneous light emission from a PMT, denoted as a flasher, was rejected efficiently [2].

IBD events were selected with the following criteria: $0.7 < E_p < 12.0$ MeV, $6.0 < E_d < 12.0$ MeV, $1 < \Delta t < 200$ μs, the prompt-delayed pair was vetoed by muons if $t_d - t_{\mu_{\rm WS}} < 600$ μs, $t_d - t_{\mu_{\rm AD}} < 1000$ μs or $t_d - t_{\mu_{\rm sh}} < 1$ s, and a multiplicity cut that requires no additional >0.7 MeV trigger in the time range $(t_p - 200\mu{\rm s}, t_d + 200\mu{\rm s})$, where E_p (E_d) is the prompt (delayed) energy and $\Delta t = t_d - t_p$ is the time difference between the prompt and delayed signals. Statistically consistent performance was achieved by an independent analysis that used different energy reconstruction, muon veto, and multiplicity cuts. The prompt-energy, the delayed-energy and the capture-time distributions for data and MC are shown in Figs. 3- 5.

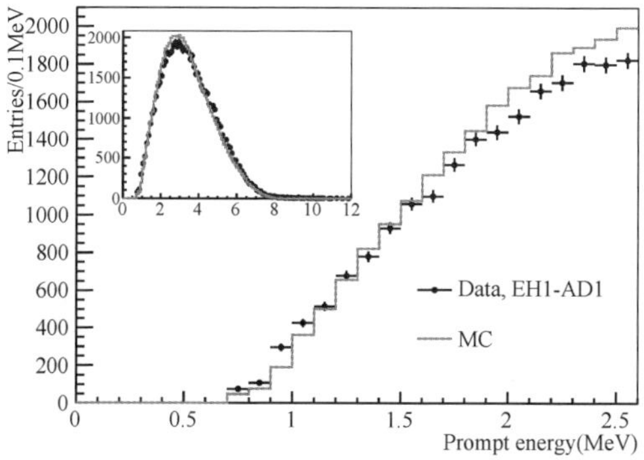

Figure 3: The prompt energy spectrum from AD1. 0.7 MeV cut is loosed to show cut and accidental backgrounds have been subtracted.

4.2 Background Subtraction

4.2.1 Accidental Backgrounds

The accidental background is defined as any pair of otherwise uncorrelated triggers that happen to satisfy the IBD selection criteria. The probability of forming an accidental background could be calculated as the product of the probability of a prompt-like trigger within 1-200 μs before the neutron-like trigger and the probability of no other trigger within 200 μs before the prompt-like trigger or 200 μs after the neutron-like trigger. Since the rate of prompt-like and neutron-like triggers changed over time, the number of

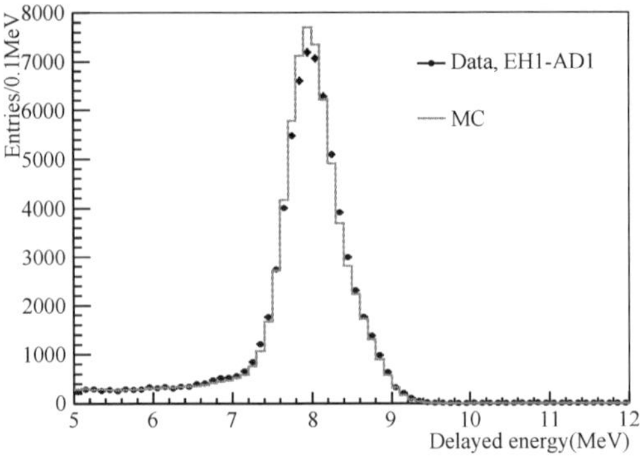

Figure 4: The delayed energy spectrum from AD1. 6 MeV cut is loosed to show cut and accidental backgrounds have been subtracted.

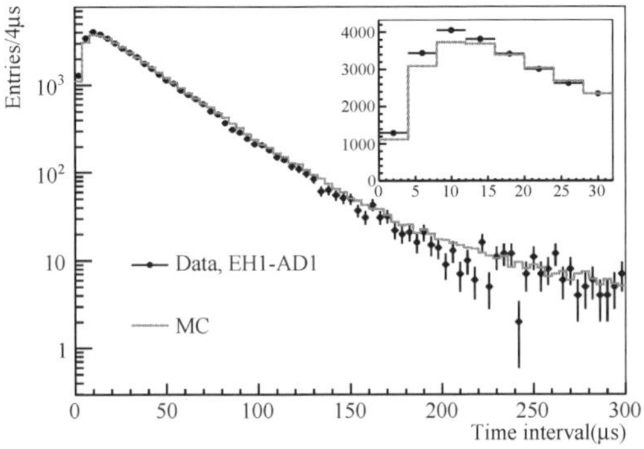

Figure 5: The neutron capture time from AD1. Cuts are loosed and accidental backgrounds have been subtracted.

accidental backgrounds was calculated every four hours then summed up as:

$$N_{acc.bkg.} = \sum_i N_i e^{-R_i \cdot 400\mu s}(1 - e^{-R_i \cdot 199\mu s}), \tag{2}$$

where N_i and R_i are the number of neutron-like singles and prompt-like single rate in the i-th four-hour period, respectively. The statistical uncertainty is dominated by N_i, and can be approximated as

$$\delta N^{(stat.)}_{acc.bkg.} = \frac{N_{acc.bkg.}}{\sqrt{\sum_i N_i}}. \tag{3}$$

An alternate method was used to estimate accidental backgrounds by counting the coincidence signals in a randomly-picked off-window. Two methods gave the consistent results.

4.2.2 Fast Neutron Backgrounds

An energetic neutron entering an AD can form a fast-neutron background by recoiling off a proton before being captured on Gd. By relaxing the $E_p < 12$ MeV criterion in the IBD selection, a flat distribution in E_p was observed up to 100 MeV as showed in Fig. 6. Extrapolation into the IBD energy region gave an estimate for the residual fast-neutron background. Two different extrapolation methods were used. First, by assuming the recoiled proton energy spectrum follows a flat distribution, the mean value for the distribution from 12 to 100 MeV was used to estimate the number of fast-neutron events from 0.7 to 12 MeV. Alternately, the data from 12 to 100 MeV was fit with a first-order polynomial function($f(E) = a + bE$), and the fit parameters were used to estimate fast-neutron backgrounds. A similar flat E_p distribution was found in the muon-tagged fast-neutron sample produced by inverting the muon veto cut. Consistent results were obtained by scaling the muon-tagged fast-neutron rate with muon inefficiency, and by MC.

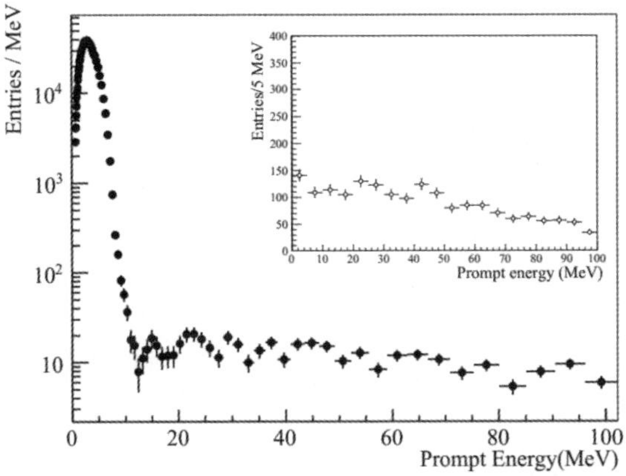

Figure 6: Prompt energy spectrum of IBD candidates with the upper limit relaxed. The energy spectrum of the fast neutron backgrounds tagged by muon is shown in the inset.

4.2.3 ^{9}Li/^{8}He Backgrounds

The rate of correlated background from the -n cascade of 8He/9Li decays was evaluated from the distribution of the time since the last muon, which can be described by the following function:

$$f(t) = N_{^9Li} \cdot \lambda_{^9Li} \cdot e^{-\lambda_{^9Li} t} + N_{^8He} \cdot \lambda_{^8He} \cdot e^{-\lambda_{^8He} t} + N_{ibd} \cdot R_\mu \cdot e^{-R_\mu t} \qquad (4)$$

where $N_{^9Li}$, $N_{^8He}$ and N_{ibd} are the number of ^{9}Li, ^{8}He and IBD events, $\lambda_i = R_\mu + 1/\tau_i^{isotope}$, R_μ is the muon rate, $\tau_i^{isotope}$ is the known decay time

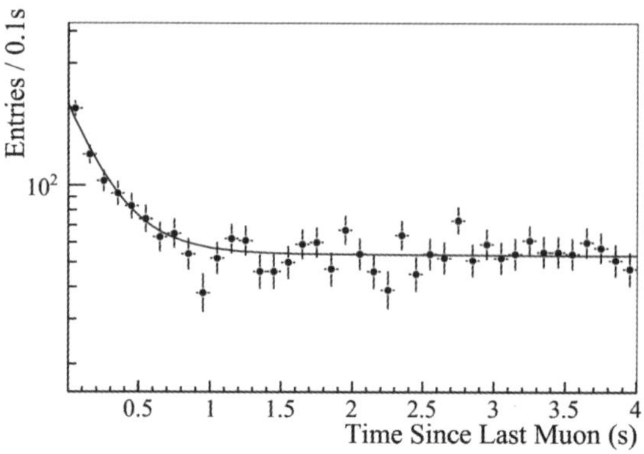

Figure 7: An example of fitting for ^{8}He/^{9}Li

constant for ^{8}He or ^{9}Li.

To reduce the number of minimum ionizing muons in the data samples, we assumed that most of ^9Li and ^8He production was accompanied with neutron generation, rejecting AD tagged muon events with no follow-on neutron(defined as $>$ 1.8 MeV trigger within a 10 - 200 μs window). Both muon samples with and without reduction were prepared for ^{9}Li and ^{8}He background estimation. A fit example is showed in Fig. 7.

4.2.4 Other Backgrounds

The ^{238}U, ^{232}Th, ^{227}Ac decay chains and ^{210}Po would bring us the ^{13}C(α n)^{16}O backgrounds. The Am-C sources sat inside the ACUs on top of each AD emit neutrons, which would occasionally ape IBD events by scattering inelastically with nuclei in the shielding material (emitting gamma rays) before being captured on a metal nuclei, such as Fe, Cr, Mn or Ni (releasing more gamma rays). Both these two backgrounds can be estimated with MC.

4.3 Uncertainties

The $\overline{\nu}_e$ rate in the far hall was predicted with a weighted combination of the two near hall measurements assuming no oscillation. Since the flux is constrained by the measurements from the near sites, this analysis is largely independent of reactor flux. The choice of flux models has a negligible impact on the result. Furthermore, any correlated detector and reactor uncertainties can be neglected.

The detailed analysis for the absolute efficiencies and the systematic uncertainties can be found in [4]. The uncertainties of the absolute efficiencies

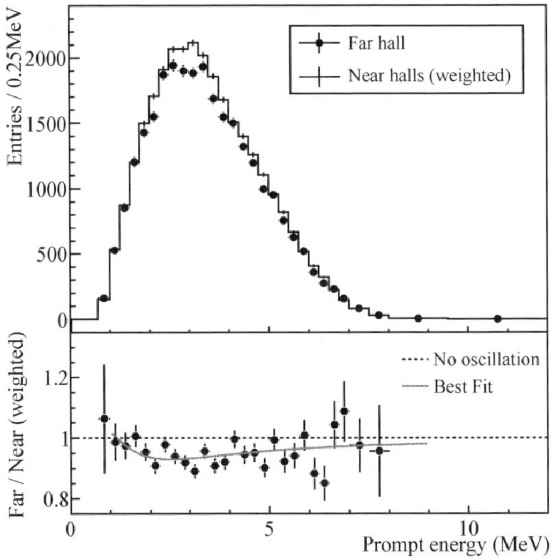

Figure 8: Measured prompt energy spectrum at the far site compared with the prediction from the near sites.

are correlated among the ADs. No relative efficiency, except the muon veto cut and multiplicity selection cut efficiencies was corrected. All differences between the functionally identical ADs were taken as uncorrelated uncertainties.

5 Results

As indicated in Fig. 8, there is a clear deficit in the far hall, expressed as a ratio of observed to expected events, $R = 0.944 \pm 0.007(\text{stat}) \pm 0.003(\text{syst})$. The value of $\sin^2 2\theta_{13}$ was determined with a χ^2 constructed with pull terms accounting for the correlation of the systematic errors. The best-fit value is $\sin^2 2\theta_{13} = 0.089 \pm 0.010(\text{stat.}) \pm 0.005(\text{syst.})$ with a χ^2/NDF of 3.4/4. The no-oscillation hypothesis is excluded at 7.7 standard deviations as showed in Fig. 9.

The recent results [4] from the Daya Bay Reactor Experiment constitute the world's most precise measurement of $\sin^2 2\theta_{13}$ to date with a precision of 12.6%. After installation of final two ADs in the summer of 2012, running for a period of three years, the Daya Bay experiment is expected to achieve a measurement of $\sin^2 2\theta_{13}$ with a precision of 5% or better.

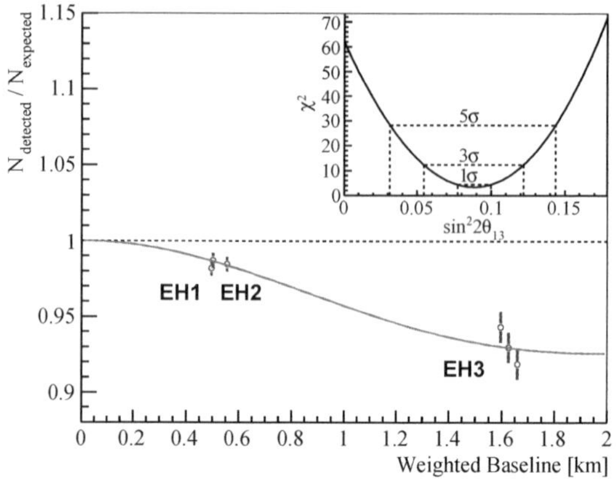

Figure 9: Ratio of measured versus expected signal in each detector, assuming no oscillation. The expected signal is corrected with the best-fit normalization parameter. Reactor and survey data were used to compute the flux-weighted average baselines. The oscillation survival probability at the best-fit value is given by the smooth curve. The AD4 and AD6 data points are displaced by -30 and +30 m for visual clarity. The χ^2 versus $\sin^2 2\theta_{13}$ is shown in the inset.

References

[1] F. P. An *et al.* (Daya Bay Collaboration), Phys. Rev. Lett. 108, 171803 (2012), arXiv:1203.1669.

[2] F. P. An *et al.* (Daya Bay Collaboration), Nucl. Instr. Meth. A685, 78-97 (2012), arXiv:1202.6181.

[3] X. H. Guo *et al.* (Daya Bay Collaboration), Proposal of the Daya Bay experiment, arXiv:hep-ex/0701029 (2007).

[4] F. P. An *et al.* (Daya Bay Collaboration), Chinese Physics C. 37, 011001 (2013), arXiv:1210.6327v2.

TTP12-036
SFB/CPP-12-69

Vacuum stability in the SM and the three-loop β-function for the Higgs self-interaction

M. F. Zoller[1]

Institut für Theoretische Teilchenphysik, Karlsruhe Institute of Technology (KIT), D-76128 Karlsruhe, Germany

Abstract

In this article the stability of the Standard Model (SM) vacuum in the presence of radiative corrections and for a Higgs boson with a mass in the vicinity of 125 GeV is discussed. The central piece in this discussion will be the Higgs self-interaction λ and its evolution with the energy scale of a given physical process. This is described by the β-function to which we recently computed analytically the dominant three-loop contributions [1].[2] These are mainly the QCD and top-Yukawa corrections as well as the contributions from the Higgs self-interaction itself. We will see that for a Higgs boson with a mass of about 125 GeV the question whether the SM vacuum is stable and therefore whether the SM could be valid up to Planck scale cannot be answered with certainty due to large experimental uncertainties, mainly in the top quark mass.

1 The Higgs potential and the stability of the SM vacuum

In the SM the Higgs potential at tree-level appears as part of the Lagrangian for a scalar SU(2)-doublet field:

$$\mathcal{L}_\Phi = \partial_\mu \Phi^\dagger \partial^\mu \Phi - \underbrace{\left(m^2 \Phi^\dagger \Phi + \lambda \left(\Phi^\dagger \Phi\right)^2\right)}_{V(\Phi)}, \qquad \Phi = \begin{pmatrix} \Phi_1 \\ \Phi_2 \end{pmatrix}. \tag{1}$$

This doublet aquires a non-zero vacuum expectation value (VEV) under spontaneous symmetry breaking (SSB) and we get the Higgs field, three Goldstone bosons[3] and the masses of the SM particles:

$$\Phi = \begin{pmatrix} \Phi_1 \\ \Phi_2 \end{pmatrix} \xrightarrow{SSB} \begin{pmatrix} \Phi^+ \\ \frac{1}{\sqrt{2}}(v + H + i\chi) \end{pmatrix}, \qquad |\Phi_{SM}| = \sqrt{\frac{-m^2}{2\lambda}} = \frac{v}{\sqrt{2}} \tag{2}$$

At tree level the mass of the Higgs boson is then given by

$$M_H^2 = -2m^2 = 2\lambda v^2. \tag{3}$$

---

[1]email: max.zoller@kit.edu

[2]In [1] we also give the dominant contributions to the β-functions for the top-Yukawa coupling, the strong coupling and the anomalous dimensions of the scalar, gluon and quark fields in the unbroken phase of the Standard Model at three-loop level.

[3]These Goldstone bosons can be absorbed by the massive W and Z bosons.

For $M_H = 125$ GeV the Higgs potential is shown in Fig. 1. If we now include radiative

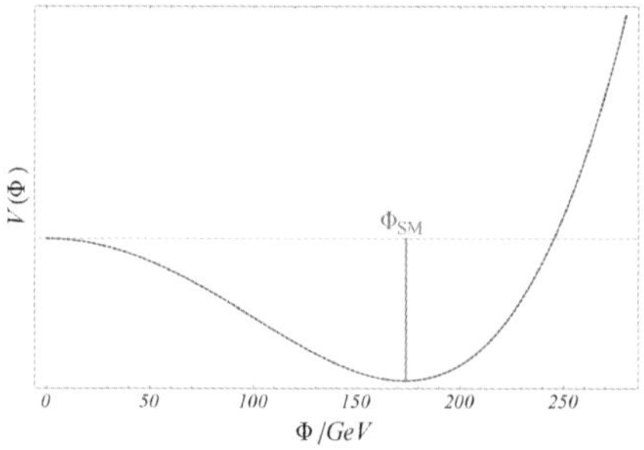

Figure 1: The SM Higgs potential at tree-level for $M_H = 125$ GeV

corrections we have to consider the effective potential $V_{eff}(\lambda(\Lambda), g_i(\Lambda), \Phi(\Lambda))$ as introduced in [2]. All couplings and fields undergo an evolution up to some scale Λ where the theory ceases to be valid.[4] Here we want to investigate whether a scenario in which the SM provides a good description of nature up to the Planck scale, i.e. $\Lambda \sim 10^{18}$ GeV, is possible.[5] The general shape of this effective potential for the SM is shown in Fig. 2

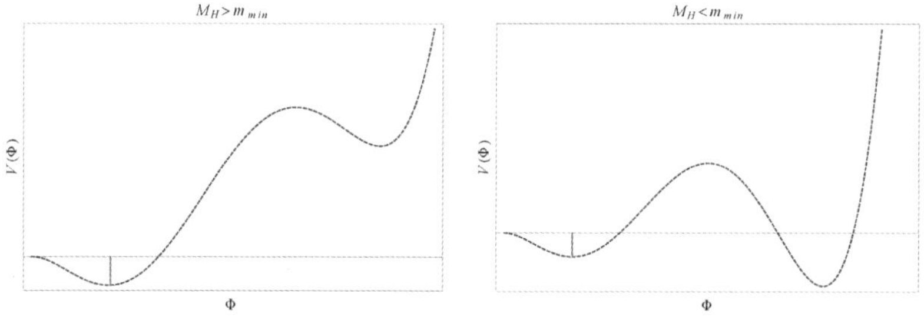

Figure 2: The effective Higgs potential

for the cases of a Higgs mass larger and smaller than a critical value m_{min}, the minimal stability bound[6] (see also [3]). If the Higgs mass is chosen below this critical value a second minimum develops which is lower than the SM one. This implies that the SM vacuum is no longer stable, i.e. it can tunnel into this energetically favoured state. Since this is in contradiction to our observation[7] we are led to the conclusion that our theory is incomplete and that new physics has to enter between Fermi and Planck scale. It has been demonstrated in [4] that for $\Phi \sim \Lambda \gg \mu_0$ a good approximation for the effective

---

[4]With $t = \log\left(\frac{\Lambda}{\mu_0}\right)$ we have the field $\Phi(\Lambda) = \Phi_{cl} \cdot \exp\left(\int_0^t dt' \gamma_\Phi(\lambda(t'), g_i(t'))dt'\right)$ where Φ_{cl} is the classical field in the absence of radiative corrections. μ_0 is the scale at which we start the running of the paramters.

[5]An extended model is expected to be needed at this scale due to gravity.

[6]There is also an upper bound m_{max} on the Higgs mass from the requirement that no Landau pole appears at energies $\mu \leq \Lambda$.

[7]It is however not possible to exclude that we live in metastable universe, i.e. the lifetime of the (local) SM minimum could be longer than the age of the universe.

potential is

$$V_{eff}[\Phi] \approx \lambda(\Lambda)\Phi^4(\Lambda) + \mathcal{O}(\lambda^2(\Lambda), g_i^2(\Lambda)), \tag{4}$$

which means that the stability of the SM vacuum is approximately equivalent to the question whether λ stays positive up to the scale Λ (see also [5, 6]). Fig. 3 shows the

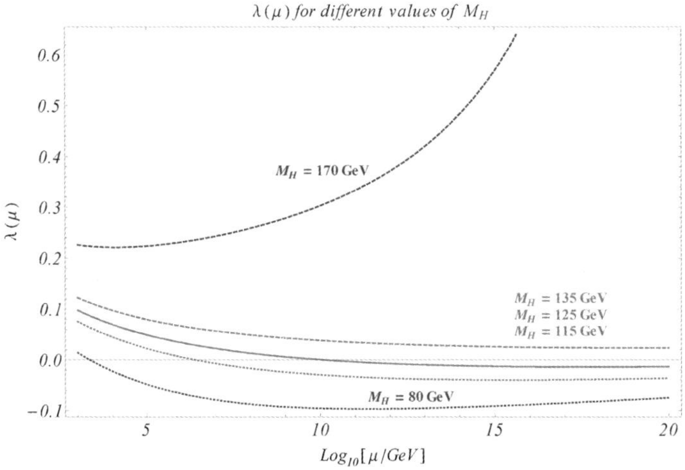

Figure 3: The behaviour of $\lambda(\mu)$ for different Higgs mass values

evolution of λ for different Higgs mass values. For a large enough Higgs mass λ diverges quickly[8] and for a small Higgs mass λ would become negative at a relatively low scale.[9] The interesting region is around 125 GeV where λ is very close to zero at the Planck scale! As the minimal stability bound has been estimated to be about 129 ± 3 GeV [3, 10], which is very close to the mass of the boson recently descovered at the LHC, the question of vacuum stability in the SM becomes one for precision physics. This serves as a strong motivation for calculating the three-loop β-function of the Higgs self-interaction which describes the evolution of this crucial coupling as well as the β-functions of the relevant SM parameters on which β_λ depends.

2 The three-loop β-function for the Higgs self-interaction

The β-function is defined as the derivative of the coupling with respect to the renormalization scale:

$$\beta_\lambda(\lambda, y_t, g_i, \ldots) = \mu^2 \frac{d}{d\mu^2}\lambda(\mu). \tag{5}$$

[8]We find a Landau pole below the Planck scale for $M_H > m_{max} \approx 175$ GeV (see also [5, 7–9]).

[9]Eventually λ will increase again and reach a Landau pole as well due to the the evolution of the gauge couplings and y_t. At large scales (above 10^{16} GeV) the electroweak couplings, especially the $U(1)$ coupling g_1, start to be the dominant contributions. In contrast y_t, which is responsible for the decrease of λ at lower scales, becomes small. In order to see this one has to extrapolate the evolution of λ well beyond the Planck scale. This is also the reason why the effective potential stays bounded from below at large field strengths Φ beyond the second minimum.

This object has been known at the two-loop level for a while including the dependence on all the gauge couplings g_1,g_2 and g_s, the quartic Higgs self-interaction λ and the Yukawa couplings which give mass to the fermions [11–14]. The one-loop and two-loop results for all SM couplings have been known for a long time [11–25] as have been partial three-loop results [26–31]. Four-loop β-functions are available for QCD [32,33] and the purely scalar part of the SM [34–36].

As there are many Feynman diagrams at the three-loop level and the treatment of γ_5 matrices in dimensional regularization poses a serious problem at this order for the Higgs and Yukawa sector we try to find the dominant contributions to the evolution of the Higgs self-interaction at the scales that we are intersted in. At the scale of the top mass $\mu = M_t = 172.9$ GeV we find the strong coupling $g_s \approx 1.17$ and the top-Yukawa coupling $y_t \approx 0.93$ to be much larger than the electroweak couplings $g_2 \approx 0.65$ and $g_1 \approx 0.36$. The Higgs self-interaction for a Higgs boson around 125 GeV is $\lambda(M_H) \approx 0.13$. [10] From this we conclude that a simplified model containing only g_s, y_t and λ will give the numerically largest terms for β_λ at three-loop level. When combinen with the full SM result at one and two-loop level this leads to the evolution of λ to the highest precision so far. The Lagrangian of our model consists of three pieces:

$$\mathcal{L} = \mathcal{L}_{QCD} + \mathcal{L}_{\Phi} + \mathcal{L}_{y_t}. \tag{6}$$

with the standard QCD Lagrangian, the Higgs part as defined in (1) and the top-Yukawa sector

$$\mathcal{L}_{y_t} = -y_t \left\{ \bar{t}_R \left(\Phi_2, -\Phi_1\right) \cdot \begin{pmatrix} t \\ b \end{pmatrix}_L + \left(\bar{t}, \bar{b}\right)_L \cdot \begin{pmatrix} \Phi_2^* \\ -\Phi_1^* \end{pmatrix} t_R \right\} \tag{7}$$

The indices L and R indicate the left- and right-handed part of the Dirac fermion fields as obtained by the projectors

$$P_L = \frac{1}{2}\left(1 - \gamma_5\right) \qquad P_R = \frac{1}{2}\left(1 + \gamma_5\right). \tag{8}$$

Hence γ_5 enters into our calculation and has to be treated carefully as described below.

2.1 Calculation

In order to obtain the β-functions for a coupling, here λ or y_t, we have to compute the renormalization constant for a vertex involving this coupling and the field renormalization constants for the external legs of this vertex. The latter is done by computing loop corrections to the respective propagators. The renormalization constant for y_t for example can be computed as[11]

$$Z_{y_t} = \frac{Z^{(\text{Yukawa vertex})}}{\sqrt{Z_L^{(\text{top field})} Z_R^{(\text{top field})} Z^{(\text{Higgs field})}}}. \tag{9}$$

[10] These couplings enter at every loop order with characteristic factors $\frac{g_s^2}{4\pi} \approx 0.11$, $\frac{g_2^2}{4\pi} \approx 0.03$, $\frac{g_1^2}{4\pi} \approx 0.01$ and $\frac{y_t^2}{4\pi} \approx 0.07$. The Higgs self-coupling is already present at tree level and enters at each order linearly as $\frac{\lambda}{4\pi} \approx 0.01$. The second-largest Yukawa coupling to be considered would be $y_b = \sqrt{2}\frac{M_b}{v} \approx 0.02$ which is negligible in comparison.

[11] Note that in general left and right-handed fields have to be renormalized separately.

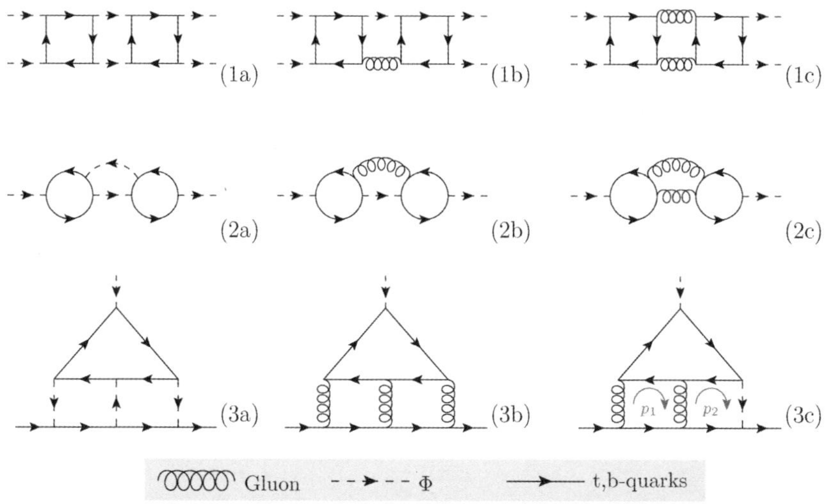

Figure 4: Sample Feynman diagrams for the calculation of renormalization constants for the quartic Higgs vertex, the Higgs self-energy and a Yukawa vertex at three-loop order.

The β-function β_{y_t} can then be computed from the requirement that the bare Yukawa coupling

$$y_t^{bare} = Z_{y_t}[g_i(\mu), y_t(\mu), \lambda(\mu)]\, y_t(\mu) \tag{10}$$

is independent of the renormalization scale μ and hence its μ^2-derivative must vanish.

For the purpose of calculating renormalization constants and β-functions we only need the UV-divergent part of all these diagrams. Nevertheless, there are two issues which have to be considered carefully in a computation of this type. The first problem is the complicated nature of our integrals when we have arbitrary momenta flowing into the external legs of our Feynman diagrams. A nice feature of the $\overline{\text{MS}}$-scheme however is that renormalization constants do not depend on external momenta[12], so we can set those to zero. Unfortunately, this introduces artificial IR-divergences into our diagrams which cannot be distinguished from the UV ones in dimensional regularization. In many cases this problem can be avoided by setting all external momenta to zero except for one which enters at one leg and exits at another. If we have no masses, as we do here, the propagator-like integrals resulting from this method can be computed with the FORM package MINCER [37] up to three-loop order. For the λ-vertex and its radiative corrections this fails, however, as there are still IR-divergent diagrams. A method to compute only the UV-divergences without having to worry about the IR ones has been described in [38, 39]. The trick is to introduce the same auxiliary mass into every propagator denominator, Taylor expand in all the external momenta and introduce all possible counterterms for the auxiliary mass in order to cancel subdivergences arising from this new mass. The resulting massive tadpole integrals can be computed with the FORM package MATAD [40]. This method will yield the correct UV-pole part of the calculated Feynman diagrams (but not the correct finite part) which is enough for the computation of β-functions. Where possible both methods have been used for the calculation in order to have an independent check.

[12]To be precise, UV-divergent terms can only depend polynomially on external momenta (and masses).

The second problem is the treatment of γ_5 matrices, which arise here from the projectors in the Yukawa sector, in d space-time dimensions. This matrix is only well-defined in four dimensions:

$$\gamma_5 = i\gamma^0\gamma^1\gamma^2\gamma^3 = \frac{i}{4!}\varepsilon_{\mu\nu\rho\sigma}\gamma^\mu\gamma^\nu\gamma^\rho\gamma^\sigma \text{ where } \varepsilon_{0123} = 1 = -\varepsilon^{0123}. \tag{11}$$

A naive treatment of γ_5, i.e. using the relations $\{\gamma_5, \gamma_\mu\} = 0$ and $\gamma_5^2 = \mathbb{1}$ to eliminate as many γ_5 as possible and then discarding all terms which still have one γ_5 in them, can only be applied to external fermion lines and closed fermion loops with less than four Lorentz indices or momenta flowing in or out of the fermion loop in question.[13] An example for a problematic diagram is shown in Fig. 4 (3c). The Lorentz indices of the two gluons connected to the closed fermion loop and the two loop momenta p_1 and p_2 offer the possibility of a non-trivial γ_5 contribution from such a diagram. And indeed, if we apply the treatment suggested by 't Hooft and Veltman in [41], i.e. using the definition with $\varepsilon_{\mu\nu\rho\sigma}$ from (11) and contracting the ε-tensor from the closed fermion loop with the one from the external fermion line or a projector $\propto \gamma_5$ acting on this external line to make the integral scalar, we find a sizable contribution. But even this treatment is not exact, in fact it is only correct up to an error of $\mathcal{O}(\varepsilon)$. Fortunately, there are only first order poles in ε from these integrals which makes the UV-divergent part and therefore our renormalization constants correct. Once again, the finite part calculated for these diagrams is unreliable but luckily not needed here.

2.2 Result

The result for the β-function of the Higgs self-interaction

$$\mu^2\frac{d}{d\mu^2}\lambda(\mu) = \beta_\lambda(g_s, y_t, \lambda) = \sum_{n=1}^{\infty}\frac{1}{(16\pi^2)^n}\beta_\lambda^{(n)}(g_s, y_t, \lambda) \tag{12}$$

in our simplified version of the Standard Model is given by

$$\begin{aligned}
\beta_\lambda^{(1)} =& 12\,\lambda^2 + 6\,y_t^2\lambda - 3\,y_t^4,\\
\beta_\lambda^{(2)} =& -156\,\lambda^3 - 72\,y_t^2\lambda^2 - \frac{3}{2}\,y_t^4\lambda + 15\,y_t^6 + 40\,g_s^2y_t^2\lambda - 16\,g_s^2y_t^4,\\
\beta_\lambda^{(3)} =& \lambda^4\,(3588 + 2016\zeta_3) + 873\,y_t^2\lambda^3 + y_t^4\lambda^2\left(\frac{1719}{2} + 756\zeta_3\right)\\
&+ y_t^6\lambda\left(\frac{117}{8} - 198\zeta_3\right) - y_t^8\left(\frac{1599}{8} + 36\zeta_3\right)\\
&+ g_s^2y_t^2\lambda^2\,(-1224 + 1152\zeta_3) + g_s^2y_t^4\lambda\,(895 - 1296\zeta_3)\\
&+ g_s^2y_t^6\,(-38 + 240\zeta_3) + g_s^4y_t^2\lambda\left(\frac{1820}{3} - 32n_f - 48\zeta_3\right)\\
&+ g_s^4y_t^4\left(-\frac{626}{3} + 20n_f + 32\zeta_3\right)
\end{aligned}$$

in the $\overline{\text{MS}}$-scheme. $\beta_\lambda^{(1)}$, $\beta_\lambda^{(2)}$ and $\beta_\lambda^{(3)}$ are the one, two and three-loop results respectively. To get an idea of the size of the individual terms and the overall result we evaluate β_λ at

[13]Four different Lorentz indices are needed to support a non-vanishing $\varepsilon_{\mu\nu\rho\sigma}$ after the trace over the fermion line has been performed. These can be the Lorentz indices of gauge boson vertices or they can be contracted with internal momenta from other loops. External momenta can be set to zero without changing the UV-divergent part and can hence be ignored in this consideration.

the scale $\mu = M_Z$ (with an assumed Higgs mass of 125 GeV and the number of fermion flavours $n_f = 6$) which yields a value of $\beta_\lambda \sim (-0.01)$ at the one-loop level. The two and three-loop contributions change this result by $\sim 1\%$ and $\sim (-0.04)\%$ respectively which is quite small e.g. in comparison with the β-function for the top-Yukawa coupling where we get a value of $\beta_{y_t} \sim (-0.023)$ and corrections of $\sim 16.6\%$ and $\sim 0.7\%$ at two and three-loop level. The full result for β_{y_t} can be found in [1]. If we have a look at the the numerically largest individual terms in (2.2) we find a curious behaviour:

$$\frac{\beta_\lambda^{(3)}}{(16\pi^2)^3}(\mu = M_Z) = (\underbrace{+7.9}_{g_s^2 y_t^6} \quad \underbrace{-4.8}_{y_t^8} \quad \underbrace{-3.1}_{g_s^2 y_t^4 \lambda} \quad \underbrace{-2.5}_{g_s^4 y_t^4} \quad \underbrace{+2.6}_{g_s^4 y_t^2 \lambda}) \times 10^{-5}. \tag{13}$$

There is a strong cancellation between these terms making the overall effect almost two orders of magnitude smaller than the largest individual contributions.[14] This significantly improves the convergence of the perturbation series for β_λ and makes the remaining theoretical uncertainty small.

3 The evolution of the Higgs self-coupling

Now we want to investigate the effect of the new three-loop result $\beta_\lambda^{(3)}$ on the running of λ and therefore its effect on the stability of the electroweak vacuum in the SM. For this we also include the electroweak contributions up to the two-loop level. For recent detailed discussions of the SM vacuum stability and its dependence on key parameters like the Higgs and the top mass see for example [1, 3, 42–44]. We start the evolution of λ at the scale of the top mass and go up to the Planck scale at 10^{18} GeV. To find the starting values in the $\overline{\text{MS}}$-scheme we have to match the physical parameters, like e.g. pole masses, to their $\overline{\text{MS}}$ counterparts.[15] These matching relations depend on the exact values of the $\overline{\text{MS}}$ parameter $\alpha_s(M_Z)$, the pole mass M_t of the top quark and of course the mass of the Higgs boson M_H. We use [48]

$$\alpha_s(M_Z) = 0.1184 \pm 0.0007, \qquad M_t = 172.9 \pm 0.6 \pm 0.9 \text{ GeV}. \tag{14}$$

Fig. 5 shows the evolution of λ for the two cases $M_H = 124$ GeV and $M_H = 126$ GeV.. The dependence of the λ-running on the parameters $\alpha_s(M_Z)$ and M_t can be estimated from the shifted curves where these parameters are changed by $\pm\sigma$ as given in eq. (14). Note that there is a considerable difference between $M_H = 124$ GeV and $M_H = 126$ GeV which means that the evolution of λ is very sensitive to the value of the Higgs mass. Given a fixed value for M_H the largest uncertainty lies in the exact value of the top mass. The second largest uncertainty comes from α_s. The total effect due to the three-loop part of the β-functions, an extension of the vacuum stability to larger scales, is a little smaller than the α_s uncertainty. Note that an analysis including higher order corrections for the matching between the pole masses and $\overline{\text{MS}}$-masses has been performed in [3, 10].

In conclusion it can be said that for a Higgs mass in the vicinity of 125 GeV the question of vacuum stability cannot be resolved with certainty. Although it looks as if λ becomes indeed negative at high scales the experimental uncertainties, mainly on the Higgs mass and the top mass, are too large at the moment. However, this is a very good motivation to attempt higher precision experiments and calculations in the near future.

[14] Note that for a significantly different Higgs mass this would not happen to such extent.

[15] We take the electroweak corrections at one-loop and the QCD ones at two-loop level from [45–47].

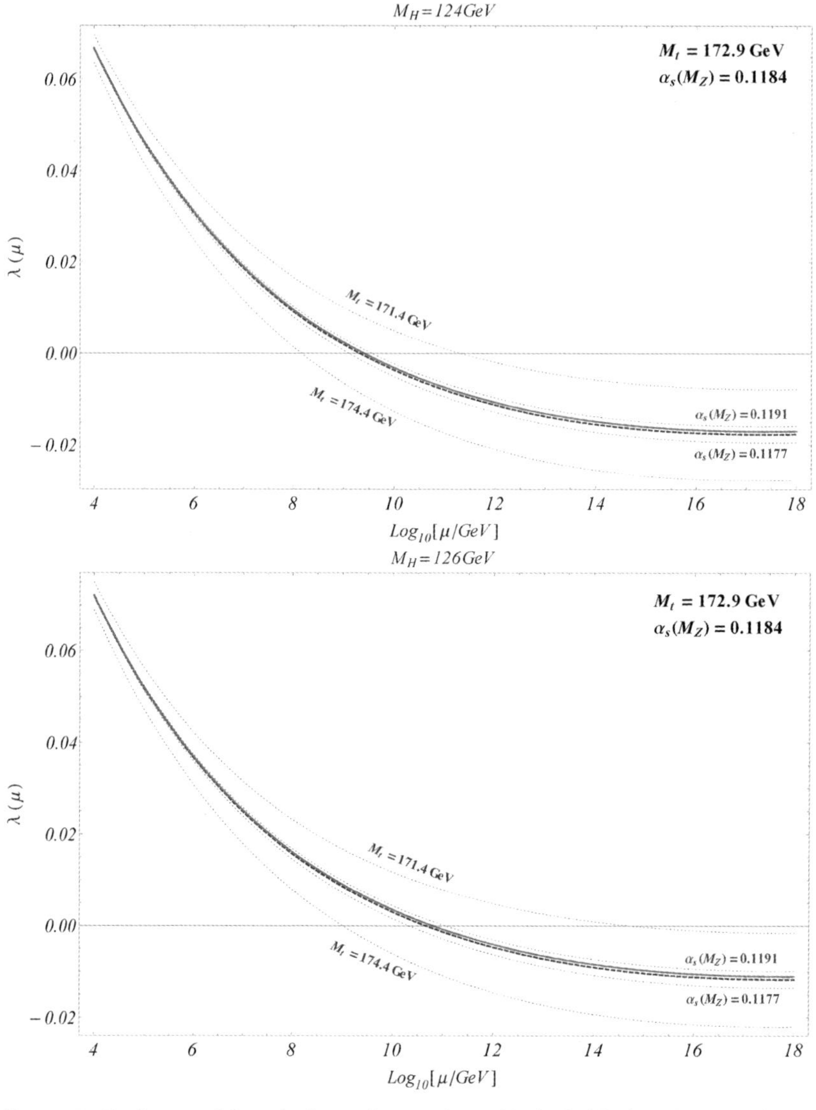

Figure 5: Evolution of λ with the scale μ: 2 loop (dashed, blue) and 3 loop (continuous, red) results; Uncertainties with respect to the two-loop result: $\pm 1\sigma_{\alpha_s}$, $\pm 1\sigma_{M_t}$ (dotted)

Acknowledgments

I want to thank my collaborator on this project K. G. Chetyrkin for invaluable discussions and J. Kühn for his support and useful comments. This work has been supported by the Deutsche Forschungsgemeinschaft in the Sonderforschungsbereich/Transregio SFB/TR-9 "Computational Particle Physics" and the Graduiertenkolleg "Elementarteilchenphysik bei höchsten Energien und höchster Präzission"

References

[1] K. Chetyrkin and M. Zoller, *Three-loop β-functions for top-Yukawa and the Higgs self-interaction in the Standard Model.* JHEP **1206** (2012) 033, arXiv:1205.2892 [hep-ph].

[2] S. Coleman and E. Weinberg, *Radiative Corrections as the Origin of Spontaneous Symmetry Breaking.* Phys. Rev. D **7** (1973) 1888–1910.

[3] F. Bezrukov, M. Y. Kalmykov, B. A. Kniehl, and M. Shaposhnikov, *Higgs boson mass and new physics.* arXiv:1205.2893 [hep-ph].

[4] G. Altarelli and G. Isidori, *Lower limit on the Higgs mass in the standard model: An update.* Physics Letters B **337** (1994) no. 1-2, 141–144.

[5] N. Cabibbo, L. Maiani, G. Parisi, and R. Petronzio, *Bounds on the Fermions and Higgs Boson Masses in Grand Unified Theories.* Nucl. Phys. **B158** (1979) 295–305.

[6] C. Ford, D. Jones, P. Stephenson, and M. Einhorn, *The Effective potential and the renormalization group.* Nucl. Phys. **B395** (1993) 17–34, arXiv:hep-lat/9210033 [hep-lat].

[7] L. Maiani, G. Parisi, and R. Petronzio, *Bounds on the Number and Masses of Quarks and Leptons.* Nucl. Phys. **B136** (1978) 115.

[8] M. Lindner, *Implications of Triviality for the Standard Model.* Z. Phys. **C31** (1986) 295.

[9] T. Hambye and K. Riesselmann, *Matching conditions and Higgs mass upper bounds revisited.* Phys.Rev. **D55** (1997) 7255–7262, arXiv:hep-ph/9610272 [hep-ph].

[10] J. Elias-Miro, J. R. Espinosa, G. F. Giudice, G. Isidori, A. Riotto, *et al.*, *Higgs mass implications on the stability of the electroweak vacuum.* Phys. Lett. **B709** (2012) 222–228, arXiv:1112.3022 [hep-ph].

[11] M. Fischler and J. Oliensis, *Two-loop corrections to the beta function for the Higgs-Yukawa coupling constant.* Phys. Lett. B **119** (1982) no. 4, 385–386.

[12] M. E. Machacek and M. T. Vaughn, *Two-loop renormalization group equations in a general quantum field theory (II). Yukawa couplings.* Nucl. Phys. B **236** (1984) no. 1, 221–232.

[13] C. Ford, I. Jack, and D. Jones, *The Standard model effective potential at two loops.* Nucl.Phys. **B387** (1992) 373–390, arXiv:hep-ph/0111190 [hep-ph].

[14] M.-x. Luo and Y. Xiao, *Two loop renormalization group equations in the standard model.* Phys. Rev. Lett. **90** (2003) 011601, arXiv:hep-ph/0207271 [hep-ph].

[15] D. J. Gross and F. Wilczek, *Ultraviolet Behavior of Non-Abelian Gauge Theories.* Phys. Rev. Lett. **30** (1973) 1343–1346.

[16] H. D. Politzer, *Reliable Perturbative Results for Strong Interactions?* Phys. Rev. Lett. **30** (1973) 1346–1349.

[17] D. Jones, *Two-loop diagrams in Yang-Mills theory.* Nuclear Physics B **75** (1974) no. 3, 531–538.

[18] O. Tarasov and A. Vladimirov, *Two Loop Renormalization of the Yang-Mills Theory in an Arbitrary Gauge.* Sov.J.Nucl.Phys. **25** (1977) 585.

[19] W. E. Caswell, *Asymptotic Behavior of Non-Abelian Gauge Theories to Two-Loop Order.* Phys.Rev.Lett. **33** (1974) 244–246.

[20] E. Egorian and O. Tarasov, *Two loop renormalization of the QCD in an arbitrary gauge.* Teor.Mat.Fiz. **41** (1979) 26–32.

[21] D. R. T. Jones, *Two-loop β function for a $G_1 \times G_2$ gauge theory.* Phys. Rev. D **25** (1982) 581–582.

[22] M. S. Fischler and C. T. Hill, *Effects of Large Mass Fermions on M_X and $\sin^2 \theta_W$.* Nucl.Phys. **B193** (1981) 53.

[23] I. Jack and H. Osborn, *General background field calculations with fermion fields.* Nucl. Phys. B **249** (1985) no. 3, 472–506.

[24] M. E. Machacek and M. T. Vaughn, *Two-loop renormalization group equations in a general quantum field theory: (I). Wave function renormalization.* Nucl. Phys. B **222** (1983) no. 1, 83–103.

[25] M. E. Machacek and M. T. Vaughn, *Two-loop renormalization group equations in a general quantum field theory: (III). Scalar quartic couplings.* Nucl. Phys. B **249** (1985) no. 1, 70–92.

[26] T. Curtright, *Three loop charge renormalization effects due to quartic scalar selfinteractions.* Phys.Rev. **D21** (1980) 1543.

[27] D. Jones, *Comment on the charge renormalization effects of quartic scalar selfinteractions.* Phys.Rev. **D22** (1980) 3140–3141.

[28] O. Tarasov, A. Vladimirov, and A. Y. Zharkov, *The Gell-Mann-Low Function of QCD in the Three Loop Approximation.* Phys.Lett. **B93** (1980) 429–432.

[29] S. Larin and J. Vermaseren, *The Three loop QCD Beta function and anomalous dimensions.* Phys. Lett. **B303** (1993) 334–336, `arXiv:hep-ph/9302208 [hep-ph]`.

[30] M. Steinhauser, *Higgs decay into gluons up to O(alpha**3(s) G(F)m**2(t)).* Phys.Rev. **D59** (1999) 054005, `arXiv:hep-ph/9809507 [hep-ph]`.

[31] A. Pickering, J. Gracey, and D. Jones, *Three loop gauge beta function for the most general single gauge coupling theory.* Phys.Lett. **B510** (2001) 347–354, `arXiv:hep-ph/0104247 [hep-ph]`.

[32] T. van Ritbergen, J. Vermaseren, and S. Larin, *The Four loop beta function in quantum chromodynamics.* Phys. Lett. **B400** (1997) 379–384, `arXiv:hep-ph/9701390 [hep-ph]`.

[33] M. Czakon, *The Four-loop QCD beta-function and anomalous dimensions.* Nucl.Phys. **B710** (2005) 485–498, `arXiv:hep-ph/0411261 [hep-ph]`.

[34] E. Brezin, J. C. Le Guillou, J. Zinn-Justin, and B. G. Nickel, *Higher oder contributions to critical exponents.* Phys. Lett. **44A** (1973) 227–228.

[35] E. Brezin, J. C. Le Guillou, and J. Zinn-Justin, *Addendum to wilson's theory of critical phenomena and callan-symanzik equations in 4-epsilon dimensions.* Phys. Rev. **D9** (1974) 1121–1124.

[36] D. I. Kazakov, O. V. Tarasov, and A. A. Vladimirov, *Calculation Of Critical Exponents By Quantum Field Theory Methods.* Sov.Phys. **JETP 50** (1979) 521.

[37] S. G. Gorishnii, S. A. Larin, L. R. Surguladze, and F. V. Tkachov, *MINCER: Program for multiloop calculations in quantum field theory for the SCHOONSCHIP system.* Comput. Phys. Commun. **55** (1989) 381–408.

[38] M. Misiak and M. Munz, *Two loop mixing of dimension five flavor changing operators.* Phys. Lett. **B344** (1995) 308–318, `arXiv:hep-ph/9409454`.

[39] K. G. Chetyrkin, M. Misiak, and M. Munz, *Beta functions and anomalous dimensions up to three loops.* Nucl. Phys. **B518** (1998) 473–494, `arXiv:hep-ph/9711266 [hep-ph]`.

[40] M. Steinhauser, *MATAD: A program package for the computation of massive tadpoles.* Comput. Phys. Commun. **134** (2001) 335–364, `arXiv:hep-ph/0009029`.

[41] G. 't Hooft and M. J. G. Veltman, *Regularization and Renormalization of Gauge Fields.* Nucl. Phys. **B44** (1972) 189–213.

[42] I. Masina, *The Higgs boson and Top quark masses as tests of Electroweak Vacuum Stability.* `arXiv:1209.0393 [hep-ph]`.

[43] M. Holthausen, K. S. Lim, and M. Lindner, *Planck scale Boundary Conditions and the Higgs Mass.* JHEP **1202** (2012) 037, `arXiv:1112.2415 [hep-ph]`.

[44] G. Degrassi, S. Di Vita, J. Elias-Miro, J. R. Espinosa, G. F. Giudice, *et al.*, *Higgs mass and vacuum stability in the Standard Model at NNLO.* JHEP **1208** (2012) 098, `arXiv:1205.6497 [hep-ph]`.

[45] J. Espinosa, G. Giudice, and A. Riotto, *Cosmological implications of the Higgs mass measurement.* JCAP **0805** (2008) 002, `arXiv:0710.2484 [hep-ph]`.

[46] R. Hempfling and B. A. Kniehl, *On the relation between the fermion pole mass and MS Yukawa coupling in the standard model.* Phys.Rev. **D51** (1995) 1386–1394, `arXiv:hep-ph/9408313 [hep-ph]`.

[47] A. Sirlin and R. Zucchini, *Dependence of the Higgs coupling hMS(M) on mH and the possible onset of new physics.* Nucl. Phys. B **266** (1986) no. 2, 389–409.

[48] K. Nakamura *et al.*, *The review of particle physics.* J. Phys. **G** (2010) no. 37, 075021.

Closing Ceremony

DIPLOMAS

Twenty five Diplomas were open for competition among the participants. They have been awarded as follows:

- John S. BELL Diploma to:

  Rafael LOPES DE SÀ
  Stony Brook University, New York, USA

- Patrick M.S. BLACKETT Diploma to:

  Max ZOLLER
  Karlsruhe Institute für Tecnologie, Karlsruhe, Germany

- Nicola CABIBBO Diploma to:

  Maria Valentina CARLUCCI
  Technische Universität München, Garching, Germany

- James CHADWICK Diploma to:

  Stefanos LEONTSINIS
  CERN, Geneva, Switzerland

- Sidney COLEMAN Diploma to:

  Anton SHEYKIN
  St. Petersburg State University, Russia

- Richard H. DALITZ Diploma to:

  Tom MELIA
  Rudolf Peierls Centre for Theoretical Physics, Oxford, UK

- Paul A.M. DIRAC Diploma to:

  Alexey SLEPTSOV
  ITEP, Moscow, Russia

- Bruno FERRETTI Diploma to:

  Alexander KRIKUN
  ITEP, Moscow, Russia

- Vladimir N. GRIBOV Diploma to:

  Scott DAVIES
  UCLA, Los Angeles, CA, USA

- Robert HOFSTADTER Diploma to:

  Fabian WISSBROCK
  DESY, Zeuthen, Germany

- Gunnar KÄLLEN Diploma to:

  Fabio COLAMARIA
  INFN, Bari, Italy

- Seymour J. LINDENBAUM Diploma to:

  Kieran BOYLE
  RIKEN BNL Research Center, Upton, NY, USA

- Yuval NE'EMAN Diploma to:

  Alexandr TUMANOV
  St. Petersburg Nuclear Physics Institute, Russia

- Giuseppe P.S. OCCHIALINI Diploma to:

  Sabine HEMMER
  University of Padova, Italy

- Oreste PICCIONI Diploma to:

  Konstantinos KARAKOSTAS
  CERN, Geneva, Switzerland

- Bruno PONTECORVO Diploma to:

  Feihong ZHANG
  Institute of High Energy Physics, Beijing, China

- Giampietro PUPPI Diploma to:

  Francesco SANFILIPPO
  Laboratoire de Physique Théorique, Orsay, France

- Isidor I. RABI Diploma to:

  Javier Mauricio G. DUARTE
  CALTECH, Pasadena, CA, US

- Giulio RACAH Diploma to:

  Rob KNEGJENS
  NIKHEF, Amsterdam, The Netherlands

- Bruno ROSSI Diploma to:

  Maxim PERFILOV
  Moscow State University, Moscow, Russia

- Julian S. SCHWINGER Diploma to:

  Deepa THOMAS
  Institute for Subatomic Physics, Utrecht, The Netherlands

- Victor F. WEISSKOPF Diploma to:

  David TWIGG
  DAMP Center for Mathematical Science, University of Cambridge, UK

- Eugene P. WIGNER Diploma to:

  Ksenia PTITSYNA
  Moscow State University, Moscow, Russia

- Bjorn H. WIIK Diploma to:

  Hayk PIRUMOV
  Physikalisches Institut, Universität Heidelberg, Heidelberg, Germany

- Chien Shiung WU Diploma to:

  Zhiqing LIU
  Institute of High Energy Physic, Beijing, China

AWARDS

The Awards for the best 'New Talent – Speakers' have been attributed to:

i) Alexey SLEPTSOV
ITEP, Moscow, Russia
for his original presentation in
MATHEMATICAL PHYSICS

ii) Max ZOLLER
Karlsruhe Institute for Technology, Karlsruhe, Germany
for his original presentation in
THEORETICAL PHYSICS

iii) Rafael LOPES DE SÀ
Stony Brook University, NY, USA
for his original presentation in
DETECTOR PHYSICS

iv) Feihong ZHANG
Institute of High Energy Physics, Beijng, China
for his original presentation in
EXPERIMENTAL PHYSICS

Award for Best Scientific Secretary given to:
Deepa THOMAS
Institute for Subatomic Physics
Utrecht, The Netherlands

Award for Best Student given to:
Maria Valentina CARLUCCI
Technische Universität München
Garching, Germany

PARTICIPANTS

Andrea AGOSTINELLI Italy	INFN and University of Bologna Bologna, Italy
Muhammad AJAZ Pakistan	COMSATS Institute of Information Technology Islamabad, Pakistan
Lasma ALBERTE Latvia	Ludwig Maximilian University, ASC Munich, Germany
Artur APRESYAN Armenia	CALTECH Pasadena, CA, USA
Richard ARNOWITT **USA**	Texas A&M University College Station, TX, USA
Michal ARTYMOWSKI **Poland**	Faculty of Physics, University of Warsaw Warsaw, Poland
Evelina ARUSHANOVA **Russia**	ITEP Moscow, Russian Federation
Paolo ASCHIERI **Italy**	University of Piemonte Orientale Alessandria, Italy
Tagir AUSHEV **Russia**	ITEP, Moscow, Russian Federation EPFL - High Energy Physics Laboratory Lausanne, Switzerland
Dario AUTIERO **Italy**	CNRS Lyon, France
Daniele BARDUCCI **Italy**	School of Physics and Astronomy University of Southampton Southampton, UK
Grigory BEDNIK **Russia**	Institute for Nuclear Research of RAS Moscow, Russia
Viktor BEGUN **Ukraine**	Bogolyubov Institute for Theoretical Physics Kiev, Ukraine
Gianpaolo BELLINI **Italy**	University of Milan, Milano, Italy
Zvi BERN **USA**	UCLA Los Angeles, CA, USA
Alessandro BETTINI **Italy**	INFN & Padoa University Padova, Italy
Thomas BLAKE **UK**	CERN Geneva, Switzerland

Philippe BLOCH **France**	CERN Geneva, CH
Kieran BOYLE **USA**	RIKEN BNL Research Center Brookhaven National Lab Upton, NY, US
Carlo BROGGINI **Italy**	INFN & Padoa University Padova, Italy
Ksenia BULYCHEVA **Russia**	ITEP Moscow, Russian Federation
Philipp BURDA **Russia**	ITEP Moscow, Russian Federation
Tian Yu CAO **USA**	Boston University MA, USA
Maria Valentina CARLUCCI **Italy**	Technische Universität München Garching, Germany
John Joseph CARRASCO **USA / Ireland**	Stanford Institute for Theoretical Physics Stanford, CA, US
Bianca Letizia CERCHIAI **Italy**	University of Milano Milano, Italy
Marcin CHRZASZCZ **Poland**	Institute of Nuclear Physics Polish Academy of Sciences Kracow, Poland
Frank CLOSE **UK**	University of Oxford, UK Oxford, UK
Eugenio COCCIA **Italy**	INFN & University of Tor Vergata, Rome, Italy
Fabio COLAMARIA **Italy**	INFN and University of Bari Bari, Italy
Màtè CSANÀD **Hungary**	Eötvös Loránd University Budapest, Hungary
Scott DAVIES **USA**	UCLA Los Angeles, CA, US
Cesareo A. DOMINGUEZ **South Africa**	University of Cape Town, ZA Cape Town, South Africa
Qader DOROSTI HASANKIADEH **Iran**	University of Groningen Groningen, The Netherlands

Tatiana DROZHZHOVA
Russia
St. Petersburg State University
Saint Petersburg, Russian Federation

Javier Mauricio G. DUARTE
USA
CALTECH
Pasadena, CA, US

Petr DUNIN-BARKOWSKI
Russia
ITEP
Moscow, Russian Federation

Ahmed FARAG ALI
Egypt
Benha University, Benha, Qaliubiya; and
ECTP, MTI University, Cairo, Egypt

Angelo Raffaele FAZIO
Italy
Universidad Nacional de Colombia
Bogotá, Colombia

Sergio FERRARA
Italy
CERN
Geneva, Switzerland and
LNF–INFN, Frascati, Italy

Antonia Micol FRASSINO
Italy
FIAS (Frankfurt Institute for Advanced Studies)
Goethe Universität
Frankfurt am Main, Germany

Harald FRITZSCH
Germany
Ludwig-Maximilians-Universitat, Muenchen,
Muenchen,DE

Ilmar GAHRAMANOV
Azerbaijan
DESY
Hamburg, Germany

Evangelos GAZIS
Greece
National Technical University, Athens, Greece and
CERN, Geneva, Switzerland

Murray GELL-MANN
USA
Institute for the Complexity, Santa Fe Institute
New Mexico, USA

Paolo GIUBELLINO
Italy
University of Torino, Italy and
CERN, Geneva, Switzerland

Ivan GNESI
Italy
University of Turin
Turin, Italy

Martin GÖRNER
Germany
Universität Hamburg
Hamburg, Germany

Barbara GUERZONI
Italy
INFN and University of Bologna
Bologna, Italy

Sergey Vladimirovich GUTS
Russia
ITEP
Moscow, Russia

Dieter HAIDT
Germany
DESY
Hamburg, Germany

Sabine HEMMER
Germany
University of Padova
Padova, Italy

Peter Ware HIGGS
UK
The University of Edinburgh
Edinburgh, UK

HUSHNUD
India
Aligarh Muslin University
Aligarh, India

Gianluca INGUGLIA
Italy
Queen Mary University of London
London, UK

Volodymyr IURASOV
Ukraine
V.N. Karazin Kharkiv National University
Kharkiv, Ukraine

Iryna IVANCHENKO
Ukraine
Taras Shevchenko National University of Kiev
Kiev, Ukraine

Peter JENNI
Switzerland
CERN,
Geneva, Switzerland

Sonia KABANA
Greece
Subatech
Nantes, France

Konstantinos KARAKOSTAS
Greece
CERN
Geneva, Switzerland

Andrey KHMELNITSKIY
Russia
Ludwig Maximilian University, ASC
Munich, Germany

Vladimir KIRILIN
Russia
ITEP
Moscow, Russian Federation

Yaroslav KLOPOT
Ukraine
Joint Institute for Nuclear Research (IJNR)
Dubna, Moscow Region, Russia

Rob KNEGJENS
The Netherlands
NIKHEF
Amsterdam, The Netherlands

Rob KNOOPS
Belgium
CERN-TH
Geneva, Switzerland

Petr KOPNIN
Russia
ITEP
Moscow, Russian Federation

Maxim KURKOV
Russia
St. Petersburg Nuclear Physics Institute
Saint Petersburg, Russia

Oleg LEMETS
Ukraine
Kharkiv Institute of Physics and Technology
Kharkiv, Ukraine

Stefanos LEONTSINIS
Greece
CERN
Geneva, Switzerland

Illia LEVANDOVSKYI **Ukraine**	V.N. Karazin Kharkiv National University Kharkiv, Ukraine
Ka Ki Carmen LI **Hong Kong**	School of Mathematics, The University of Edinburgh Edinburgh, Scotland, UK
Lev N. LIPATOV **Russia**	PNPI St. Petersburg, Russia
Zhiqing LIU **China**	Institute of High Energy Physics Beijing, People's Republic of China
Rafael LOPES DE SÀ **Brazil**	Dept Physics and Astronomy, Stony Brook University Stony Brook, NY, US
Árpád László LUKÁCS **Hungary**	Institute for Particle and Nuclear Physics Budapest, Hungary
Patricia McBRIDE **USA**	FNAL Batavia, IL, USA
Magnus MAGER **Germany**	CERN, Geneva, Switzerland Technische Universität Darmstadt Darmstadt, Germany
Mohammed Salim MALIYEKKAL **India**	Aligarh Muslim University Aligarh, India
Elisa MANONI **Italy**	INFN and University of Perugia Perugia, Italy
Alessio MARRANI **Italy**	CERN Geneva, Switzerland
David MARZOCCA **Italy**	SISSA Trieste, Italy
Alexey S. MATVEEV **Russia**	P.N. Lebedev Physical Institute of the RAS Moscow, Russia Federation
Tom MELIA **UK**	Rudolf Peierls Centre for Theoretical Physics University of Oxford Oxford, UK
Thomas METCALF **UK**	The University of Edinburgh Edinburgh, Scotland, UK
Andrey MOROZOV **Russia**	ITEP Moscow, Russian Federation
Edvard MUSAEV **Russia**	Queen Mary University of London London, UK
Francesco NOFERINI **Italy**	Enrico Fermi Centre, Rome; and University of Bologna, Italy

Oleg NOVIKOV **Russia**	St. Petersburg State University Saint Petersburg, Russian Federation
Vladyslav PAUK **Ukraine**	Nuclear Physics Institute Johannes Gutenberg-Universität Mainz Mainz, Germany
Maxim PERFILOV **Russia**	Moscow State University Moscow, Russian Federation
Giacinto PIACQUADIO **Italy**	CERN Geneva, Switzerland
Hayk PIRUMOV **Armenia**	Physikalisches Institut, Universität Heidelberg Heidelberg, Germany
Roberto PREGHENELLA **Italy**	Enrico Fermi Centre, Rome; and University of Bologna, Italy
Olga PTITSYNA **Russia**	Moscow State University Moscow, Russia
Ksenia PTITSYNA **Russia**	Moscow State University Moscow, Russia
Bartlomiej RACHWAL **Poland**	Kracow University of Technology Kracow, Poland
Ninfa RADICELLA **Italy**	Universidad Autonoma de Barcelona Barcelona, Spain
Martti RAIDAL **Estonia**	National Institute of Chemical Physics and Biophysics Tallin, Estonia
William REECE **UK**	CERN Geneva, Switzerland
Alessandro ROSSI **Italy**	INFN and University of Perugia Perugia, Italy
Igor RUDENOK **Ukraine**	Taras Shevchenko National University of Kiev Kiev, Ukraine
Riccardo RUSSO **Italy**	INFN and University of Turin Turin, Italy
Andrey SADOFYEV **Russia**	ITEP Moscow, Russian Federation
Francesco SANFILIPPO **Italy**	Laboratoire de Physique Théorique Université de Paris-Sud Orsay, France

Petr SATUNIN
Russia
Institute for Nuclear Research of RAS
Moscow, Russian Federation

Dieter SCHILDKNECHT
Germany
Universität Bielefeld, DE; and
Max-Planck-Institute for Physics
Munich, Germany

Eike SCHLIECKAU
Germany
Universität Hamburg
Hamburg, Germany

Kgotlaesele Johnson SENOSI
South Africa
University of Cape Town/iThemba Labs Somerset West
Cape Town, South Africa

Volodymyr SHAPOVAL
Ukraine
Bogolyubov Institute for Theoretical Physics
Kiev, Ukraine

Andriy SHCHERBAKOV
Ukraine
LNF–INFN
Frascati, Italy

Olga SHEKHOVTSOVA
Ukraine
IFIC, University of Valencia
Valencia, Spain

Anton SHEYKIN
Russia
St. Petersburg State University
Saint Petersburg, Russian Federation

Kirill SKOVPEN
Russia
Budker Institute of Nuclear Physics
Novosibirsk, Russian Federation

Alexey SLEPTSOV
Russia
ITEP
Moscow, Russian Federation

Valentin SLEPUKHIN
Russia
ITEP
Moscow, Russian Federation

Lev SPODYNEIKO
Russia
Landau Institute for Theoretical Physics
Moscow, Russian Federation

Horst STOECKER
Germany
GSI
Darmstadt, Germany

Michael J. TANNENBAUM
USA
Brookhaven National Laboratory
Upton, NY, USA

Daniel Jesus TAPIA TAKAKI
Mexico
IPN – Université de Paris-Sud
Orsay, France

Andrei TARASOV
Russia
St. Petersburg State University
Saint Petersburg, Russian Federation

Astkhik TER-MINASYAN
Russia
RRC Kurchatov Research Center
Moscow, Russian Federation

Deepa THOMAS
India
Institute for Subatomic Physics (SAP)
Utrecht, The Netherlands

Gerardus 't HOOFT **The Netherlands**	Utrecht University Utrecht, Netherlands
Alexandr TUMANOV **Russia**	St. Petersburg Nuclear Physics Institute Saint Petersburg, Russian Federation
David TWIGG **UK**	DAMP Center for Mathematical Science University of Cambridge, UK
Karoly ÜERMÖESSY **Hungary**	Institute for Particle and Nuclear Physics Budapest, Hungary
Tejinder VIRDEE **India**	CERN Geneva, Switzerland
Benedikt VORMWALD Mr. **Germany**	DESY – FLC Hamburg, Germany
Lucia VOTANO **Italy**	LNGS L'Aquila, Italy
Alan WALKER **UK**	School of Physics and Astronomy The University of Edinburgh, UK
Horst WENNINGER **Germany**	CERN Geneva, Switzerland
Fabian WISSBROCK **Germany**	DESY Zeuthen, Germany
Cigdem YALCIN **Turkey**	Istanbul University Istanbul, Turkey
Ivan YELETSKIKH **Ukraine**	Joint Institute for Nuclear Research (JINR) Dubna, Moscow Region, Russian Federation
Danylo YEROKHIN **Ukraine**	Kharkiv Institute of Physics and Technology Kharkiv, Ukraine
Egor ZENKEVICH **Russia**	Institute for Nuclear Research of RAS Moscow, Russian Federation
Feihong ZHANG **China**	Institute of High Energy Physics Beijing, People's Republic of China
Antonino ZICHICHI **Italy**	CERN, Geneva, Switzerland and INFN & University of Bologna, Italy
Max ZOLLER **Germany**	Karlsruhe Institute für Tecnologie Karlsruhe, DE